AF790901

The Integrated Electro-Mechanical Drive

A mechatronic approach

The Integrated Electro-Mechanical Drive

A mechatronic approach

B T Fijalkowski and J Tutaj
Krakow University of Technology, Krakow, Poland

IOP Publishing, Bristol, UK

© IOP Publishing Ltd 2019

All rights reserved. No part of this publication may be reproduced, stored in a retrieval system or transmitted in any form or by any means, electronic, mechanical, photocopying, recording or otherwise, without the prior permission of the publisher, or as expressly permitted by law or under terms agreed with the appropriate rights organization. Multiple copying is permitted in accordance with the terms of licences issued by the Copyright Licensing Agency, the Copyright Clearance Centre and other reproduction rights organizations.

Permission to make use of IOP Publishing content other than as set out above may be sought at permissions@iop.org.

B T Fijalkowski and J Tutaj have asserted their right to be identified as the authors of this work in accordance with sections 77 and 78 of the Copyright, Designs and Patents Act 1988.

ISBN 978-0-7503-2048-1 (ebook)
ISBN 978-0-7503-2046-7 (print)
ISBN 978-0-7503-2047-4 (mobi)

DOI 10.1088/2053-2563/aae7d7

Version: 20190201

IOP Expanding Physics
ISSN 2053-2563 (online)
ISSN 2054-7315 (print)

British Library Cataloguing-in-Publication Data: A catalogue record for this book is available from the British Library.

Published by IOP Publishing, wholly owned by The Institute of Physics, London

IOP Publishing, Temple Circus, Temple Way, Bristol, BS1 6HG, UK

US Office: IOP Publishing, Inc., 190 North Independence Mall West, Suite 601, Philadelphia, PA 19106, USA

To our families

Contents

Appendices

Preface

To make progress, science should not accept the limitations placed on discovery by traditional methods, conventional approaches, or existing infrastructure.

Moselio Schaechter, Roberto Kolter and Merry Buckley 2004
Microbiology in the 21st Century: Where Are We and Where Are We Going?
(Washington DC: AMS)

The real voyage of discovery consists not in seeking new technologies but in having new eyes.

Marcel Proust

A particular new scientific result does not usually gain a victory in a way that the opponents suffer a defeat and declare that they are converted but much rather the opponents gradually die out and the new generations grow ab ovo familiar with the truth.

Max Planck

The resistance to a new idea increases as the square of its importance.

Bertrand Russell

The title of this textbook requires some explanation. 'Mechatronics' is the integration of physics—mechanics, fluidics, electrics and electronics, etc—and computer technologies into the research and development (R&D) of cyber-physical heterogeneous continuous dynamical hypersystems (Fijalkowski 2010, 2011, 2016, Tutaj 2012). Thus, mechatronics consists of the synergistic combination of different physical disciplines like mechanics, fluidics, electrics and electronics, as well as informatics, etc. Synergy is a very creative and therefore dynamic process, far more so than the usual co-operation of the different physical disciplines, and even more so than a close integration of machine hardware and software.

Mechatronics offers new solutions and unprecedented flexibility in transportation systems, industrial production processes, and aerospace, aviation, automotive and traction cyber-physical heterogeneous dynamical hypersystems, containing homogenous dynamical systems, hyposystems and components, etc. Its economic success is based on functional integration, i.e. the multiple use of mechano-mechanical (M-M), fluido-mechanical (F-M) and/or electromechanical (E-M) actuators (machines); the decentralisation of intelligence into cyber-physical heterogeneous continuous dynamical hypersystems, e.g. aerospace, automotive and industrial integrated electromechanical drives (IEMD) and/or traction direct-drive (DD) propulsion cyber-physical heterogeneous continuous or discrete dynamical hypersystems with their commutation, control, communication and automation homogenous dynamical systems, hyposystems and components, as well as the inherent options for sensorless self-monitoring and system protection.

Mechatronics is a rapidly developing interdisciplinary field of engineering. It deals with the synergistic integration of mechanical engineering, macro- and microelectronic engineering (macrocommutators and microprocessors), control engineering, computer technology, in the development of E-M products, e.g. an IEMD through a unified design, dynamical systems approach.

A mathematical model is as a rule a simplification of reality. In physics and engineering, the author discriminates three fundamental aims for a formulation of mathematical models of cyber-physical heterogeneous continuous or discrete dynamical hypersystems: analytical study, design and control.

In an identification of some real (existing) cyber-physical heterogeneous continuous or discrete complex and/or simple dynamical hypersystems there is usually a great degree of indeterminacy, connected with the existence of stochastic disturbances and often an inaccurate knowledge of the cyber-physical heterogeneous continuous or discrete dynamical hypersystem's structure. This indeterminacy often limits statistical identification methods (Manczak and Nahorski 1983).

In this textbook, the authors have confined themselves exclusively to 'deterministic identification methods', because in the considered mathematical models of cyber-physical heterogeneous continuous or discrete dynamical hypersystems, the stochastic disturbances may be neglected. Deterministic identification methods for cyber-physical heterogeneous continuous or discrete dynamical hypersystems are interesting for practitioners and designers.

Dynamical process identification is a complex work that comprises (Wegrzyn 1974): the formulation of mathematical models, experimental studies (collection of measurement data), analytical studies (physical-parameter computer-simulation of mathematical models) and the verification (and inspection) of mathematical models.

The most important and most difficult aspect of physically continuous dynamical hypersystems is the formulation of mathematical models, which is why this is a principal aim of this textbook.

In principle, the problems that exist upon the formulation of a simple mathematical model of the physical homogeneous continuous or discrete dynamical system may be considered to have been sufficiently studied.

In the case of the formulation of a synthetic mathematical model or a functional mathematical model of the cyber-physical heterogeneous continuous dynamical hypersystem, a suitable methodology of its formulation should be applied.

In the holor theory of abstract functional heterogeneous continuous dynamical hypersystems (including physically continuous dynamical hypersystems, among others) two modes of mathematical modelling methodology may be selected: an analytical mode with decomposition methods and a synthetic mode comprising aggregation methods.

The full advantages of both modes are taken into consideration in this textbook. That is why when considering cyber-physical heterogeneous continuous dynamical hypersystems two kinds of mathematical models may be created. Namely, synthetic mathematical models for a structure of the cyber-physical heterogeneous continuous dynamical hypersystem, as well as simple mathematical models for individual, functional and structural dynamical hyposystems or components of cyber-physical heterogeneous continuous dynamical hypersystems, that is, elemental homogeneous continuous dynamical systems, hyposystems and components.

The formalisation of cyber-physical heterogeneous continuous dynamical hypersystems provides everything about the mathematical models' methodology, comprising different operations, phases and connections in a synthetic dynamical process. However, this is not the case for identification, which most often resolves itself in the identification of the homological, functional and structural dynamical components of physically continuous dynamical hypersystems, that is, physically homogenous continuous dynamical systems, or dynamical hyposystems, by means of well-known methods.

Stating a generality of knowledge on a considered problem precisely in the domain of mathematical formalism is termed 'mathematical model formulation', and concrete mathematical relationships are termed 'mathematical models'. Because the mathematical model only simulates these features, which specify a purpose for which it has been created, a given cyber-physical heterogeneous continuous dynamical hypersystem may have no single mathematical model, but instead several, simulating heterogeneous viewpoints. It may be affirmed, going out from most generalised assumptions, that each of the simple 'component' mathematical models may be led out from a certain synthetic mathematical model of the physical heterogeneous continuous dynamical hypersystem, in which they are connected to each other.

A synthetic mathematical model that is solely in the domain of abstract mathematical models and uses the Euler–Lagrange second-order differential equations of dynamics for its formation may be formatted in the dynamical systems approach and holor matrix notation. In compliance with this, it is good to determine the physical heterogeneous continuous dynamical hypersystem by means of a structure, representing the so-called 'synthetic mathematical model'. The cyber-physical heterogeneous continuous dynamical hypersystem as a whole may be concerned or it may be divided into cyber-physical homogeneous continuous dynamical systems on the grounds of which form of energy is concerned (e.g. kinetic energy: radiant, thermal, motion, sound and electrical homogeneous continuous dynamical systems; potential energy: chemical, nuclear, stored mechanical, fluidic, and electrical and gravitational homogeneous continuous systems). Next, they may be divided as cyber-physical homogeneous continuous dynamical hyposystems and, as follows, they may be divisible as functional and structural cyber-physical homogeneous continuous dynamical components.

The authors' intention in writing this textbook is to submit 'generalised physical commutation matrixer holor analyses' in the easiest practicable dynamical systems approach (systems thinking).

The author believes that this can be best realised through the application of physical commutation matrixers and the functional and structural cyber-physical homogeneous continuous dynamical components. Nearly all mechanical and electrical engineering students have been educated concisely regarding physical commutation matrixers during their university physics courses.

In view of this, the physical model represents an exceptional structure with which to form a junction between static and dynamical analysis. It is probably particularly important that the generalised physical commutation matrixer concept lends itself systematically to analogy.

Owing to the application of the physical commutation matrixer method, one may confirm that it is practicable to consider the area under discussion in mechatronics (kinetic energy—radiant, thermal, motion, sound and electrical—as well as potential energy—chemical, nuclear, stored mechanical, fluidic, electrical and gravitational). For instance, the analogy between charging an electrical capacitor and saturating a fluidic container of water is made evident by equating the physical and mathematical models of these physical processes. Naturally, one might have identified this analogy instinctively.

In spite of this, one becomes aware that these physical processes are also comparable to the variation in linear/angular velocity of a mass/moment of inertia when a force/torque, respectively, is relevant. This is not practically so evident.

As a result, the physical commutation matrixer concept confirmation is exceptionally convenient in the absence of physical situations in various cyber-physical heterogeneous continuous dynamical hyposystems for a generalised cyber-physical heterogeneous continuous dynamical hyposystem from which a universal type of solution may be completed.

A new definition of the 'physical commutation matrixer', based on matrix interconnection cyber-physical heterogeneous continuous dynamical hypersystem theory concepts, is proposed. The definition may be used to evaluate physical commutation matrixers, which perform multivalent logical functions, and continuously operating physical commutation matrixers. The serial and parallel connections of two physical commutation matrixers and simplified formulae are defined, which are valid for high component physical commutation matrixers.

A comparison is made between results arrived at using the new and conventional definitions. It is found that the results differ little in the case of equal probability output letters. For instance, a physical commutation matrixer is a general term referring to a cyber-physical heterogeneous continuous dynamical hypersystem or part of a cyber-physical heterogeneous continuous dynamical hypersystem of matrixery conductive parts and their matrixery inter-connections through which an energy-transfer holor is intended to flow.

A physical commutation matrixer, i.e. a configuration of physically (electrically, magnetically, optically, or radiationally) connected dynamical components or analogue and/or digital devices, is made up of active and passive physical, functional and structural dynamical components or an assemblage of physical, functional and structural dynamical components, and their matrix-interconnected row and column conductive collectors. Thus, a physical commutation matrixer is a physical device powered by physical energy.

The active physical, functional and structural dynamical components are the sources of physical energy for the physical commutation matrixer; for instance, they may be chemo-electrical/electro-chemical (ChE/ECh) storage batteries, direct current (DC) or alternating current (AC) mechano-electrical/electromechanical (M-E/E-M) dynamotors (generators/motors), photovoltaic cells, or fuel cells.

The passive physical, functional and structural dynamical components are impeders or admitters, i.e. resistors, inductors and capacitors.

The physical commutation matrixer described by a matrixer physical model (matrixery signal-flow diagram or map) shows the active and passive physical, functional and structural dynamical components and their matrixery interconnected row and column conductive collectors (conductors). For the purposes of analysis, apparatus (equipment) and devices with an individual physical identity are often represented by equivalent physical commutation matrixers. These equivalent physical commutation matrixers are made up of the basic passive and active physical, functional and structural dynamical components listed above. For instance, mechano-electrical (M-E) split-ring/flat or electrical commutation matrixers are used not only to convert electro-electrical (E-E) electrical energy as ME split-ring/flat commutators or macroelectronic commutators (macrocommutators) of electrical machines which are used in IEMDs, but also to transmit electrical energy in high-voltage power lines and E-E transformers or in low-voltage distribution in factories and homes. They are used to convert energy from or to its electrical form, for example, as in E-M motors, M-E generators, microphones, loudspeakers and lamps; to communicate information, as in telephones, radios, televisions and internet systems; to process and store data and make logical decisions, as in computers; and to form systems for the automatic control of equipment.

Physical commutation matrixer theory includes the study of all aspects of matrixers, including analysis, design and application. In it, the fundamental quantities are the 'energy-potential-difference holors' (i.e. generalised force or torque holors) between various points, the 'energy-transfer holors' (i.e. generalised translational or angular velocity holors) flowing in a number of row and column conductive collectors, and the parameters, which describe the passive physical, functional and structural dynamical components.

Other important physical commutation matrixer quantities, such as power, energy and time constants, may be computed from the fundamental physical variables. For a discussion of these parameters, physical commutation matrixer theory is often divided into special topics. This can be based on how the energy-potential-difference holors (i.e. voltage holors and current holors) in the physical commutation matrixer vary with time (i.e. DC or AC, sinusoidal, non-sinusoidal, digital and transient physical commutation matrixer theory). Moreover, they can be divided based on the arrangement or configuration of the energy-transfer holors' (i.e. electrical current holors) row and column conductive collectors (series, parallel, series parallel, parallel series, coupled, open circuited and short-circuited physical commutation matrixers).

The physical commutation matrixer can be divided into special topics according to which physical devices form the matrixer, or the application and use of physical commutation matrix theory can also be divided (power, communication, macro- and microelectronic, solid-state, integrated computers and programmable commutation matrixers). This textbook provides a comprehensive and careful development of the physical commutation matrixer base and the use of holor theory formulations of analytical methods.

No attempt is made to provide the required mathematical tools: matrixer theory, the functions of a complex variable, Laplace transforms, numerical methods and programming. However, some references in these areas have been included in the bibliography.

Holor analysis has been shown beyond doubt to be of prime importance in physics and engineering. Every current scientist and engineer must be systematically well acquainted with the symbolism and methods of manipulation.

Because both the nomenclature and routine manipulation are quite uncomplicated and may be learned in a short time, the question emerges as to whether an independent one-term subject on holors is adequate for a programme of scientific study.

If the subject is on the experiential level, it may be integrated into one of the necessary physics subjects, such as electronics and magnetics, mechanics, fluidics, photonics, electronics, etc. However, if one expects to use matrices as a systematically dependable tool, one must really understand them. This not only necessitates that an independent subject is taught in the programme of study, but also that there is a textbook that is dissimilar from the old school one.

An examination of the handbooks on matrix analysis since 2000 confirms that virtually all of them are founded on the same physical models. One considers these physical models to be inadequate on two counts (Moon and Spencer 1965):

- they do not refer to 'invariance', which is in reality the imperative factor that makes a matrix a matrix, a vector a vector and a scalar a scalar;
- they give too much weight to rectangular coordinates.

Matrix analysis as column matrix or vector analysis is the infant of quaternions and Ausdehnungslehre. The former are principally algebraic, and in them questions of invariance do not arise. The latter deals with geometrical figures that are tacitly assumed to be unconcerned by coordinate transformations. However, precise consideration of invariance under coordinate transformation (which has assumed such importance in the 20th century, especially after the advent of relativity) is not present in either parents or offspring. This is why some mathematicians have considered column matrix analysis or vector analysis to be a marginal subject: an ad hoc combinatory logic foundation.

The inexperienced confidence that a column matrix or a vector maintains its form and size is transformed when the coordinates are not adequate. One must also enquire, 'under what group of transformations'? Because matrix analysis has disregarded this question, it is concerned by imprecision. The unacceptable dissimilarity among free, bound and sliding column matrices or vectors is a paradigm. In addition, the old school treatment of matrices gives too much weight to orthogonal Cartesian coordinates. Even the definitions of 'gradient', 'divergence' and 'curl' are normally only provided for particular cases of rectangular coordinates, as if no other coordinates ever emerge in mathematics or physics.

Over the past couple of decades, the academic community has made considerable progress in developing educational materials and laboratory exercises for elementary mechatronics education. Students learn mathematical motion-control theory, board-level macro- and microelectronics, interfacing, and microprocessors supplemented with educational laboratory equipment. As new electrical and mechanical engineering graduates become practicing engineers, many are engaged in projects where knowledge of industrial electromechanical drive technology is an absolute must since industrial automation is designed primarily around specialised electromechanical drive hardware and software.

This textbook introduces new development trends and is in line with current trends in industry and technology, it will also help to design a modern integrated electromechanical drive (IEMD), which in the near future will replace the conventional electromechanical drive (CEMD). The textbook presents the methodology of creating mathematical models of integrated electrical machines, which is the basis for further computer-simulation studies of this IEMD.

With every passing year, it is getting more difficult to recognise the current crop of an IEMD as the descendants of Ward-Leonard, Kraemer and Scherbius systems.

At present, the IEMD is correctly stuffed with macroelectronic commutators (macrocommutators) and microelectronic controllers (microcontrollers) and human–machine interfaces (HMI) designed to take on functions once performed by these earlier Ward-Leonard, Kraemer and Scherbius systems.

IEMD users' demands are constantly varying in the market. Their challenges will be on the increase year on year. Minimising their risk as a manufacturing industry, saving energy and reducing downtime are always at the focus of all users' demands.

To continue to be competitive, IEMD suppliers often look towards technology and the latest innovations to enable their demands while reinforcing risk minimisation.

Now if IEMD users take these challenges into account and focus on IEMD cyber-physical dynamical hypersystem demands, with EM engineering complexities they will see that optimising the IEMD cyber-physical homogeneous continuous dynamical systems, hyposystems and components can be critical. IEMD users necessitate combining highly complex solutions with commutation, control, communication and automation cyber-physical homogeneous continuous dynamical systems that are high quality, reliable and are fully supported throughout the lifecycle. The IEMD cyber-physical homogeneous continuous dynamical systems, hyposystems and components need to be optimally combined to encounter the application demands, reduce transmission losses and ensure the IEMD cyber-physical heterogeneous continuous dynamical hypersystem is highly maintainable in the future. Therefore, implications further down the IEMD cyber-physical heterogeneous continuous dynamical hypersystem can be very critical for suppliers' operations as a whole.

IEMD cyber-physical heterogeneous continuous dynamical hypersystem is a concept that brings together a suite of products that connect to each other and integrate not only with each other, but also with the commutation, control, communication and automation homogenous continuous dynamical systems, hyposystems and components, all from a one-stop shop. For example, an integrated E-M motor coupled onto a gearbox and an adjustable-velocity IEMD doing the commutation, control, communication and automation integrate together in a way that it seamlessly adds value to the user.

Designing an IEMD cyber-physical heterogeneous continuous dynamical hypersystem means taking a step back to look at the bigger picture, listening to the user's demands and developing a solution that hits their exceptional demands, their future requirements and adds value way beyond just connecting IEMD cyber-physical heterogeneous continuous dynamical systems, hyposystems and components together.

The IEMD cyber-physical heterogeneous continuous dynamical hypersystem gives users more flexibility as the solution is viewed end to end rather than component by component, configured once and delivered as a package.

The benefit for the user is the reassurance that support on the IEMD cyber-physical heterogeneous continuous dynamical hypersystem is a single phone call away.

At suppliers, they bring together all their research and development (R&D) and innovation, all their engineering expertise for the given application in order to make a difference to the user. For the authors, optimisation is taking a step back and seeing the bigger picture; offering a solution that is better for the user.

At suppliers, they are in an exceptional position to have the breadth of portfolio that allows their scientists and engineers to propose a less biased perspective when offering a solution. For example, a conveyor solution could have centralised or decentralised control, which in turn would affect the type of IEMD cyber-physical heterogeneous continuous dynamical systems, hyposystems and components offered, which therefore has an impact on IEMD cyber-physical heterogeneous continuous dynamical hypersystem solution.

Most users currently are interested in gaining rich data quality from their plants. So again not only do they minimise the risk of putting the IEMD heterogeneous continuous dynamical systems, hyposystems and components together in an optimised solution so it meets their application demands, but they minimise risk by integrating cyber-physical heterogeneous continuous dynamical hypersystem seamlessly with their proven IEMD platform, which enables their users to make operational decisions based on the data presented to them.

Efficiency and productivity are decisive success factors for manufacturing industries. Engineering plays a central role in this especially as it relates to ever more complex machinery and plants. For that reason, a high level of efficiency is already demanded at the engineering stage, as the first step toward better production: faster, more flexible, and more intelligent. The supplier has an intelligent answer to this: uniform hardware and software interfaces. These shared characteristics minimise engineering time. The result: lower costs, reduced time to market, and greater flexibility.

Suppliers have suffered with engineering resource over the years, and often their users are looking towards suppliers to support them in the design and implementation of solutions IEMD. IEMD is the name given to efficient interoperability of all IEMD heterogeneous continuous dynamical systems, hyposystems and components. The open IEMD cyber-physical heterogeneous continuous dynamical hypersystem architecture covers the entire production process and is based on the consistent presence of shared characteristics: consistent data management, global standards, and IEMD cyber-physical heterogeneous continuous dynamical hypersystem gives them access not only to their breadth of portfolio but also engineering resource and experience across the IEMD, from both suppliers and their partner network. Putting the risk back to companies who have the suite of products and the best in class engineering knowledge is a real advantage.

Users do not have to invest in the infrastructure in place to be able to maintain these complex solutions. They can then focus on their products and their R&D and they will not need to worry about everything else.

What they are trying to do here for the user is to say take the helicopter view with them; let us look at the overall solution, end to end. Let us design a solution that surpasses user demands. Then they ask their users: 'What else do you need? What other value add would you like to see for this application?' listening to the market and driving innovation and collaboration.

In conclusion, the major benefit of IEMD is risk minimisation. If anything fails, users know they need to only go to one company to sort it out. The authors really want the user to be feeling that, actually it is a supplier they can rely on. With IEMD, they aim to be the trusted partner for their users, focusing on a solution that really works for them, really advances their processes.

Ultimately, IEMD will minimise their risk, recover their efficiencies and increase their profitability. A real win–win for both partners.

The treatment presented in this interdisciplinary textbook differs from the old school one by introducing invariance into the theory and providing general definitions that are sensible in all coordinate systems. In this approach, matrix analysis is endowed with a concrete logical foundation. Irrespective of the forward thinking aspects of the textbook, the authors do not sense that the treatment is too difficult for ordinary readers. This interdisciplinary subject has been taught to undergraduates and postgraduates at the Cracow University of Technology in Krakow and the State Higher Vocational School in New Sandec (Nowy Sacz), Poland. This textbook has been written primarily for undergraduate students of physics and electrical, mechanical, fluidic and thermal engineering. Its logical structure, however, should also make it valuable for mathematicians, and it may serve as a helpful review source for graduate students.

The first few chapters of the book introduce a variety of concepts that may be unknown to some readers, but mastery of these concepts may provide a much deeper knowledge of matrix analysis than could otherwise be acquired. Index notation is employed where necessary, but applications to electronics and mechanics, etc, are performed in the well-known physical commutation matrixer nomenclature. Various problems are known to provide readers with satisfactory preparation in using holors.

We are the authors of this textbook and all the text contained herein is of our own conception unless otherwise indicated. Any text, figures, theories, results, or designs that are not of our own devising are appropriately referenced in order to give acknowledgement to the original authors. All sources of assistance have been assigned due acknowledgement. In this textbook, all the information has been obtained and presented in accordance with academic rules and conduct. We have fully cited and referenced all the material and results that are not original to this book.

We are also indebted to our international and national colleagues who contributed indirectly to this book. In addition, we are grateful to all of you who have adopted this text for your interdisciplinary classes or for your own use. Without you we would not be in business. We hope that you find this textbook to be a valuable learning tool and reference for students.

The authors are also grateful to their friends for helpful comments. The above declaration should not be interpreted as a suggestion that these benevolent friends concur with all the iconoclastic propositions presented in the text; quite the contrary!

For all the radicalisms and imperfections, we assume total responsibility. We would like to conclude this preface by sharing our life motto—with focus, motivation and concerted efforts, one can make the seemingly impossible, possible.

Bogdan Thaddeus Fijalkowski
Jozef Tutaj
Krakow, 31 August 2018

References

Fijalkowski B T 2010 *Automotive Mechatronics: Operational and Practical Issues. Vol I. International Series on Intelligent Control, and Automation Science and Engineering* vol 47 (Berlin: Springer)

Fijalkowski B T 2011 *Automotive Mechatronics: Operational and Practical Issues. Vol II. International Series on Intelligent Control, and Automotive Science and Engineering* vol 52 (Berlin: Springer)

Fijalkowski B 2016 *Mechatronics: Dynamical Systems Approach and Theory of Holors* (Bristol, UK: IOP Publishing)

Manczak K and Nahorski Z 1983 *Komputerowa identyfikacja obiektow dynamicznych* (Computerised Identification of Dynamical Objects) (Warszawie: Panstwowe Wydawnictwo Naukowe) (in Polish)

Moon P H and Spencer D E 1965 *Vectors* (Princeton, NJ: Van Nostrand)

Tutaj J 2012 *Ujecie systemowe dynamiki wielofunkcyjnego prądnico-rozrusznika silnika spalinowego (Dynamic systems approach of the polyfunctional generator-starter for a combustion engine of the automotive vehicle)* Monografia 409, Seria Mechanika (Krakow: Politechnika Krakowska im. Tadeusza Kosciuszki) (in Polish)

Wegrzyn S 1974 *Systemy automatyki kompleksowej* Teoria i doswiadczenie Prace VI Krajowej Konferencji Automatyki 1974 (in Polish)

Acknowledgements

The authors give their sincere thanks to the many friends and students who not only inspired their initial interest in the subject, but also gave many helpful suggestions during the writing of the book. Particularly, the authors wish to express their gratitude to their families, who have been indispensible in the writing of this book, and whose enthusiasm, advice, and encouragement helped make all of it possible. Special thanks go to Bogdan's grandson Marcel for his very helpful discussions and for the preparation of many of the figures in this book. We are grateful to the many authors referenced in this book from whom, during the course of writing, we learned so much regarding the subjects covered. We are also indebted to our international and national colleagues who contributed indirectly to this book. We would like to take this opportunity to also thank the IOP Publishing crew for their persistence in making this book a reality. In addition, we are grateful to all of you who have adopted this text for your interdisciplinary classes or for your own use. Without you we would not be in business.

Author biographies

B T Fijalkowski, Professor, Dr;

Professor Dr Bogdan T Fijalkowski was born in 1932 in Poland. He obtained an MSc in Electrical Engineering, PhD in Power Electronics and DSc in Physical Heterogeneous Continuous and Discrete Dynamical Systems from Szczecin University of Technology in 1959, Academy of Mining and Metallurgy in 1965 and Poznan University of Technology in 1988, respectively. He began his professional and electrical engineering carrier in 1955. He has been working for five years in industry and 50 years in academia; he was the Director of the Electrotechnics & Industrial Electronics Institute, Faculty of Electrical and Computer Engineering and he was the Head of Automotive Mechatronics Institution, Faculty of Mechanical Engineering at the Cracow University of Technology, Poland. He was a visiting professor at several well-known universities. He serves as Consultant to several organisations in Poland and the United States. He is the signatory of the MoU for the establishment of World Electric Vehicles Association (WEVA). He was Guest Editor of *Journal of Circuits, Systems and Computers*—Special Issue on Automotive Electronics, reviewer and referee, *IEEE Transactions on Circuits and Systems*—Part I, *IEEE Transactions on Fuzzy Systems*, *International Journal of Vehicle Design*, *USA*, *International Journal on Advanced Robotic Systems* and *International Journal of Technology Management (IJTM),* USA. He is a reviewer of books and book chapters on automotive electrics and electronics, electric drives systems dynamics and wireless information transmission, as well as information security, and is listed in *Who's Who in the World, Who's Who in Science and Engineering* and so on. He has published 30 books and book chapters as well as over 200 technical papers and 25 patents on mining and automotive electrics and electronics as well as mechatronics. Recent publications by him include books (monographs), book chapters, journal articles, and conference proceedings on topics such as mathematical models of selected aerospace and automotive discrete dynamical hypersystems as well as the civil and military, wheeled and tracked all-electric and hybrid-electric vehicles and also nanomagneto-rheological fluid (NMRF) mechatronic commutator, 'crankless' internal combustion engines with energy storage, termed the Fijalkowski engines and automotive gas turbines that are based on the Fijalkowski turbine boosting (FTB) system. Recently, Springer published two volumes of his book entitled *Automotive Mechatronics: Operational and Practical Issues* and IOP Publishing published his interdisciplinary book entitled *Mechatronics: Dynamical Systems Approach and Theory of Holors*. He made valuable contributions to several of ASME, ISATA, ISTVS, SAE, IEEE SMC, as well as EVS and WEVA, conferences as a track and/or session chair and speaker. His hobby is yachting. He has the rank of Ocean-Going Yacht Master.

J Tutaj, Professor, Dr;

Professor Dr Jozef Tutaj was born in 1957 in Poland. He obtained an MSc in Electrical Engineering, PhD in Electrical Engineering and DSc in Electro-Mechanical Drive Systems from Academy of Mining and Metallurgy in 1981 and 1996 as well as Cracow University of Technology in 2013, respectively. He began his professional and electrical engineering career in 1981. He has been working for one year in industry and 30 years in academia; he was the Head of Mechatronics Institution, Faculty of Mechanical Engineering at the Cracow University of Technology, Poland. He has published three books and book chapters as well as over 50 technical papers and five patents on automotive electrics and electronics as well as mechatronics. Recent publications by him include books (monographs), book chapters, journal articles, and conference proceedings on topics such as mathematical models of electrical machines and automotive drive systems. Recently, he published his DSc dissertation entitled *Dynamical Systems Approach of the Polyfunctional Generator-Starter for a Combustion Engine of the Automotive Vehicle*, which has received the Minister of Economy Award in 2015. He made valuable contributions to several SAE, IEEE, as well as other conferences as a session chair and speaker. His hobby is classical music.

Acronyms

ABW	absorb-by-wire
AC	alternating current
AEV	all-electrical vehicle
AI	artificial intelligence
AV	adjustable velocity
AWA	all-wheel-absorbed
AWB	all-wheel-braked
AWD	all-wheel-driven
AWS	all-wheel steered
BBW	brake-by-wire
BEMF	back electromotive force
BLG	brush-lifting and split-ring short-circuiting gear
CASE	computer-aided system engineering
CHINT	charge injection transistor
CHP	combined heat and power
ChE	chemo-electrical
CMOS	complementary MOS
COTS	commercially-off-the-shelf
CPU	central processing unit
CS	computer system
CS	current source
CUPL	computer for universal programmable logic
D	differential
DBW	drive-by-wire
DC	direct current
DD	direct-drive
DDPWM	direct duty-ratio PWM
DoF	degrees-of-freedom
DOL	direct-on-line
DSP	digital signal processor
DTC	direct torque control
ECE	external combustion engine
ECL	emitter-coupled logic
ECM	electrical commutation matrixer
ECU	electronic control unit
E2CMOS	electrically erasable complementary metal-oxide semiconductor
E2D	electrical energy distribution
EFD	electro-fluido-dynamical
EIC	electrical integrated circuit
EIM	electrical integrated matrixer
EMD	electro-mechano-dynamical
EMF	electromotive force
EOF	electro-osmotic flow
EPD	electro-plasmo-dynamical
EPROM	erasable PROM
EMU	electrical multiple unit
EE	electro-electrical

ECh	electro-chemical
EM	electromechanical
EMI	electromagnetic interference
FBW	fly-by-wire
FEM	finite elements method
FET	field effect transistor
FF	ferro-fluid
FFET	flow field effect transistor
FL	fuzzy logic
FLC	full-load current
FM	frequency modulation
FMMEA	failure mode and effect analysis
F-M	fluido-mechanical
FPGA	field programmable gate array
FPLA	field programmable logic array
FSLESL	freescale embedded software libraries
GaN	gallium nitride
GCM	generalised commutation matrixer
GERF	giant electro-rheological fluid
GTO	gate-turn-off
HBT	hetero-junction bipolar transistor
HD	heavy duty
HDI	high-pressure direct injection
HDL	hardware description language
HE	hybrid electrical
HEMT	high electron injection transistor
HEV	hybrid electrical vehicle
HMC	human–machine communication
HMI	human–machine interface
HOLM	holor modulation
HOLPWM	holor PWM
HPS	hybrid power source
HV	high voltage
HVDC	high-voltage direct current
I	integral
ICE	internal combustion engine
IEEE	Institution of Electrical and Electronics Engineers
IEC	International Electrotechnical Commission
IEMD	integrated electromechanical drive
IGBT	insulated-gate bipolar transistor
IGCT	integrated-gate commutated transistor
IP	internet protocol
IPM	interior permanent magnet
ISO	International Standard Organisation
IT	information technology
I/O	input/output
JFET JUGFET	junction-gate FET
KFL	Kirchhoff's first law
KSL	Kirchhoff's second law
LAN	local area network

LCI	load commutated inverter
LED	light-emitting diode
LTQ	light triggered and quenched
MIMO	multi-input/multi-output
MCM	mechanical commutation matrixer
MCT	MOS controlled thyristor
MCU	microcontroller unit
MEMS	micro-electromechanical system
MESFET	metal–semiconductor FET
MFD	magneto-fluido-dynamical
MFOC	magnetic-field-oriented control
MIMO	multiple-input multiple-output
MISFET	metal–insulator–semiconductor FET
MMC	man–machine communication
MMD	magneto-mechano-dynamical
MMF	magnetomotive force
MNOS	metal-nitride-oxide silicon
MOS	metal-oxide semiconductor
MOSFET	metal-oxide semiconductor FET
MPD	magneto-plasmo-dynamical
MRV	Martian Roving Vehicle
MSI	medium scale integration
MSI	mild soft iron
ME	mechano-electrical
M-F	mechano-fluidic
MM	mechano-mechanical
M-P	mechano-pneumatic
M-V	mechano-vacuum
MPWM	multi PWM
N	north
NEMA	National Electrical Manufacturer Association
NF	neural network and fuzzy logic
NMOS	non-complementary metal-oxide semiconductor
NMRF	nanomagneto-rheological fluid
NN	neural network
OFET	organic field effect transistor
OTS	off-the-shelf
P	proportional
PBC	passively based control
PC	personal computer
PC	predictive control
PC	process control
PCC	point of common coupling
PF	power factor
PHEV	plug-in hybrid-electrical vehicle
PI	polarisable interference
PI	proportional-integral
PID	proportional-integral-derivative
PLC	programable logic controller
PLM	programmable logic matrixer

PML	programmable matrix logic
PROM	programmable read-only memory
PWM	pulse-width modulation
P-M	pneumo-mechanical
QUIT	bipolar quantum interference transistor
RB	reverse blocking
RBSOA	reverse-biased safe-operating area
RBW	ride-by-wire
RCGTO	reverse conducting GTO
RHC	real holor control
R&D	research and development
RMS	Root mean square
R–C	resistor–capacitor
R–L	resistor–inductor
R–L–C	resistor–inductor–capacitor
S	south
SCR	semiconductor controlled rectifier
SET	single electron transistor
SF	service factor
Si	silicon
SiC	silicon carbide
SINM	sinusoidal modulation
SINPWM	sinusoidal PWM
SISO	single-input/single-output
SIT	static induction transistor
SMES	superconducting magnetic energy storage
SM&GW	steered, motorised and/or generatorised wheel
SOP	sum-of-product
SPWM	single PWM
SR	semiconductor rectifier
SSI	standard scale integration
SUPER-HET	bipolar superconductor hot-electron transistor
TAB	tape automated bonding
TTL	transistor–transistor logic
UAS	unnamed aircraft system
UGV	unnamed ground vehicle
UPS	uninterruptible power supply
UV	ultra-violet
VAP	very advanced propulsion
VAT	very advanced technology
VECM	vector modulation
VECPWM	vector PWM
VDU	video display unit
VLSI	very large scale integration
VoIP	voice over Internet protocol
VPI	vacuum pressure impregnate
VR	variable reluctance
VS	voltage-source
V/F	voltage per frequency
V/Hz	voltage per hertz

XBW	X-by-wire
Δ2Y	delta-to-wye
Π2T	pi-to-tee
2-D	two dimensional
3-D	three dimensional
2 × 1	two-by-one
2 × 2	two-by-two
3 × 2	three-by-two
3 × 3	three-by-three
3 × 5	three-by-five
5 × 5	five-by-five
5 × 6	five-by-six

IOP Publishing

The Integrated Electro-Mechanical Drive
A mechatronic approach
B T Fijalkowski and J Tutaj

Chapter 1

General considerations

'... Absolutely amazing! I wish I had a tool like this when I was learning about motors. Five stars!'

Nikola Tesla
Inventor of the AC induction motor

1.1 Introduction

Modern **electro-mechanical drive** (EMD) users are more likely to choose suppliers who are able to provide them with a comprehensive delivery. Such a solution gives them a sense of security because it avoids problems resulting from a possible incompatibility.

In the case of a modern EMD, it is often the case that the EMD suppliers do not recommend using **direct current** (DC) and **alternating current** (AC) commutator synchronous or asynchronous (induction) motors with macroelectronic commutators (macrocommutators), micro-electronic controllers (microcontrollers) and **human–machine interfaces** (HMI) from different manufacturers. This is all the more justified when the authors talk about an **integrated electro-mechanical drive** (IEMD) with macrocommutators, microcontrollers and HMIs as well as components of precision mechanics such as gears, rotary tables, cross tables, electric cylinders, etc. Buying an IEMD in a set is therefore completely natural. The only exception is when a very simple solution is sought, e.g. a stepper motor operating in an open feedback loop. Then the user can allow the EM motor and its driver to come from different suppliers. In the case of more advanced IEMDs, this solution has no reason to exist, because the **electro-mechanical** (EM) motor and driver are inseparable.

In this day and age, selling any automation components without providing adequate technical support is very difficult. These are the realities of the market and it is what users expect. An IEMD is more and more technologically advanced, new functions are constantly being added. New suppliers also appear on the market. It is

doi:10.1088/2053-2563/aae7d7ch1

© IOP Publishing Ltd 2019

difficult for users to keep up with all these new products. Especially because in the IEMD or other equipment some details are not always obvious, there are a lot of nuances. Of course, users can read all this in the detailed technical documentation, but it is much easier to ask for a technical advisor, which is very useful in this situation. Considerable competition also means that the suppliers themselves strive to provide their users with as much as possible within the so-termed added value. It happens that this is no longer just a help in the selection of components, after-sales technical support, or an efficiently operating service. Sometimes the suppliers participate in the implementation of the project, putting into practice and finally in commissioning.

Such a cooperation hypothesis means that on the one hand the recipient can count on comprehensive support at all stages of the implementation of the project, and on the other hand, the supplier will win the contractor's loyalty on subsequent projects. In short: without good technical support there is no good sale.

Over the past two decades there is have been several attempts to bring a modern EMD closer and more tightly integrated with novel **magneto-mechano-dynamical** (MMD) electrical machines. Such activities have intensified over the past several years. The drivers for this are technology breakthroughs in integrated power electronic devices, termed by the authors' macroelectronic commutators (macro-commutators), new materials as well as the increasingly cyber-physical heterogeneous dynamical hypersystem requirements for a wide range of applications.

This textbook will cover a wide range of applications highlighting the advantages and challenges of achieving such integration. The textbook will also highlight the current research trends.

The IEMD is a concept that brings together an arrangement of EMD electrical machines: macrocommutators, microcontrollers, and HMIs as well as gearboxes, clutches and actuators that connect to each other and integrate not only to each other, but also to the control and automation cyber-physical homogeneous continuous dynamical system, all from a one-stop shop. An EM motor coupled with a gearbox and an adjustable-velocity IEMD doing the control integrate together in a way that it seamlessly adds value.

Designing an IEMD means taking a step back to look at the bigger picture, listening to user requirements and developing a solution that hits the users unique requirements, user emerging and future IEMD cyber-physical heterogeneous dynamical hyposystem needs and adds value way beyond just connecting cyber-physical homogeneous dynamical systems, hyposystems and components together. The IEMD gives you more flexibility as the solution is viewed end-to-end rather than component-by-component, configured once and delivered as a package.

In general, an IEMD cyber-physical heterogeneous dynamical hyposystem (as illustrated in figure 1.1), can be defined as a power conversion means characterised by its capability to efficiently convert electrical energy from an electrical energy source (voltage and current) into mechanical energy (torque and velocity) to control a mechanical load or process. In some cases, this mechanical energy flow is reversed or can even be bilateral as regards the energy-flow direction.

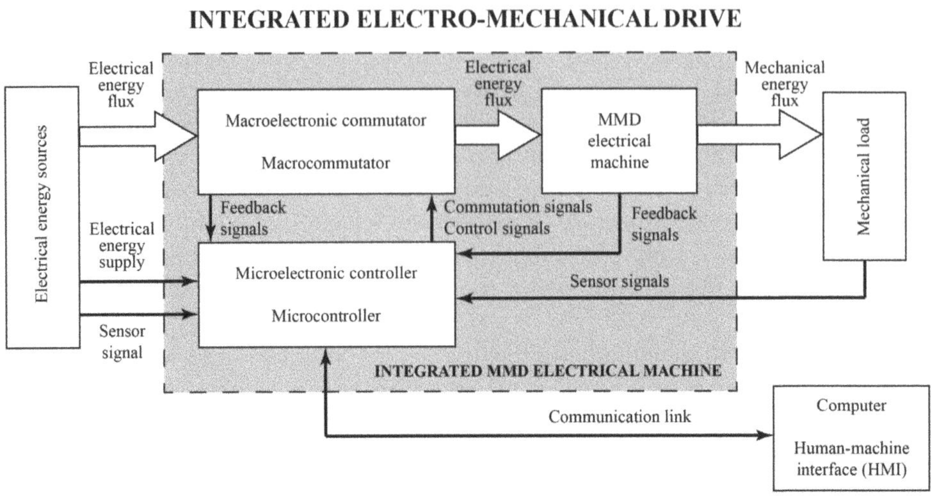

Figure 1.1. IEMD cyber-physical heterogeneous dynamical hypersystem.

At present an IEMD makes use of a macroelectronic commutator (macro-commutator) to (digitally) control this EM energy conversion process. In addition, as an IEMD is being integrated more and more in cyber-physical heterogeneous dynamical hypersystems, communication links to higher level computer networks are essential to support commissioning, initialisation, diagnostics and higher level process control.

Consequently, the main IEMD components consist of an EM energy converter (usually an MMD electrical machine or actuator), an embedded macroelectronic **electrical-to-electrical** (EE) energy converter, i.e. a macrocommutator and an embedded digital control unit. The digital control unit directly controls the macro-electronic semiconductor electrical valves (electronic switches) of the macroelectronic converter. To this end not only suitable control hardware, sensors, high-speed digital logic devices and processors are needed but also suitable control algorithms. From this perspective, IEMD technology is a fairly modern development. Indeed, although MMD electrical machines were first developed over 160 years ago, power electronic converters have been available for only 45 years, dynamic torque control algorithms for AC–DC–AC commutator induction motors (magnetic-field oriented control) have been around for about 40 years and high-speed digital control using **digital signal processors** (DSP) have been available for less than 60 years. Even now with all components (integrated electrical machine, macroelectronics, control hardware and software) being developed, IEMD technology is still evolving at a rapid pace. Over the past two decades, new integrated electrical machine types have been developed, optimised and investigated, such as surface **interior permanent magnet** (IPM) and buried IPM electrical machines, commutated-reluctance electrical machines, transversal magnetic-flux electrical machines, axial magnetic-flux electrical machines, linear electrical machines, etc.

Each electrical machine type requires its specific control and sensors. During the past 10 years, the position of sensorless IEMDs have been investigated to eliminate expensive sensors and make an IEMD more robust (reliable). The power range of a modern IEMD spans many decades, from milliwatts up to hundreds of megawatts, which demonstrates the flexibility and the broad application of this technology.

In the following several technology trends of state-of-the-art EMD are being discussed. An attempt is made to derive future trends based on the development of an IEMD over the past 20 years.

IEMD technology represents growing markets, albeit less impressive than recent **information technology** (IT) and nanotechnologies, but has proven to be a robust market segment which has been affected less by speculation and global market fluctuations or crises. One can say that an IEMD literally is a robust cyber-physical heterogeneous dynamical hypersystem which keeps the world's economy moving towards higher prosperity (more work done by machines) and more efficient use of primary energy (as an adjustable-velocity IEMD is more efficient when production rates need to be adapted).

The needs of users are varying in the market all the time. The challenges users deal with ought to be increasing year on year. For original suppliers, reducing users risk as a business, minimising downtime, saving energy, and decreasing engineering time are always a requirement of users'. To remain competitive, suppliers routinely look towards technology and the most recent innovations to supply their needs while minimising risk.

In this textbook the authors bring together all their years of innovation and engineering expertise for the given application in order to make a difference to you, the user.

The IEMD cyber-physical heterogeneous dynamical hypersystems is a trend-setting answer to the high degree of complexity that characterises the IEMD and its automation technology currently.

The world's only proper one-stop answer for the whole IEMD cyber-physical heterogeneous dynamical hypersystems is particularly characterised by threefold integration: horizontal, vertical, and lifecycle integration demonstrate that every dynamical system, hyposystem and/or component fits seamlessly into the whole IEMD cyber-physical heterogeneous dynamical hypersystem, into any automation environment, and even into the entire lifecycle of a plant.

The vision of the IEMD treatment is to develop the necessary technology so that IEMD capabilities can be economically embedded inside emerging and future integrated EM motors with minimal impact on their size, mass, and environmental robustness.

The long-term goal is to develop integrated EM motors with adjustable-velocity capabilities that, from their external appearance, show minimal evidence of the internally-packaged IEMD macroelectronic commutators (macrocommutators) and microelectronic controllers (microcontrollers).

Equally important, this IEMD must be manufacturable with a minimal cost premium while demonstrating environmental robustness and reliability characteristics that match those of conventional EM motors currently.

Consistent with these minimal impact objectives, the input power quality and **electromagnetic interference** (EMI) characteristics of an emerging and future IEMD must approach those of the integrated EM motor fed directly from the utility grid.

One of the most important rewards accompanying the success of emerging and future IEMDs will be major energy savings resulting from the cyber-physical heterogeneous dynamical hypersystem efficiency improvements made possible by introducing adjustable-velocity capabilities into applications that use fixed-velocity conventional EM motors today.

As the cost of electrical power inevitably increases during the coming years, the lifetime cost savings generated by the introduction of an IEMD will make the integrated EM motor increasingly attractive for new applications.

If the demanding technical challenges associated with the development of an IEMD can be successfully surmounted, the day will arrive when integrated EM motors will be promoted with the baseline expectation that they have adjustable-velocity capabilities.

Since the IEMD macrocommutators and microcontrollers ought to be embedded inside the electrical machine, many users may not even be aware that it is there. That is, adjustable-velocity capabilities will become an assumed inherent feature of integrated EM motors.

Development of mature, low-cost IEMDs will greatly accelerate the penetration of the IEMD into a wide variety of applications ranging from aerospace, aviation, automotive, metallurgy, mining and cement, and home appliances. Early steps towards achieving this vision can already be seen in industry today, and the objective of the IEMD is to develop technology that will accelerate the practical realisation of this ambitious vision. Thus the effective integration of emerging and future IEMDs require the development of technology that allows for volume and mass reduction of critical components.

The **research and development** (R&D) teams are studying the potential for volume and mass reduction through the integration of macrocommutators and microcontrollers into an integrated EM motor. Integration of macrocommutators and microcontrollers into the integrated EM motor frame offers space saving advantages, allowing the EM motor macrocommutators and microcontrollers to share the same housing and cooling cyber-physical homogeneous dynamical system. Accordingly, significant volume and mass reductions are possible in the macroelectronics and microelectronics housing and cooling auxiliaries.

The aim of the integrated high-power electronic (macroelectronic) commutators (macrocommutators) is to develop technologies that will enable emerging and future, macroelectronic-based electrical energy processing units. The state-of-the-art electrical energy processing unit is still largely the mechanical split-ring or flat (rotary disc) commutators, developed in the late 19th century.

Although it has long been argued that macroelectronic commutators can help improve IEMD controllability, reliability, and overall energy and power efficiency, their penetration in electrical energy processing units is still quite low.

The often-cited barriers of higher cost and lower reliability of the macrocommutators are quite high if high-power macroelectronics is used as a direct, one-to-one, replacement for the existing mechanical split-ring or flat commutators. However, if the whole IEMD was designed as a cyber-physical heterogeneous dynamical hypersystem of controllable macrocommutators, the overall cyber-physical heterogeneous dynamical hypersystem cost and reliability could actually improve, as is currently the case at low-power microelectronics within computer and telecommunications equipment.

The vision of the macrocommutators is to develop concepts for macroelectronic-based electrical energy processing units that can impact applications in an IEMD. The new macrocommutators thrust inherits the previously existing mechanical split-ring or flat commutators thrust, with scope expanding to a wider power and application range.

With the new vision, there are four major research focuses:

- macrocommutator and microcontroller architecture design and optimisation;
- electrical energy processing and control;
- high-density macrocommutator integration;
- physical and mathematical modelling, analysis, simulation and management.

The application focus will be on autonomous, electrical energy processing units, and on emerging and future applications such as portable, alternative, and sustainable energy sources.

The emerging and future of AC or DC motor IEMDs is the driving force behind all cyber-physical heterogeneous dynamical hypersystems used in industry, commerce and buildings. They are used in a wide range of applications in many industries such as aerospace, aviation, automotive, metallurgy, mining and cement, and home appliances, improving the efficiency and reliability of these processes while at the same time improving safety and energy savings.

As many technologies continue to evolve, R&D teams continue to work on making the AC or DC motor IEMDs even smaller and more affordable. However, it is not only size that matters.

Scientists and engineers are designing an AC and DC motor IEMD that is more intelligent, has better communications and is easier to install and control. Such an AC or DC motor EMD will open the door to many new markets, such as chemical, pulp and paper, metal and oil and gas, and contribute enormously to increasing applications and provide manufacturers with a whole host of new market IEMD opportunities (Barnes 2003, Wilamowski and Irwin 2011, Chan and Shi 2011, Wach 2011, Hughes and Drury 2013, Holmes and Lipo 2003, Geyer 2017, Rashid 2018).

Three main torque and energy conservation mechanisms, for integrated AC and DC MMD electrical machines, exist.

- *Synchronous:* Electromagnetic torque results because of the interaction of a time varying electromagnetic rotational field generated in the stator windings and a stationary electromagnetic or magnetoelectric field established by the windings or interior permanent magnets (IPM), respectively, in the EM motor.
- *Synchronous (variable reluctance):* Electromagnetic torque produced to minimise the reluctance of the electromagnetic system. Thus the torque is created in an attempt to align the minimum reluctance path of the rotor with the time varying rotating air gap.

- *Asynchronous (induction):* Electromagnetic torque is the result of a time varying electromagnetic rotational field present due to time varying voltage or motion of the rotor with respect to the stator.

The R&D teams' most recent innovation, the vision of the AC or DC motor IEMD is considered a 'revolution' in terms of size and simplicity, with 'extensive' performance and overall functionality.

An AC and a DC motor IEMD ought to consist of its individual units, as a microcontroller, a macrocommutator and an EM motor windings, fully integrated as a single intelligent EM power module, which is an integrated AC–AC or AC–DC–AC or DC–AC commutator synchronous or asynchronous (induction) motor.

The macrocommutator and microcontroller are compactly built into the integrated AC–AC or AC–DC–AC or DC–AC commutator synchronous or asynchronous (induction) motor, effectively saving space in the equipment due to the small amount of installation and wiring space required.

No cables are required to connect the integrated AC–AC or AC–DC–AC or DC–AC commutator synchronous or asynchronous (induction) motor, macrocommutator and microcontroller. Problems caused by electrical noise can also be expected to decrease.

The compactness and optional mounting are made possible by using the latest technologies, such as the newest-generation of electrical valves, for instance, **reverse blocking** (RB) **insulated gate bipolar transistors** (IGBT) or **integrated gate commutated thyristors** (IGCT) and an innovative **ferro-fluid** (FF) cooling homogeneous dynamical system.

Technologies such as macroelectronics, microelectronics and nanoelectronics, software, sensors, industrial communication and materials science are making these individual units smaller and smarter. The overall result is a more complex, technologically advanced, not to mention cost effective IEMD family with a broad range of industrial and consumer applications.

The use of very advanced semiconductor or superconductor electrical valves (electronic switches) or intelligent EM power modules enable the application of control techniques that, a few decades ago, seemed only a vision.

In the last decades, the integrated AC–AC or AC–DC–AC or DC–AC commutator synchronous or asynchronous (induction) motors themselves have improved in efficiency by an average of 5%. Moreover, the journey does not end here. The integrated AC–AC or AC–DC–AC or DC–AC commutator synchronous or asynchronous (induction) motor and IEMD technologies such as high-energy **interior permanent magnets** (IPM), semiconductor or superconductor commutated EM motors, silicon micromotor technology and soft magnetic materials are developing at a record pace. In fact, scientists and engineers have further developed IPM and very high-voltage integrated EM motor technology to satisfy various user requirements.

In industry, conventional EM motors and EMD powering mechanical equipment account for about 65% of the total electrical energy consumed. Reducing this figure is therefore of prime importance when it comes to electrical energy savings. The longer EMD and EM motors are in operation, the higher the savings. Over the total running time, more than 97% of the total cost of an IEMD cyber-physical heterogeneous dynamical hypersystem is accounted for by its power consumption and only 3% by the capital investment, hence the vital importance for high standard EM motor efficiency.

With years of experience and know-how, the authors are showing the way in the development of AC and DC IEMD technology. One such development has been a radical new control technique termed **holor control** (HC).

The HC contributes directly to energy efficiency by an integrated AC–AC or AC–DC–AC or DC–AC commutator synchronous or asynchronous (induction) EM motor's magnetic-flux optimisation. Sinusoidal waveform on voltage and current ensures that will be absolutely no interferences from the IEMD to other sensitive electronic equipment, such as radars, echo sounders, and seismic research instrumentation, as well as guaranteeing no bearing damages, and an absolute minimum of acoustic noise from the integrated EM motor and IEMD.

Original closed loop HC technology maintains positioning operation even during abrupt load fluctuations and accelerations. The rotor position detection sensor monitors the rotation. When an overload condition is detected, the HC will instantaneously regain control using the closed loop mode.

When an overload condition continues the HC will output an alarm signal, thereby providing reliability equal to that of an integrated AC–AC or AC–DC–AC or DC–AC commutator synchronous or asynchronous (induction) motor.

Another striking development affecting the AC or DC IEMD is miniaturisation. Increasing individual unit integration means **integrated matrixer** (IM) and **integrated circuit** (IC) boards are becoming smaller, which in turn leads to more cost- and energy-efficient manufacturing. On top of this, environmentally-friendly and energy-efficient technologies combined with sound manufacturing processes and increased recycling of resources contributes enormously to the environmental health of the planet.

AC and DC IEMD technology is a field that inspires scientists and engineers to develop better cyber-physical heterogeneous dynamical hypersystems with innovative technology. In addition, it contributes to the efficient use of electrical energy, which in turn improves the economy of the users. Finally, yet importantly, IEMD technology supports a sustainable development for all people.

In this textbook, let the authors 'drive' the readers through the interesting world of the IEMD and integrated AC–AC or AC–DC–AC or DC–AC commutator synchronous and asynchronous (induction) motors. The authors wish the readers of this textbook an interesting journey.

1.2 Definitions of integrated electro-mechanical drive (IEMD) cyber-physical heterogeneous dynamical hypersystems

Whenever the term of an EM motor or ME generator is used, one tends to think that the translational or angular velocity of these MMD electrical machines are totally controlled only by the applied voltage and frequency of the AC power source.

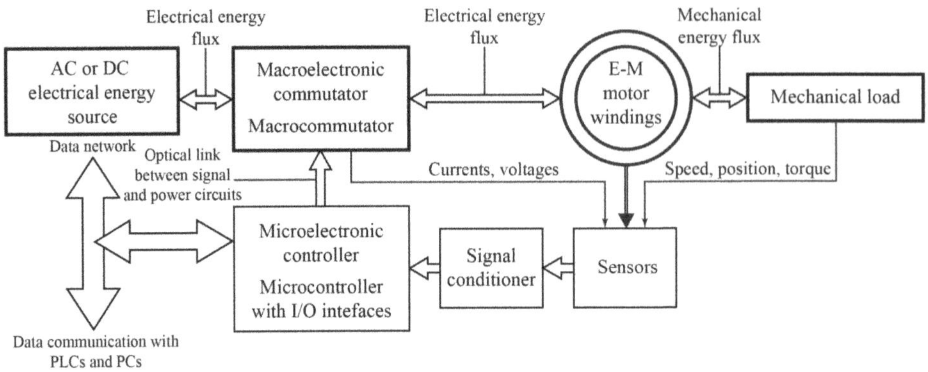

Figure 1.2. General structural and functional diagram of the emerging and future AC or DC motor IEMD cyber-physical heterogeneous dynamical hypersystem.

However, the translational or angular velocity of an electrical machine can be controlled precisely, also by implementing the vision of the AC or DC motor IEMD.

The main advantage of this concept is that the motion control is easily optimised with the help of the AC or DC motor IEMD. In very simple words, the cyber-physical heterogeneous dynamical hypersystem, which controls the motion of the electrical machines, is known as the AC or DC motor IEMD.

A typical cyber-physical IEMD dynamical hypersystem is assembled with an integrated EM motor (there may be several) and a sophisticated control cyber-physical dynamical system that controls the translational or angular velocity of the EM motor's mover or shaft, respectively.

Currently, this control can be done easily with the help of software. Therefore, the controlling becomes more and more accurate and this concept of AC or DC motor IEMD provides the ease of use. This IEMD cyber-physical heterogeneous dynamical hypersystem is widely used in a large number of industrial and domestic applications like factories, transportation systems, textile mills, fans, pumps, motors, robots etc. The IEMD is employed as the prime mover (starter) for diesel or petrol combustion engines, gas or steam turbines.

At present, almost everywhere the application of the IEMD ought to be seen. The very basic general structural and functional diagram of the emerging and future AC or DC motor IEMD cyber-physical heterogeneous dynamical hypersystem is shown in figure 1.2. The mechanical load in figure 1.2 represents various types of equipment, which consists of an integrated AC–AC or AC–DC–AC or DC–AC commutator synchronous or asynchronous (induction) motor, like fans, pumps, washing machines etc.

Integrated AC–AC or AC–DC–AC or DC–AC commutator motor and micro-controller packages are designed to simplify IEMD cyber-physical dynamical system integration, minimise interconnection cabling, and reduce or eliminate noise and EM motor/IEMD compatibility issues. The AC or DC motor IEMD is a cyber-physical heterogeneous dynamical hypersystem, in particular IEMD cyber-physical heterogeneous dynamical hypersystems convert electrical energy into mechanical energy and control the converted mechanical-energy flux according to a specific law.

Technically, a cyber-physical heterogeneous dynamical hypersystem is a smooth action of the reals or the integers on another object (usually a manifold). When the reals are acting, the cyber-physical heterogeneous dynamical hypersystem is termed a cyber-physical heterogeneous continuous dynamical hypersystem and when the integers are acting, the cyber-physical heterogeneous dynamical hypersystem is termed a cyber-physical heterogeneous discrete dynamical hypersystem.

Cyber-physical heterogeneous dynamical hypersystems integrate computing, communication and storage capabilities with the monitoring and/or control of entities in the physical world dependably, safely, securely, efficiently and in real-time:

- cyber-physical heterogeneous dynamical hypersystems is an exciting prospect for the next decades!
- involves multi-disciplinary R&D works;
    - high confidence software;
- cyber-physical heterogeneous dynamical hypersystems have the potential to change the way people interact with their surroundings;
- applications in the emerging and future for cyber-physical heterogeneous dynamical hypersystems are limited only by human imagination;
- affordability and ease of application will drive adoption.

The AC or DC motor IEMD in general plays a key role in power generation, aerospace, aviation, automotive metallurgy, mining, industrial cement applications and household appliances. The rapidly expanding area of IEMDs as used in robotics, wind turbines and all-electric or hybrid-electric vehicles is driven by innovations in electrical machine design, power semiconductors and superconductors, **digital signal processors** (DSP) and simulation software.

Cyber-physical heterogeneous dynamical hypersystems are expected to play a major role in the R&D of emerging and future physical heterogeneous dynamical hypersystems, in particular IEMD cyber-physical heterogeneous dynamical hypersystems with new capabilities that far exceed today's levels of autonomy, functionality, usability, reliability, and cyber security.

Advances in their R&D can be accelerated by close collaborations between academic disciplines in computation, communication, control, and other engineering and computer science disciplines, coupled with grand challenge applications. Selected recommendations for R&D in cyber-physical heterogeneous dynamical hypersystems:

- Standardised abstractions and architectures that permit modular design of cyber-physical heterogeneous dynamical hypersystems are urgently needed.
- Cyber-physical heterogeneous dynamical hypersystems' applications involve components that interact through a complex, coupled physical environment. Reliability and security pose particular challenges in this context—new frame-works, algorithms, and tools are required.
- Emerging and future cyber-physical heterogeneous dynamical hypersystems will require hardware and software components that are highly dependable, reconfigurable, and in many applications, certifiable and trustworthiness must extend to the system level.

In this textbook, the authors will focus on the cyber-physical heterogeneous dynamical hypersystems, in particular IEMD cyber-physical dynamical hypersystems of the application area of AC or DC motor IEMDs, and measure their performance to compare with the theoretical holor analysis.

An AC or a DC motor IEMD comprises an EM motor, i.e. an EM energy converter, which is a mechanical split-ring or flat commutator or macrocommutator, i.e. an EE converter operating as an energy processing unit, and a microcontroller and communication unit. An AC or a DC motor IEMD may be also used as propulsion and/or dispulsion systems in elevators, escalators, rolling mills, mine winders, as well as high-speed trains, all-electric and hybrid-electric ships, all-electric forklift and platform trucks, all-electric or hybrid-electric vehicles. Advanced control algorithms (mostly digitally implemented) allow translational/angular velocity and/or force/torque control over a high bandwidth. Hence, precise motion control can be achieved. Examples are an IEMD in robots, pick-and-place machines, factory automation hardware, etc.

Principally an AC or DC motor IEMD can operate in motoring and generating mode. Wind turbines use an AC or DC motor IEMD to convert wind energy into electrical energy. More and more, AC or DC motor IEMDs are used to save energy for example, in air-conditioning units, compressors, blowers, pumps and home appliances. Procedures to ensure stable operation of an IEMD in the aforementioned applications are translational/angular velocity and/or force/torque control algorithms.

In a very advanced AC or DC motor IEMD, a unique approach is followed to derive model-based translational/angular velocity and/or force/torque microcontroller/communication units for all types of Lorentz-force MMD electrical machines, i.e. DC and AC synchronous and asynchronous (induction) electrical machines.

The rotating-transformer physical model forms the basis for this generalised modelling approach that finally leads to the development of universal field-oriented control algorithms. In the case of variable-reluctance (commutated-reluctance) MMD electrical machines, force/torque observers are proposed to implement direct force/torque algorithms.

Changes in engineering are transforming the AC or DC motor IEMD from purely mechanical machines into hubs of complex macroelectronics and microelectronics. From an engineering perspective, this means that in many respects, the IEMD is undergoing a major redesign, and the rate of change is not likely to decelerate anytime soon. If anything, it will probably accelerate even more over the years ahead.

Formerly, a variety of terms have been used to describe an IEMD cyber-physical heterogeneous dynamical hypersystem that permits a mechanical load to be driven at user-selected translational or angular velocities.

An 'adjustable-velocity IEMD' is the abbreviated form of terms, which include, but are not limited to:

- adjustable-speed IEMD;
- variable-speed IEMD;

- adjustable-frequency IEMD;
- variable-frequency IEMD.

As the comprehensive explanation better conveys, it allows one to adjust the translational or angular velocity of an integrated EM motor (by varying the voltage and frequency of the supply power delivered to the integrated EM motor).

The term variable means a change that may or may not be under the control of the user. Adjustable is the chosen term since this relates to a change directly under control of the user. The term frequency can only be attached to an IEMD with an AC output, while the term velocity is preferred since this includes both AC or DC motor IEMDs. Thus, the term most universally known is adjustable-velocity IEMD.

Just as angle or distance and displacement or position have distinctly different meanings (despite their similarities), so do speed and velocity.

Speed is a scalar quantity that refers to 'how fast an EM motor's shaft or mover is rotating or moving'. Speed can be thought of as the rate at which an EM motor's shaft or mover covers angle or distance, respectively.

A fast rotating or moving EM motor's shaft or mover has a high speed and covers a relatively large angle or distance in a short amount of time. Contrast this to a slow rotating or moving EM motor's shaft or mover that has a low speed; it covers a relatively small amount of angle or distance in the same amount of time. An EM motor's shaft or mover with no movement at all has a zero speed.

Velocity is a holor or vector quantity that refers to 'the rate at which an EM motor's shaft or mover changes its displacement or position'. Imagine an EM motor's shaft or mover rotating or moving rapidly—one-step forward and one-step back—always returning to the original starting displacement or position. While this might result in a whirl of activity, it would result in a zero velocity.

A holor or vector is quantitative, it has magnitude, direction and sense of direction. The magnitude represents the holor or vector size or physical quantity. The direction represents the holor or vector position with respect to a holor elements (merates) or reference axis, respectively. The sense of direction represents the holor or vector orientation and its arrowhead represents it. This contrasts with the definition of a scalar, which has only magnitude. Examples of scalar quantities include temperature, resistivity, voltage and mass.

In comparison, examples of holor or vector quantities would include velocity, force, acceleration and position. The most familiar and intuitive use of holors or vectors is in the two-dimensional (holor merates or x, y coordinates) or **three-dimensional** (3-D) holor merates (x, y and z coordinates) Cartesian coordinate system.

Because the EM motor's shaft or mover always returns to the original displacement or position, the motion would never result in a change in displacement or position. Since velocity is defined as the rate at which the displacement or position changes, this motion results in zero velocity. If an EM motor's shaft or mover in motion desires to maximise their velocity, then that EM motor's shaft or mover must make every effort to maximise the amount that they are displaced from their original displacement or position. Every step must go into moving that EM motor's shaft or mover further from where it started.

The EM motor's shaft or mover should never change senses of direction and begin to return to the starting displacement or position. Thus, velocity is a holor or vector quantity. As such, velocity is 'sense of direction aware'. When evaluating the velocity of an EM motor's shaft or mover, one must keep track of direction. It would not be enough to say that an EM motor's shaft or mover has a velocity of 25 rad s^{-1} or 75 m s^{-1}.

One must include sense of direction information in order to fully describe the velocity of the EM motor's shaft or mover. For instance, one must describe an EM motor's shaft or mover velocity as being 25 rad s^{-1} or 75 m s^{-1}, right- or leftwards. This is one of the essential differences between speed and velocity. Speed is a scalar quantity and does not 'keep track of direction'; velocity is a holor or vector quantity and is 'sense of direction aware'.

The task of determining the sense of direction of the velocity holor or vector is easy. The sense of direction of the velocity holor or vector is simply the same as the sense of direction that an EM motor's shaft or mover is rotating or moving. It would not matter whether the EM motor's shaft or mover is speeding up or slowing down.

If an EM motor's shaft or mover is rotating or moving forwards, then its velocity is described as being forwards. If an EM motor's shaft or mover is rotating or moving downwards, then its velocity is described as being downwards. Therefore, an EM motor's shaft or mover rotating or moving in a forward direction with a speed of 25 rad s^{-1} or 75 m s^{-1} has a velocity of 25 rad s^{-1} or 75 m s^{-1}, forwards.

Note that speed has no sense of direction (it is a scalar) and the velocity at any instant is simply the speed value with a sense of direction. As an EM motor's shaft or mover rotates or moves, it often undergoes changes in speed. For example, during an average EM motor operation, there are many changes in speed. Rather than the tachometer or speedometer maintaining a steady reading, the indicator constantly moves up and down to reflect the stopping, starting, the accelerating, and decelerating. One instant, the EM motor's shaft or mover may be rotating or moving at 25 rad s^{-1} or 75 m s^{-1} and another instant, it might be stopped (i.e. 0 rad s^{-1} or m s^{-1}).

Yet during the EM motor operation the EM motor's shaft or mover might average 25 rad s^{-1} or 75 m s^{-1}. The average speed during an entire motion can be thought of as the average of all tachometers or speedometer readings. If the tachometer or speedometer readings could be collected at 1 s intervals (or 0.1 s intervals or ...) and then averaged together, the average speed could be determined.

The average value of speed during the course of a motion is often computed using the following formula:

$$\text{Average value of angular speed} = \frac{\text{Angle}}{\text{Time}}$$

$$\text{Average value of translational speed} = \frac{\text{Distance}}{\text{Time}}$$

In contrast, the average value of velocity is often computed using this formula:

$$\text{Average value of angular velocity} = \frac{\text{Angular displacement}}{\text{Time}}$$

$$\text{Average value of translational velocity} = \frac{\text{Translational displacement}}{\text{Time}}.$$

Since a rotating or moving EM motor often changes its speed during its rotation or motion, it is common to distinguish between the average value of speed and the instantaneous value of speed.

The distinction is as follows:

- instantaneous value of speed—the speed at any given instant in time;
- average value of speed—the average of all instantaneous values of speed; found simply by a distance/time ratio.

One might think of the instantaneous value of speed as the speed that the tachometer or speedometer reads at any given instant in time and the average speed as the average of all the tachometer or speedometer readings during the course of the EM motor operation. Since the task of averaging tachometer or speedometer readings would be quite complicated (and maybe even dangerous), the average speed is more commonly calculated as the angle or distance/time ratio.

Rotating or moving EM motors do not always operate with variable and changing speeds. Occasionally, an EM motor will rotate or move at a steady rate with a constant speed. That is, the EM motor will cover the same angle or distance every regular interval of time. For instance, an EM motor's shaft or mover might be rotating or moving with a constant speed of 25 rad s^{-1} or 75 m s^{-1} for several minutes. If EM motor's shaft or mover speed is constant, then the angle or distance rotated or moved every second is the same.

The EM motor's shaft or mover would cover an angle or a distance of 25 rad or 75 m every second, respectively. If one could measure the EM motor's shaft or mover displacement or position (angle or distance from an arbitrary starting point) each second, then we would note that the displacement or position would be changing by 25 rad or 75 m each second. This would be in stark contrast to an EM motor that is changing its speed. An EM motor with a changing speed would be rotating or moving a different angle or distance each second.

To sum up, speed and velocity are kinematic quantities that have distinctly different definitions. Speed, being a scalar quantity, is the rate at which an EM motor's shaft or mover covers angle or distance. The average speed is the angle or distance (a scalar quantity) per time ratio. Speed is 'ignorant of direction'. On the other hand, velocity is a holor or vector quantity; it is 'sense of direction aware'. Velocity is the rate at which the EM motor's shaft angle or mover position changes. The average velocity is the displacement or position change (a holor or vector quantity) per time ratio.

1.3 Emerging and future AC or DC motor integrated electro-mechanical drives (IEMD)

Emerging and future AC or DC motor IEMDs are impressively smaller than their counterparts from the 20th century, meaning that installing them is now easier than ever before. For example, control rooms have become more compact and less costly because panel builders are now able to fit other IEMDs into a standard cubicle. Suppliers have also benefited in that it is now much easier for them to fit an IEMD into their equipment. With many R&D teams working to make AC or DC motor IEMDs smaller, the question arises as to just how small AC or DC motor IEMDs can get. All suppliers believe that there are few restrictions, particularly in the lower power range, and that over the next ten years, AC or DC motor IEMDs in this range will shrink by another 60%–70%.

So how is all of this possible? To begin with, there seems to be no end to how small macroelectronics, microelectronics and nanoelectronics can get, and these developments are rapidly finding their way into the high-power semiconductor and superconductor industry. In addition, lower losses are being achieved from the same area of **silicon** (Si) or **silicon carbide** (SiC) and **gallium nitride** (GaN). These two factors combined not only mean smaller semiconductors, but also the amount of heat generated within the AC or DC motor IEMD is reduced, so smaller heatsinks (radiators) are now possible. There is one limitation though: the cable terminations have to be big enough to accommodate the electrical power-carrying cables.

The development of high-power semiconductors and superconductors is an important factor that influences IEMD miniaturisation, but so too is the technology used for cooling.

Even though air-cooling is likely to become the dominant technique, a considerable amount of R&D effort is being invested in developing new cooling techniques as well as in reducing the need for cooling:

- Developments in numerical modelling mean that advanced computer flow modelling techniques are used to design heatsinks that achieve more effective cooling.
- Scientists and engineers are looking at new materials, integrating the heatsink with the high-power module for better cooling performance and improving fan performance with variable-velocity control.
- Liquid-cooling, especially the innovative FF method, is finding increasing use in wind power, transportation and marine.

  Scientists and engineers developed a new way of pumping FFs without the use of any mechanical components. They claim that their technique, dubbed 'ferro-hydrodynamic pumping', can be easily scaled up or down to be used in micro-fluidic devices or industrial-scale pumping devices, and anything in between.

  Using a FF can provide significant compactness while retaining original (or better) cooling performance:

  - The thermomagnetic effect can completely replace the natural convection driving force, while still retaining a cooling performance enhancement of

50% or more. This eliminates the required rig height, enabling more compact solutions.

- Using a FF can allow reductions of heatsink size to about 25% of original size while retaining original performance.

In addition to the ongoing developments mentioned above, new cooling technologies, such as heat pipes and thermosyphons may be applied over the next few years. Thermosyphons use evaporation followed by condensation to transfer heat directly out of the AC or DC motor IEMD. Even though the principles of these devices are well known, cost and performance issues must be solved before they can be commercially applied.

Another area that holds much promise for the emerging and future of AC or DC motor IEMDs is the 'cool chip'. The cool chip is an early application of nanotechnology that uses electrons to transfer heat from one side of a vacuum diode to the other. It uses the principle of electron tunnelling in which a voltage bias is applied to make energetic electrons 'jump' across a tiny gap between two surfaces. These electrons transfer heat energy between the two layers, and because of the gap, the heat cannot be conducted back.

Applied to IEMD technology, the cool chip principle could be used to carry heat from the semiconductor directly to the heatsink, thereby vastly improving the heatsink's efficiency. This would mean smaller active power devices, generating a lot less heat than would be expected for the rated power. To achieve this, relatively large surface areas with a gap of less than 10 nm need to be manufactured. In addition, the manufacturers must ensure no contact between the surfaces at any point.

Reducing the cost of an IEMD is a goal for all suppliers, and miniaturisation contributes enormously in achieving this goal. Three smaller and cheaper AC or DC motor IEMDs will find new applications as diverse as running machines and small centrifuges used in honey production. Not only is it intended for small industrial applications, but also for user products such as air conditioners, exercise machines and washing machines.

Component integration also contributes to a cheaper AC or DC motor IEMD. Suppliers predict that over the next 25 years, a combination of tighter semiconductor and mechanical part integration will lead to even fewer parts within a modular adjustable-velocity IEMD. Fewer parts mean fewer interfaces and fewer mechanical fixings, and this means improved reliability. In the future of the AC or DC motor IEMD, another form of integration, that of the IEMD and integrated EM motor with the application will have its place. This is already happening in some specialised applications.

One supplier, for example, has developed a fully integrated tubular submersible pump. This form of integration is also seen as being important in the field of robots where true mobility will be obtained with a fully-integrated AC or DC motor IEMD.

Naturally, software has a big part to play in the future. As software continues to develop, the AC or DC motor IEMD can expect to have increased capability with less hardware. All suppliers play a major part in the overall cost-reduction process. They do this by looking at ways of improving every aspect of their products.

For example, improvements can be made by means of:

- improved components;
- more integration;
- up-to-date design techniques;
- very advanced and efficient manufacturing processes;
- better logistics.

As the IEMD market continues to grow, economies of scale in volume production will be needed to cover the substantial investments needed in R&D to maintain the steep decline in prices seen in recent years.

The intelligent AC or DC motor IEMD are certain to benefit from the growth of ethernet communications by becoming an integral part of control, maintenance and monitoring cyber-physical dynamical systems. Decentralised control cyber-physical dynamical hypersystems will be created in which multiple IEMDs share control functions, with one taking over in the event of a fault or error in another IEMD. The advantage of this is that reliance on costly **programmable logical controllers** (PLC) would be greatly reduced and automation reliability would improve dramatically.

The authors think that ethernet-based AC or DC motor IEMDs will become a valuable source of data for preventive maintenance programs. Taking advantage of ethernet's wide bandwidth, these intelligent IEMDs would be able to communicate greater amounts of monitoring information than would standard web-based cyber-physical dynamical hypersystems.

In addition to this type of information, the IEMD would also collect data that describes the state of the process being controlled.

If each IEMD had its own **internet protocol** (IP) address, it would be easy to gather a log of every IEMD on a central server via ethernet, and build-up a highly detailed picture of the entire process and its performance. A detailed analysis of this data could be used to adjust the process and improve productivity. It could also be used to increase process availability through proactive fault management and asset optimisation.

Taking the intelligent AC or DC motor IEMD a step further; it could even have the capability of detecting the cause of a fault and providing a course of action for its resolution. All of this fits properly with the intelligent technology concept, in that the AC or DC motor IEMD with advanced communication capabilities can be seamlessly integrated into larger real-time automation and information cyber-physical homogeneous dynamical systems.

The increase in AC or DC motor IEMD intelligence will meet a growing demand from users for IEMDs that are easier to set up and control. As reliability is now taken for granted, ease of use and ease of commissioning are becoming the most important demands of emerging and future IEMD users.

The ultimate goal of all suppliers is to have a completely self-commissioning AC or DC motor IEMD, requiring no manual setting of parameters. They believe that achieving this goal is getting closer with advanced set-up wizards installed in the latest AC or DC motor IEMD.

The dynamic performance of the AC or DC motor IEMD in general has improved dramatically over the years. However, with **real holor control** (RHC) technology, the authors believe it has reached the ultimate in control performance. Using RHC applications that were only feasible with other IEMD technologies, such as the DC motor IEMD and servo EMD, are now routine for the AC motor IEMD. For example, the control of the new low-speed AC–AC or AC–DC–AC or DC–AC commutator IPM synchronous motors using new developments in RHC technology is likely to find increasing use in a variety of industries.

To control the EM motor, the authors have adapted the control algorithms in its RHC technology to achieve highly accurate control at low speeds without encoder feedback.

Standard AC asynchronous (induction) motors, normally designed to run at 750–3000 rpm, have poor efficiency at low speeds and often cannot deliver sufficiently smooth torque across the speed range. This problem is normally overcome by using a gearbox, but gearboxes are complex and take up valuable space and maintenance resources.

A **direct drive** (DD) IEMD cyber-physical dynamical hypersystem, using the integrated AC–AC or AC–DC–AC or DC–AC commutator IPM synchronous motor with the macrocommutator, provides a high-torque IEMD directly coupled to the driven application, thus eliminating the need for a gearbox. This cyber-physical dynamical hypersystem saves on integrated EM motor maintenance because the integrated AC–AC or AC–DC–AC or DC–AC commutator IPM synchronous motor is robust, and in maintenance terms, similar to standard AC asynchronous (induction) motors.

The DD IEMD cyber-physical dynamical hypersystem has already been applied in the paper industry, as paper machines require large numbers of a high-accuracy, low-speed IEMD. Another application is in ship propulsion dynamical systems. Suppliers, designed to give ships extreme manoeuvrability, uses a DD IEMD with a fixed-pitch propeller mounted directly onto the integrated EM motor shaft. The integrated EM motor's small size enables the outer diameter of the pod to be reduced, thereby improving hydrodynamic efficiency. The DD IEMD cyber-physical dynamical hypersystem is well suited to smaller vessels. Overall, the future looks very good for AC or DC motor IEMD users. It will be possible to buy an AC or DC motor IEMD that is smaller, more intelligent, easier to install and suitable for many applications, particularly at low power and low speed. However, the best news of all is that this IEMD will be cheaper than ever before.

References

Barnes M 2003 *Practical Variable Speed Drives and Power Electronics Automated Control Systems* (Perth, Australia: IDC Technologies)

Chan T-F and Shi K 2011 *Applied Intelligent Control of Induction Motor Drives* (New York: Wiley)

Geyer T 2017 *Model Predictive Control of High Power Converters and Industrial Drives* (New York: Wiley)

Holmes D G and Lipo T A 2003 *Pulse Width Modulation for Power Converters Principles and Practice* (Piscataway, NJ: Institute of Electrical and Electronics Engineers)

Hughes A and Drury W 2013 *Electric Motors and Drives Fundamentals, Types and Applications* 4th edn (Amsterdam: Elsevier)
Rashid M H 2018 *Power Electronics Handbook* 4th edn (Amsterdam: Elsevier)
Wach P 2011 *Dynamics and Control of Electrical Drives* (Berlin, Heidelberg: Springer)
Wilamowski B M and Irwin D J 2011 *The Industrial Electronics Handbook - Power Electronics and Motor Drives* (Boca Raton, FL: CRC Press)

IOP Publishing

The Integrated Electro-Mechanical Drive
A mechatronic approach
B T Fijalkowski and J Tutaj

Chapter 2

Integrated DC or AC motors with the mechanical split-ring flat and or macroelectronic commutator

'I would not give my rotating field discovery for a thousand inventions, however valuable… A thousands years hence, the telephone and the motion picture camera may be obsolete, but the principle of the rotating magnetic field will remain a vital, living thing for all time to come.'

Nikola Tesla

'A Famous Prophet of Science Looks into the Future'
(*Popular Science Monthly,* November, 1928)

2.1 Introduction

Scientists and engineers at present are tasked with understanding, evaluating and applying many **direct current** (DC) and **alternating current** (AC) integrated **electro-mechanical** (EM) motor technologies because EM motors ultimately power the majority of translational or angular motion. The key trends in industry related to the integrated EM motors, and what users and the market value, are energy efficiency, size/footprint reduction, and reliability.

In this textbook, the authors outline the capabilities of integrated AC–AC or AC–DC–AC or DC–AC commutator synchronous or asynchronous (induction) motors with the mechanical split-ring/flat commutator and/or macroelectronic commutator (macrocommutator), the major technologies with partially overlapping functionalities for larger, higher-end applications requiring precisely metered torque, velocity or positioning. They then expound on some engineering requirements and compare all options for specific situations.

Integrated AC–AC or AC–DC–AC or DC–AC commutator synchronous or asynchronous (induction) motors offer good performance in a number of

doi:10.1088/2053-2563/aae7d7ch2

© IOP Publishing Ltd 2019

important areas. Many of the EM motor's key attributes were previously considered mutually exclusive, i.e. it was not possible to combine them in the same electrical machine.

Thanks to the integrated AC–AC or AC–DC–AC or DC–AC commutator synchronous or asynchronous (induction) motors, users have been taking advantage of high power, a wide translational or angular-velocity range, high force or torque, small dimensions, and high overload capacity or low mass or moment of inertia simultaneously for some years.

The improvements were important enough to justify calling integrated AC–AC or AC–DC–AC or DC–AC commutator synchronous or asynchronous (induction) motors the beginning of a new era in integrated AC or DC motor **integrated electro-mechanical drive** (IEMD) technology.

As was shown above, the majority of IEMDs sold today are based on integrated AC–AC or AC–DC–AC or DC–AC commutator IPM synchronous or asynchronous (induction) motors and increasingly **variable reluctance** (VR) motors.

2.2 MCM or ECM AC–AC or AC–DC–AC or DC–AC commutator motors—a basic application

Macro- and microelectronics technology led to the real breakthrough in the changeover from the today's power electronics static converter torque and velocity control systems for **mechanical commutation matrixer** (MCM) DC–AC commutator synchronous motors (DC motors) and AC–AC commutator asynchronous (induction) motors to tomorrow's **electrical commutation matrixer** (ECM) AC–AC or AC–DC–AC or DC–AC commutator motors.

In the past, power electronics technology led to the changover from the rotary dynamo-electric Ward-Leonard torque and velocity control systems for MCM DC–AC commutator synchronous motors (DC motors) to static converter Ward-Leonard torque and speed control systems. This commenced in the 1930s with the introduction of mercury-arc converters and in the 1950s with the introduction of transductor-controlled **semiconductor rectifiers** (SR), i.e. power diodes.

Contemporary, **semiconductor controlled rectifiers** (SCR), i.e. thyristor converters for MCM DC–AC commutator synchronous motor (DC motor) advanced EMD and traction propulsion cyber-physical heterogeneous continuous or discrete dynamical hypersystems have matured, and this technology is well established in many applications. Because of their simple and robust design AC synchronous and asynchronous motors, and then above all AC squirrel-cage-rotor asynchronous (induction) motors, invented by Polish electrician Dolivo-Dobrovolsky (1862–1919) in 1888 would be preferable to the MCM DC–AC commutator synchronous motor (DC motor) for many torque and speed control applications, particularly in difficult and hazardous environments (Hoseman *et al* 1999). However, the torque and angular-velocity control of AC synchronous or asynchronous (induction) motors has only been a dream for many years. It is true that this technique has existed for a long time and indeed suppliers have also been manufacturing static converters (static frequency changers) for AC synchronous or asynchronous (induction) motors for today's industrial EMD and traction propulsion cyber-physical heterogeneous continuous or discrete dynamical hypersystems.

In recent years, the cost trend for ECM commutators has now made it increasingly interesting from the commercial viewpoint to use ECM commutators more widely, in ECM AC–AC. AC–DC–AC and AC–DC commutator synchronous or asynchronous (induction) motors for variable-speed industrial EMD and traction propulsion cyber-physical heterogeneous continuous or discrete dynamical hypersystems. This trend is still in its infancy and this technique should experience its definite breakthrough during the next years.

The design of ECM commutators is closely tied to the complexity of the MMD electrical machine. Beyond doubt, the electrical energy conversion is technically the most advanced today. It involves the static conversion of electrical energy.

It has been, and still is, a challenge for the application of modern newly designed energy-saving ECM AC–AC, AC–DC–AC and DC–AC commutator synchronous or asynchronous (induction) dynamotors (generators/motors), particularly if the industrial EMD and traction propulsion cyber-physical heterogeneous continuous or discrete dynamical hypersystems are considered. It is consequently not an exaggeration to state that ECM AC–AC or AC–DC–AC or DC–AC commutator synchronous or asynchronous (induction) dynamotors are technically in advance of those for the other MMD electrical machines.

The authors have been very active during the past years in the field of ECM AC–AC or AC–DC–AC or DC–AC commutator synchronous or asynchronous (induction) motors. Viewed from a historical perspective, the authors have kept up with the technical developments in this area and on many occasions have quickly and successfully exploited new technical advances (Fijalkowski 1985a, 1985b, 1985c).

As far back as 1881, an electrical machinery industry supplied the first MCM DC–AC commutator synchronous motors (DC motors) for industrial EMD and traction propulsion cyber-physical heterogeneous continuous or discrete dynamical hypersystems.

The technology was based on an EM energy conversion. Development then continued with the introduction of mercury-arc and semiconductor converters. The first fully static ECM commutators and ECM AC–AC. AC–DC–AC and AC–DC commutator motors were launched at the beginning of the 1930s (see appendix A).

Recently there has been considerable interest in the use of ECM AC–AC, AC–DC–AC, AC–DC, DC–AC, DC–DC and DC–AC–DC commutators technology for industrial EMD and traction propulsion mechatronic control systems as well as for aerospace and automotive applications. The aerospace and automotive industries have also shown considerable interest in ECM AC–AC, AC–DC–AC, AC–DC, DC–AC, DC–DC and DC–AC–DC commutators due to the inherent potential advantages they have in aircraft and automotive applications. This textbook will give an in-depth introduction to the full range of ECM AC–AC, AC–DC–AC, AC–DC, DC–AC, DC–DC and DC–AC–DC commutator technologies and investigate the suitability and design of ECM AC–AC, AC–DC–AC, AC–DC, DC–AC, DC–DC and DC–AC–DC commutators.

Table 2.1. Mechanical commutator versus electronic commutator.

Mechanical commutator	Electronic commutator
Made up of commutator copper segments and mica insulation. Brushes are made up of carbon or graphite.	Macroelectronic electrical valves (electronic switches) are used in the commutator.
Shaft position sensing is inherent in the arrangement.	It requires a separate rotor position sensor.
Number of commutator copper segments is very high.	Number of electrical valves is limited to 6, 9 or 12.
Sliding contact between commutator copper segments and carbon brushes.	No sliding contacts.
Sparking takes place.	There is no sparking.
It requires regular maintenance.	It requires less maintenance.
Difficult to control the voltage available across armature tappings.	Voltage available across armature tappings can be controlled by PWM techniques.
Commutator arrangement is located in the rotor.	Commutator arrangement is located in the stator.
Highly reliable.	Reliability can be improved by specially designed devices and protecting circuits.

The potential advantages of ECM AC–AC, AC–DC–AC, AC–DC, DC–AC, DC–DC and DC–AC–DC commutator technology will be examined, and the factors that have so far limited commercial exploitation of the ECM are discussed (Fijalkowski 2016; Tutaj 1996, 2012).

We will use appropriate practical results from various prototype MIMO ECM AC–AC, AC–DC–AC, AC–DC, DC–AC, DC–DC and DC–AC–DC commutators to illustrate the topics under discussion. These examples will focus on application areas where the advantages of the ECM AC–AC, AC–DC–AC, AC–DC, DC–AC, DC–DC and DC–AC–DC commutator carry a large premium, for example industrial EMD and traction propulsion mechatronic control systems, especially for aerospace, aviation and automotive applications (table 2.1).

This textbook is aimed at anybody interested in the area of MIMO ECM AC–AC, AC–DC–AC, AC–DC, DC–AC, DC–DC and DC–AC–DC commutators.

It will start with a basic introduction to the topology and lead into in-depth discussion on the critical research developments and topical issues.

At present, macro- and microelectronics technology has become widely accepted for new concept integrated ECM AC–AC or AC–DC–AC or DC–AC commutator synchronous and asynchronous (induction) motors. Such technology has resulted in considerably improved industrial EMD and traction propulsion cyber-physical heterogeneous continuous or discrete dynamical hypersystems, providing fast, smooth and precise torque and angular-velocity control. For instance, for railway and automotive vehicles this means better split controls, which have made it possible to take out high tractive efforts close to the limit of adhesion.

2.3 New concept ECM AC–AC or AC–DC–AC or DC–AC commutator motors

The development of modern ECM AC–AC or AC–DC–AC or DC–AC commutator synchronous or asynchronous (induction) motors is closely connected with further development of macro- and microelectronics technology. Moreover, it is necessary to pay attention to the interesting fact, that contemporary MCM AC–AC or DC–AC commutator synchronous or asynchronous (induction) motors have not changed their basic construction and appearance since the late 1980s (Barnes 2003, Wilamowski and Irwin 2011, Chan and Shi 2011, Wach 2011, Hughes and Drury 2013, Holmes and Lipo 2003, Geyer 2017, Rashid 2018).

R&D works on ECM AC–AC or AC–DC–AC or DC–AC commutator synchronous or asynchronous (induction) motors concentrated initially on design of their details (component parts) and deeper understanding of physical phenomena appearing during their working.

Modern ECM AC–AC or AC–DC–AC or DC–AC commutator synchronous or asynchronous motors with the ECM commutators are, up to now, very rarely used, mainly in aerospace, aviation and automotive industrial EMD and traction propulsion cyber-physical heterogeneous continuous or discrete dynamical hypersystems (Fijalkowski 1985b).

Another factor that contributed to this state of affairs in power electronics was the fact, that in the 1940s a rapid development of mercury-arc converters, and in the 1950s—thyristors converters took place. Thus, the scientists were interested only in the large scale development of electronic commutators, i.e., polyvalvular, ECM AC–AC or AC–DC–AC or DC–AC commutators with the unipolar ECM, realised on ‘continuous’ power diodes and thyristors, up to now called in the field of power electronics—‘static converters’.

The new concept integrated ECM AC–AC or AC–DC–AC or DC–AC commutator synchronous or asynchronous (induction) motors conceived by the authors, can be used, e.g. in advanced cradled dynamometer simple continuous dynamical systems for testing of combustion engines, electrical, fluidical and pneumatical machines as well as other rotary machines etc, and industrial EMD and traction propulsion cyber-physical heterogeneous continuous or discrete complex dynamical hypersystems operate in **sinusoidal-pulse width modulation** (SINPWM) mode with integral ECM commutators just like conventional MCM AC–AC or DC–AC commutator asynchronous (induction) motors. A dynamometer that is designed to be driven is called an ‘absorption dynamometer’. A dynamometer that can either drive or absorb is called a ‘universal dynamometer’. A dynamometer can also be used to determine the torque and power required to operate a driven machine, for example, such as a **mechano-fluidical** (M-F) pump or a **mechano-pneumatical** (M-P) compressor. In that case, a motoring or driving dynamometer is used.

A motoring dynamometer acts as an MCM AC–AC or a DC–AC/AC–DC commutator synchronous or asynchronous (induction) motor that drives the equipment under test. It must be able to drive the **external combustion engine** (ECE) or **internal combustion engine** (ICE), EM, FM or PM motor or other rotating prime mover at any angular speed and develop any level of torque that the test requires. Only torque and angular speed can be measured directly.

In most dynamometers, power is not measured directly, it must be computed from the torque and angular-velocity holors or force and linear-velocity holors according to the following holor product:

$$\boldsymbol{P}^{\kappa} = \boldsymbol{T}_i \otimes \boldsymbol{\Omega}^j = \gamma^{\kappa i}_{..j} \boldsymbol{T}_i \boldsymbol{\Omega}^j;$$

or

$$\boldsymbol{P}^{\kappa} = \boldsymbol{F}_i \otimes \boldsymbol{V}^j = \gamma^{\kappa i}_{..j} \boldsymbol{F}_i \boldsymbol{V}^j.$$

where:

$\boldsymbol{P}^{\kappa}$ is the power holor (W);
$\boldsymbol{T}_i$ is the torque holor (Nm);
$\boldsymbol{\Omega}^j$ is the angular-velocity holor (rad s^{-1});
$\boldsymbol{F}_i$ is the force holor (N);
$\boldsymbol{V}^j$ is the linear-velocity holor (m s^{-1}).

An increasing feature of the industrial EMD and traction dynamical systems is that the sense of a power-flow direction in the ECM commutator of the MMD electrical machine is arbitrary, that is, it can be used for both EM motoring and M-E generating (regenerative braking). The change-over of the power-flow direction can take place without any reconnections.

Other characteristics that can be achieved with the industrial EMD and traction propulsion cyber-physical heterogeneous continuous or discrete dynamical hypersystems are a power factor, which can be controlled in an optimum way and low machine armature-windings phase current harmonics. These characteristics are perhaps more important than the industrial EMD and traction propulsion cyber-physical heterogeneous continuous or discrete dynamical hypersystem as such.

In unison, with reasonable forecasts of an ever more widespread application of new concept integral ECM AC–AC or AC–DC–AC or DC–AC commutator synchronous or asynchronous (induction) motors in industrial EMD and traction propulsion heterogeneous continuous or discrete dynamical hypersystems, a constant action has been developed to study and update these initial new concept integral ECM AC–AC or AC–DC–AC or DC–AC commutator synchronous or asynchronous (induction) motors and integral lower- and higher-level microprocessor controllers (microcontroller) broken down into integral MMD electrical machine types (Fijalkowski 1985a, 1985b, 1985c). These new concept integral MMD electrical machines with ECM commutators will have a fully laminated stator and all the electrical insulation is of class H.

The material utilisation that is the ratio between the mass and the output power will be better for the new concept ECM AC–AC or AC–DC–AC or DC–AC commutator synchronous or asynchronous (induction) motors compared with the preceding concepts of MMD electrical machines. This improvement has been achieved without having to increase either the electrical stresses or the relative level of the material used. This means that, for example, the margin to harmful temperature rise in connection with overloads is at least as great as for the older designs.

ECM AC–AC or AC–DC–AC or DC–AC commutator synchronous or asynchronous (induction) motors with their accessories and above all the 'intelligent' MMD electrical machines represent one of the main themes of this textbook. Concepts such as 'artificial intelligence' and microcomputer capacity are factors that are increasingly setting their mark on our world.

The use of industrial EMD and traction propulsion cyber-physical heterogeneous continuous or discrete dynamical hypersystems having integral ECM AC–AC or AC–DC–AC or DC–AC commutator synchronous or asynchronous (induction) motors is rapidly growing. This is largely due to the increasingly severe demands being made on these cyber-physical heterogeneous continuous or discrete dynamical hypersystems.

Emerging and future industrial EMD and traction propulsion cyber-physical heterogeneous continuous or discrete dynamical hypersystems must be capable of being adjusted to cope the varying operation conditions and handle a wide range of current and voltage/torque and velocity controls.

In addition, the must be flexible to meet operation modifications, where new operation conditions can be needed in existing EMD and traction propulsion cyber-physical heterogeneous continuous or discrete dynamical hypersystems. To meet these demands, integrated ECM AC–AC or AC–DC–AC or DC–AC commutator synchronous or asynchronous (induction) motors are becoming more advanced and their range of accessories is growing. These EM motors will be available in the near future as complete 'function packages' for different applications. Their accessories will include flexible microprocessor-based ECM commutator microcontrollers and a new **artificial intelligence** (AI) cyber-physical heterogeneous continuous or discrete dynamical hypersystem.

Advantages and disadvantages of integrated ECM AC–AC or AC–DC–AC or DC–AC commutator motors—Here follows a short list, pointing out the advantages and disadvantages of integrated ECM AC–AC or AC–DC–AC or DC–AC commutator synchronous or asynchronous motors.

Some of the pros of integrated ECM AC–AC or AC–DC–AC or DC–AC commutator synchronous or asynchronous (induction) motors are:

- Variable/adjustable voltage and/or current as well as torque and/or angular-velocity controls, which may increase the efficiency of the load. Increased efficiency leads to reduced electrical energy costs.
- Easy installation, leading to reduced installation costs.
- Easy commissioning, leading to reduced commissioning costs.
- No space for an ECM AC–AC or AC–DC–AC or DC–AC commutator cabinet is required.
- Reduced EMC problems, both radiated and line-carried, due to the containment of the ECM and the cable from the ECM AC–AC or AC–DC–AC or DC–AC commutator to the EM motor.

- Reduced stock inventory, since one integrated ECM AC–AC or AC–DC–AC or DC–AC commutator synchronous or asynchronous (induction) motor can replace, for example, several AC asynchronous (induction) motors with magnetic-pole numbers for a certain torque.

Some of the cons of integrated ECM AC–AC or AC–DC–AC or DC–AC commutator motors are:

- Novel integrated ECM AC–AC or AC–DC–AC or DC–AC commutator synchronous or asynchronous (induction) motor concept scares potential buyers.
- Slightly more expensive to buy than an AC or a DC motor with a separate static converter.
- The AC–AC or AC–DC–AC or DC–AC commutator's ECM is exposed to the same environment as the EM motor, regarding vibrations and heat.

Modern industrial EMD and traction propulsion cyber-physical heterogeneous continuous or discrete dynamical hypersystems having as the chief and only component the integrated ECM AC–AC or AC–DC–AC or DC–AC commutator synchronous motor are generally built up as hierarchical cyber-physical heterogeneous continuous or discrete dynamical hypersystems (figure 2.1).

The EM motor's microprocessor-based ECM commutator microcontrollers (lower-level microcontrollers) communicate with higher-level EMD and propulsion microcontrollers, which handle the co-ordination within the industrial EM drive and traction propulsion cyber-physical heterogeneous continuous or discrete dynamical hypersystem, in their turn, the EMD or propulsion microcontrollers communicate

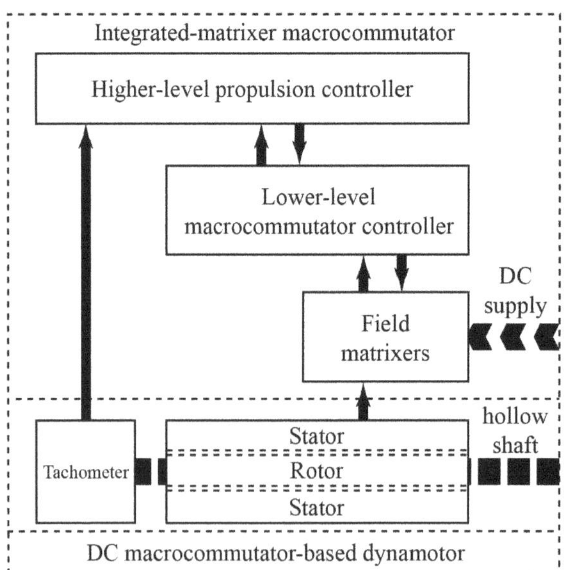

Figure 2.1. Main functions of an ECM AC–AC or AC–DC–AC or DC–AC commutator synchronous motor (Fijalkowski 1985a, 1985b, 1985c).

with the ECM commutator microcontrollers. Further down in the hierarchy is a relatively small volume of data for the different AC–AC and AC–DC/DC–AC commutator synchronous motors' ECM commutator microcontrollers.

On the other hand, greater demands are frequently made on the rate of the signal processing than in the higher-level EMD or traction propulsion microcontrollers. The need for rate of the signal processing in the control functions on the higher level decreases.

In the case of higher-level control and monitoring with operator presentation, the microcomputer must be able to handle large volumes of data, while the demand made on the rate of processing decreases.

There are three main function categories in each type of process computer system:

- **process control** (PC) functions;
- **human–machine communication** (HMC) functions;
- **computer system** (CS) functions.

Figure 2.1 shows the main functions and signal routes of the industrial EMD or traction propulsion cyber-physical heterogeneous continuous or discrete dynamical hypersystem. The control equipment is divided into a 'higher-level EM drive or traction propulsion microcontroller' and 'lower-level ECM commutator microcontroller'. Both sets of control equipment are located in the ECM commutator compartment.

The ECM commutator microcontroller incorporates circuits for controlling and monitoring the armature and field ECMs. In principle, the input signal to the ECM commutator microcontroller is a torque reference (set point), which is compared with a torque signal obtained by means of suitable control of armature and field ECMs.

The EMD or traction propulsion microcontroller incorporates circuits for controlling and monitoring the object driven by the integral ECM AC–AC or AC–DC–AC or DC–AC commutator synchronous dynamotor. The control mode is generally angular-velocity control, but position or some other variables, for example, tension, pressure, flow and loop may also be controlled. The reference (set point) for the controlled variable may be formed in the EMD or traction propulsion microcontroller or be obtained from the higher-level industrial EMD or traction process commander (control system). Today's microprocessor-based ECM commutator microcontrollers consist of microcomputers working real time, of varying degrees of complexity. They are frequently linked in hierarchical structures to satisfy different functional requirements in the best possible way. The modern industrial EMD or traction propulsion cyber-physical heterogeneous continuous or discrete dynamical hypersystems impose demands on advanced data communication methods.

The operator controls and supervises their control process with the aid of a **human–machine communication** (HMC) system, which consists essentially of multi-colour **video display units** (VDU) and keyboards. These VDUs reproduce in a quick, correct and easily understandable way what is happening in the integral ECM AC–AC, AC–DC–AC and AC–DC/DC–AC commutator synchronous or asynchronous dynamotors, i.e. in the industrial EMD or traction propulsion cyber-physical heterogeneous continuous or discrete dynamical hypersystems. This highlights current requirements and demands on new concept integral MMD electrical machines with ECM AC–AC, AC–DC–AC and/or AC–DC/DC–AC commutators.

Placement of the ECM AC–AC, AC–DC–AC and/or AC–DC/DC–AC commutators—Some of the possibilities may be chosen to place the ECM AC–AC, AC–DC–AC and/or AC–DC/DC–AC commutator on top of the integrated AC–AC and/or AC–DC/DC–AC commutator synchronous or asynchronous (induction) motor, some of them also allow it to be placed on the left or the right side of the EM motor.

A top placement conserves the footprint of the integrated ECM AC–AC, AC–DC–AC and/or AC–DC/DC–AC commutator synchronous or asynchronous (induction) motor but increases the height. Other possibilities for a placement of the ECM AC–AC, AC–DC–AC and/or AC–DC/DC–AC commutator at the non-drive end. This placement extends the axial length of the ECM AC–AC, AC–DC–AC and/or AC–DC/DC–AC commutator synchronous or asynchronous (induction) motor but keeps the height and reduces the risk of harmful vibrations.

In this case, the AC–AC, AC–DC–AC and/or AC–DC/DC–AC commutator's ECM is better protected from heat, especially the rising heat of the integrated ECM AC–AC, AC–DC–AC and/or AC–DC/DC–AC commutator synchronous or asynchronous (induction) motor after it is turned off.

In the industrial EMD and traction propulsion cyber-physical heterogeneous continuous or discrete dynamical hypersystems, conformal coatings are used to protect components from the increasingly harsh environments in which they must operate. As technologies continue to advance, often becoming smaller and more complex and/or utilising advanced materials in their design, many surface treatment options struggle to provide reliable protection.

An ultra-thin parylene conformal coating which ensures the reliability of critical components when they fail is an option.

Ultra-thin parylene conformal coatings can enhance the reliability of components, including circuit boards, sensors, MEMS, LEDs, and elastomers that are used in a wide variety of industrial EMD and traction propulsion cyber-physical heterogeneous continuous or discrete dynamical hypersystems such as EMD data and recording, communications and monitoring. Ultra-thin parylene's properties are: excellent chemical and moisture barrier, high dielectric strengths, thermal stability up to 350 °C (short-term), and superior **ultra-violet** (UV) stability.

Modern integrated ECM AC–AC, AC–DC–AC and/or AC–DC/DC–AC commutator synchronous or asynchronous (induction) motors should have ECMs together with single-chip microcontrollers (designed for their controlling and monitoring), built in the newly designed terminal boxes of integrated MMD electrical machines.

2.4 Integrated DC–AC commutator synchronous motors with the mechanical split-ring or flat commutator (DC motors)

Integrated DC–AC/AC–DC commutator synchronous motors with the mechanical split-ring or flat commutator (formerly known as conventional DC motors) have existed for more than a century, and at least twice during that period they have been

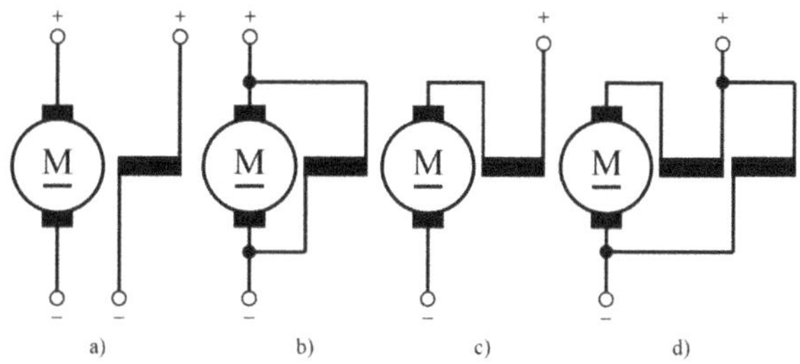

Figure 2.2. Physical model of the integrated DC–AC commutator synchronous motor (formerly known as a conventional DC motor): (a) separate field; (b) shunt field; (c) series field; (d) compound field.

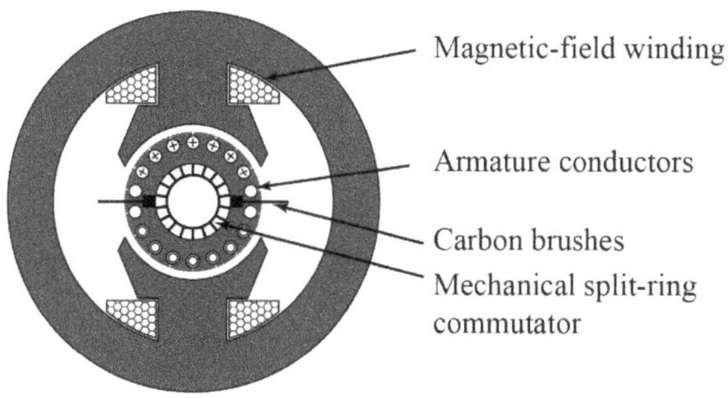

Figure 2.3. Construction of the integrated DC–AC commutator synchronous motor.

popularly regarded as doomed to extinction. The first time this happened was when three-phase AC supply came onto the scene around the turn of the 20th century. An integrated DC–AC commutator synchronous motor (a conventional DC motor) converts electrical energy into mechanical energy. An integrated AC–DC commutator synchronous generator (formerly known as a conventional DC generator) converts mechanical energy into electrical energy.

A physical model of the integrated AC–DC commutator synchronous motor is shown in figure 2.2, while a construction of the integrated AC–DC commutator synchronous motor is shown in figure 2.3.

An integrated DC–AC commutator synchronous motor (DC motor) and an integrated AC–DC commutator synchronous generator (DC generator) can be realised in the same DC electrical machinery, but in two reversible processes. This DC electrical machine usually contains a long conducting wire wrapped around an armature (or a rotor), forming a set of armature-winding coils. The armature (rotor) is placed inside a magnetic field. When the electric current passes through the armature-winding coils, the armature current forces on the armature-winding coils will exert a torque on the rotor and rotate it.

An integrated AC–DC commutator synchronous generator is a reversed DC–AC commutator synchronous motor. An external mechanical energy source (wind, engine, or water flow) is used to rotate the rotor. Therefore, magnetic flux through the armature-winding coils are changing rapidly due to the rotation of the armature-winding coils in the magnetic field. Thus an induced voltage of rotation (formerly known as an electromotive force—EMF) is generated in the armature-winding coils and it will supply a voltage to the load.

The angular velocity of AC–AC or AC–DC–AC or DC–AC/AC–DC commutator synchronous or asynchronous (induction) motor, depends on two factors: the number of EM motor magnetic poles and the frequency of the applied AC power. In an adjustable-frequency IEMD, variable speed of an EM motor is achieved by varying the frequency of the voltage applied to the EM motor. The lower the frequency, the slower the operating rotational speed of the EM motor. AC–AC or AC–DC–AC or DC–AC/AC–DC commutator synchronous or asynchronous (induction) motors can be detrimentally affected when operated by variable-frequency IEMD. 'Commutator duty' and 'holor or vector duty' describe a class of EM motors that is capable of operation from a variable-frequency IEMD. Low temperature rise in this class of EM motor is accomplished with better insulation systems, additional active material (iron and copper), and/or external fans for better cooling at low speed operation. A particular design includes an independent cooling fan to cool down the EM motor so that it can operate within a wide speed range without any heating problems.

A **mechanical commutation matrixer** (MCM) DC–AC commutator is a component part of an integrated DC–AC commutator synchronous motor and is fitted on the armature (rotor). An MCM DC–AC commutator alternates the sense of direction of the flow of electric current in the armature winding so as to retain the sense of direction of armature spinning in the integrated DC–AC commutator synchronous motor. The voltage generated in the armature, placed in a rotating magnetic field, of an AC–DC commutator synchronous generator is alternating in nature.

Thanks to their good regulatory properties, conventional DC–AC/AC–DC commutator synchronous motors (DC motors) nevertheless survived and, during the last few decades, have become more popular than ever, largely due to mechanical split-ring or flat MCM AC–DC/DC–AC commutators acting as AC–DC rectifiers or DC–AC inverters making it easy to capitalise on the advantages of DC technology.

However, in the mid to late of the XX century, the sceptics' came knocking again, this time predicting the demise of the DC motor IEMD because it was possible to approach its precision and responsiveness with ECM AC–AC or AC–DC–AC macroelectronic commutators (macrocommutators) acting as cycloconverters or frequency changers and standard AC squirrel-cage asynchronous (induction) motors.

Once more, the prophecy did not materialise; sales of standard DC motors in recent years have remained stable; and what is more, thanks to the introduction of the integrated DC–AC/AC–DC commutator synchronous or asynchronous (induction) motors with the AC–AC or AC–DC–AC macrocommutator, suppliers experienced significant growth in this sector.

The stiff competition in the field of integrated DC–AC/AC–DC commutator synchronous motors (DC motors) with the mechanical split-ring or flat commutator has led to ever more 'stressed' designs. While performance in terms of output power, torque and maximum velocity has increased over the years, it has been at the expense of the electrical margins and velocity range at rated output power.

This in turn has led to a greater need for maintenance of carbon brushes and mechanical split-ring or flat commutators and to a reduction in lifetime.

The challenge for the integrated DC–AC/AC–DC commutator synchronous motor (DC motors) with the mechanical split-ring or flat commutator project scientists and engineers was therefore to turn the development of integrated DC–AC/AC–DC commutator synchronous motor (DC motors) back towards ideal characteristics, including a wide speed range at constant output power, low electrical stress levels and, as a result, less need for maintenance. Some time ago, sales of the standard DC motor's forerunner, the integrated DC–AC/AC–DC commutator synchronous motor, gave suppliers reason to contemplate the future of their involvement in DC electrical machines. They decided they wanted to reverse the market trend towards increasingly restrictive designs and, based on their broad know-how, experience and the state-of-the art design tools, develop an ideal integrated EM motor with respect to its mechanical and electrical properties. Specifically, this meant higher output power, higher torque, higher velocity, a wider speed range, lower moment of inertia, better low-speed properties and lower electrical and mechanical stresses.

Two possible ways were open: start with the current EM motor series and improve it, or design an entirely new electrical machine. The first was less risky and less costly, but also less promising; the second, although it obviously entailed more risk and higher costs, held more promise. After careful analysis, suppliers chose the second path and laid plans to design an entirely new series of EM motors. This was done because its potential by far outweighed the greater risks and higher development costs.

The result was the integrated DC–AC/AC–DC commutator synchronous motor with the mechanical split-ring or flat commutator, this proved to be the right choice.

Compared with the forerunner (standard DC motors), this integrated DC–AC/AC–DC commutator synchronous motor with the mechanical slip-ring or flat commutator exhibits approximately:

- 50% higher output power;
- 90% higher torque;
- 30% higher maximum speed;
- 30% wider speed range.

The development project spawned a number of novel technical solutions. The most important development was a novel principle for cooling the armature windings. In this principle, the cooling channels are moved through the armature laminations closer to the winding slots, while at the same time significantly widening the overall cooling area. Because of this, the cooling capacity is considerably better. In addition, the magnetic balance of the armature is improved by the fact that the cooling channels are located symmetrically in relation to the winding slots.

The symmetry of the magnetic flux in the armature circuit, which ensures that no armature winding coil is subjected to higher electrical stress than any other, was one of the aspects examined during the computerised optimisation of the magnetic circuit (figure 2.3), shows an example of the way the results of magnetic-flux calculations were presented.

Other improvements to the armature were:

- The manufacturing process for core laminations was refined to reduce the variations in magnetic polarisation of the electrical steel used. This measure has resulted in smoother running at low angular velocities.
- The armature winding coils are skewed, resulting not only in smoother running at low angular velocities but also in lower noise levels.
- The winding-coil ends of the armature are mounted on rigid support rings made of aluminium, thus reducing vibrations and contributing to the high angular-velocity limit of the EM motor. The arrangement also improves the cooling of the winding-coil ends, a weakness in many older designs which directly affects the durability of the insulation.
- The EM motor's shaft is over dimensioned to further reduce vibrations and contribute to the high angular-velocity capability of the EM motor.

The EM motor's stator was also improved. More space was created for the windings. The magnetic balance is now better, as is the temperature distribution. All of this was made possible by computerised optimisation of the magnetic circuit, more compact winding coils and improved manufacturing methods.

The mechanical split-ring or flat commutator was also redesigned: its diameter was reduced and made sturdier. Consequently, it too contributes to the EM motor's high angular-velocity capability and wide commutation margin (i.e. its capacity to conduct current without sparking).

The above mentioned improvements were introduced to boost the output power of the EM motor without compromising the commutation margin. However, the results far exceeded the expectations. Despite a boost in power of up to 70%, the commutation stress level fell significantly.

Commutation margin: the key to high reliability. The commutation margin is a measure of the ability of integrated DC–AC/AC–DC commutator synchronous motor (DC motors) with the mechanical split-ring or flat commutator to conduct currents between the carbon brushes and the mechanical split-ring or flat commutator without sparking. Consequently, it is one of the most important properties governing operational reliability and maintenance requirements.

The commutation margin is affected by factors such as current ripple, vibration, temperature, load and contaminants in the cooling air—all factors which, in different ways, have been attended to in the EM motor.

The commutation ability of the integrated DC–AC/AC–DC commutator synchronous motor with the mechanical split-ring or flat commutator has been verified with EE converter supply (ECM commutator), which is significantly more demanding than a smoothed DC power supply.

An extremely wide commutation margin ensures high operational reliability and minimal maintenance for the EM motor. In reasonably stable operating environments, the mechanical split-ring or flat commutator and the carbon brushes do not need servicing more often than the bearings. Suddenly the standard DC motor is on par with standard AC squirrel-cage asynchronous (induction) motors in terms of maintenance intervals. Different materials are used for the insulation of the integrated DC–AC/AC–DC commutator synchronous motor to extend its life for as long as possible. For example, materials with high temperature indices—far higher than Class H—are used where the temperature is high and mechanically stronger materials where temperatures are lower.

In this context it is important to point out that the supplier calculates temperature margins for the integrated DC–AC/AC–DC commutator synchronous motor on the basis of the actual temperature during operation, not the temperature some time after shut-down, which the IEC 34-1 standard permits. The supplier's temperature tests are also based on the more demanding **electrical–electrical** (EE) converter (ECM commutator) supply. At the same time, the risk of hot spots developing is eliminated thanks to near ideal electromagnetic dimensioning and improved cooling.

A systematic approach to identifying the measures required to reach the set quality objectives is offered by **failure mode and effect analysis** (FMEA). This method was therefore used throughout the development and production-engineering phases of the integrated DC–AC/AC–DC commutator synchronous motor with the mechanical split-ring or flat commutator (DC motor).

Designing EM motors, like so many other design assignments, is largely a matter of finding the best balance between desirable, but conflicting, properties. In the case of integrated DC–AC/AC–DC commutator synchronous motor (DC motors) with the mechanical split-ring or flat commutator, the desirable properties include high output power, high torque, high angular velocity, a wide angular-velocity range, small dimensions, low mass and minimal maintenance. Different EM motors, not least from different suppliers, have traditionally been good at different things; none has been good at all of them.

Thanks to unique human resources in its design department, up-to-date computer-based design tools and a good measure of innovative systems thinking. All suppliers have, with the integrated DC–AC/AC–DC commutator synchronous motor (DC motors) with the mechanical split-ring or flat commutator, a product that is significantly better on all counts. This is good news for all involved, and for the ability of DC technology to remain viable in the long term.

Thanks to creative innovations and state-of-the-art computerised optimisation of technical solutions that earlier were considered to have reached the *'design limits'*, a completely new generation of integrated DC–AC/AC–DC commutator synchronous motor (DC motors) with the mechanical split-ring or flat commutator has evolved.

They embody a number of positive characteristics, which previously could not be combined:

- high output and small dimensions;
- high output over a wide angular-velocity range;

- smooth running at low angular velocities and low moment of inertia;
- benefits of physical and mathematical modelling;
- reduced maintenance.

The emerging and future generation of integrated DC–AC/AC–DC commutator synchronous motor (DC motors) with the mechanical split-ring or flat commutator offers completely new opportunities for improving productivity as a result of the substantially faster angular-velocity control. At the same time, the investment costs are lowered.

Thanks to the precise optimisation of the electrical and mechanical characteristics and the wide angular-velocity range, oversizing of a DC motor IEMD to achieve the desired angular-velocity range is unnecessary. It is no exaggeration to say that the integrated DC–AC commutator synchronous motors (standard DC motors) with the mechanical split-ring or flat commutator belong to a generation offering a completely new performance and new opportunities.

A low level of maintenance is an important user requirement. However, it is just as important for the user to be able to draw up reliable plans for the maintenance intervals without any unforeseen disturbances occurring between them.

The integrated DC–AC commutator synchronous motor (DC motors) with the mechanical split-ring or flat commutator has not only been designed to reduce the maintenance needs. Its built-in margins increase the reliability and the possibility to predict with certainty the maintenance needs.

The integrated DC–AC commutator synchronous motors (DC motors) with the mechanical split-ring or flat commutator have the following features:

- Large commutation margin. Thereby optimum carbon-brush grade can be selected.
- Long carbon-brush wear length.
- Very good mechanical split-ring or flat commutator shape stability which decreases the carbon-brush wear.
- Small mechanical split-ring or flat commutator diameter which decreases the carbon-brush wear.
- Small sealing gaps in the bearings. No risk of grease leakage into the EM motor (provided the correct grade of grease used). In addition the bearings' resistance to contaminants increases.
- Bearings with long lubrication intervals.
- Filter with high degree of filtration in the cooling cyber-physical homogeneous dynamical system which results in long cleaning intervals.
- High resistance to contaminants which results in long cleaning intervals.

Salient features of the integrated DC–AC commutator synchronous motors (DC motors) with the mechanical split-ring or flat commutator are their very high market.

The development work on the stator, for example, output and torque, in relation to the frame size. A number of new solutions have together given the integrated DC–AC commutator synchronous motors (DC motors) with the mechanical split-ring or flat commutator such a superior performance that they are in a class of their own in the DC has provided more space for the windings and led to a more uniform temperature distribution due to a more compact winding-coil design and improved manufacturing processes.

The biggest difference, however, has been achieved in the armature. Arranging the cooling ducts below the winding slots gives first and foremost a fully symmetrical distribution of the magnetic-flux path in the armature (see figure 2.3).

In addition, there is a minimal intrusion of the cooling ducts into the magnetic-flux path. The efficiency of the cooling ducts could consequently be strikingly improved as a result of them being moved closer to the winding slots. At the same time, their cooling area could be enlarged. Combining in a way to increase cooling with a simultaneous improvement of the magnetic-flux circuit was impossible in the past. The improvement of the magnetic-flux paths in the armature was also followed up with a careful optimisation of the EM motor stator.

2.5 Integrated AC–AC or AC–DC–AC or DC–AC commutator synchronous motor with the macroelectronic commutator (macrocommutator)

Integrated AC–AC or AC–DC–AC or DC–AC commutator synchronous or asynchronous (induction) motors with the macroelectronic commutator (macro-commutator) are delivering high performance in industrial processes, the marine and offshore sectors, utilities and specialised applications all over the world.

In every case, suppliers work with their users, using their expertise in different applications to tailor the optimum cost-effective solution based on their modular and standardised platforms. Designed for outstanding levels of reliability and efficiency, integrated AC–AC or AC–DC–AC or DC–AC commutator synchronous motors not only help their users to cut operating, maintenance and energy costs, but to reduce their environmental impact. Extensive service programs and their global organisation ensure that they can support users over their EM motors' entire lifecycle.

Physical models of the integrated AC–AC or AC–DC–AC or DC–AC commutator synchronous motor are shown in figures 2.4–2.6, while a construction of the integrated AC–AC or AC–DC–AC or DC–AC commutator synchronous motor is shown on figure 2.7.

Fixed-speed standard AC synchronous or asynchronous (induction) motors are typically used in applications such as compressors, fans, pumps, wood grinders and refiners. In each case suppliers utilise advanced software tools to optimise the EM motor design for the specific application and ensure that the chosen starting method is appropriate. The results of this work are processed into an easy-to-understand format and incorporated into the technical specification they provide with each proposal.

Suppliers can engineer the EM motor to develop sufficient torque for smooth starting and acceleration with the starting current limited typically to 3.5–5 times of the rated value. If this results in unacceptable line voltage drops then they can investigate alternative starting methods. The main factors to be considered in

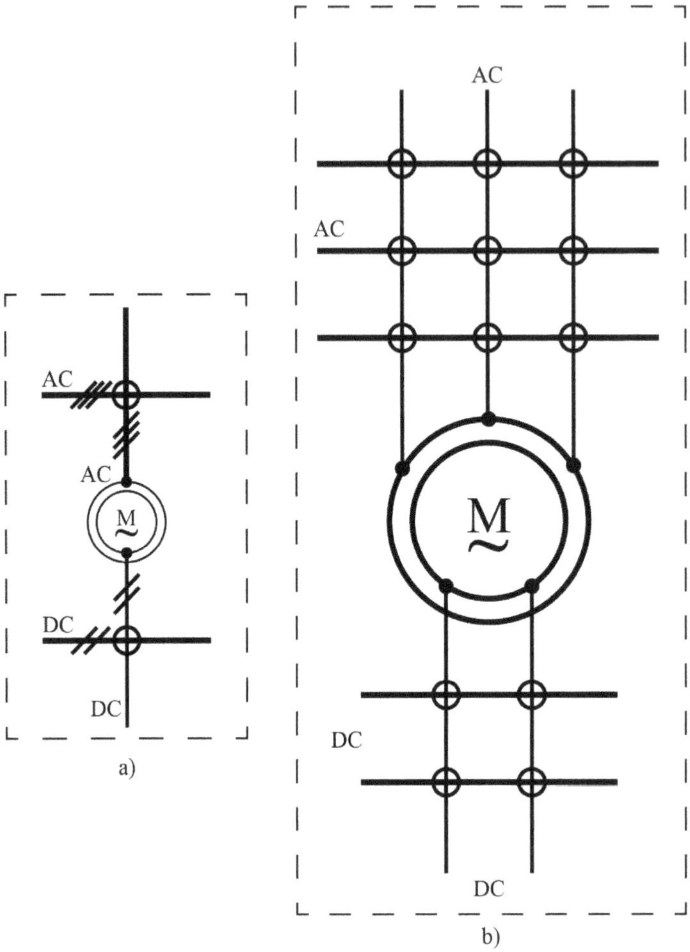

Figure 2.4. Physical model of the integrated AC–AC commutator synchronous motor.

selecting the starting method are the user's demands, network capability and the requirements of the process. The starting methods most commonly used for fixed-speed standard AC synchronous motors are **direct-on-line** (DOL), reactor, auto-transformer, **load commutated inverter** (LCI) or 'soft starter', capacitor and reactor capacitor. Rolling mills, mine winders, hoists, pumps, and compressors are examples of variable-speed applications.

Integrated AC–AC or AC–DC–AC or DC–AC commutator synchronous motors with the AC or DC motor IEMD is also commonly used in the main propulsion system in ships or vessels, aircraft, automotive vehicles etc.

In applications such as extruders, compressors and pumps, optimised AC and DC motor IEMD cyber-physical dynamical hypersystems based on integrated AC–AC or AC–DC–AC or DC–AC commutator synchronous motors can provide considerable energy savings. Suppliers deliver integrated EM motors and an IEMD that meet torque requirements over the entire operating range, from zero to maximum

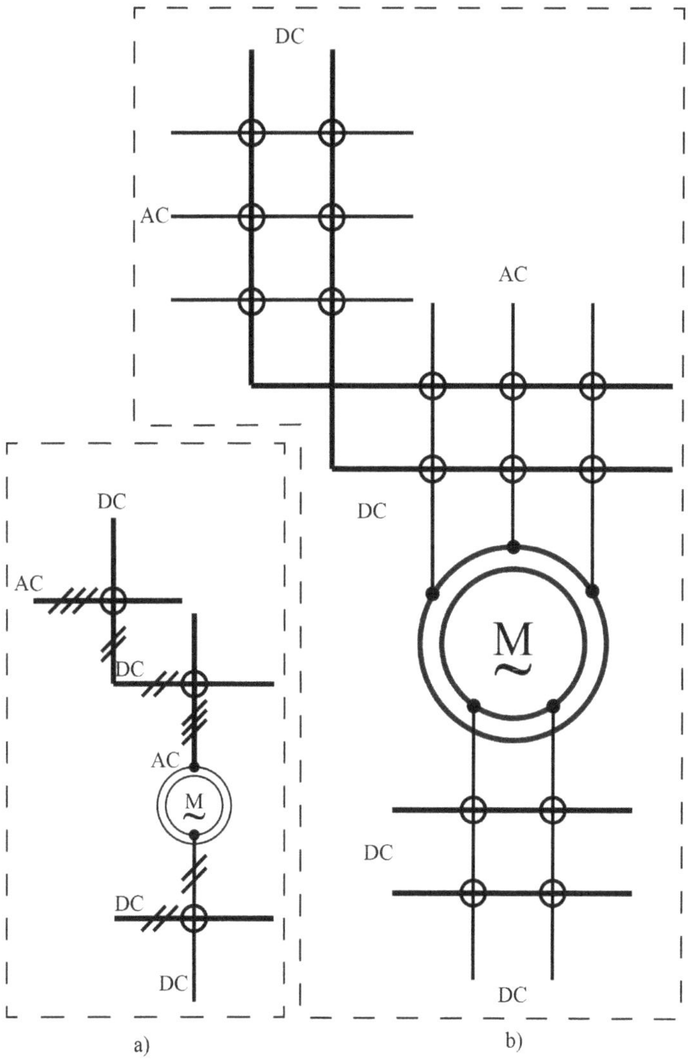

Figure 2.5. Physical model of the integrated AC–DC–AC commutator synchronous motor.

process speed. This ensures smooth starting, acceleration and operation. When suppliers deliver IEMD packages they apply their engineering and application know-how to ensure that all components, particularly the integrated AC–AC or AC–DC–AC or DC–AC commutator synchronous motor's interface, are optimally integrated to meet the needs of the process. Benefits include special integrated AC–AC or AC–DC–AC or DC–AC commutator synchronous motor's designs, increased efficiency and improved torque production capabilities.

The integrated DC–AC/AC–DC or AC–AC or AC–DC–AC commutator synchronous motor's rotor plays a crucial part in achieving the best possible electrical and mechanical performance.

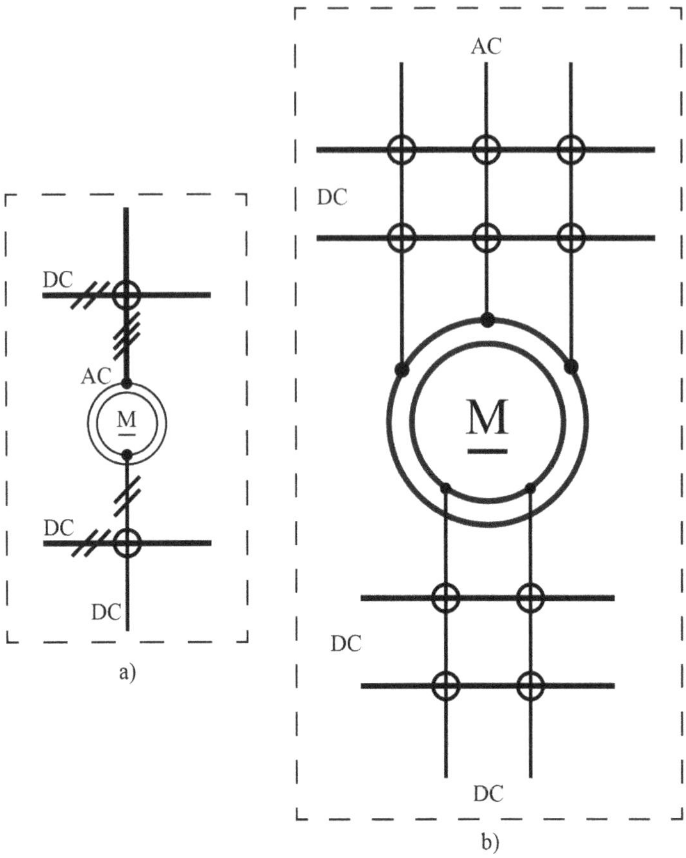

Figure 2.6. Physical model of the integrated DC–AC commutator synchronous motor.

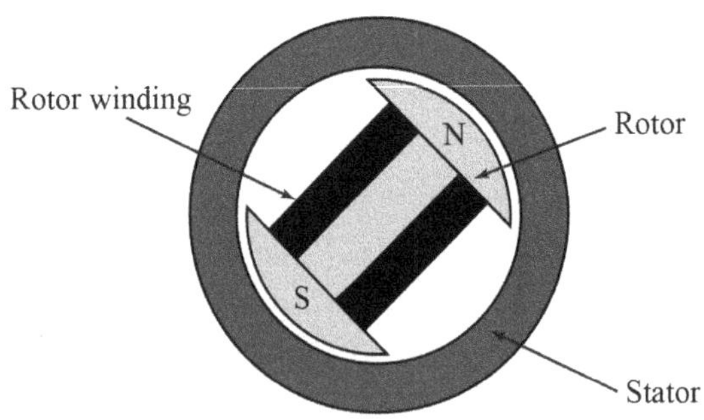

Figure 2.7. Construction of the integrated AC–AC or AC–DC–AC or DC–AC commutator synchronous motor.

Higher-speed EM motors (four and six magnetic poles) have solid rotors with integrated magnetic poles made from a one-piece steel forging. A solid magnetic-pole plate is attached to the magnetic pole, and this design has good overload capability and low harmonics. At the same time, this rotor configuration contributes to the EM motors' excellent starting characteristics, high starting torque and low starting current. The large cooling surfaces and effective flow of cooling air mean that the rotor temperature remains low and uniform, helping to ensure reliability and a long operating life. The winding coils are class H insulated for extra thermal margins. The rigid rotor construction and minimum distance between bearings ensure that the EM motor operates below the first critical speed, keeping vibration levels low.

Lower-speed EM motors (eight or more magnetic poles) typically have rotors with the magnetic poles manufactured from 2 mm laminated steel sheet. The sheets are pressed together with inserted steel bars, which are welded to the endplates. The magnetic poles are secured to the shaft or rotor centre by bolts from above or below, or by means of dovetails. A copper or brass damper winding is often fitted.

For lower-speed EM motors, the technically preferred option is to **vacuum pressure impregnate** (VPI) the rotor assembly after it is wound to achieve excellent insulation and mechanical strength. After impregnation, the complete rotor assembly is dynamically checked for balance.

The EM motor shaft is manufactured of forged or rolled steel and machined to exact specifications. The EM motor shaft ends can be cylindrical, conical or flanged, or two shaft ends can be provided (double end IEMD).

The rotor windings are made either of preformed enamelled rectangular copper wire or of flat copper. Proper supports between adjacent windings are used to ensure stability up to the rated over-speed. The rotor windings are made to match the insulation class of the stator. This ensures outstanding reliability and a long service life, even with asymmetric loads or under exceptional conditions.

Each application has its own specific requirements for EM motor performance. Modular EM motor stator design allows designers to precisely match the performance of every integrated AC–AC or AC–DC–AC or DC–AC commutator synchronous motor with the macro-commutator to the needs of its application.

The stator core itself is built of stacked, high-grade, low-loss laminated electrical steel, insulated on both sides with high quality insulating coating (heat-resistant inorganic coating is also available). The use of high-grade core material increases efficiency and therefore reduces operating costs. Radial cooling ducts ensure uniform and effective cooling. The rigid stator construction transmits all forces via the frame to the foundation to minimise vibrations.

The windings are insulated with Mica based tape. When the windings are in place, the complete stator undergoes VPI. The windings are class F insulated, which gives good thermal margins. This insulation homogeneous dynamical system has been used very successfully over many years and in several thousand EM motors, and it provides reliability and a long operating life. Long-term reliability is also assured by the use of well-proven methods for locking the winding coils into the slots and bracing the winding coil ends.

For maximum flexibility, integrated AC–AC or AC–DC–AC or DC–AC commutator synchronous motors are designed for horizontal, inclined or vertical mounting (vertical mounting available for EM motors with eight or more magnetic poles). The robust frame transfers dynamic and static stresses directly to the foundation, reducing vibration and contributing to the overall excellent performance of the EM motor.

Suppliers can supply integrated AC–AC or AC–DC–AC or DC–AC commutator synchronous motors—even up to the largest sizes and outputs—as complete, ready-to-install units (in some cases the heat exchanger or terminal box has to be removed prior to shipment).

This means that no further on-site assembly is necessary which substantially reduces installation times and cuts the risk that something could go wrong during installation.

R&D team works closely with suppliers of driven equipment, using **finite element method** (FEM) and dynamic animation techniques, to analyse vibration mathematical models, test critical parts and verify that hazardous mechanical resonances will not occur in the EM motor.

Integrated AC–AC or AC–DC–AC or DC–AC commutator synchronous motors are designed and built to withstand the relevant environmental conditions. All surfaces made of steel, aluminium alloy or cast iron are treated in accordance with the chosen paint system. Selection of a suitable paint system gives reliable anti-corrosion protection even under the most severe environmental conditions. For moderate indoor conditions, the standard finish is moisture-proof in accordance with the relevant standards. Solvent free paints are used wherever possible in order to minimise environmental impacts.

At suppliers, they take care to ensure that the excitation homogeneous dynamical system meets the very high overall standards of reliability that they have set for their integrated AC–AC or AC–DC–AC or DC–AC commutator synchronous motors. For fixed-velocity applications and adjustable-velocity applications with less demanding dynamic control requirements, a magnetic-field exciter and automatic voltage regulator are generally provided. The excitation homogeneous dynamical system has no wearing parts, and the external excitation power requirement is low. The magnetic-field exciter is a separate AC–DC commutator generator with the macrocommutator acting as AC–DC rectifier mounted on the EM motor shaft at the non-drive end.

In most fixed-velocity AC synchronous motors, the field winding is DC fed; in adjustable-velocity EM motors (and fixed-velocity EM motors with LCI starting) the magnetic field winding is AC supply power fed. The exciter is VPI using the MCI method, ensuring that the windings are sealed and secured. The advanced and yet straightforward design has a low component count and effective protection functions, and it offers high reliability and easy access for maintenance.

For adjustable-velocity IEMD applications where very fast and accurate angular velocity or torque control are required, the EM motor is generally equipped with carbon brushes and an M-E split-ring unit to allow excitation and control of the EM motor from the AC–AC commutator acting as an AC–AC cycloconverter.

The M-E split-rings are mounted on the EM motor shaft with access via removable inspection covers. In general, the split-rings and mounting flange or hub are made of steel, and they are normally mounted as a single unit. Mechanical split-ring units with brass rings, as well as split flange mounted units, are available on request. The mechanical split-ring unit is fitted with brass connection pins to facilitate installation. No excitation homogeneous dynamical system is needed for adjustable-velocity EM motors with IPM rotors.

Higher-velocity EM motors (four and six magnetic poles) use a compact, magnetic-field exciter unit mounted on the rotor shaft outboard of the bearings. No independent support or alignment is required. The high level of magnetic field forcing delivers improved dynamical system performance, which increases the production of reactive power and is beneficial when faults arise in the supply network.

Lower-velocity EM motors (eight or more magnetic poles) typically have a magnetic-field exciter unit mounted inside the EM motor enclosure. The EM motor excitation control panel can be supplied in a variety of basic formats. It houses the excitation equipment, protection homogeneous dynamical system and logic functions for starting. Various options are available on user request.

Suppliers offer a wide range of instrumentation and control equipment to protect integrated AC–AC or AC–DC–AC or DC–AC commutator synchronous motors with the macrocommutator and ensure excellent reliability and availability, and extended product lifetimes.

The EM motor excitation control panel can, as an option, be adapted for immediate integration into a superior management and supervisory system. Communication via modem can also be supplied to facilitate remote support.

2.6 Integrated AC–AC or AC–DC–AC or DC–AC commutator IPM synchronous motor with the macroelectronic commutator (macrocommutator)

While it would be difficult to find the difference between an integrated DC–AC commutator IPM synchronous motor (DC motors) with the mechanical split-ring or flat commutator EM motor described above and an integrated DC–AC IPM synchronous motor with the macrocommutator by just looking at them, the concept of operation is quite different as is the analysis.

Physical models of the integrated AC–AC or AC–DC–AC or DC–AC commutator IPM synchronous motor are shown in figures 2.8–2.10, while a construction of the integrated AC–AC or AC–DC–AC or DC–AC commutator IPM synchronous motor is shown in figure 2.11.

The windings in the EM motor stator in an integrated DC–AC IPM synchronous motor with the macrocommutator are not sinusoidally distributed but instead they are concentrated, each occupying one third of the magnetic-pole pitch. The magnetic-flux density on the magnet surface and in the air gap is also not sinusoidally distributed over the magnet but almost uniform in the air gap. As the stator currents interact with the magnetic flux coming from the magnet torque is developed.

It should be clear that for the same sense of direction of magnetic flux, currents in opposite sense of directions result in opposite forces, and therefore in reduction of

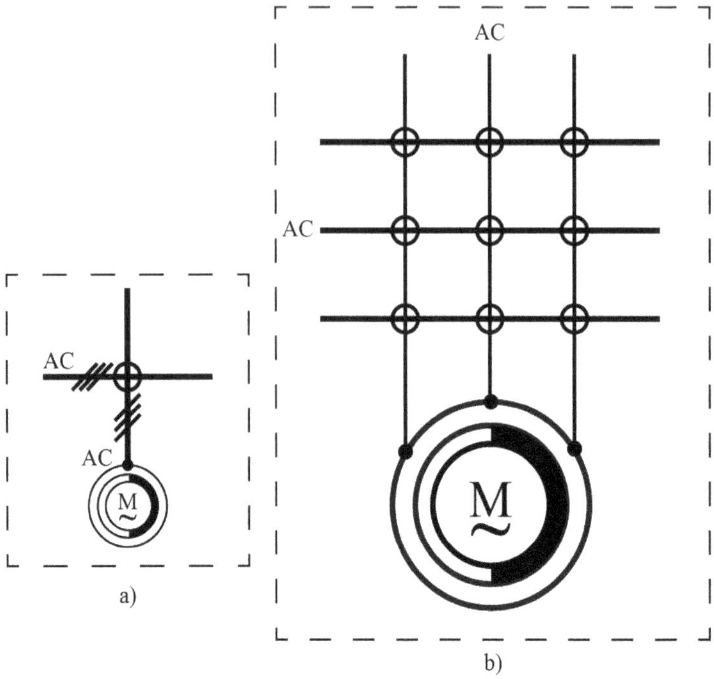

Figure 2.8. Physical model of the integrated AC–AC commutator IPM synchronous motor.

total torque. This in turn makes it necessary that all the current in the stator above the rotor is in the same sense of direction.

To accomplish this following are needed:

- Sensors on the stator that sense the direction of the magnetic flux coming from the rotor.
- A fast supply that will provide currents to the appropriate stator windings as determined by the magnetic-flux sense of direction.
- A way to control these currents, e.g. through pulse width modulation (PWM).
- A microcontroller with inputs the desired angular velocity, the magnetic-flux sense of direction in the stator and the stator currents, and outputs the desired currents in the stator.

Interior permanent magnet (IPM) poles are used in certain adjustable-velocity applications. Stators with the IPM poles are straightforward constructions and do not necessitate an electromagnetic excitation homogeneous dynamical system. When necessary, high magnetic-pole numbers can be used in low angular-velocity applications. The formulae that describe the operation of the cyber-physical dynamical hypersystem are quite simple.

The developed EM torque holor $\boldsymbol{T}_i^e$ is proportional to the stator-current holor $\boldsymbol{I}_s^j$:

$$\boldsymbol{T}_i^e = \boldsymbol{C}_{ij}\boldsymbol{I}_s^j \tag{2.1}$$

where: $\boldsymbol{C}_{ij}$ —constant coefficient holor.

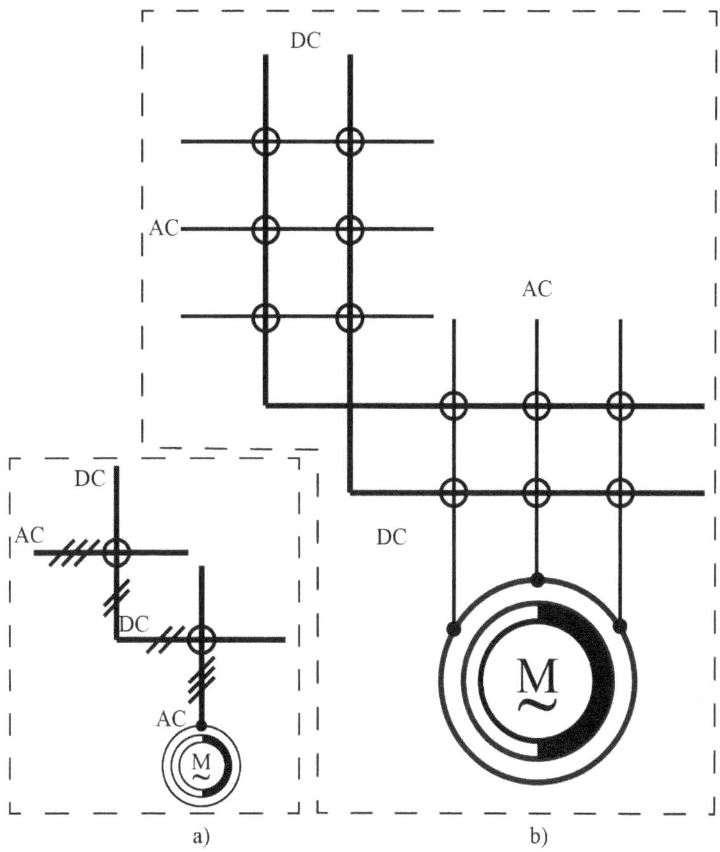

Figure 2.9. Physical model of the integrated AC–DC–AC commutator IPM synchronous motor.

At the same time, the rotating magnetic-flux induces an induced voltage holor $\boldsymbol{E}_i$ in the energised windings:

$$\boldsymbol{E}_i = \boldsymbol{C}_{ij}\,\boldsymbol{\Omega}^j \tag{2.2}$$

where: $\boldsymbol{\Omega}^j$—angular-velocity holor.

Finally, the terminal DC voltage holor $\boldsymbol{V}_i^{DC}$ differs from the induced-voltage holor by a resistive voltage-drop holor $\Delta \boldsymbol{V}_i^R = \boldsymbol{R}_{ij}\boldsymbol{I}_s^j$:

$$\boldsymbol{V}_i^{DC} = \boldsymbol{E}_i + \boldsymbol{R}_{ij}\boldsymbol{I}_s^j \tag{2.3}$$

where: $\boldsymbol{R}_{ij}$ —stator resistance holor.

This is the reason that although this electrical machine is entirely different from a standard DC motor, it is termed an integrated DC–AC commutator IPM synchronous motor with the macrocommutator.

These equations are similar to those of an integrated DC–AC IPM synchronous motor with the mechanical split-ring or flat commutator (DC motor).

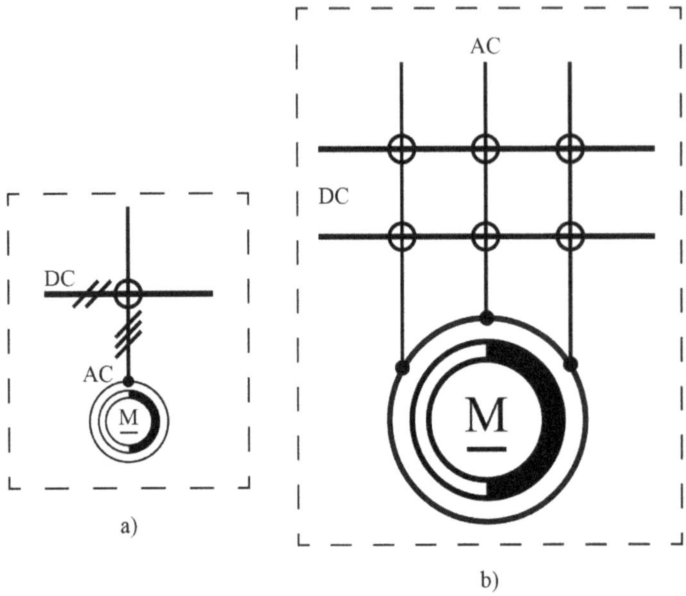

Figure 2.10. Physical model of the integrated DC–AC commutator IPM synchronous motor.

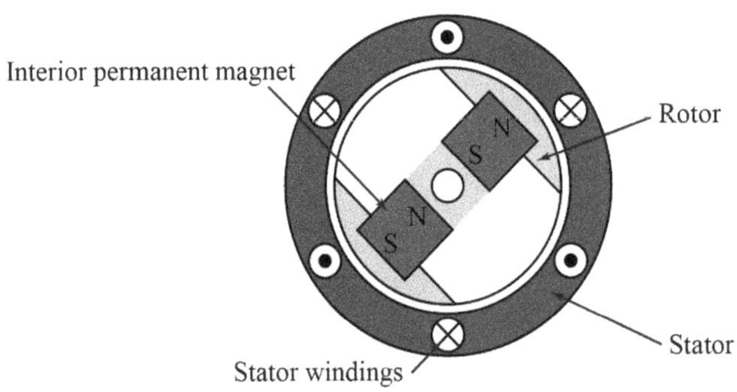

Figure 2.11. Construction of the integrated AC–AC or AC–DC–AC or DC–AC commutator IPM synchronous motor.

Integrated AC–AC or AC–DC–AC or DC–AC commutator IPM synchronous motors with the macroelectronic commutator (macrocommutator) do not rely entirely on current for magnetisation. IPM fields are, by definition, constant and not subject to failure, except (as we will explore) in extreme cases of magnet abuse and demagnetisation by overheating. Until recently, integrated AC–AC or AC–DC–AC or DC–AC commutator IPM synchronous motors were available but not widely distributed; now these EM motors are proliferating. All integrated AC–AC or AC–DC–AC or DC–AC commutator IPM synchronous motors require a matched IPM IEMD and must be operated with such; integrated AC–AC or AC–DC–AC or DC–AC commutator IPM synchronous motors are not to be started for across the AC supply-line starting.

Rare-earth components are those 30 metals found in the Mendeleyev's periodic table's oft-omitted long centre two rows; they are used in many modern applications. Magnets made of rare-earth metals are particularly powerful alloys with crystalline structures that have high magnetic anisotropy—which means they readily align in one sense of direction, and resist it in others. Discovered in the 1940s and identified in 1966, rare-earth IPMs are one third to two times more powerful than traditional ferrite magnets—generating magnetic fields up to 1.4 T in some cases. Their IPMs are used in MMD machines, portable electronic devices, hysteresis clutches, accelerometers and—last but not least—IPM rotary and linear EM motors. IPM poles are used in certain adjustable-velocity applications. Rotors with the IPM magnetic poles are straightforward constructions and do not need an electromagnetic excitation homogeneous dynamical system. When necessary, high magnetic-pole numbers can be used in low angular-velocity IEMD applications.

An integrated AC–AC or AC–DC–AC or DC–AC commutator IPM synchronous motor has a sinusoidally distributed stator winding to produce sinusoidal induced voltage of rotation waveforms. Induced voltage of rotation is voltage that opposes the current that causes it.

In fact, induced voltage of rotation arises in any EM motor when there is relative motion between the current-carrying armature (whether rotor or stator) and the external magnetic field. As the rotor spins (with or without power applied to the windings), the mechanical rotation generates a voltage—so, in effect, becomes an M-E generator. Typical units are (V $krpm^{-1}$).

IPM-compatible AC or DC motor IEMD (termed as IPM IEMD) substitute the more conventional trapezoidal waveform's flat tops with a sinusoidal waveform that matches IPM induced voltage of rotation more closely, so torque output is smoother. Each commutation of phases must overlap, selectively commuting more than one pair of power-commutation AC–AC or AC–DC–AC or DC–AC commutators at a time. These IEMD setups can be operated as open-loop systems used in midrange performance applications requiring speed and torque control. In this case, integrated AC–AC or AC–DC–AC or DC–AC commutator IPM synchronous motors are placed under holor-type control.

In fact, though IPMs require an IEMD specifically designed to drive integrated AC–AC or AC–DC–AC or DC–AC commutator IPM synchronous motors, the IPM IEMD setup is most similar to magnetic-flux holor IEMD—for integrated AC–AC or AC–DC–AC or DC–AC commutator asynchronous (induction) motors, in that the IEMD uses current-commutation techniques to control EM motor torque—and simultaneously controls both torque and magnetic-flux current via mathematically intensive transformations between one coordinate homogeneous dynamical system and another. This IPM IEMD uses EM motor data and current measurements to calculate rotor position; the **digital signal processor** (DSP) calculations are quite accurate. During every sampling interval, the three-phase AC cyber-physical homogeneous dynamical system—dependent on time and velocity—is transformed into a rotating two-coordinate cyber-physical homogeneous dynamical system in which every current is expressed and controlled as the sum of two holor's merates.

Many scientists and engineers associate IPM construction with DC servomotors; however, integrated DC–AC/AC–DC or AC–AC or AC–DC–AC commutator IPM synchronous motors are now an option, and they exceed the power density efficiency of standard AC asynchronous (induction) motors. For instance, IEMD's IPM technology reduces rotor losses to save energy for many fractional and integral, mechanical power EM motors; a radial magnet-pole design greatly improves EM motor efficiency and specific output power. An adjustable-velocity IEMD operation in constant and variable-torque applications is also possible.

In integrated AC–AC or AC–DC–AC or DC–AC commutator IPM synchronous motors, angular velocity is a function of frequency—the same as it is with integrated AC–AC or AC–DC–AC or DC–AC commutator asynchronous (induction) motors. However, integrated AC–AC or AC–DC–AC or DC–AC commutator IPM synchronous motors rotate at the same velocity as the magnetic field produced by the stator windings; it is a synchronous electrical machine. Therefore, if the magnetic field is rotating at 3000 rpm, the rotor also turns at 3000 rpm—and the higher the input frequency from the IEMD, the faster the EM motor rotates.

Most OEMs of standard AC synchronous motors hold magnetic-pole count constant so input frequency dictates the EM motor's speed. To calculate required input frequency f (Hz) when the number of magnetic poles p and rotational speed n (rpm) are known

$$f = pn/120.$$

Integrated AC–AC or AC–DC–AC or DC–AC commutator IPM synchronous motors are suitable for variable or constant-torque applications: The IEMD and application parameters dictate to the EM motor how much torque to produce at any given angular velocity. This flexibility makes IPMs suitable for adjustable-velocity IEMD operation requiring ultra-high EM motor efficiency. Now a word on a common misconception:

Cogging is the unwanted jerking during EM motor spinning from repeatedly overcoming the attraction of IPMs and stator's steel structure—is often associated with integrated AC–AC or AC–DC–AC or DC–AC commutator IPM synchronous motors. Particularly at start-up, cogging arises from the interaction of the rotor IPMPoles and stator winding when it is energised, due to harmonics. Cogging, in turn, causes noise, vibration and non-uniform rotation. Many methods for reducing cogging can be leveraged to eliminate torque and angular-velocity ripple.

Some integrated AC–AC or AC–DC–AC or DC–AC commutator IPM synchronous motors are designed with more poles than equivalent integrated AC–AC or AC–DC–AC or DC–AC commutator asynchronous (induction) motors, which helps reduce these issues. With reference to integrated AC–AC or AC–DC–AC or DC–AC commutator IPM synchronous motors, saliency refers to the difference in EM motor inductance at the EM motor terminals as the EM motor's rotor is rotated.

This difference corresponds to alignment and misalignment of the stator's rotor—a characteristic that an EM motor's drive tracks to monitor rotor position during operation.

In special cases, integrated AC–AC or AC–DC–AC or DC–AC commutator IPM synchronous motors are used in closed-loop configurations using angular-velocity feedback. Feedback allows the IEMD to track the exact rotor position—to provide true infinite angular-velocity range, including full torque at zero angular velocity. The angular-velocity reference required from an external source can be an analogue signal and encoder feedback, or a serial command from a feedback device on an axis one wishes to follow—or a programmable limit switch, potentiometer or any external device that can create and communicate a value to the IEMD. This normally is an angular-velocity signal, sometimes further processed in the IEMD before it is used as a command.

The EM motor's angular velocity is limited by induced voltage of rotation because the latter increases directly with the EM motor angular velocity. The EM motor is connected to the AC or DC motor IEMD and its macroelectronic and microelectronic units are designed for a maximum voltage above the rated IEMD voltage.

Normally, the EM motor and its controls are designed to operate well below the maximum voltage of the macroelectronic and microelectronic units. However, if the EM motor's angular velocity exceeds the design angular-velocity range (either being powered from the control or being driven by the mechanical load), it is possible to exceed the maximum voltage of the IEMD units—and cause failures.

Note that an AC or a DC motor IEMD is capable of limiting the EM motor's induced voltage of rotation when operating properly. However, if the IEMD faults and loses control during over-speed, it cannot protect itself. In addition, integrated AC–AC or AC–DC–AC or DC–AC commutator IPM synchronous motor control requires some technical knowledge for implementation.

All commercially available integrated AC–AC or AC–DC–AC or DC–AC commutator IPM synchronous motors require an IPM-compatible IEMD to operate, although there is ongoing research in the development of a line-start integrated AC–AC or AC–DC–AC or DC–AC commutator IPM synchronous motor. This makes them indispensable for producing high magnetic-flux levels but also means they must be handled with care.

Catching one's finger between two of these magnets (during servicing, for example) poses a serious pinching hazard, so the EM motor supplier or an authorised shop should generally execute any magnet-related maintenance tasks on IPM EM motors. Those with pacemakers or other medically implanted devices (including hearing aids) should exercise extra caution when working around the strong magnetic fields of these devices; cell phones and credit cards may also be at risk.

That said, when an IPM EM motor's rotor is secured within the enclosure, radiated magnetic energy is no higher than that of an AC asynchronous (induction) motor.

Not every AC or DC motor IEMD is suitable for operation of integrated AC–AC or AC–DC–AC or DC–AC commutator IPM synchronous motors; only an IEMD specifically designed for IPM EM motor compatibility is suitable. Often, a parameter in the IEMD programming allows an operator to set the IEMD for an IPM EM motor. As a rule, the IEMD not specifically designed for it can run and control IPM EM motors, though performance is degraded—and one can damage the EM motor or IEMD if they are mismatched.

Finally, high current or operating temperatures can cause the magnets in integrated AC–AC or AC–DC–AC or DC–AC commutator IPM synchronous motors to lose their magnetic properties. IPMs, once demagnetised, cannot recover, even if the current or temperatures return to normal levels.

An IPM IEMD reduces the risk of high-current demagnetisation, as these are equipped with over-current protection.

Some EM motor designs further minimise the possibility of excessive-temperatures IPM failure with high-temperature magnets, integrated thermostats and restricted EM motor operating temperature. As with other EM motor designs that include IPMs, axial magnetic-flux integrated AC–AC or AC–DC–AC or DC–AC commutator IPM synchronous motors exist. In these EM motors, magnetic force (through the air gap) is along the same plane as the EM motor shaft—along the EM motor length. Axial magnetic flux can be thought of as having the same orientation as disc brakes on a regular automotive vehicle, as the disc rotates like the rotor in an axial magnetic-flux design.

Radial magnetic-flux integrated AC–AC or AC–DC–AC or DC–AC commutator IPM synchronous motors are the more conventional design, in which the magnetic force is perpendicular to the length of the EM motor shaft. A design's form factor determines which orientation is most suitable. Does the electrical machinery require a longer, skinnier radial EM motor or is a 'pancake' axial design more appropriate? The final determining factor may be cost as the axial design, once tooled for production, provides equivalent torque but uses less active material for better power density. Though not yet suitable for elevator applications, scientists and engineers are developing radial integrated AC–AC or AC–DC–AC or DC–AC commutator IPM synchronous motors to incorporate axial air-gap IPM designs into elevators without machine rooms.

2.7 Integrated AC–AC or AC–DC–AC or DC–AC commutator variable-reluctance synchronous motor with the macroelectronic commutator (macrocommutator)

Combining the advantages of integrated AC–AC or AC–DC–AC or DC–AC commutator IPM synchronous motors and integrated AC–AC or AC–DC–AC or DC–AC–AC commutator asynchronous (induction) motors, innovative integrated AC–AC or AC–DC–AC or DC–AC commutator variable-reluctance (commutated-reluctance) synchronous motors offers potentially breakthrough technology for modular AC or DC motor IEMD and EM motor packages. These EM motors use an innovative rotor design that eliminates the split losses associated with standard AC asynchronous (induction) motors while offering the benefits of IPM operation.

The key trends in industry related to EM motors, and what users and the marketplace value, are energy efficiency, size/footprint reduction, and reliability. Innovative integrated AC–AC or AC–DC–AC or DC–AC commutator variable-reluctance synchronous motor IEMD technology offers benefits in precisely these three areas.

It provides improved efficiency compared to standard AC asynchronous (induction) motors, the ability to go to higher power density, and higher reliability that includes longer service intervals. The fundamental technology makes it possible to reach these benefits. To create a smaller, more efficient EM motor with a long lifetime and lower maintenance needs, as well as a new EM motor type that could be adapted to modular AC or DC motor IEMD operation, suppliers say they considered all technology options.

An integrated AC–AC or AC–DC–AC or DC–AC commutator IPM synchronous motor is a synchronous EM motor meaning that its rotor spins at the same speed as the EM motor's internal rotating magnetic field. Other AC synchronous technologies include hysteresis EM motors, larger DC-excited EM motors and common reluctance EM motors. The latter includes both a stator and rotor with multiple projections; the stator's magnetic poles are wrapped with windings that are energised, while the rotor's magnetically permeable steel projections act as salient magnetic poles that store magnetic energy by reluctance averaging the tendency of magnetic flux to follow the path of least magnetic reluctance in order to repeatedly align the rotor and stator magnetic poles.

A physical model of the integrated AC–AC or AC–DC–AC or DC–AC commutator variable-reluctance synchronous motor is shown in figures 2.12–2.14, while a construction of the integrated AC–AC or AC–DC–AC or DC–AC commutator variable-reluctance synchronous motor is shown on figure 2.15.

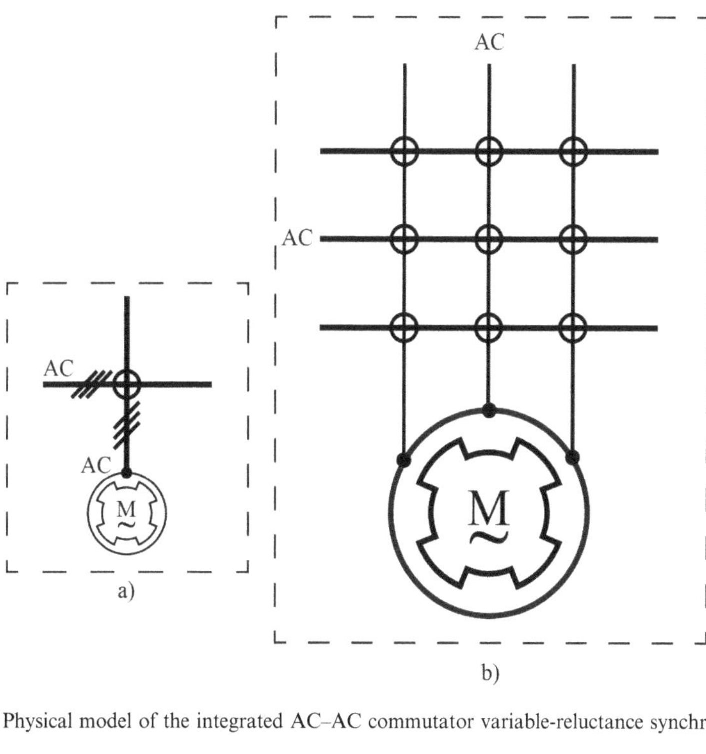

Figure 2.12. Physical model of the integrated AC–AC commutator variable-reluctance synchronous motor.

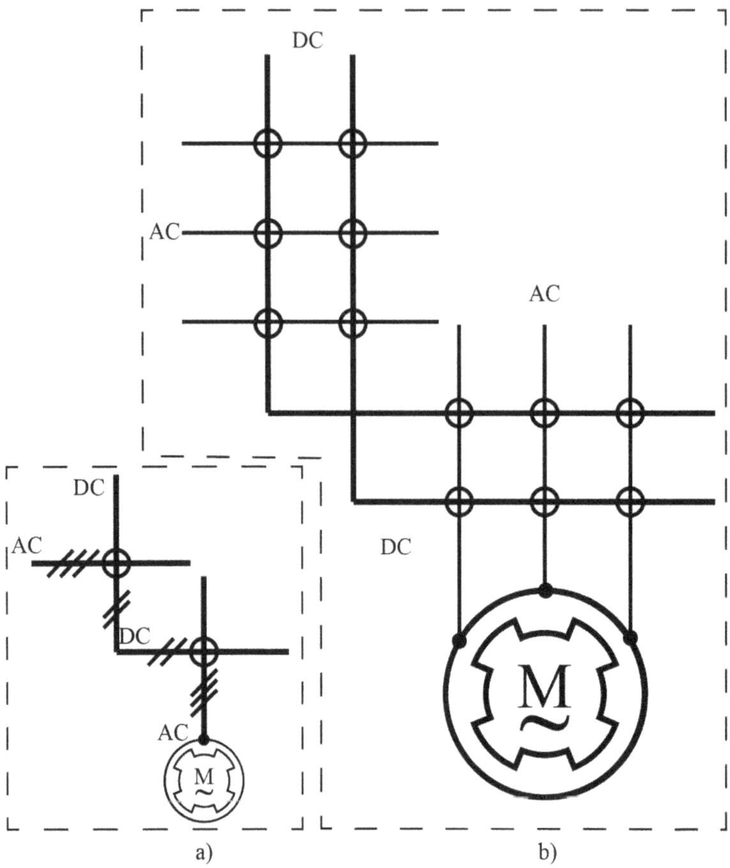

Figure 2.13. Physical model of the integrated AC–DC–AC commutator variable-reluctance synchronous motor.

Though extended coverage of reluctance EM motor variants is not within the scope of this textbook, it is helpful to know a bit about its variable-reluctance iteration. This EM motor can be built to deliver up to 150 kW, overlapping with AC asynchronous (induction) and IPM AC synchronous motor capabilities. In variable-reluctance (commutated-reluctance) EM motors, the stator winding coils are synchronously energised with rotor rotation, with overlapping phases. While variable-reluctance EM motors are typically used as open-loop steppers, their variable-reluctance derivative (also sometimes termed commutated-reluctance) is typically operated under closed-loop control. In fact, step EM motors are somewhat similar to variable reluctance, and step to each defined rotor position, resulting in high repeatability, and accuracy. Variable-reluctance EM motors produce high efficiency and control, and produce 100% torque at stall indefinitely—useful for applications that require holding. Finally, although torque ripple must be overcome, variable-reluctance EM motors can be operated at higher speeds than integrated

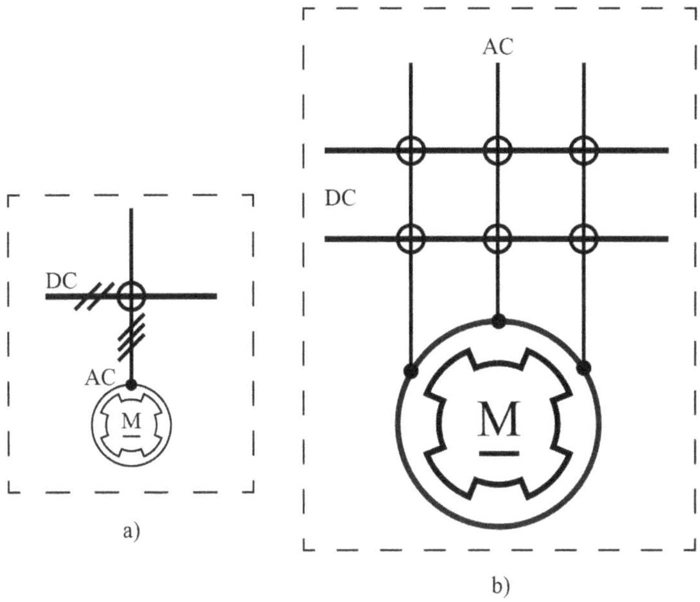

Figure 2.14. Physical model of the integrated DC–AC commutator variable-reluctance synchronous motor.

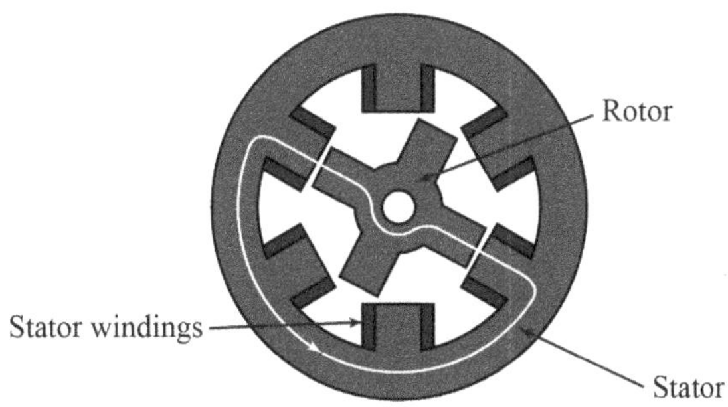

Figure 2.15. Construction of the integrated AC–AC or AC–DC–AC or DC–AC commutator variable-reluctance synchronous motor.

AC–AC or AC–DC–AC or DC–AC commutator IPM synchronous motors, as they lack induced voltage of rotation constraints.

Starting an integrated AC–AC or AC–DC–AC or DC–AC commutator variable-reluctance synchronous motor is very different from a direct-line connection start, but this highlighted the opportunities to simplify the EM motor design and improve efficiency.

As a high-efficiency solution, integrated AC–AC or AC–DC–AC or DC–AC commutator variable-reluctance synchronous motor technology is based on the concept of magnetic resistance or reluctance. The rotor is designed to produce the smallest possible reluctance in one sense of direction and the highest reluctance in the perpendicular sense of direction.

The EM motor produces torque as the rotor attempts to align the magnetically conducting sense of direction to the stator magnetic field. What this means in practice is a rotor technology where the rotor is rotating at the same frequency as the stator magnetic field.

On the rotor side of the EM motor, suppliers are able to eliminate so-called split-page losses that exist in AC asynchronous (induction) motors. These rotor-side losses are an essential part of the total losses of the EM motor, so eliminating them creates higher efficiency.

They can reduce the losses by up to 40%, depending on the power rating of the EM motor, and increase the efficiency of the EM motor up to 5% in smaller power EM motors and 0.5% in the biggest frame EM motors, which puts them in compliance with the upcoming IE4 efficiency class in Europe.

Because of the lower losses on the rotor side, the design makes the integrated AC–AC or AC–DC–AC or DC–AC commutator variable-reluctance synchronous motors smaller and lets them produce up to 40% more power at any EM motor size. These are the two main benefits of this technology.

Other benefits relate to reliability. The rotor construction itself is robust, and there is no squirrel-cage or IPMs on the rotor side.

Thanks to the low losses, EM motor operating temperatures are also lower, which results in longer life for the insulation and bearings.

If one compares the innovative integrated AC–AC or AC–DC–AC or DC–AC commutator variable-reluctance synchronous motors to other integrated AC–AC or AC–DC–AC or DC–AC commutator IPM synchronous motor technologies, there are no IPMs and no risk of losing performance due to demagnetisation from overheating or failure. There are no induced voltages of rotation caused by the spinning EM motor, which can be a safety concern, and, because of the synchronous rotation, the speed control is by nature accurate without needing additional encoders.

High-speed operation with the innovative integrated AC–AC or AC–DC–AC or DC–AC commutator variable-reluctance synchronous motors can eliminate use of mechanical power transmission components such as gearboxes in some applications.

The core innovation is the rotor design, since the stator side of the EM motor is identical to an AC asynchronous (induction) motor. Suppliers' knowledge and expertise helped it construct a robust, efficient rotor capable of high-speed IEMD operation. However, IEMD technology and the matched AC or DC motor IEMD packages and algorithms make the cyber-physical dynamical hypersystem breakthrough possible from a technology perspective.

In many ways, integrated AC–AC or AC–DC–AC or DC–AC commutator variable-reluctance synchronous motor technology combines the benefits of AC asynchronous (induction) motors and AC IPM synchronous motors. It provides the robustness of an AC asynchronous (induction) motor and the size, efficiency, and synchronous speed operation benefits of AC IPM synchronous motor technology while eliminating concerns related to IPM technology.

The primary benefits of the technology—including size, efficiency, and reliability—are generic and are applicable for a wide array of areas, but pumps, fans, and compressors are some of the mainstream applications. For example, the low-inertia rotor makes it possible to provide high-speed ramp rates that can be useful for harbour cranes and mine winders.

2.8 Integrated AC–AC or AC–DC–AC or DC–AC commutator asynchronous (induction) motor with the macroelectronic commutator (macrocommutator)

One third of the world's electrical energy consumption is used for running standard AC asynchronous (induction) motors driving pumps, fans, compressors, elevators and machinery of various types.

The integrated AC–AC or AC–DC–AC or DC–AC commutator asynchronous (induction) motor with the macrocommutator is a common form of an AC asynchronous (induction) motor whose operation depends on three electromagnetic phenomena:

- ***EM motor action***—When an iron rod (or other magnetic material) is suspended in a magnetic field so that it is free to move or rotate, it will align itself with the magnetic field. If the magnetic field is moving or rotating, the iron rod will move or rotate with the moving or rotating magnetic field so as to maintain alignment.
- ***Rotating magnetic field***—A rotating magnetic field can be created from fixed-stator magnetic poles by driving each magnetic-pole pair from a different phase of the AC power supply.
- ***AC–AC transformer action***—The current in the rotor windings is induced from the current in the stator windings, avoiding the need for a direct connection from the power source to the rotating windings.

The integrated AC–AC or AC–DC–AC or DC–AC commutator asynchronous (induction) motor with the macrocommutator can be considered as an AC–AC transformer with a rotating secondary winding. Rotating magnetic fields are created by polyphase excitation of the stator windings. In the example below of a three-phase commutator asynchronous (induction) motor, as the current applied to the winding of magnetic-pole pair a (phase 1) passes its peak and begins to fall, the magnetic-flux associated with the winding also begins to weaken, but at the same time the current in the winding of the next magnetic-pole pair b (phase 2) and its associated magnetic-flux is rising.

Simultaneously the current through the winding of the previous magnetic-pole pair (phase 3) and its associated magnetic-flux will be negative and rising (towards positive). The net effect is that a magnetic-flux wave is set up as the magnetic-flux created by the stator magnetic-poles rotates from one magnetic pole to the next, about the axis of the EM motor, at the frequency of the applied voltage. In other words, the rotating magnetic-flux field appears to the stator as the north and south magnetic poles of a magnet rotating about the stator.

The magnitude of the rotating magnetic-flux wave is proportional applied induced **magnetomotive force** (MMF). Ignoring the effect of the induced voltage of rotation also termed **back electromotive force** (EMF) set up by the induced

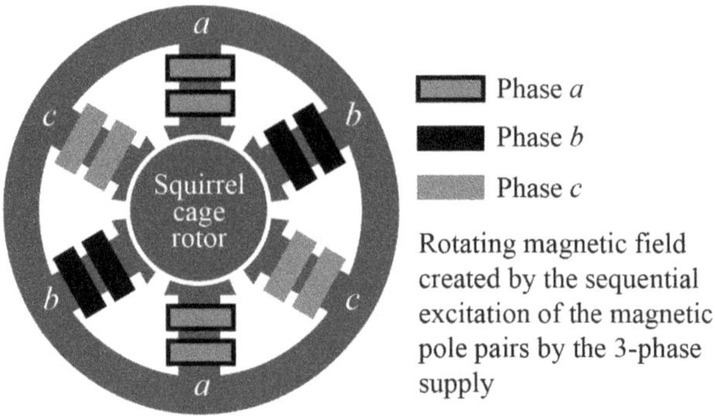

Figure 2.16. AC squirrel-cage asynchronous (induction) motor.

Figure 2.17. Squirrel-cage rotor.

currents in the rotor windings, the magnetic-flux density B will be proportional to the applied voltage.

The stator carries the EM motor primary windings and is connected to the AC power source through a macrocommutator acting as the DC–AC inverter or AC–AC cycloconverter or AC–DC–AC frequency changer by AC–AC transformer action. The magnitude of the rotor current will be proportional to the magnetic-flux density B in the air gap and the relative motion, termed the *'split'*, of the rotor with respect to the rotating magnetic field (as shown in figure 2.17).

Torque is produced by the reaction between the induced rotor currents and the air-gap magnetic flux created by the stator currents. Many rotor types are used.

The most popular integrated AC–AC or AC–DC–AC or DC–AC commutator asynchronous (induction) motors use 'squirrel-cage' rotors (figure 2.16) which are constructed from copper or aluminium bars fixed between conducting end rings which provide the short circuit path for the currents induced in the bars.

When the EM motor is first commuted (switched) on and the rotor is at rest, a current is induced in the rotor windings (conductors) by AC–AC transformer action. Another way of seeing this is that the relative motion of the rotating magnetic-flux passing over the slower moving (initially static) rotor windings causes a current to flow in the windings by mechanical generator action.

Once current is flowing in the rotor windings, the EM motor action due to the Lorentz force on the conductors comes into effect. The reaction between the current flowing in the rotor conductors and the magnetic flux in the air gap causes the rotor to rotate in the same sense of direction as the rotating magnetic-flux as if it was being dragged along by the magnetic-flux wave.

Similar to the integrated DC–AC commutator synchronous induction motor with the mechanical split-ring or flat commutator (DC motor), the torque in an integrated AC–AC or AC–DC–AC or DC–AC commutator asynchronous (induction) motor with the macrocommutator, the torque holor $\boldsymbol{T}_i$ is proportional to the magnetic-flux density holor $\boldsymbol{B}_{ij}$ and the induced rotor current holor $\boldsymbol{I}^j$.

Thus

$$T_i = c\boldsymbol{B}_{ij}\boldsymbol{I}^j$$

where c is a constant depending on the number of stator turns, the number of phases and the configuration of the magnetic circuit.

The rotor angular velocity builds up due to the EM motor action described above, but as it does so, the relative motion between the rotating stator magnetic field and the rotating rotor conductors is reduced.

This in turn reduces the M-E generator action and thus the current in the rotor conductors and the torque on the rotor.

As the angular velocity of the rotor approaches the angular velocity of the rotating field, (termed the 'synchronous angular velocity'), then the torque of the rotor drops to zero.

Thus the angular velocity of a standard AC asynchronous (induction) motor can never reach the synchronous angular velocity. A standard AC induction motor is therefore an 'AC asynchronous motor'.

The relative motion between the rotating magnetic field and the rotating rotor is termed the 'split' and is given by:

$$s = \frac{\Omega_s - \Omega_r}{\Omega_s}$$

where s is the slip, Ω_s is the synchronous angular velocity (s^{-1}), and Ω_r is the rotor angular velocity.

Since the rotor current is proportional to the relative motion between the rotating magnetic field and the rotor angular velocity, the rotor current and hence the torque are both directly proportional to the slip. The rotor current is proportional to the rotor resistance. Increasing the rotor resistance will reduce the current and increase the split; hence a form of angular velocity and torque control is possible with AC wound-rotor asynchronous (induction) motors.

Increased rotor resistance also has the added benefit of reducing the input surge current and increasing starting torque on switch on, but all of these benefits are at the expense of more complex rotor designs and unreliable split-rings to give access to the rotor windings.

In all its iterations, the integrated AC–AC or AC–DC–AC or DC–AC commutator asynchronous (induction) motor with the macroelectronic commutator (macrocommutator) induces magnetism that is leveraged to output translational or angular motion. The stationary outer stator is connected to an external DC power source via the macrocommutator acting as a DC–AC inverter or an AC–AC cycloconverter or AC–DC–AC frequency changer, respectively; this is fed to the mover's or rotor's poles in a translational or angular progression that causes travels or revolutions of the magnetic field within the EM motor's stator.

Conducting bars in the mover or rotor interact with the stator's magnetic fields; current is induced in those bars, which in turn generate magnetic fields that are attracted to those of the stator.

As the mover's or rotor's induced current and magnetism cause it to follow the magnetic field generated by the stator, translational or angular motion is output. Because, an integrated AC–AC or AC–DC–AC or DC–AC commutator asynchronous (induction) motor with the macrocommutator increases the magnetic-flux enclosed by its stationary single- or polyphase stator-winding coils, it is a transformer with a travelling or rotating secondary (mover or rotor). The mover or rotor current's effect on the air-gap magnetic-flux causes force or torque, respectively.

Standard AC asynchronous (induction) motors are built by suppliers according to established **International Standards Organisation** (ISO) or **National Electrical Manufacturers Association** (NEMA) standards in many fractional and integral mechanical power ratings and associated frame sizes build a standard asynchronous (induction) motor. These standard AC asynchronous (induction) motors are quite common—'the workhorse of industry'. Electric motors are often referred to as 'the workhorse of industry'. This phrase has probably been around for almost a century and it is still in use today.

A physical model of the integrated AC–AC or AC–DC–AC or DC–AC commutator squirrel-cage-rotor asynchronous (induction) motor is shown in figures 2.18–2.20, while a construction of the integrated AC–AC or AC–DC–AC or DC–AC commutator squirrel-cage-rotor asynchronous motor is shown in figure 2.21.

An integrated DC–AC/AC–DC or an AC–AC or AC–DC–AC commutator asynchronous (induction) motor's stator consists of a stack of thin, highly permeable steel laminations with slots; the laminations are either secured in a steel or cast-iron frame that provides, a mechanical support. The windings that accept the external power supply are run through the slots.

The AC inductor mover or rotor assembly resembles a squirrel-cage consisting of aluminium or copper conducting bars connected by short-circuiting end rings. The rotor also has laminations; radial slots around the laminations contain the bars.

As mentioned, the mover or rotor turns when the travelling or rotating magnetic field induces current in the shorted conductors, and the rate at which it travels or rotates is the EM motor's synchronous speed—determined by AC power-supply frequency and the number of stator magnetic poles.

$$n_s = 120f/p$$

where: n_s—synchronous speed (rpm);

f—frequency (Hz);

p—number of magnetic poles.

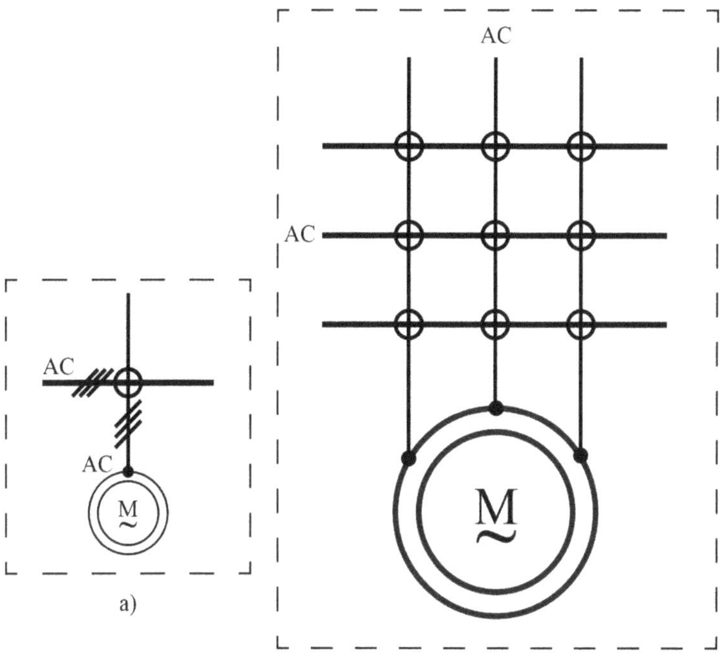

Figure 2.18. Physical model of the integrated AC–AC commutator asynchronous (induction) motor.

Synchronous speed is the fastest theoretical speed an EM motor can spin—when the mover or rotor spins at the same speed as the EM motor's internal travelling or rotating magnetic field.

In practice, a standard AC induction motor is an EM motor (in which the mover or rotor lags magnetic-field speed), so its mover or rotor must spin more slowly than the magnetic field, or split. This allows the induction of mover or rotor current to flow, and production of force or torque to drive the attached mechanical load while overcoming internal losses.

Single-phase AC asynchronous (induction) motors power many low mechanical power commercial and industrial IEMD applications where three-phase power is impractical; they are not efficient, but can last a lifetime. They are also classified by how they are started, as these AC asynchronous (induction) motors alone develop no starting torque, but require external means for initial actuation.

The simple split-phase (induction-start-induction-run) integrated EM motor has a small-gage-start winding with fewer turns than the main winding to create more resistance and put the start winding's field at a different electrical angle than that of the main—causing the EM motor to rotate until it reaches 75% of rated speed. Then the main winding of heavier wire keeps the integrated EM motor running. This inexpensive design develops high starting current (700%–1000% of rated), so prolonged starts cause overheating.

Suitable applications include small grinders, blowers and low-starting-torque applications requiring up to 1/3 kW. For instance, this single-phase, capacitor-start,

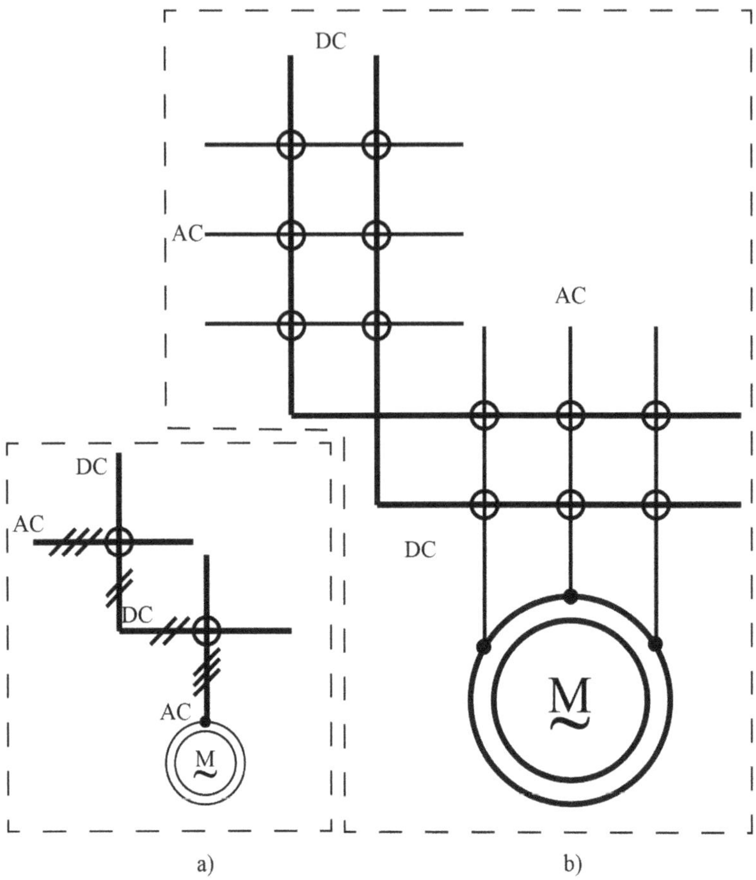

Figure 2.19. Physical model of the integrated AC–DC–AC commutator asynchronous (induction) motor.

induction-run EM motor form is an instant reversing EM motor designed for parking gates and door operators requiring up to 1 kW.

This type of general-purpose EM motor has a beefed-up start winding plus a capacitor for boost. Starting torque is typically 200%–400% of rated mechanical load; starting current to 575% of rated current allows higher cycle rates and thermal resilience.

Split-capacitor AC asynchronous (induction) motors are common but are slowly being replaced with more efficient EM motors and an IEMD, which we will explore later. Split-capacitor AC asynchronous (induction) motors have a run-type capacitor permanently connected in series with the start winding, making the latter an auxiliary winding once the EM motor reaches running speed. Starting torques are 30%–150% of rated mechanical load, unsuitable for hard-to-start applications. However, starting current is less than 200% of rated load current, making them suitable for cycling or frequent reversals—in fans, blowers, intermittent adjusting mechanisms and quick-reversing garage door openers.

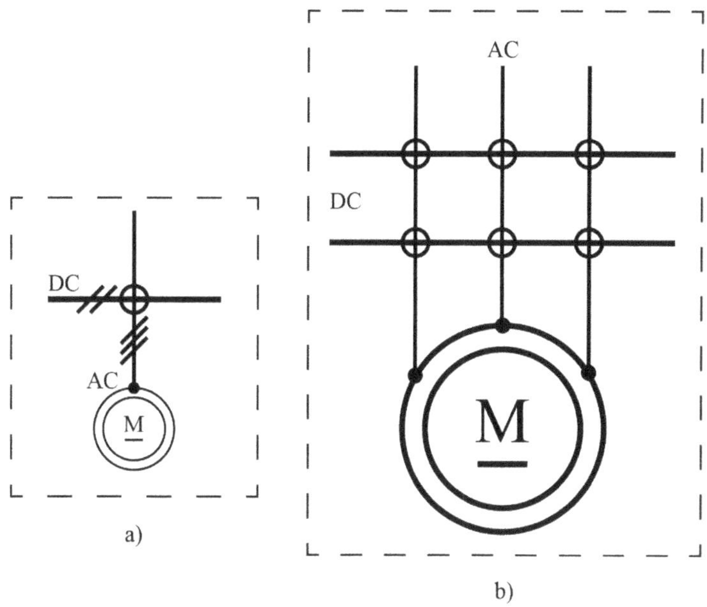

Figure 2.20. Physical model of the integrated DC–AC commutator asynchronous (induction) motor.

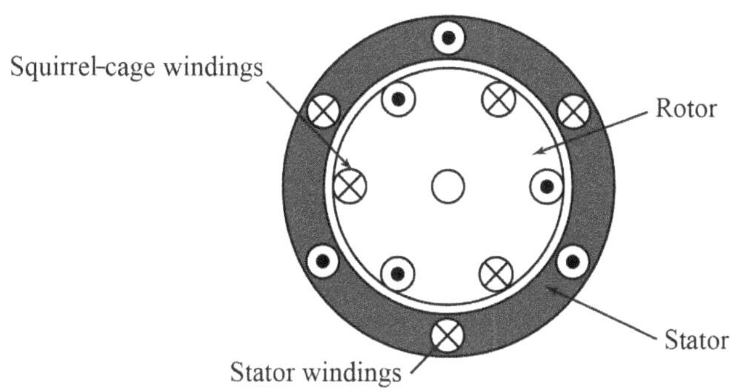

Figure 2.21. Construction of the integrated AC–AC or AC–DC–AC or DC–AC commutator asynchronous (induction) motor.

Most powerful of all the single-phase types, capacitor-start capacitor-run AC asynchronous (induction) motors have a start capacitor in series with auxiliary winding, plus a run-type capacitor in series with the auxiliary winding for high overload torque. Some have lower full-load current and higher efficiency—so operate more coolly than other single-phase motors of comparable mechanical power.

Cost is higher but these EM motors are indispensable in wood-working machinery, air compressors, vacuum pumps and other 1–7.5 kW high-torque applications. Lastly, inexpensive, shaded-pole, single-phase AC induction motors have only one main winding. Starting is through a copper loop partially covering a portion of each EM motor magnetic-pole—causing the magnetic field in the ringed area to lag behind that in the unringed portion. The reaction of the two magnetic fields causes shaft rotation; varying voltage controls speed. These are disposable EM motors, and are common in household fans; efficiency is 20% or lower. By far the most common industrial EM motor is the three-phase AC asynchronous (induction) motor. Approximately 90% of all three-phase AC induction motors are used in industrial applications. Why? The standard utility has three-phase power at 50 or 60 Hz.

Single-phase AC asynchronous (induction) motors are versatile but three-phase (and other polyphase) EM motors offer higher power and efficiency, and require no electrical valve (electronic switch), capacitor or relays. More specifically, three-phase AC asynchronous (induction) motors have high starting torque because six poles (only $\pi/3$ rad per phase apart) operate in pairs to boost the power factor—also useful in large industrial applications. (For example, 4 Nm of torque from a four-pole EM motor yields 1 kW.)

NEMA classifies general-purpose, three-phase AC asynchronous (induction) motors as A, B, C or D according to their electrical design. For example, NEMA Design C AC asynchronous (induction) motors have higher starting torque with normal start current and less than 5% split. Some of these EM motors are connected directly to line power with a basic switch; others are fitted with wye/delta windings or soft starters. However, in these designs, roughly one third of all industrial designs are variable-load applications.

Integrated EM motors are increasingly being driven by an IEMD—which have (thanks to increasingly sophisticated, yet affordable, semi- or superconductor electrical valves and integrated electrical commutation matrixers) proliferated over the last two decades.

In these integrated **electrical commutation matrixers** (ECM), **pulse width modulation** (PWM) is used to vary EM motor voltage. In turn, solid-state electrical valves such as, for example, **reverse blocking** (RB) **insulated gate bipolar transistors** (IGBTs) or **integrated gate commutated thyristors** (IGCT) or **gate turn off** (GTO) thyristors (SCRs) execute PWM. For instance. AC line voltage is converted by the AC–DC–AC macroelectronic commutator (macrocommutator) unit acting as an AC–DC rectifier to DC and then reshaped by the AC–DC–AC macrocommutator unit acting as a DC–AC inverter so that EM motor speed varies with the frequency of the pulses in the output voltage.

When a modular PWM AC or DC motor IEMD were developed using RB IGBTs, this new technology was used to control integrated AC–AC or AC–DC–AC or DC–AC commutator asynchronous (induction) motors in industrial and commercial applications.

The advantages were obvious:

- Force or torque and EM motor translational or angular velocity could now be precisely controlled to optimise both process and energy requirements.
- The potential to save energy by operating EM motors at only the needed translational/angular velocity while maintaining force/torque requirements could potentially result in 20%–50% energy savings.
- Processes could be optimised and controlled by computer processing systems to achieve productivity increases or energy savings.

A PWM modular AC or DC motor IEMD allows wide speed ranges, programmable acceleration and deceleration ramps, and good energy efficiency; force/torque and translational/angular velocity precision can, in some cases, match that of DC motor IEMD cyber-physical heterogeneous dynamical hypersystems.

- In its simplest iteration, voltage-per-frequency AC or DC motor IEMD operation holds the ratio of voltage and frequency constant by tracking voltage magnitude. This prevents magnetic saturation (at which an EM motor's rotor cannot be magnetised further, causing high currents); voltage to be applied is calculated from the applied frequency required to maintain air-gap magnetic flux—a method that provides passable speed control, though no direct control of EM motor torque.
- Sensorless holor or vector control also modulates frequency but measures (and compensates for) split by determining the amount of current in phase with the voltage for approximated torque current—for both magnitude and phase-angle between current and voltage. This helps to keep the EM motor running at the desired speed even under varied mechanical load.
- Slightly more sophisticated, a magnetic-flux holor or vector IEMD leverage the fact that in integrated AC–AC or AC–DC–AC or DC–AC commutator asynchronous (induction) motors, some current magnetises or magnetic fluxes the rotor to magnetically couple it to the stator. A magnetic-flux holor or vector IEMD hold this magnetic-flux current at the minimum required to induce a magnetic field, while independently modulating torque-producing current pulsing through the stator. Finally, the AC or DC motor IEMD iteration, termed **magnetic-field oriented control** (MFOC), pairs an IEMD's current regulators with an adaptive microcontroller to independently meter and control EM motor's torque and EM motor's magnetic flux. This kind of IEMD can be paired with an encoder for closed-loop servo-control but its consistent performance does not typically require feedback. Torque output is consistent from zero to full load over many speeds.

Even under sophisticated AC or DC motor IEMD control, integrated AC–AC or AC–DC–AC or DC–AC commutator asynchronous (induction) motors exhibit inherent efficiency limitations and can require an encoder for feedback if low-speed accuracy is required. In addition, retrofitting an existing design with a modular AC or DC motor IEMD can be troublesome, particularly when equipped with older standard AC asynchronous (induction) motors. Why? Its AC–AC or AC–DC–AC or DC–AC commutator's synthesised AC waveform accelerates heating (although advances continue to improve the waveform to more closely approximate an AC sinewave).

Extended operation of an AC or DC motor IEMD-powered integrated AC–AC or AC–DC–AC or DC–AC commutator asynchronous (induction) motor at less than 50% of base speed also is unacceptable; modern AC–AC or AC–DC–AC or DC–AC commutator-duty EM motors have higher insulation ratings but extreme cases require a separately powered cooling fan. In fact, the wasted heat generated by any integrated AC–AC or AC–DC–AC or DC–AC commutator induction motor is capable of degrading the insulation so essential to EM motor operation. Stator insulation prevents short-circuits, winding burnout and failure. Magnet wire coating insulates wires within a winding coil from each other; slot cell and phase insulation (composite sheets installed in stator slots) shield phase-to-ground; stator varnish dip boosts moisture resistance and overall insulation.

ISO and NEMA sets specific temperature standards for standard AC induction motors of various enclosures and service factors (of 1.5 or more in most cases). These standards are based on thermal insulation classes—often B, F and H. Maximum winding temperature ratings are total temperatures, based on integrated AC–AC or AC–DC–AC or DC–AC commutator asynchronous (induction) motor 40 °C (104 °F) maximum ambient plus the temperature rise generated by EM motor operation. 4 kW and greater, premium efficiency AC–AC or AC–DC–AC or DC–AC commutator-duty EM motors typically have Class F insulation. Beyond that, many manufacturers design their EM motors to operate more coolly than their thermal class definitions. Class H insulation is reserved for heavy-duty, hot or high-altitude conditions. Another consideration is cycling: EM motors built for frequent reversals can withstand it but start-stop cycles in others can cause overheating. This is because a typical EM motor under these conditions draws five to six times the rated running current, which accelerates heating. ISO or NEMA limits three-phase, continuous-duty standard AC asynchronous (induction) motors to two starts in succession before allowing the EM motor to stabilise to its maximum continuous operating temperature. Finally, the AC or DC motor IEMD commonly used to drive integrated AC–AC or AC–DC–AC or DC–AC commutator asynchronous (induction) motors are sensitive to inertia, mechanical power, EM motor lead length and power quality, so they must be programmed with full-load and no-load currents, base angular velocity and frequency, and EM motor voltage when initially connected to a new EM motor.

Typically, an AC or a DC motor IEMD also requires tuning, during which EM motor response and electrical characteristics are logged.

Integrated AC–AC or AC–DC–AC or DC–AC commutator wound-rotor (split-ring) asynchronous (induction) motors are ideal for heavy load inertia applications or weak network conditions:

- high starting torque, high inertia;
- low starting current;
- high torque over the entire speed range;
- suitable for starting in weak networks;
- can be adapted for use with modular AC or DC motor.

These EM motors are widely used in applications requiring high starting torque or low starting current, including a mill EMD, cement plants, mines, utilities, water works and many more. These rugged EM motors with fully braced and vacuum pressure impregnated windings are made for heavy-duty IEMD operation. Their high efficiency level is mainly achieved through sophisticated EM motor design software, the effective use of high quality materials, and advanced ventilation technology. High efficiency means substantial energy savings over the life of the EM motor.

Integrated DC–AC/AC–DC or an AC–AC or AC–DC–AC commutator wound-rotor (split-ring) asynchronous (induction) motors can be supplied with:

- permanent contact carbon brushes, or
- carbon-brush-lifting gear.

In both cases, the split-rings are mounted at the N-end and enclosed in a separate housing from the EM motor. This arrangement provides easy access and maintenance, and keeps carbon dust out of the EM motor. It also makes it possible to have different cooling methods for the split-rings and EM motor.

The split-rings with permanent contact carbon brushes are manufactured from highly corrosion resistant Cu–Sn–Ni alloy and are helical grooved as standard. Mechanical split-rings with carbon-brush-lifting gear manufactured from stainless steel have a smooth, non-grooved surface.

A physical model of the integrated AC–AC or AC–DC–AC commutator split-rings or wound rotor asynchronous (induction) double-fed motor is shown in figures 2.22, and 2.23, while a construction of the integrated AC–AC or AC–DC–AC commutator split-rings or wound rotor asynchronous double-fed motor is shown in

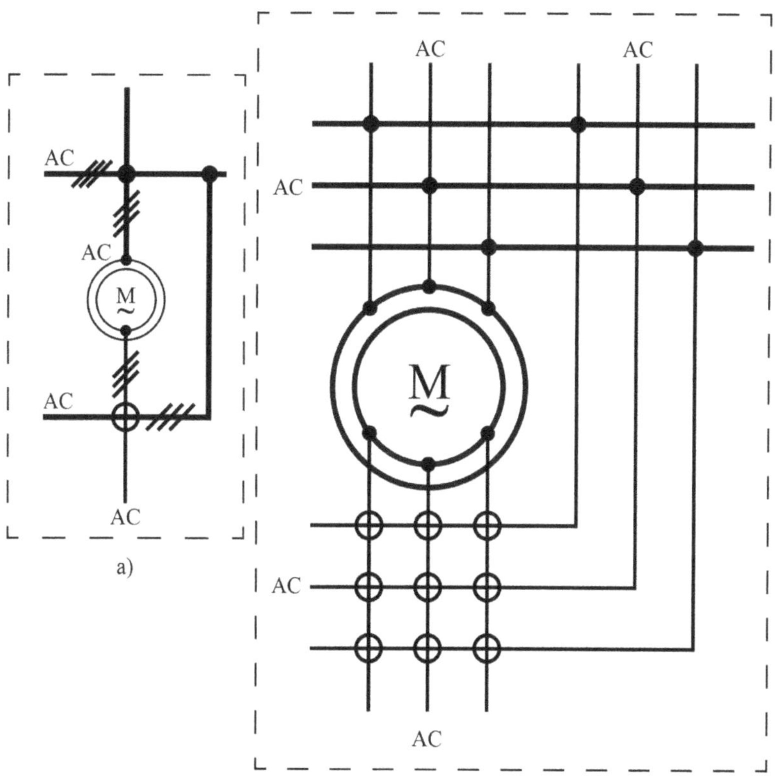

Figure 2.22. Physical model of the integrated AC–AC commutator split-rings or wound-rotor asynchronous (induction) motor.

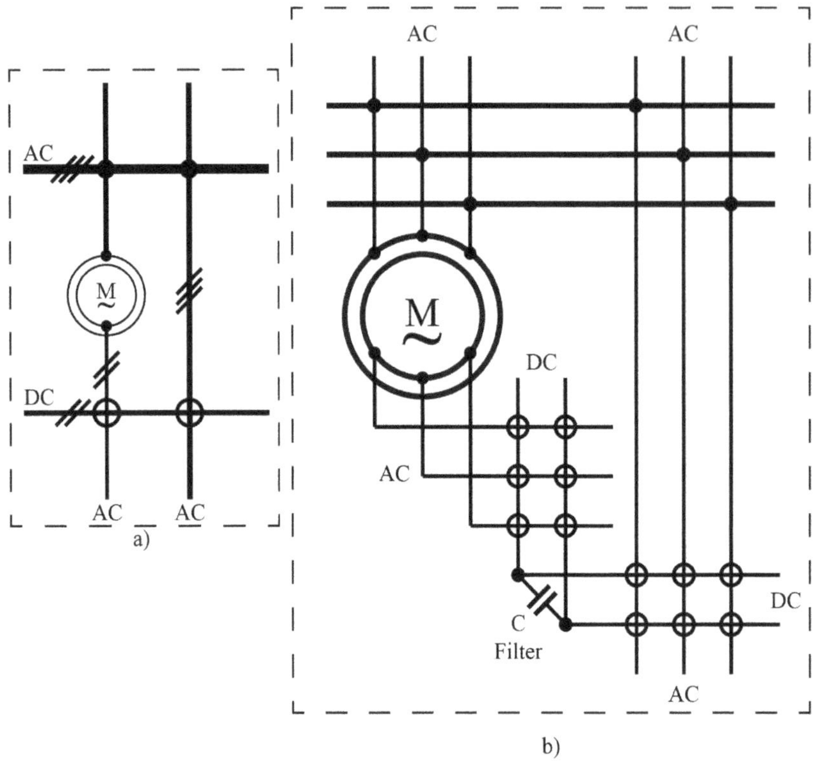

Figure 2.23. Physical model of the integrated AC–DC–AC commutator wound-rotor asynchronous (induction) double-fed motor.

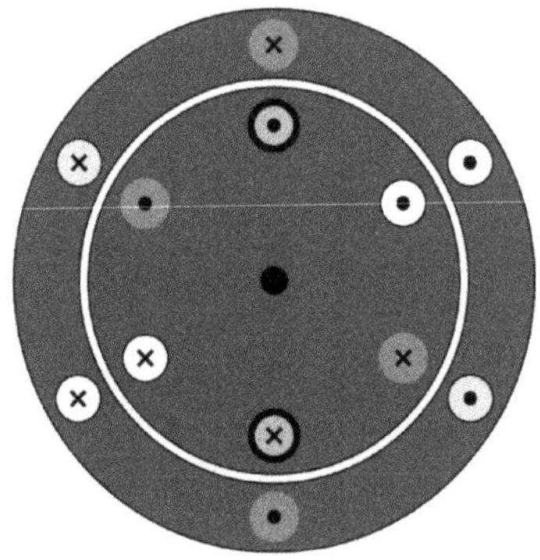

Figure 2.24. Construction of the integrated AC–AC or AC–DC–AC commutator split-rings or wound-rotor asynchronous (induction) double-fed motor.

figure 2.24. After the EM motor has achieved full speed, the **brush-lifting and split-ring short-circuiting gear** (BLG) first short-circuits the rotor winding and then raises the carbon brushes from the split-rings. Effective cooling for safe, reliable operation by means of internal air circulation is provided by a shaft-mounted fan or separate blower.

In air-to-air cooled EM motors, the external cooling air is circulated by a shaft-mounted fan or separate blower. Standard versions are always self-ventilated.

Cooling cyber-physical homogeneous dynamical systems in weather protected EM motors operate so the cooling air is drawn through the EM motor as a silencer. Radial air ducts in the stator core and spacers between the winding-coil ends ensure uniform and efficient cooling of the stator.

Mechanical split-ring units with permanent contact split-ring carbon brushes are equipped with a heat exchanger or protective cover similar to that on the EM motor.

A shaft-mounted fan circulates cooling air through the split-ring unit, filter, and upper cover. This cooling cyber-physical heterogeneous dynamical system is isolated from the EM motor's cooling cyber-physical heterogeneous dynamical system, thus providing easy maintenance and inspection.

2.9 Integrated AC–AC or AC–DC–AC or DC–AC commutator servomotor with the mechanical ring/disc or macroelectronic commutator (macrocommutator)

Emerging and future EM servomotors are integrated AC–AC or AC–DC–AC or DC–AC commutator servomotors with the mechanical split-ring/flat or macro-electronic commutator (macrocommutator) that use feedback for closed-loop control of cyber-physical heterogeneous dynamical hypersystems in which the operation is the variable.

Standard AC asynchronous (induction) motors designed for servo operation are wound with two phases at right angles. A fixed reference winding is excited by a fixed voltage source, while a variable control voltage from a servo-amplifier excites the winding. The windings often are designed with the same voltage-to-turns ratio, so that power inputs at maximum fixed phase excitation, and at maximum control phase signal, are in balance. Any EM motor designed for servo use is typically 25%–50% smaller than other EM motors with similar output and the reduced rotor inertia makes for quicker response. For instance, AC servomotors are used in applications requiring rapid and accurate response characteristics—so these AC asynchronous (induction) motors have a small diameter for low inertia and fast starts, stops and reversals. High resistance provides nearly linear speed-torque characteristics for accurate control.

Wound magnetic-field integrated AC–AC or AC–DC–AC or DC–AC commutator synchronous motors with the mechanical split-ring or flat commutator, previously termed DC motors (in various iterations, with mechanical split-ring or flat commutator's copper bars or segments in the rotor connected by magnetic-wire windings and stator windings) are another option.

More often, however, compact DC motors (which employ IPMs affixed to the inside of the EM motor frame, plus a rotating wound armature and commutating carbon brushes) are used as EM servomotors because speed control is easy: the only variable is voltage applied to the rotating-armature mechanical split-ring or flat commutator. There are no magnetic-field windings to excite, so these EM motors use less energy than wound DC designs, and have better power density than wound magnetic-field EM motors. Servo-built DC motors also include more wire wound onto the laminations, to boost torque. Integrated AC–AC or AC–DC–AC or DC–AC commutator IPM synchronous motors with macrocommutator also are commonly used for servo applications.

Commutation refers to how current is precisely routed to a carbon-brushed DC rotor's array of armature winding coils, to generate the torque needed via carbon brushes and a mechanical ring or disc commutator. A mechanical split-ring or flat commutator is mounted on the rotor shaft and includes pads, on which the carbon brushes rest. Current is conducted from the carbon brushes to the mechanical split-ring or flat commutator and then to the connected armature winding coils as the rotor spins.

Currently, the term commutation ought to be used to refer to integrated AC–AC or AC–DC–AC or DC–AC commutator asynchronous (induction) or synchronous motor operation, even though macrocommutator and a sensor on the rotor shaft replace any carbon brushes or mechanical ring/disc commutator. In this case, current is still commutated, but by macro-electronic commutator. The term loses meaning in the world of AC motors, though sometimes refers incorrectly to how AC voltage is generated in the IEMD.

Most DC motor windings are interconnected in an array and most units are fitted with a trio of Hall sensors at one stator end. These Hall sensors output low and high signals when the rotor's south and north IPM poles pass—to allow the following of energising sequence and rotor position. In its most basic form, the IEMD for a servomotor receives a voltage command that represents a desired EM motor current. The EM servomotor is modelled in terms of inertia (including EM servomotor and mechanical-load inertia) damping, and a torque constant. The mechanical load is considered rigidly coupled so that the natural mechanical resonance is safely beyond the servo-microcontroller's bandwidth.

EM motor position is usually measured by an encoder or resolver coupled to the EM motor shaft. A basic servo-control generally contains both a trajectory generator and a **proportional-integral-derivative** (PID) microcontroller: the former provides position set point commands; the latter uses position error to output a corrective torque command that sometimes is scaled to the EM motor's torque generation for a specific current (torque constant). Reliable speed controls for DC motors abound.

Many use solid-state AC–DC/DC–AC commutators with semiconductor electrical valves (for example IGCTs, MOSFETs or IGBTs), acting as controlled AC–DC rectifiers/DC–AC inverters are common, converting AC line voltage to controlled DC voltage that is applied to the DC motor's armature.

Increasing voltage increases speed—so this is sometimes termed 'armature-voltage control'. It is highly effective for DC motors up to approximately 3 kW, allowing 60:1 velocity regulation and constant torque even at reduced speeds. Servo-control, on the other hand, takes control to the next level with feedback—and is suitable for larger designs.

High-voltage DC motors are typically used with a DC–AC commutator and PWM micro-controller in applications requiring adjustable velocity and constant torque throughout the speed range. The brush holder design provides easy access, while the carbon brushes themselves are large for extended life.

Servo-control exhibits less steady-state error, transient responses and sensitivity to mechanical-load parameters than open-loop control dynamical systems. Improving transient response increases control dynamical system bandwidth, for shorter settling times and higher throughput. Minimising steady-state errors boosts accuracy. Finally, reducing mechanical-load sensitivity allows a motion homogeneous dynamical system to tolerate fluctuations in voltage, torque and mechanical-load inertia. Typically, a profile is programmed for instructions that define the operation in terms of time, position and angular velocity: a digital servo-controller sends angular-velocity command signals to an amplifier, which drives the servomotor, EM motor. With the help of resolvers, encoders or tachometers for feedback (mounted in the EM motor or on the mechanical load), the controller then compares actual position and speed to the target motion profile, and differences are corrected.

Currently, many IPM EM motors are DC and used in servo applications requiring adjustable velocity. For quick stops, these can minimise mechanical brake size (or eliminate the brake) by leveraging dynamic braking (EM motor generated energy fed to a resistor grid) or regenerative braking (EM motor-generated energy returned to the AC supply).

In addition, IPM DC motor's angular velocity can be controlled smoothly down to zero, followed immediately by acceleration in the opposite sense of direction without power circuit switching. In typical integrated AC–AC or AC–DC–AC or DC–AC commutator IPM synchronous motors, energisation is controlled macro-electronically. In some designs, IPMs are installed on the stator. Designs that are more common include stators with stacked steel laminations and windings through axial slots; IPMs are installed on the rotor. Here, the stator winding is trapezoidally wound to generate a trapezoidal induced voltage of rotation waveform with six-step commutation.

DC electrical valves (electronic switches) energise changing pairs of EM motor phases in a predefined commutation sequence. Most units are fitted with a trio of Hall sensors at one stator end to allow the following of energising sequence and rotor position. Output torque has considerable torque ripple, which occurs at each step of the trapezoidal commutation. However, due to a high torque-to-inertia ratio, integrated AC–AC or AC–DC–AC or DC–AC commutator IPM synchronous motors AC quickly to control-signal changes—making them useful in servo applications.

Most importantly, the increased performance of EM servomotor designs comes at dramatically increased cost. In addition, there are two situations in which EM servomotor efficiency declines—low voltage and high torque.

In short, EM servomotors are most often employed because of their ability to produce high peak torque, thus providing rapid acceleration—but high torque often requires that servomotors run two to three times their normal torque range, which degrades efficiency. Finally, EM servomotors are designed to operate over a wide range of voltages (as this is how their speed is varied) but efficiency drops with voltage.

2.10 Comparisons and conclusions

Designers and EM motor personnel benefit from finding a supplier that is an experienced resource of information to help in pragmatic EM motor selection. Involve application specialists as early as possible, as they can help develop prototypes, custom electrical and mechanical designs, mountings and gearboxes. This also reduces costs associated with shorter lead times and rush delivery. In the end, all industrial EM motor subtypes have strengths and weaknesses, plus application niches for which they are most suitable. For example, many industrial applications are essentially constant torque, such as conveyors. Others, such as centrifugal blowers, require torque to vary as the square of the speed.

In contrast, machine tools and centre winders are constant mechanical power, with torque decreasing as speed increases. Which EM motors are most suitable in these situations? As the authors will explore, the speed-torque relationship and efficiency requirements often determine the most appropriate EM motor.

EM servomotors are utilised in motion control applications where low inertia and dynamic response are important. In fact, many EM motors used for servo applications are similar to integrated AC–AC or AC–DC–AC or DC–AC commutator IPM synchronous motors but use special microcontrollers (in digital technology) or operational amplifiers (in analogue technology) and feedback to control position rather than just angular velocity. However, the price for EM servo-systems is quite high—often 10 to 20 times than that of an equivalently rated AC asynchronous (induction) motor.

Applications requiring near-servo performance are excellent candidates for integrated AC–AC or AC–DC–AC or DC–AC commutator IPM synchronous motors, benefitting from their cost-to-performance ratio. Case in point: integrated AC–AC or AC–DC–AC or DC–AC commutator IPM synchronous motors are quite suitable for typical **mechano-hydraulical** (MH) pump operations, which typically run at variable velocity between 75%–85% of maximum angular velocity.

Where are IPM AC synchronous motor cyber-physical heterogeneous dynamical hypersystems not suitable? The angular velocity of some EM servomotor applications sometimes reaches 10 000 rpm, which is out of the integrated AC–AC or AC–DC–AC or DC–AC commutator IPM synchronous motor range. In addition, without feedback for the IPM AC synchronous motor, designers can find it difficult to locate and position to pinpoint accuracy that EM servomotors must often deliver. Let us compare integrated AC–AC or AC–DC–AC or DC–AC commutator IPM synchronous motors to those most commonly used for servo applications—integrated DC–AC commutator synchronous motors.

A DC motor IEMD waveform is trapezoidal; two of the EM motor's three leads are used for the phases, and the third is used for hunting—so it is regularly changing magnetic fields. In brief, the three leads of the IPM AC synchronous motor are all actively used; input waveforms are sinusoidal, which is useful because these waveforms boost efficiency while minimising noise and vibration.

As mentioned, integrated DC–AC commutator synchronous motors (DC motors with the mechanical split-ring or flat commutator), and integrated AC–AC or

ACDC–AC or DC–AC commutator IPM synchronous motors' stator winding patterns are specialised for microcontrollers that produce trapezoidal and sinusoidal current waveforms, respectively.

One cannot differentiate the two by visual inspection. A microcontroller that produces trapezoidal waveforms is less costly than those that produce sinusoidal waveforms. However, sinusoidal microcontrollers and EM motors produce more consistent shaft rotation than trapezoidal—and rotor inertia, EM motor rating and specific microcontroller characteristics magnify the difference in performance.

One requirement: in low-voltage applications (anything below 250 V), conventional DC motors or AC asynchronous (induction) motors are still better choices than integrated AC–AC or AC–DC–AC or DC–AC commutator IPM synchronous motors—although there is work being done to address the issues that arise in these situations.

Concisely, integrated DC–AC commutator synchronous motors are commonly built for voltages down to 12, 24, 36 or 42 V. However, to wind an IPM AC synchronous motor for this voltage is (in effect) taking a 175 or 250 kW and winding it for 250 V. Here, lead sizes can grow to the size of an average coffee cup (an inane result) and winding such an EM motor's magnet wire (with an electrical machine or by hand) is problematic, as suppliers in this case must redesign the stator and rotor fairly extensively to ensure that the setup is physically possible.

For an apples-to-apples comparison of AC–AC or AC–DC–AC or DC–AC commutator asynchronous (induction) motors to AC–AC or AC–DC–AC or DC–AC commutator IPM synchronous motors, one must consider both with an IEMD —as the latter requires an IEMD for operation, and cannot connect directly to supply power as typical AC motors can.

System efficiency is higher for an integrated AC–AC or AC–DC–AC or DC–AC commutator IPM synchronous motor/IEMD setup from 40% to beyond 120% load. In addition, an integrated AC–AC or AC–DC–AC or DC–AC commutator IPM synchronous motor exhibits higher power density than an equivalent AC induction motor.

Rare-earth IPMs produce more magnetic flux for their physical size than the magnetic energy (and resultant torque) produced by an integrated AC–AC or AC–DC–AC or DC–AC commutator asynchronous (induction) motor's squirrel-cage rotor.

In the latter, the effect of induced voltage of rotation also is more pronounced: induced voltage of rotation reduces current and operates to slow the EM motor—and gets larger as speed increases. When a mechanical load is not present, it approaches the input voltage magnitude, reducing efficiency.

Consider that, in general, IPM AC motors are rated for variable or constant torque to 20:1 without feedback (open loop) or 2000:1 for closed loop (with an encoder).

Angular velocity (input frequency) has less effect on integrated AC–AC or AC–DC–AC or DC–AC commutator IPM synchronous motor efficiency than it does on AC asynchronous (induction) motors, which translates into more energy savings at reduced angular velocities.

Integrated AC–AC or AC–DC–AC or DC–AC commutator IPM synchronous motor losses (the inverse of efficiency) are 15%–20% lower than ISO or NEMA AC asynchronous (induction) motors. Each efficiency index represents 10% fewer (or greater) losses than its neighbour, so efficiency rating is one to three indices higher.

Depending on EM motor size, electric utility rate and duty cycle, designers can realise full return on an IPM AC motor purchase in one year. IPM AC motor efficiency ratings are one to three indexes above ISO or NEMA, which translates to 10%–30% fewer losses than a conventional EM motor.

Electricity is estimated to comprise approximately 95%–97% of the total lifecycle cost of EM motors, so energy savings significantly reduce the total investment. Concisely, IPM AC motors are inherently more efficient due to elimination of rotor conductor losses, lower resistance winding and flatter efficiency curve.

Due to their synchronous operation, IPM AC synchronous motors also offer better dynamic performance and speed control precision—a major benefit in high-inertia positioning applications. Although in some cases the dynamical system power factor with an IEMD may not be as high as an EM motor-only asynchronous (induction) machine, IPM AC synchronous motors generally provide higher power density due to higher magnetic flux. This means more torque can be produced in a given physical size, or equal torque produced in a smaller package. Finally, IPM AC synchronous motors generally operate more coolly than AC asynchronous (induction) motors, resulting in longer bearing and insulation life. Because an IPM rotor lacks conductors (rotor bars), there are no I^2R losses—so everything else being equal, an IPM AC synchronous motor is inherently more efficient.

On one specific integral mechanical power line (of IPM AC synchronous motors), the winding design has shorter end turns and a concentrated bobbin-type winding. Unlike a distributed winding, used in AC asynchronous (induction) motors, there are no shared slots—so the potential for phase-to-phase shorts is eliminated.

Shorter end turns reduce waste and make more room in the housing for more active material, enhancing power density (as end turns do nothing to generate torque).

IPM AC synchronous motor sound and vibration is often comparable to that of an AC asynchronous (induction) motor, though the sound and vibration of IPM AC synchronous motors varies widely from manufacturer to manufacturer and models designed for quiet operation exist. This tends (as with most other EM motor types) to depend on the type of application for which a specific EM motor is designed. Although the term **service factor** (SF) often is misunderstood and not recognised by the **International Electrotechnical Commission** (IEC), it is still commonly applied to describe the maximum output of NEMA EM motors.

IPM AC synchronous motors have an SF of 1.0 on AC–AC or AC–DC–AC or DC–AC commutator power, which is comparable to that of integrated AC–AC or AC–DC–AC or DC–AC commutator asynchronous (induction) motors. Operating any EM motor beyond its rated power results in additional (possibly detrimental) heating. Intermittent operation above rated power is most normally acceptable, as long as its components can withstand the additional thermal stress.

On a similar note, reserve torque capability is an expression of an EM motor's ability to safely deliver increased torque, due to higher peak torque capability, and is subject to the IEMD's ability to deliver increased current. An IPM AC synchronous motor has a reserve torque capability of 150% for 60 s.

Most applications compatible with AC asynchronous (induction) motors can utilise integrated AC–AC or AC–DC–AC or DC–AC commutator IPM synchronous motors. In centrifugally loaded variable-velocity applications (pumps, fans and blowers), integrated AC–AC or AC–DC–AC or DC–AC commutator IPM synchronous motors boost efficiency—and, in many instances, can **direct drive** (DD) these designs. Fans are unique in that they are typically sized by torque; yet here, DD integrated AC–AC or AC–DC–AC or DC–AC commutator IPM synchronous motors can eliminate the need for belts, pulleys and sheaves. This in turn simplifies maintenance, which is particularly helpful where fans are installed on roofs. In applications that incorporate belts, chains or gearboxes, integrated AC–AC or AC–DC–AC or DC–AC commutator IPM synchronous motors boost power density—and it is these applications in which reducing or eliminating power transmission devices makes the most improvement in efficiency and reduced maintenance cost. There are situations for which direct driving is not possible or desirable.

Consider conveyors driven by gearbox-fitted EM motors. Here, an integrated AC–AC or AC–DC–AC or DC–AC commutator IPM AC synchronous motor may not be able to eliminate the need for a gearbox, but can typically help designers reduce the gearbox by a size or two—which then allows integrated downsizing of other equipment as well. For example, a 48 frame, integrated AC–AC or AC–DC–AC or DC–AC commutator IPM AC synchronous motor can carry 100 Nm of torque, which equates to roughly 3 kW at 3000 rpm. One requirement: in integrated AC–AC or AC–DC–AC or DC–AC commutator IPM synchronous motors are not particularly suitable in fixed-velocity applications, as an integrated AC–AC or AC–DC–AC or DC–AC commutator IPM synchronous motor require an IPM IEMD. If an average integrated AC–AC or AC–DC–AC or DC–AC commutator asynchronous (induction) motor is replaced or retrofitted with an integrated AC–AC or AC–DC–AC or DC–AC commutator IPM synchronous motor cyber-physical heterogeneous dynamical hypersystem, typically the IEMD also must be replaced.

The IEMD topologies and algorithms are different; divergent, too, are the software and ladder logic, particularly concerning how the two IEMD types handle induced voltage of rotation. In addition, the EM motor must be able to communicate with the IEMD and vice versa. Stated another way, integrated AC–AC or AC–DC–AC or DC–AC commutator IPM synchronous motors are controlled by a PWM AC IEMD similar to those used with integrated AC–AC or AC–DC–AC or DC–AC commutator IPM synchronous or asynchronous (induction) motors, but with software to control an IPM synchronous motor.

In most situations, replacing existing, integrated AC–AC or AC–DC–AC or DC–AC asynchronous (induction) motors with integrated AC–AC or AC–DC–AC or DC–AC commutator IPM synchronous motors requires no mechanical changes to the equipment.

The effective integration of electric power in emerging and future IEMD requires the development of technologies that allow for volume and mass reduction of critical components.

The authors of this textbook are studying the potential for volume and mass reduction through the integration of the AC–AC or AC–DC–AC or DC–AC macroelectronic commutator (macrocommutator) with its microelectronic controller (microcontroller) into an EM motor. Two conceptual designs of an IEMD with integrated AC–AC or AC–DC–AC or DC–AC macrocommutators, namely with AC motors and DC motors, may be presented.

The authors compared two integration approaches. The different approaches result in trade-offs on shaft and bearing design.

The first approach (external to rotor) is to simply integrate the AC–AC or AC–DC–AC or DC–AC macrocommutator components into the stator housing. Though this approach does not take advantage of the large space available inside the rotor, it does allow for space sharing inside the stator housing.

The AC–AC or AC–DC–AC or DC–AC macrocommutator shares the same cooling cyber-physical dynamical system with the EM motor and eliminates the need for AC–AC or AC–DC–AC or DC–AC macrocommutator housing and long cables. The external design allows for a smaller shaft using standard bearings; however, the shaft and stator housing must be extended axially to account for more components inside the stator housing.

The second integration approach (internal to rotor) is to place the AC–AC or AC–DC–AC or DC–AC macrocommutator components inside the rotor. While the apparent advantage of the design is the space savings, the disadvantages include difficulty in assembly and maintenance.

The internal design allows for a shorter shaft and overall space savings, but a larger hollow shaft is necessary for access, and larger endplates are required for supporting the internal AC–AC or AC–DC–AC or DC–AC macrocommutator structure.

Integration of the AC–AC or AC–DC–AC or DC–AC macrocommutator into the EM motor frame offers space saving advantages, allowing the EM motor and macroelectronics to share the same housing and cooling cyber-physical homogeneous dynamical system. Accordingly, significant mass and volume reductions are possible in the macroelectronics housing and cooling auxiliaries.

One of the advantages of integrating the AC–AC or AC–DC–AC or DC–AC macrocommutator with the EM motor housing is that the same air-to-air, air-to-water, water or innovative FF cooling cyber-physical homogeneous dynamical system used for the EM motor windings can also be used to cool the AC–AC or AC–DC–AC or DC–AC macrocommutator.

Due to the inherent characteristics of a high-power, low-speed IEMD, there is large potential for integrating AC–AC or AC–DC–AC or DC–AC macrocommutator into the EM motor. Whether the AC–AC or AC–DC–AC or DC–AC macrocommutator is external or internal to the rotor, significant space savings are achievable. The external approach has more significant mass savings with the elimination of AC–AC or AC–DC–AC or DC–AC macrocommutator enclosure, a shared cooling cyber-physical dynamical system, and reduced cable length. The internal approach has more significant volume savings with only minor additions to the endplates from the baseline design. Both designs contribute to the optimisation of an IEMD train.

Even though this R&D was performed for a high-power, low-speed EM motor, the same techniques apply to larger EM motors. There are plenty of challenges facing the further development of integrated macroelectronics, such as tooling, assembly, and maintainability; however, the increased power density of the overall IEMD train is a technological development with payoffs in the optimisation of the IEMD.

Thermal management of integrated EM motors is a complicated process because of the multiple heat transfer paths within the EM motor and the different materials and thermal interfaces through which the heat must pass to be removed.

With the reduction in packaging volumes and the significant increase in power demands placed on a **high voltage** (HV) IEMD users can expect to see exciting new technologies being applied to thermal management over the next few years.

References

Chan T-F and Shi K 2011 *Applied Intelligent Control of Induction Motor Drives* (New York: Wiley)

Fijalkowski B 1985a New concept MACRO- and MICRO-electronics cradled dynamo-meter systems for testing of combustion engines, *Proc. ISATA 85: International Symposium on Automotive Technology and Automation* , *Graz, Austria* 23-27 September 1985 **vol 2** pp 587–616

Fijalkowski B T 1985b On the new concept hybrid and bimodal vehicles for the 1980s and 1990s, *Proc. DRIVE ELECTRIC Italy '85* , *Sorrento (Naples), Italy* 1–4 October 1985 pp 4.04.2–8

Fijalkowski B T 1985c On the new concept MMD electrical machines with integral macro-electronic commutators—Development for the future, *Proc. First European Conference on POWER ELECTRONICS AND APPLICATIONS* , *Brussels, Belgium* 16–18 October 1985 **vol 2** pp 3.377–84

Fijalkowski B 2016 *Mechatronics: Dynamical Systems Approach and Theory of Holors* (Bristol, UK: IOP Publishing)

Geyer T 2017 *Model Predictive Control of High Power Converters and Industrial Drives* (New York: Wiley)

Holmes D G and Lipo T A 2003 *Pulse Width Modulation for Power Converters Principles and Practice* (Piscataway, NJ: Institute of Electrical and Electronics Engineers)

Hoseman G, Krolikowski L, Raatz E, Shapilov E D, Turek-Kwiatkowska L and Vieselovsky O N 1999 Zhizn i delyatelnost Mikhaila Dolivo-Dobrovolskogo—vydaiu-shchegosia inzhenera elektrika evropeiskogo masshtaba, *Proceedings of the 4th Int. Conf., on Unconventional Electromechanical and Electrical Systems, 21–24 June 1999, St. Petersburg, Russia* **vol 1** pp 59–70 In Russian

Hughes A and Drury W 2013 *Electric Motors and Drives Fundamentals, Types and Applications* 4th edn (Amsterdam: Elsevier)

Rashid M H 2018 *Power Electronics Handbook* 4th edn (Amsterdam: Elsevier)

Tutaj J 1996 Estymacja jakosci energii samochodowych komutatorowych pradnic mechano-elektrycznych pradu stalego z komutatorami elektronicznymi (Energy quality estimation of

DC mechano-electrical generators with electronic commutators) *PhD thesis* Promotor: Bogdan Fijalkowski, Automotive Vehicles and Combustion Engines Institute, Cracow University of Technology, Krakow (In Polish)

Tutaj J 2012 *Ujecie Systemowe Dynamiki Wielofunkcyjnego Prądni Rozrusznika Silnika spalinowego* (Dynamic Systems Approach of the Polyfunctional Generator-Starter for a Combustion Engine of the Automotive Vehicle) Monografia 409, Seria Mechanika (Krakow: Politechnika Krakowska im. Tadeusza Kosciuszki) (In Polish)

Wach P 2011 *Dynamics and Control of Electrical Drives* (Berlin, Heidelberg: Springer)

Wilamowski B M and Irwin D J 2011 *The Industrial Electronics Handbook—Power Electronics and Motor Drives* (Boca Raton, FL: CRC Press)

IOP Publishing

The Integrated Electro-Mechanical Drive

A mechatronic approach

B T Fijalkowski and J Tutaj

Chapter 3

Advanced AC and DC motor IEMD control

3.1 Introduction

The perfect combination of high performance together with ease of use allows even the most demanding applications to be tackled easily by the AC or DC motor IEMD. The authors of this textbook have presented advanced mathematical algorithms and the very latest hardware and software technology to provide exceptional AC or DC motor IEMD control with a simple interface to help users easily apply the benefits to their applications.

Control modes:

- V_i/f voltage holor or vector to frequency.
- Energy optimised V_i/f.
- Sensorless angular velocity holor or vector control.
- Sensorless torque holor or vector control.
- Variable-reluctance synchronous motor control.
- Static AC motor control.
- Closed loop (encoder) angular velocity holor or vector control.
- Closed loop (encoder) torque holor or vector control.
- Open loop IPM synchronous motor control.

Designed for fast installation and commissioning, IEMD provides the most cost effective solution for industry.

All IEMD units provide 150% overload for 60 s as standard, ensuring each IEMD is suitable for **heavy duty** (HD) applications, whilst the IP55 enclosed versions ensure the IEMD is tough enough to survive in industrial environments. Extensive **input/output** (I/O) and communications interface capabilities ensure the IEMD can be integrated quickly and efficiently into a wide variety of control systems with the minimum commissioning time, ensuring rapid start up.

The dynamic torque holor (torque-vector) control system performs high-speed calculation to determine the required EM motor power for the load status. Very

 © IOP Publishing Ltd 2019

advanced technology is optimal control of voltage and current holors (vectors) for maximum output torque holor.

- A high starting torque of 200% at 0.5 Hz (e.g. 180% for 30 kW or larger EM motors).
- Achieves smooth acceleration/deceleration in the shortest time for the mechanical load condition.
- Using a high-speed **central processing unit** (CPU) quickly responds to an abrupt mechanical load change, detects the regenerated power to control the deceleration time. This automatic deceleration function greatly reduces the DC–AC macrocommutator acting as a DC–AC inverter tripping.

EM motors connected to an IEMD receive electrical energy including a changeable fundamental frequency, a carrier frequency, and very rapid voltage build-up. These factors can have negative impacts, especially when existing EM motors are used. There are a number of potential problems that can become real when an IEMD is used to power an existing AC asynchronous (induction) motor. (Barnes 2003, Wilamowski and Irwin 2011, Chan and Shi 2011, Wach 2011, Hughes and Drury 2013, Holmes and Lipo 2003, Geyer 2017, Rashid 2018).

As such, users should carry out a careful study to determine if these problems could be sufficiently bad to cause reconsideration of such an installation. With an IEMD, an existing EM motor normally having a number of useful years left in it could abruptly fail. Existing EM motors are designed for 50 Hz or 60 Hz only, or 50/60 Hz service. As such, one has to question whether or not a new IEMD can be matched to existing EM motor and still have the EM motor perform reasonably well. In other words, will the EM motor be able to handle additional factors that may cause greater vibration, heat rise, etc, and a possible increase in audible noise?

3.2 Problems cause by high frequencies

One should be aware of possible side effects caused by high pulsing frequency when applying an IEMD to an existing EM motor. These negative effects include additional heat, audible noise, and vibration. Also, **pulse width modulation** (PWM) circuitry, which causes a high rate of voltage rise of the carrier frequency, can cause insulation breakdown of the end turns of EM motor windings as well as feeder cable insulation.

The carrier frequency, a by-product of obtaining current at a variable fundamental frequency, is the cause of having additional power in the EM motor; this power is essentially wasted energy that adds heat to the EM motor. The amount of such loss varies, depending upon the EM motor's stator and rotor designs and frequency of the carrier wave.

With frequencies other than the fundamental, an EM motor runs at very high split-page and, therefore, is running somewhat inefficiently. Split is the difference between the rotational speed of the stator's magnetic field, i.e. the synchronous rotational speed of the AC asynchronous (induction) motor and the rotational speed of the rotor. Also, numerous lines of magnetic flux are being cut by the rotor; this phenomenon produces additional power and additional heat. (Note that the high-frequency ripples in the current are at low magnitude, and the additional heat is of the order of 5%–10% above that produced by a pure sine wave.)

The synchronous rotational speed of a four magnetic-pole EM motor served by 50 Hz power is 1500 rpm. This same EM motor, when considering 'overtones' or ripples in the current's fundamental frequency caused by a voltage carrier frequency of 4 kHz, will have current flowing through it based on that high frequency. Thus, the rotor of a four magnetic-pole, 50 Hz designed EM motor (with a rated full-load rotational speed of 1450 rpm) being supplied power by an IEMD adjusted to 10 Hz output will be turning at 1/5 rated rotational speed. If the load's torque requirements are constant at low—through full-rated rotational speeds, split rpm remains constant. For the EM motor above, which is operating at 10 Hz, the shaft will be turning at 250 rpm.

The rotor, while turning at 250 rpm and crossing lines of magnetic flux (the magnetic field) based on the 10 Hz fundamental frequency and the synchronous rotational speed of 300 rpm (1/5 of 1500 rpm), is also crossing lines of magnetic flux due to the carrier frequency voltage of 4 kHz. The synchronous speed at 4 kHz is 120 000 rpm. Based on a synchronous rotational speed of 120 000 rpm and a shaft speed of 250 rpm, the user can see that the lines of magnetic flux being cut due to the carrier frequency (4 kHz) are substantial compared with a synchronous speed of 300 rpm caused by the 10 Hz frequency. This additional current, which is transmitted to the rotor bars by the cutting of additional magnetic flux caused by carrier frequency, produces very little useful power. Most of this current is dissipated as heat, adding to the temperature rise of the EM motor.

This additional heat represents about another 5%–10% thermal build-up in the EM motor and can place an additional thermal strain on the EM motor's rotor bars and stator windings, if it is running at full load. This high frequency power is an inefficient producer of torque. Because of these and other conditions mentioned, one may wish to derate an existing EM motor when it is connected to an IEMD. The amount of energy from the carrier frequency that is dissipated by the EM motor depends upon the amplitude and the frequency of the voltage, and the reactance and resistance of the EM motor at the resultant frequency. The amplitude of the current is determined by the ratio of the voltage over the impedance, while the watts lost is a product of the current squared times the resistance.

3.3 Supplementary undesired side effects

Users also should be aware of supplementary potential side effects caused by high frequency. These include undesirable audible noise, harmful vibration, and bearing problems.

Vibration and noise problems. To avoid noise and vibration problems, it is recommended that the EM motor being used not have components that can resonate at the frequencies the EM motor (and its load) will generate. This is possible on control systems where the frequency of the power is known, such as with 50 or 60 Hz. However, a modern IEMD has no standard carrier frequency, and the fundamental frequency can range from less than 10% of 50 or 60 Hz to 100% of 50 or 60 Hz, and beyond. Depending on which brand and model number of IEMD is mated to the existing EM motor, and other factors such as the characteristics of the electrical system at the site, resonances in certain components may or may not be excited.

Users must also consider that when a 50 or 60 Hz designed EM motor is operating at a different electrical frequency, various components of the EM motor might go into mechanical resonance, such as the fan or shaft. Each component has its own natural mechanical frequency, and an electrical frequency going through the coils and rotor bars can cause mechanical vibrations that are different from the initial design parameters. When an electrical frequency matches the natural frequency of a mechanical component, serious problems may occur. This may include the disintegration of a component.

Bearing problems. Another possible problem, which still isn't fully understood, is the slow disintegration of the roller/ball (antifriction) bearings that support the shaft. It appears this is caused by bearing current and static discharge. What happens is that pitting occurs on the roller/ball surface and, when accumulated, causes the bearing to make noise. If not addressed, vibration will begin to develop.

Air flow problems. An additional factor users should consider when operating a standard 50 Hz EM motor at very low speed is that the fan, which is fixed and attached to the rotor, may not create enough air flow to effectively cool the EM motor. This is because air flow is proportional to shaft speed. Thus, at half shaft speed, the air flow is half normal flow. To compensate for low-volume air flow at low EM motor speeds, if installation is possible, the attachment of a constant velocity air blower package to the back of the motor will usually provide adequate cooling.

3.4 Insulation of conductors breakdown

As mentioned, PWM circuitry, which causes the high rate of voltage rise at the carrier frequency, can cause insulation breakdown of the end turns of the EM motor windings, as well as possible breakdown of the feeder cable insulation. This relates to the very high rate of rise of the voltage (rate of voltage change with respect to time) in combination with the very rapidly repeating voltage pulse caused by the IEMD. Conductor insulation failures in EM motors have occurred because of this phenomenon. This subject is not completely understood and is presently being researched.

The known facts are summarised as follows.

- Electrical valves (electronic switches) in the DC–AC commutator, acting as a DC–AC inverter of AC or DC motor IEMD used currently cause instantaneous turn-to-turn voltage inside an EM motor's windings to be significantly higher than what an equivalent normal sine wave supply produces.
- Each cycle of the fundamental voltage consists of numerous pulses of voltage.
- Long distance between an unintegrated EM motor and its EMD causes the turn-to-turn voltage to get even higher.

There are different approaches in explaining why there's an increase of voltage at the EM motor terminals. Some explain it in terms of resonant **capacitance/**

inductance (LC) circuits; others explain it in terms of standing wave theory. Both approaches end up with a similar result. When the distance between an unintegrated EM motor and its EMD exceeds a critical distance (which may be as low as 10 m), there is a voltage overshoot that may exceed twice the amplitude of the voltage pulse originally delivered at the EMD output terminals. This higher voltage comes at the EM motor at such a high rate of change for each of the PWM pulses, from zero volts to its peak value, that it is unevenly distributed across the winding, causing high turn-to-turn voltages in the turns connected closest to the power leads. The result places very high stress on the conductor insulation, which can cause early breakdown of the insulation.

Integrated AC–AC or AC–DC–AC or DC–AC commutator synchronous or asynchronous (induction) motors with the ECM DC–AC commutator, acting as a DC–AC inverter duty EM motors are available that are designed to meet or exceed the voltage amplitudes and rise times defined in NEMA Standard MG1, EM Motors and ME Generators, Section 31.40.4.2, Voltage Spikes. When connecting unintegrated EM motors to EMD with long cable lengths, users should consider using a filter to reduce the effects caused by the long cable.

3.5 Losses caused due to the skin effect

In addition to the problems described above, there is yet another loss component users should be aware of: the skin effect. The skin effect induces the current in an AC system to crowd to the outside surface of a conductor. This phenomenon causes resistance to be directly related to the square root of the frequency of the current. In other words, the greater the frequency, the greater the resistance due to skin effect. The carrier frequencies are usually between 0.8 to 15 kHz, and currents at these high frequencies will cause I^2R losses. While the high frequency currents are relatively nominal, the loss relates to the current's square power. And the carrier frequency, even at its square root, can be somewhat effective because of its basic high value. The geometry of the rotor bars also determines the degree to which the skin effect impacts rotor losses.

3.6 Conclusions

In applying integrated AC–AC or AC–DC–AC or DC–AC commutator synchronous or asynchronous (induction) motors to loads requiring constant mechanical power over wide speed ranges, users will often find it beneficial to work with a person knowledgeable in EM motors. When, using an existing standard AC synchronous or asynchronous (induction) motor for such use, a compromise is often made between the EM motor's capability and the actual mechanical power output, in other words derating of the EM motor. In these situations, it would probably be better to purchase a modern EM motor having the requirements users need.

When users use an integrated EM motor for an application where torque requirements remain constant or reduced over the total speed range being applied, an IEMD will be a good means to achieve angular velocity control, providing the integrated EM motor is able to handle the distorted electrical power being delivered to it by the IEMD.

Applications where torque requirements remain constant or reduced over an integrated EM motor's total speed range include fans, pumps, and conveyor belts. There are certain mechanical loads, such as centrifugal pumps and fans, whereas angular velocity decreases, torque will usually decrease as the square of the angular velocity, and mechanical power will decrease with the cube of the angular velocity. Thus, if the mechanical power is established at the low end of the speed requirement (e.g. 50% rated angular velocity at 7.5 kW), the mechanical power requirement at full angular velocity will be eight times as much, or 60 kW. As users can see in this type of situation, the deciding factor for mechanical power requirements must be based at full-rated mechanical load.

References

Barnes M 2003 *Practical Variable Speed Drives and Power Electronics, Automated Control Systems* (Perth, Australia: IDC Technologies)

Chan T-F and Shi K 2011 *Applied Intelligent Control of Induction Motor Drives* (New York: Wiley)

Geyer T 2017 *Model Predictive Control of High Power Converters and Industrial Drives* (New York: Wiley)

Holmes D G and Lipo T A 2003 *Pulse Width Modulation for Power Converters Principles and Practice* (Piscataway, NJ: Institute of Electrical and Electronics Engineers)

Hughes A and Drury W 2013 *Electric Motors and Drives Fundamentals, Types and Applications* 4th edn (Amsterdam: Elsevier)

Rashid M H 2018 *Power Electronics Handbook* 4th edn (Amsterdam: Elsevier)

Wach P 2011 *Dynamics and Control of Electrical Drives* (Berlin, Heidelberg: Springer)

Wilamowski B M and Irwin D J 2011 *The Industrial Electronics Handbook - Power Electronics and Motor Drives* (Boca Raton, FL: CRC Press)

IOP Publishing

The Integrated Electro-Mechanical Drive
A mechatronic approach
B T Fijalkowski and J Tutaj

Chapter 4

AC motor IEMD modus operandi

4.1 Introduction

The primary function of any mechatronically-controlled AC motor **integrated electro-mechanical drive** (IEMD) is to control the angular velocity, torque, acceleration, deceleration, and sense of direction of rotation of an electrical machine. Unlike constant-velocity dynamical systems, the IEMD permits the selection of an infinite number of angular velocities within its operating range. The use of the AC motor IEMD in **mechano-hydraulical** (MH) pump and **mechano-pneumatical** (MP) fan dynamical systems can greatly increase their efficiency. Outdated technology most often used throttles or dampers to interrupt the flow as a means to control it. The fluid or air was held back by the throttle or damper but the energy used to move the fluid or air was then dissipated uselessly. That wasted energy was still accounted for and paid for. Running an IEMD cyber-physical heterogeneous dynamical hypersystem this way is like driving an automotive vehicle with the accelerator pressed to the floor while controlling velocity with the brake. An IEMD, on the other hand, allows precise control of EM motor output. In the case of centrifugal fans and pumps, there are significant savings in the power required to handle the mechanical load. While single-phase supplies power to the IEMD, the output to the EM motor is three-phase. Control wiring consists of inputs and outputs connected to the control terminal strip. Various control wiring configurations are used, depending on the make of microcontroller and specific application (Barnes 2003, Wilamowski and Irwin 2011, Chan and Shi 2011, Wach 2011, Hughes and Drury 2013, Holmes and Lipo 2003, Geyer 2017, Rashid 2018).

4.2 Adjustable-velocity integrated electro-mechanical drive

AC–AC or AC–DC–AC commutator synchronous or asynchronous (induction) motors are the most common three-phase EM motors used in commercial and industrial applications. The preferred method of angular velocity control for EM motors is to alter the frequency of the supply voltage.

doi:10.1088/2053-2563/aae7d7ch4

© IOP Publishing Ltd 2019

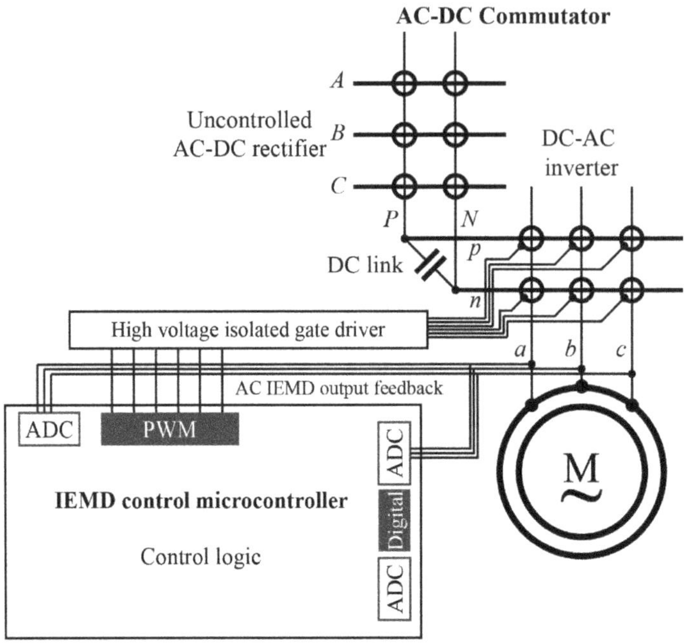

Figure 4.1. Structural and functional diagram of an integrated ECM AC–DC–AC commutator synchronous or asynchronous (induction) motor IEMD cyber-physical heterogeneous dynamical hypersystem.

An AC motor IEMD controls the angular velocity, torque, and sense of direction of an AC–AC or AC–DC–AC commutator synchronous or asynchronous (induction) motor. It takes fixed voltage and frequency AC input and converts it to a variable voltage and frequency AC output.

Figure 4.1 shows the structural and functional diagram of an integrated AC–DC–AC commutator synchronous or asynchronous (induction) motor IEMD cyber-physical heterogeneous dynamical hypersystem.

The function of each **electrical commutation matrixer** (ECM) is as follows:

- ***AC–DC rectifier:*** A full-wave ECM AC–DC rectifier that converts the applied AC to DC.
- ***DC link:*** Also referred to as a DC bus, connects the ECM AC–DC rectifier output to the input of the ECM DC–AC inverter. The DC link functions as a filter to smooth the uneven, rippled output to ensure that the rectified output resembles as closely as possible pure DC.
- ***DC–AC inverter:*** The ECM DC–AC inverter takes the filtered DC from the DC link and converts it into a pulsating DC waveform. By controlling the output of the ECM DC–AC inverter, the pulsating DC waveform can simulate an AC waveform at different frequencies.
- ***Control logic:*** The control logic system generates the necessary pulses used to control the triggering and quenching of the power semiconductor electrical valves such as IGBTs, IGCTs or MOSFETs. Fairly involved control circuitry coordinates the commutation (switching) of power electrical valves, typically

through a control board that dictates the firing of power components in the proper sequence.

An embedded microprocessor is used for all internal logic and decision requirements. Sometimes termed the front end of the IEMD, the AC–DC rectifier is commonly a three-phase, full-wave AC–DC rectifier. However, one of the advantages of an IEMD is being able to operate a three-phase AC–DC–AC or AC–AC commutator synchronous or asynchronous (induction) motor from a three-phase AC supply. The key to this is process is the rectification of the AC input to a DC output. At this rectification point, the DC voltage has no phase characteristics; the IEMD is simply producing a filtered pulsating DC waveform. The AC motor IEMD inverts the DC waveform into three different pulse-width modulated waveform signatures that duplicate an AC three-phase waveform.

Figure 4.2 shows three-phase and single-phase ECM AC–DC rectifier input connections. AC input voltage levels that are different from that required to operate the AC–DC–AC commutator synchronous or asynchronous (induction) motor require the ECM AC–DC rectifier section to raise or lower the voltage to the proper operating level of the EM motor. As an example, an AC motor IEMD supplied with 115 V AC that must deliver 230 V AC to the EM motor requires a transformer capable of stepping up the input voltage.

The AC motor IEMD offers an alternative to other forms of power conversion in areas where three-phase power is unavailable. Since it converts incoming AC power to DC, the IEMD really does not care if its source is single or three-phase.

Regardless of the input power, its output will always be three-phase. Drive sizing, however, is a factor since it must be capable of rectifying the higher-current, single-phase source. As a rule of thumb, most suppliers recommend doubling the normal three-phase capacity of an IEMD that will be operating on a single-phase input Single-phase operation is limited to smaller mechanical power motors.

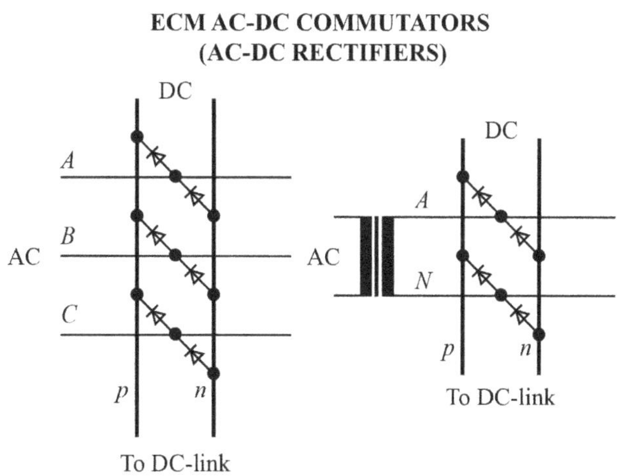

Figure 4.2. Three-phase and single-phase ECM AC–DC rectifier input connections.

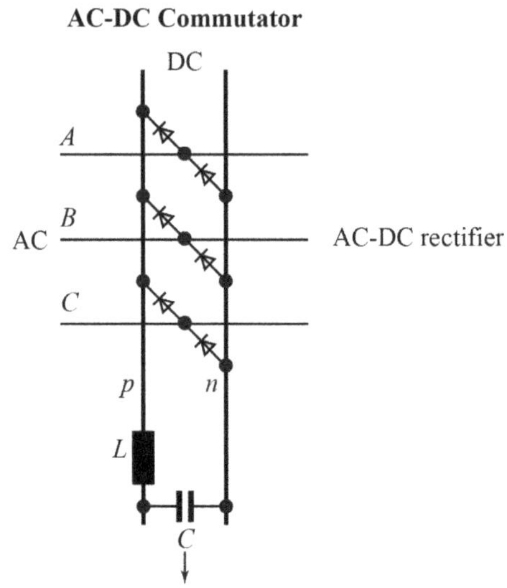

Figure 4.3. Inductor L and capacitor C connections within the DC link.

Some suppliers offer some exemplars for single-phase input only and others that are fully rated for both single-phase and three-phase input. After full-wave rectification of an AC supply into an IEMD, the DC output passes through a DC link.

Figure 4.3 shows the inductor L and capacitor C connections within the DC link. They act together to filter out any AC component of the DC waveform.

The principal energy storage component is the DC link capacitors. Any ripple that is not smoothed out will show up as distortion in the EM motor output waveform. As a rule, IEMD suppliers provide a special terminal block for DC link voltage measurement. For example, with a 400 V_{AC} input you should read an average DC link voltage of about 650–680 V_{DC}. The DC value is calculated by taking the **root mean square** (RMS) value of the line voltage and multiplying it by $\sqrt{2}$. AC voltage readings of more than 4 V_{AC} on the DC link may indicate a possible capacitor filtering problem or a problem with the diode AC–DC rectifier section.

The DC–AC inverter is the final output section of an AC motor IEMD. This is the point where the DC link voltage is switched on and off at specific intervals. In doing so, the DC energy is changed into three channels of AC energy that an AC motor uses to operate. Up-to-date DC–AC inverters use **integrated-gate commutated transistors** (IGCTs) to commute (switch) the DC link on and off.

Figure 4.4 shows a simplified structural and functional diagram of the three sections of an adjustable-velocity IEMD cyber-physical heterogeneous dynamical hypersystem. Six electrical valves (IGCT) are used in the DC–AC inverter section. The control logic uses a microcontroller to commute (switch) the electrical valves (IGCT) on and off at the proper time. The main objective of the AC motor IEMD is to vary the singular velocity of the AC–DC–AC commutator asynchronous (induction) motor while providing the closest approximation to a sine wave for current. In the simplest circuit implementation, two electrical valves (IGCT) are

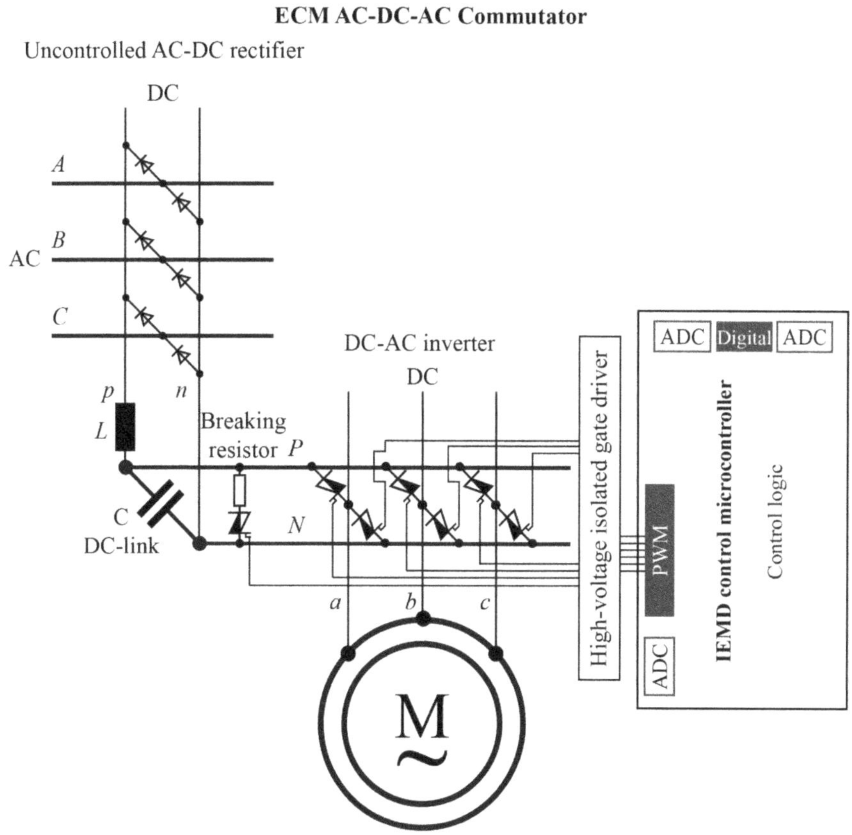

Figure 4.4. Simplified structural and functional diagram of the three sections of an adjustable-velocity IEMD cyber-physical heterogeneous dynamical hypersystem.

placed in series across the DC supply and are commuted (switched) on and off to generate one phase of the three phases for the EM motor. Two other identical circuits generate the other two phases.

Figure 4.5 shows a simplified circuit of a **pulse-width modulation** (PWM) DC–AC inverter. Electronic switches are used to illustrate the way that the electrical valves (IGCT) are commuted (switched) to produce one phase (A to B) of the three-phase output. The output voltage is switched from positive to negative by opening and closing the electrical valves (electronic switches) in a specific sequence of steps.

The operation can be summarised as follows:

- During steps 1 and 2, electrical valve (IGCT) V1 and V4 are closed.
- The voltage from phase A to B is positive.
- During step 3, electrical valve (IGCT) V1 and V3 are closed.
- The difference in voltage between phase A and phase B is zero, resulting in zero output voltage.
- During steps 4 and 5, electrical valves (IGCT) V2 and V3 are closed.
- This results in a negative voltage between phases A and B.

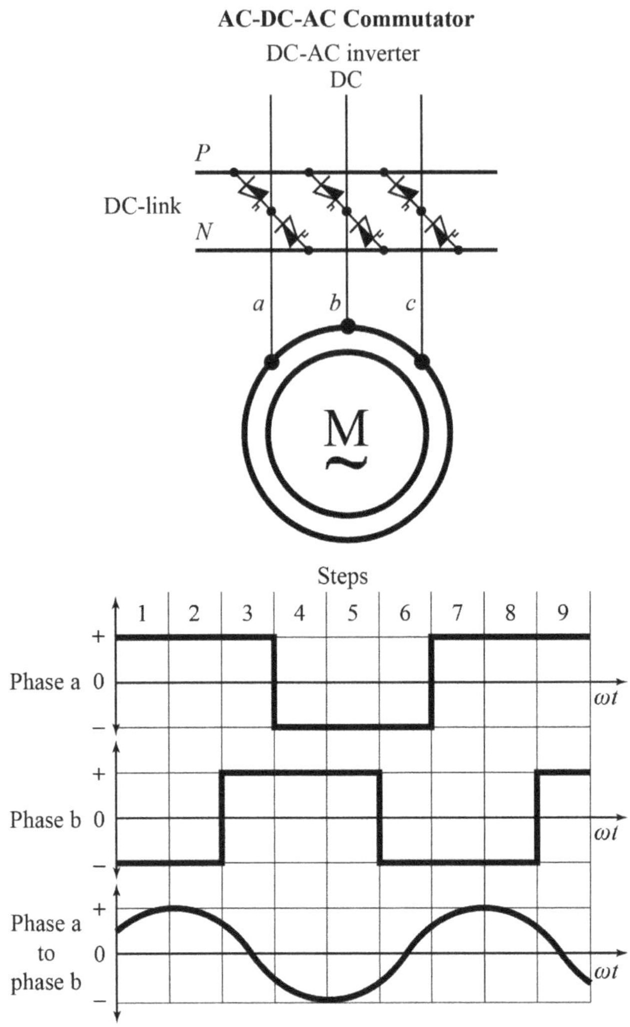

Figure 4.5. Simplified structural and functional diagram of a pulse-width modulation (PWM) DC–AC inverter.

- The other steps continue in a similar manner.
- Output voltage is dependent on the state of the electrical valves (open or closed), and the frequency is dependent on the speed of commutation (switching).

Figure 4.6 shows the sine wave (AC) line voltage, superimposed on the pulsed DC–AC inverter output, or simulated AC. Notice that the pulses are the same height for each pulse.

This is because the DC link voltage the IEMD uses to create these pulses is constant. Output voltage is varied by changing the width and polarity of the switched pulses. Output frequency is adjusted by changing the switching cycle time.

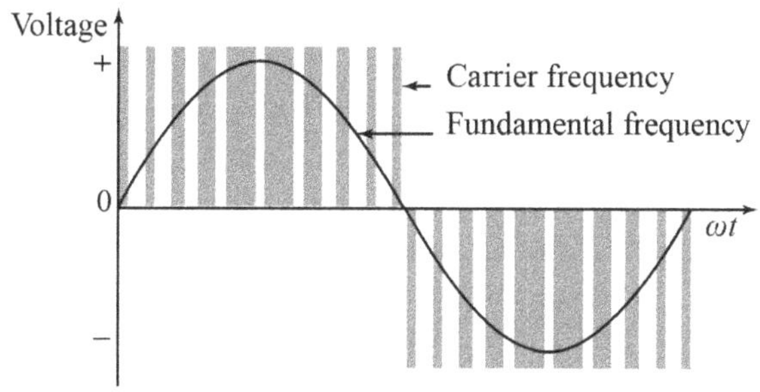

Figure 4.6. Sine wave AC line voltage, superimposed on the pulsed DC–AC inverter output, or simulated AC.

The resulting current in an AC–DC–AC commutator asynchronous (induction) motor simulates a sine wave of the desired output frequency.

Most true RMS-measuring multimeters are fast enough to measure the RMS value of the PWM voltage and current. There are two frequencies associated with a PWM IEMD: the fundamental frequency and the carrier frequency. The fundamental frequency is the variable frequency an AC–DC–AC commutator asynchronous (induction) motor uses to vary angular velocity. In a typical IEMD, the fundamental frequency will vary from a few Hz up to a few hundred Hz. The inductive reactance of an AC magnetic circuit is directly proportional to the frequency. Therefore, when the frequency applied to an AC–DC–AC commutator asynchronous (induction) motor is reduced, the applied voltage must also be reduced to limit the current drawn by the AC–DC–AC commutator induction motor at reduced frequencies. The microcontroller control adjusts the output voltage waveform to simultaneously change the voltage and frequency to maintain the constant volts/hertz ratio.

The carrier frequency (also known as the commutation frequency) is one at which the pulses in pulse-width modulation commutation occurs. The carrier frequency is a fixed frequency substantially higher than the fundamental frequency. This high commutation (switching) speed produces the classic whine associated with an IEMD. Higher carrier frequency allows a better approximation to the sinusoidal form of the output current. However, higher commutation frequencies decrease the efficiency of the IEMD because of increased heat in the power electrical valves (IGCT). The carrier frequency for the IEMD is in the 2–16 kHz range. Adjusting the carrier frequency automatically in accordance with the changing mechanical load and temperature will result in quieter operation. The DC–AC inverter-duty EM motor is designed for optimised performance to operate in conjunction with AC motor IEMD. A DC–AC inverter duty EM motor can withstand the higher voltage spikes produced by the IEMD and can run at a very slow value of the angular velocity without overheating.

DC motor IEMDs are most commonly used to control DC–AC commutator synchronous motors (DC motors), but the dynamical system is also used in some older AC motor IEMD. Earlier types of the IEMD used thyristors or **semiconductor controlled rectifiers** (SCRs) to do the commutation (switching). As they have become available in higher voltage and current ratings, faster-switching electrical valves (IGBTs) became the preferred commutation (switching) components for use in

ECMs. Angular-velocity control can be 'open loop', where no feedback of actual EM motor angular velocity is used, or 'closed loop', where feedback is used for more accurate angular-velocity regulation. How an EM motor reacts is very dependent on the mechanical load conditions. An open-loop IEMD knows nothing about mechanical load conditions; it only tells the EM motor what to do. If for example it provides 43 Hz to the EM motor, and the EM motor spins at an angular velocity equivalent to 40 Hz, the open loop does not know. With closed-loop control the microcontroller tells the EM motor what to do, then checks to see if it did it, then changes its command to correct for any error. Often a tachometer is used to provide the necessary feedback in a closed-loop system. The tachometer is coupled to the EM motor and produces a speed feedback signal that is used by the microcontroller. With closed-loop control, a change in mechanical load demand EM motor, acts to maintain a constant angular velocity. In general, AC motor IEMD controls the EM motor angular velocity by varying the frequency of the current supplying the EM motor. Although frequency can be varied in different ways, the two most common angular-velocity control methods in use today are **voltage per frequency** (V/F) and magnetic-flux holor.

4.3 Voltage per frequency AC motor IEMD

Of the angular velocity control methods, V/F technology is the most economical and easiest to apply. The V/F IEMD controls shaft angular velocity by varying the voltage and frequency of the signal powering the AC–DC–AC or AC–AC commutator synchronous or induction motor. V/F control in its simplest form takes an angular-velocity reference command from an external source and varies the voltage and frequency applied to the EM motor. By maintaining a constant V/F ratio, the IEMD can control the angular velocity of the connected EM motor. V/F IEMD acts well on applications in which the mechanical load is predictable and does not change quickly, such as MP fan and MH pump loads.

In order to prevent overheating, the voltage applied to the AC–DC–AC or AC–AC commutator synchronous or asynchronous (induction) motor must be decreased by the same amount as the frequency. V/F control runs in open loop without a feedback device.

The ratio between voltage and frequency is termed V/F. To find the **Volts per Hertz** (V/Hz) ratio, simply divide the rated nameplate by the rated nameplate frequency. For example the V/Hz ratio for a 380 V, 50 Hz EM motor is calculated as follows:

$$V/F = \text{Voltage/Frequency} = 380\text{ V}/50\text{ Hz} = 7.6\text{ V Hz}^{-1}$$

V/F control provides a linear (straight-line) voltage ratio to the frequency of an EM motor from zero to base angular velocity. This is illustrated in figure 4.7 using a 380 V_{AC}, 50 Hz EM motor as an example. The V/F ratio of 7.67 is supplied to the EM motor at any frequency between 0 and 50 Hz. If applied frequency is reduced to 25 Hz, the shaft will slow to half its original speed.

In this situation, a V/F IEMD also halves the voltage (here, to 220 V_{AC}) in order to maintain the 7.6 V/F ratio, which allows the EM motor to continue producing its

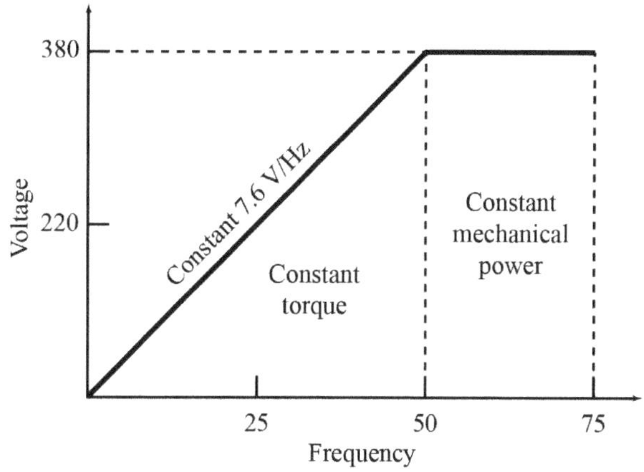

Figure 4.7. V/F control provides a linear (straight-line) voltage ratio to the frequency of an EM motor from zero to base angular velocity.

rated torque. Mechanical power increases and torque remains constant up to the base angular velocity; however, above base angular velocity (i.e. above 50 Hz frequency), the torque decreases while the mechanical power remains constant. This can easily be understood by the simple relationship between mechanical power holor $\boldsymbol{P}^{\kappa}$, torque holor $\boldsymbol{T}_i$ and angular-velocity holor $\boldsymbol{\Omega}^j$respectively:

$$\boldsymbol{P}^{\kappa} = \boldsymbol{T}_i \otimes \boldsymbol{\Omega}^j$$

V/F control of an AC–DC–AC or AC–AC commutator synchronous or asynchronous (induction) motor is based on the principle that to maintain constant magnetic flux in the EM motor, the terminal voltage magnitude must increase roughly proportionally to the applied frequency. This is only an approximate relationship, and V/F IEMD designs may include the following refinements:

- Low-frequency voltage boost (also referred to as IR compensation)—below 15 Hz the voltage applied to the EM motor is boosted to compensate for the power losses EM motors experience at low values of the angular velocity and increase the starting torque capability.
- Steady-state split compensation—increases frequency on the basis of a current measurement to give better steady-state speed regulation.
- Stability compensation—to overcome mid-frequency instabilities evident in high-efficiency EM motors.

4.4 Magnetic-flux holor AC motor IEMD

A magnetic-flux holor AC motor IEMD uses feedback from what is happening at the AC–DC–AC or AC–AC commutator synchronous or asynchronous (induction) motor to make changes in the output of the IEMD. However, it still relies on the basic V/F core for controlling the EM motor. These combined techniques control not only the magnitude of EM motor magnetic flux but also its orientation, thus the magnetic-flux holor name. The magnetic-flux holor method provides more precise

EM motor angular velocity and torque control. Magnetic-flux holor control improves on the basic V/F control technique (by providing both a magnitude and angle control) only the magnitude.

Holor AC motor IEMD comes in two types, open loop and closed loop based on the way they get their feedback information. Open loop is actually a misnomer because it is actually a closed loop system, but the feedback loop comes from within the IEMD itself instead of an external encoder. For this reason there is a trend to refer to open-loop IEMD as sensorless holor IEMD. Sensorless holor control removes a major source of complexity and simplifies the IEMD installation.

Its operation can be summarised as follows:

- Split is the difference between the rotor angular velocity and the synchronous angular velocity of the magnetic field and is required to produce EM motor torque. The split estimator lock keeps actual EM motor rotor angular velocity close to the desired set angular velocity.
- The torque current estimator determines the percent of current that is in phase with the voltage, providing an approximate torque current. This is used to estimate the amount of split, providing better angular velocity control under mechanical load.
- Holor angle controls the amount of total EM motor current that goes into EM motor magnetic flux enabled by the torque current estimator. By controlling this angle, low-angular velocity operation and torque control are improved over those of the standard V/F AC motor IEMD.
- The magnetic-flux holor control retains the V/F core and adds additional blocks around the core to improve the performance of the IEMD.
- The current resolver attempts to identify the magnetic flux and torque producing currents in the EM motor and makes these values available to other components in the IEMD.
- The current limit component monitors EM motor current and alters the frequency command when the EM motor current exceeds a predetermined value.

A true closed-loop holor AC motor IEMD uses an EM motor-mounted encoder or similar sensor to give positive shaft position indication back to the micro-controller. The EM motor's rotor position and angular velocity are monitored in real time via a digital encoder to determine and control the EM motor's actual angular velocity, torque, and power produced. Thus, a typical magnetic-flux holor AC motor IEMD and EM motor-mounted encoder can be used in IEMD applications. The encoder operates by sending digital pulses back to the IEMD indicating both angular velocity and its sense of direction. The processor counts the pulses and uses this information along with information about the EM motor itself in order to control the EM motor torque and associated operating angular velocity. The most common encoders deliver 1024 pulses per revolution. To guard against **electromagnetic interference** (EMI), the cable assembly between the encoder and IEMD should be one continuous shielded cable.

The structural and functional diagram of a sensorless magnetic-flux holor control AC motor IEMD is shown in figure 4.8.

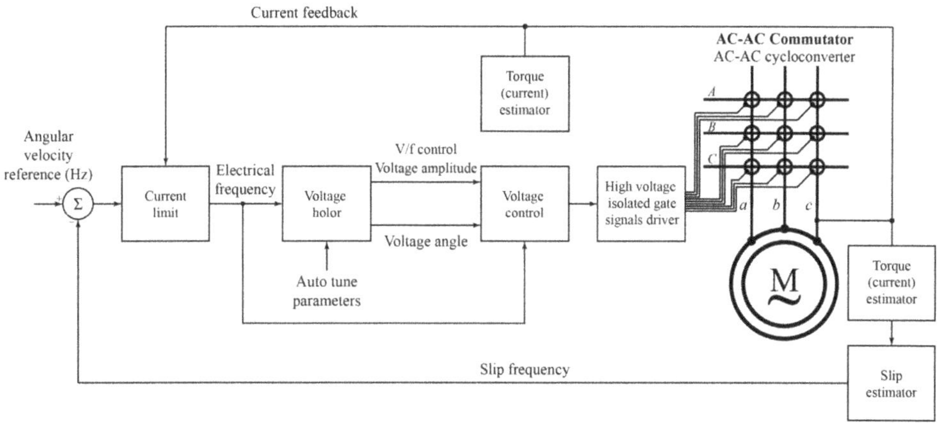

Figure 4.8. Structural and functional diagram of a sensorless magnetic-flux holor control AC motor IEMD.

True closed-loop holor AC motor IEMD can also make an AC–DC–AC or AC–AC commutator synchronous or asynchronous (induction) motor develop continuous full torque at zero angular velocity, that previously only DC–AC commutator synchronous motors (DC motors) with the mechanical split-ring/flat commutator were capable of. That makes them suitable for crane and hoist applications where the EM motor must produce full torque before the brake is released or else the mechanical load begins dropping and cannot be stopped.

4.5 AC motor IEMD installation and programming parameters

Careful planning of AC motor IEMD installation will help avoid many problems. Follow the instructions from the IEMD supplier for required and optional installation requirements. Important considerations include temperature and line power quality requirements, along with electrical connections, grounding, fault protection, EM motor protection, and environmental parameters.

4.6 AC motor IEMD selection

In selecting an AC motor IEMD, consideration must be given to the load characteristics of the driven machinery. The three basic load categories can be summarised as follows:

- Constant-torque loads require a constant EM motor torque throughout the operational angular-velocity range. Loads of these types are essentially friction mechanical loads such as a traction IEMD and conveyors.
- Variable-torque loads require much lower torque at low angular velocities than at high angular velocities. Loads that exhibit variable torque characteristics include centrifugal fans, pumps, and blowers.
- Shock (impact) loads require an EM motor to operate at normal load conditions followed by a sudden, large mechanical load applied to the EM motor. An example would be the sudden shock mechanical load that results from engaging a clutch that applies a large mechanical load to the EM motor

(as it would during a hard start). This current spike could cause the IEMD to trip as a result an excessive EM motor current fault.

4.7 AC motor IEMD line and load reactors

An AC motor IEMD reactor, as shown in figure 4.9, is basically an inductor installed on the input or output of the IEMD AC–DC–AC commutator. Line reactors stabilise the current waveform on the input side of an IEMD, reducing harmonic distortion and the burden on up-stream electrical equipment.

Harmonics are high-frequency voltage and current distortions within the power dynamical system normally caused by non-linear mechanical loads that do not have a constant current draw, but rather draw current in pulses. AC motor IEMDs create harmonics when they convert AC–DC and back DC–AC. By absorbing line spikes and filling some voltage sags, line and load reactors can prevent over-voltage and under-voltage tripping problems.

Load reactors, connected between the IEMD AC–DC–AC commutator and EM motor, will dampen overshoot peak voltage and reduce EM motor heating and audible noise. A load reactor helps to extend the life of the EM motor and increase the distance that the EM motor can be from the IEMD AC–DC–AC commutator.

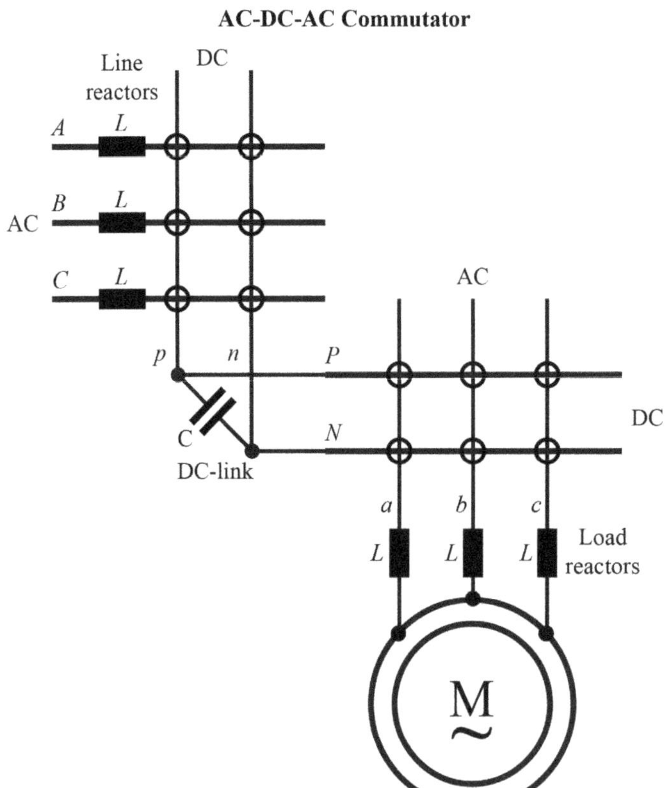

Figure 4.9. AC motor IEMD reactor.

4.8 AC motor IEMD location

Location is an important consideration in installing an IEMD as this can have a significant effect on the IEMD's performance and reliability. Location considerations are summarised as follows:

- Mount the IEMD near the EM motor windings. Excessive cable length between the IEMD and the EM motor windings can result in extremely high voltage spikes at the EM motor leads. It is important to verify the maximum cable distance stated in the IEMD specifications, when users are installing an IEMD onto AC–DC–AC commutator synchronous or asynchronous (induction) motors. Excessive voltages can reduce the expected life of the insulation system, especially non-commutator-duty EM motors.
- The enclosure surrounding the IEMD should be well ventilated or in a climate-controlled environment, as the build-up of excess heat may damage the IEMD components over time. Large fluctuations in ambient temperatures can result in condensation forming inside the IEMD enclosures and possible damage to components.
- Locations in dusty, wet, and corrosive environments, constant vibration, and direct sunlight should be avoided.
- The location should have adequate lighting and sufficient working space to carry out maintenance of the IEMD.

4.9 AC motor IEMD enclosures

Once a suitable location is chosen, it is then important to select the appropriate IEC- or NEMA-type enclosure which should have a rating appropriate to the level of protection for the environment.

4.10 AC motor IEMD mounting techniques

As a rule, a small AC motor IEMD is mounted in rack slots or on a DIN rail. The clips for mounting to the DIN rail are usually built into the fins of the heat sink to which the IEMD is mounted. This makes them easily installable in control cabinets. Larger AC motor IEMD units usually have through-hole mounting to accommodate individual fasteners. The fastening method should be adequate to support the mass of the IEMD and allow the free flow of air across the heat sink; airflow in some applications is aided by a cooling fan.

4.11 AC motor IEMD operator interface

The typical AC motor IEMD operator interface provides the means for an operator to start and stop the EM motor and adjust the operating angular velocity. Additional operator control functions might include reversing and switching between manual angular velocity adjustment and automatic control from an external process control signal. The operator interface often includes an alphanumeric display and/or indication lights and meters to provide information about the operation of the IEMD cyber-physical heterogeneous dynamical hypersystem.

When mounted within another enclosure, a remote operator keypad and display may be cable-connected and mounted a short distance from the controller.

A communications port is normally available to allow the IEMD to be configured, adjusted, monitored, and controlled using a **personal computer** (PC). PC-based software offers greater flexibility, as more detailed information on the IEMD parameters can be viewed simultaneously on the monitor. Modes of operation may include program, monitor, and run.

Typical data accessible in real time include:

- frequency output;
- voltage output;
- current output;
- EM motor angular velocity;
- EM motor kilowatts;
- DC bus voltage;
- parameter settings;
- faults.

4.12 AC motor IEMD electromagnetic interference

EMI, also termed electrical noise, is the unwanted signals generated by electrical and electronic equipment. EMI IEMD problems range from corrupted data transmission to IEMD damage. A modern IEMD using electrical valves (electronic switches) IGBTs for EM motor frequency control are very efficient because of their high commutation (switching) speed.

Unfortunately the high-speed commutation (switching) also results in much higher EMI being generated. All IEMD suppliers detail installation procedures that must be followed in order to prevent excessive noise on both sides of the IEMD.

Some of these noise suppression procedures include the following:

- Use a shielded power cable to connect the IEMD AC–DC–AC commutator to the EM motor windings.
- Use a built-in or external EMI filter.
- Use twisted control wiring leads to provide a balanced capacitive coupling.
- Use shielded cable to return the noise current flowing in the shield back to the source, instead of through the signal leads.
- Maintain at least 20 cm separation between control and power wires in open air, conduit, or cable trays.
- Use a common-mode choke wound with multiple turns of both signal and shield.
- Use optical isolation modules for control signal communications.
- Inherent in all EM motor cables is line-to-line and line-to ground capacitance. The longer the cable, the greater is this capacitance.

Electrical spikes occur on the outputs of a PWM AC–DC–AC motor IEMD due to currents charging the cable capacitances. Higher voltages, such as 400 V_{AC}, along with higher capacitances, result in larger current spikes. These spikes can shorten the lives of AC–AC, AC–DC–AC, DC–AC/AC–DC, DC–DC and DC–AC–DC commutators and EM motors windings. For this reason cable length must be limited to that recommended by the supplier.

4.13 AC motor IEMD grounding

Proper grounding plays a key role in the safe and reliable operation of the IEMD cyber-physical heterogeneous dynamical hypersystem. As a rule, IEMD, EM motor windings, and related equipment must meet the grounding and bonding requirements. The AC motor IEMD's safety ground must be connected to system ground. Ground impedance must conform to the grounding requirements in order to provide equal potential between all metal surfaces and a low-impedance path to activate overcurrent devices and reduce electromagnetic interference.

4.14 AC motor IEMD bypass contactor

A bypass contactor is intended for use in case of an IEMD failure for short-time emergency service. A structural and functional diagram of the power circuit connection of an AC motor IEMD bypass contactor is shown in figure 4.10. The isolation contactor electrically isolates the IEMD during bypass operation and is mechanically and electrically interlocked with the bypass contactor to ensure that both cannot be closed at the same time. Upon a sensed malfunction of the IEMD, the control circuit automatically opens the AC motor IEMD isolation contactor and closes the bypass contactor to keep the EM motor connected to the source. When automatic transfer to bypass operation occurs, the EM motor continues to operate at full speed. The IEMD isolation contactor must be opened during closing of the bypass contactor so that AC power is not fed into the output of the IEMD, causing damage. The automatic switch to bypass ensures no downtime and no interruption of service to critical loads. For example, in HVAC applications, this allows heating or cooling to be maintained at all times.

4.15 AC motor IEMD disconnecting means

Safety in operation and maintenance dictates that all EM motor-operated equipment have a means of fully disconnecting the power supply. As with starters, to reduce cost and size, most IEMD suppliers do not provide a disconnect switch as part of their standard IEMD package. If the optional input disconnect is not specified, a separate switch or circuit breaker must be installed. Requirements for disconnecting means for the AC motor IEMD itself and for the EM motor microcontroller; both sets of requirements must be satisfied.

The general rules that apply are as follows:

- Under 600 V, the microcontroller disconnecting means must be within sight (and less than 125 m according to definitions) of the EM motor microcontroller as specified in requirements.
- The microcontroller is not required to be in sight from the EM motor.
- The microcontroller disconnecting means shall also be permitted as the EM motor disconnecting means according to requirements.

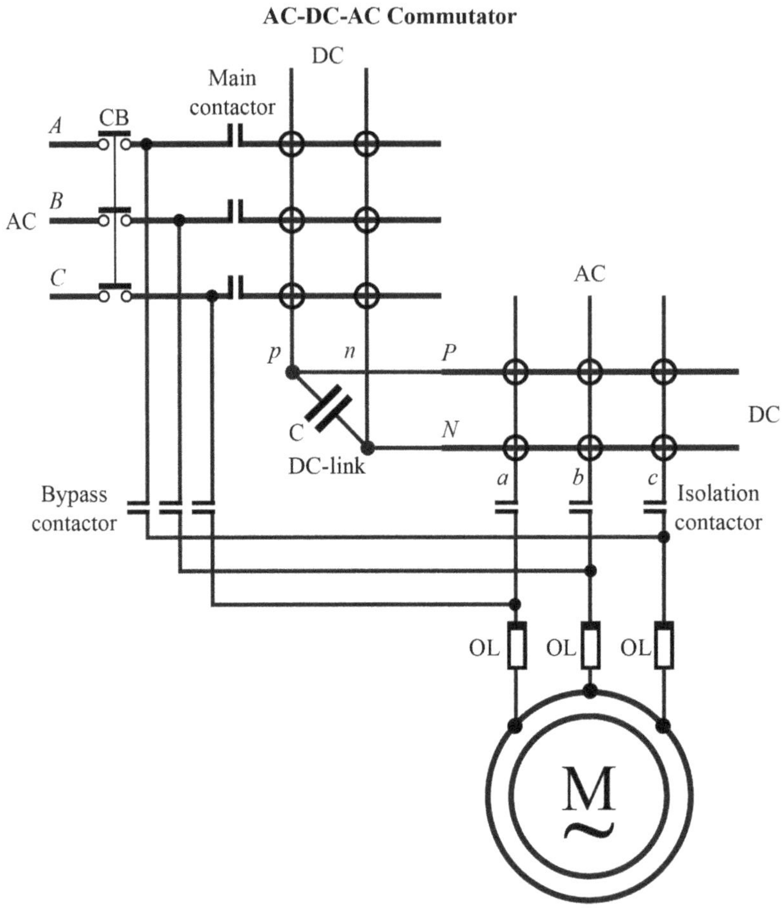

Figure 4.10. Structural and functional diagram of the power circuit connection of an AC motor IEMD bypass contactor.

- The EM motor disconnecting means shall be in sight of the EM motor. See exceptions under requirements that would allow EM motor disconnecting means to be out of sight of the EM motor. These exceptions, if applicable, would allow one lockable disconnect to serve as both microcontroller and EM motor disconnect while not in sight of the EM motor.

4.16 AC motor IEMD protection

An AC motor IEMD can operate as EM motor protection devices along with their role as EM motor angular-velocity microcontrollers. As a rule, AC motor IEMD have short-circuit protection (usually in the form of fuses) already installed by the supplier in the IEMD package.

The selection and sizing of these fuses is critical for semiconductor electrical valves protection in the event of a fault. The supplier's recommendations must be followed in installing or replacing fuses for the IEMD to assure fast operation of fuses in case of a fault.

In most IEMD applications the IEMD AC–DC–AC commutator itself provides overload protection for the EM motor. However, the feeder cable cannot be protected by IEMD built-in protection. The AC motor IEMD provides protection based on EM motor nameplate information that is programmed into it.

Microcontrollers incorporate many complex protective functions, such as:

- stall prevention;
- current limitation and overcurrent protection;
- short-circuit protection;
- under-voltage and over-voltage protection;
- ground fault protection;
- power supply phase failure protection;
- motor thermal protection through sensing of the motor winding temperature.

When an AC motor IEMD is not approved for overload protection, or if multiple EM motors are fed from the AC–DC–AC commutator, one or more external overload relays must be provided. The most common practice is to use an EM motor overcurrent relay that will protect all three phases and protect against single phasing.

4.17 AC motor IEMD braking

With AC–DC–AC commutator synchronous or asynchronous (induction) motors, there is excessive energy generated when the mechanical load drives the EM motor during deceleration, instead of the EM motor driving the mechanical load. This energy goes back into the IEMD and will result in an increasing DC link voltage. If the DC link voltage goes too high, the IEMD will be damaged. Depending on design, an AC motor IEMD can redirect this excess energy through resistors or back to the AC supply.

When dynamic braking is used, the drive connects a braking resistor across the DC link, as shown in figure 4.11, to absorb the excess energy. For smaller mechanical power EM motors, the resistance is built into the IEMD. External resistance banks are used for larger-mechanical power EM motors to dissipate the increased heat load.

Regenerative braking is similar to dynamic braking, except the excess energy is redirected back to the AC supply. IEMDs designed to use regenerative braking are required to have an active front end to control regenerative current. With this option the diodes in the AC–DC rectifier are replaced with IGBT electrical valves.

The IGBT electrical valves are commutated (switched) by the control logic, and operate in both EM motoring and M-E regenerative modes. DC-injection braking is a standard feature on a IEMD. As the term implies, DC injection braking generates electromagnetic forces in the EM motor when the microcontroller, in stop mode, injects direct current into the stator windings—after it has cut off alternating current supply to two of the stator phases—thus turning off the normal rotating magnetic field.

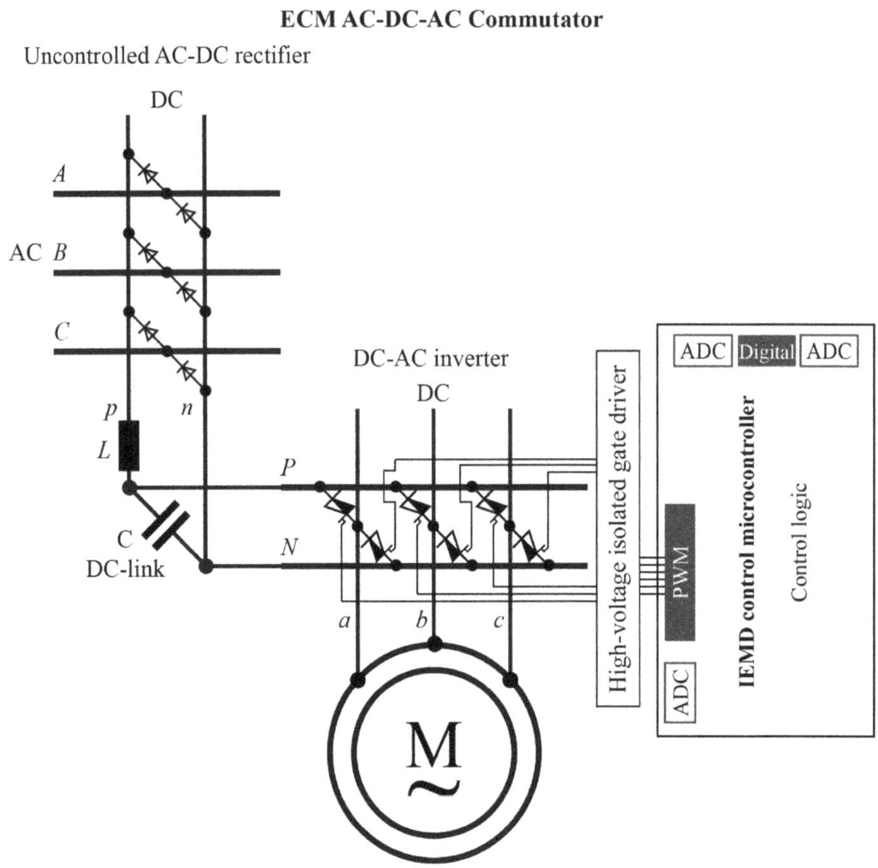

Figure 4.11. Dynamic braking applied to an AC motor IEMD.

Most DC injection braking systems have the ability to adjust the length of time they will operate and the maximum torque they will apply. They generally begin braking when they detect that the EM motor is no longer receiving its run command and come equipped with hardware to prevent the EM motor from receiving another run command until the braking is finished.

4.18 AC motor IEMD ramping

An AC motor IEMD offers many of the same advantages as reduced-voltage and soft start starters. The timed speed ramp-up feature found in an IEMD is similar to the soft start function of starters. However, the AC motor IEMD timed speed ramp-up generally has a much smoother acceleration than soft starting, which is usually done in steps. Soft starting with an IEMD reduces the frequency initially supplied to the EM motor and steps up the frequency over a pre-set period of time. An IEMD with soft start capabilities has replaced many of the older types of reduced-voltage starters. While an IEMD offers soft start capabilities, an IEMD is not considered as a true soft start starter.

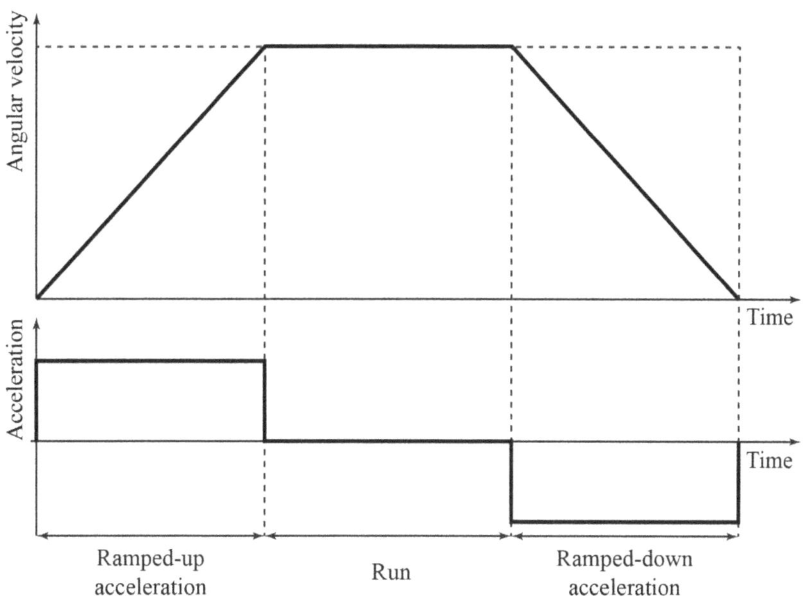

Figure 4.12. AC motor IEMD ramping functions.

Ramping is the ability of an AC motor IEMD to increase or decrease the voltage and frequency to an AC–DC–AC commutator synchronous or asynchronous (induction) motor gradually. This accelerates and decelerates the EM motor smoothly, as shown in figure 4.12, with less stress on both the EM motor and connected mechanical load.

Ramp-up is generally a smoother acceleration than the stepped increases used in soft starters. The length of time pre-set for the angular-velocity ramp-up can be varied from a few seconds to 120 s or more, depending on the IEMD capabilities.

Timed ramp-down is a function of an IEMD that provides smooth deceleration, bringing the EM motor to a full stop in a pre-set time. Acceleration and deceleration are separately programmable. Depending on IEMD parameters, ramp-down times can vary from fractions of a second (when used with dynamic braking) to more than 120 s.

The ramp-down function is applied in processes that require smooth stops, but also require the process to stop within a given period of time.

4.19 AC motor IEMD control inputs and outputs

Typical power and control inputs and output termination points found on an AC motor IEMD.

The three-phase power source is connected at the line input terminals *A*, *B*, and *C* and the EM motor feed conductors are connected to the output EM motor terminals *a*, *b* and *c*. The line and EM motor terminals pass through electronic circuitry so there is no direct connection between them, as with an across-the-line starter.

As a rule, AC motor IEMDs contain control terminal strips for external connection for both digital and analogue inputs and outputs.

The number and types of inputs and outputs will vary with the complexity of the IEMD and serve as a means of comparison between suppliers of the IEMD.

IEMD inputs and outputs are either digital or analogue signals. Digital inputs and outputs have two states (either ON or OFF), while analogue inputs and outputs have many states that vary across a range of values.

4.20 AC motor IEMD digital inputs

Digital inputs are used to interface the IEMD with devices such as push buttons, selector switches, relay contacts, and **programmable logical controller** (PLC) digital output modules. Each digital input may have a pre-set function assigned to it, such as start/stop, forward/reverse, external fault, and pre-set angular-velocity selections. For example, if an EM motor has to operate at three different values of the angular velocity, a relay or switch contact could be made to close and send signals to separate digital inputs points that would change the EM motor angular velocity to the pre-set value. Typical input digital connections are programmed an IEMD are macro- and microelectronic devices, they can have only one phase rotation output at a time. Therefore, interlocking, as required on EM devices, is not required for IEMD forward/reverse operations. Inputs can also be programmed for two- or three-wire control to accommodate either maintained or momentary start methods. Note that the control logic is determined and executed by the program within the IEMD and not by the hard-wiring arrangement of the input control devices.

4.21 AC motor IEMD digital/relay outputs

Digital/relay outputs are two-position signals (ON/OFF) sent by the IEMD to devices such as pilot lights, alarms, auxiliary relays, solenoids, and PLC digital input modules. Digital outputs have a voltage potential (e.g. 24 V_{DC}) coming from them. Relay outputs, which are known as ‘dry’ contacts, switch something external, closing or opening another potential. Relay outputs are normally rated for both AC and DC voltages.

4.22 AC motor IEMD analogue inputs

Analogue inputs are used to interface the IEMD with an external 0 to 10 V_{DC} or 4 to 20 mA signal—for example, a rotational speed set-point from an external speed control potentiometer.

4.23 AC motor IEMD analogue outputs

Analogue outputs are modulating signals sent by the IEMD to a device such as a meter that could display angular velocity or current.

4.24 EM motor nameplate

Data EM motor specifications are programmed into the IEMD to ensure optimum drive performance as well as adequate fault and overload protection. This may include the following items found on the nameplate or derived through measurements:

- ***Frequency*** (Hz)—Nameplate frequency required by the EM motor to achieve base angular velocity. The default value is normally 50 Hz or 60 Hz.
- ***Angular velocity*** (rad s^{-1})—Nameplate maximum angular velocity or rotational speed at which the EM motor should be rotated.
- ***Full-load current*** (A)—Nameplate maximum current that the EM motor may use. **Full-load current** (FLC) are the same as the EM motor nameplate current.
- ***Supply voltage*** (V)—Nameplate voltage required by the EM motor to achieve maximum torque.
- ***Power rating*** (kW)—Nameplate rating of EM motors is generally rated in kilowatts (kW).
- ***EM motor magnetising current*** (A)—Current that the EM motor draws when operating with no load at nameplate rated voltage and frequency. If not specified, it can be measured using a true-RMS clamp-on ammeter.
- ***EM motor stator resistance*** (Ω)—DC resistance of the stator between any two phases. If not specified, it can be measured with an ohmmeter.

4.25 AC motor IEMD derating

Derating an AC motor IEMD means using a larger than normal EMD in the application. Derating is required when an AC motor IEMD is to be operated outside the normal operating range specified by the supplier. Most suppliers provide derating data when the AC motor IEMD is to be operated outside the specified temperature, voltage, and altitude. As an example, derating must be considered when the AC motor IEMD is installed at high altitude, greater than 1000 m. The cooling effect of the DC motor IEMD is deteriorated because of the reduced density of the air at high altitudes.

4.26 Types of an AC motor IEMD

The evolution of AC motor IEMD technology has seen many changes in a relatively small time frame. As a result, a newer AC motor IEMD with greater functionality is now available. As a rule, an AC motor IEMD supplied currently is a PWM AC motor IEMD that convert the 50 or 60 Hz AC line power to DC link power, then pulse the output voltage for varying lengths of time to imitate an AC at the frequency desired.

An older AC motor IEMD was distinguished by the type of ECM AC–DC–AC commutator used in the AC motor IEMD. Two earlier types of IEMD were the ECM AC–DC–AC commutator with the DC–AC commutator, acting as a **voltage-source** (VS) DC–AC inverter or a **current-source** (CS) DC–AC inverter.

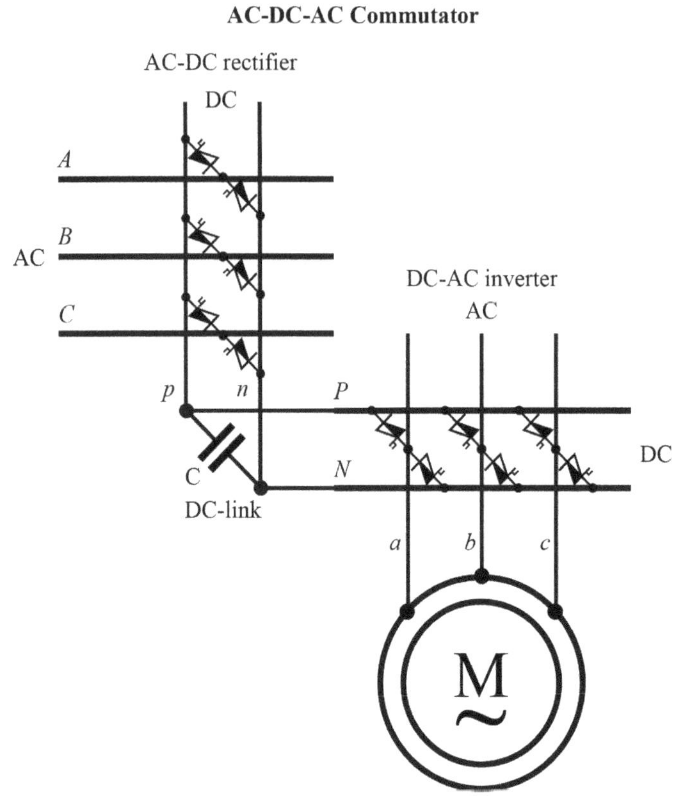

Figure 4.13. Simplified ECM AC–DC–AC commutator of one type of VS DC–AC inverter.

Figure 4.13 shows a simplified ECM AC–DC–AC commutator of one type of VS DC–AC inverter, also termed **variable-voltage** (VV) DC–AC inverter. This DC–AC inverter uses AC–DC rectifier with the controlled (triggered and/or quenched) electrical valves, e.g. IGCTs to convert the incoming AC voltage into rectified DC voltage. The electrical valves provide a means of controlling the value of the rectified DC voltage. The energy storage in the DC link between the AC–DC rectifier and DC–AC inverter consists of capacitors. The DC–AC inverter that is the AC–DC–AC commutator's section utilises six electrical valves (electronic switches). The control logic uses a microcontroller's high-voltage insulation gate drivers to commute (switch) the electrical valves ON and OFF, providing a variable voltage and frequency to the EM motor. This type of commutation (switching) is often referred to as six-step because it takes six $\pi/3$ steps to complete one 2π cycle. Although the EM motor prefers a smooth sine wave, a six-step output can be satisfactorily used. The main disadvantage is torque pulsation, which occurs each time an electrical valve is commuted (switched). The pulsations can be noticeable at low angular velocities as angular-velocity variations in the EM motor.

These angular-velocity variations are sometimes referred to as cogging. The non-sinusoidal current waveform causes extra heating in the EM motor, requiring an EM motor derating.

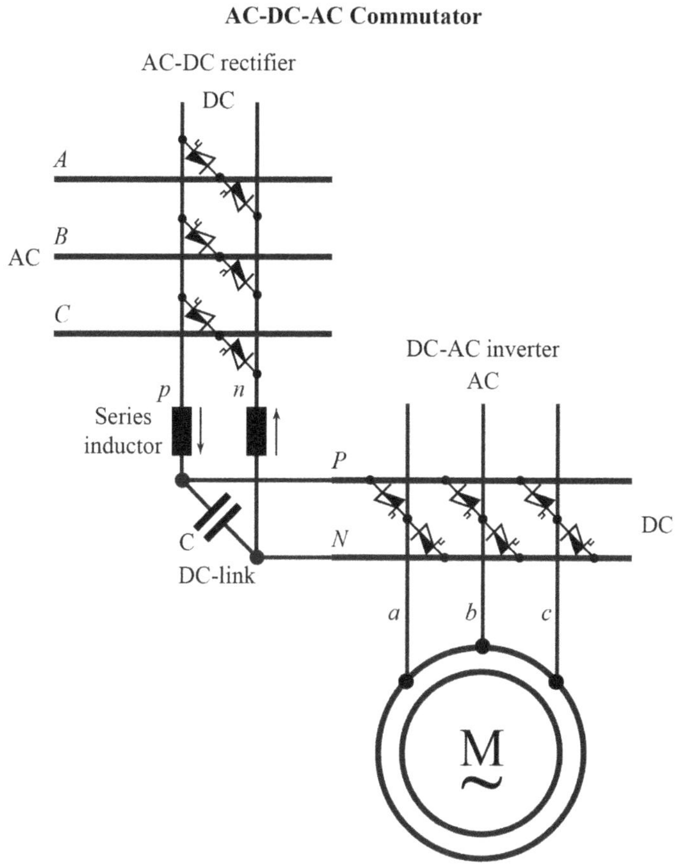

Figure 4.14. Simplified ECM AC–DC–AC commutator of a CS DC–AC inverter.

An ECM DC–AC commutator, acting as the VS DC–AC inverter IEMD can operate any number of EM motors up to the total rated mechanical power of the IEMD. With an ECM DC–AC commutator, acting as the CS DC–AC inverter, the DC power supply is configured as a current source rather than a voltage source. This IEMD employers a closed-loop system that monitors the actual angular velocity of the EM motor and compares it with the pre-set reference angular velocity, creating an error signal that is used to increase or decrease the current to the EM motor.

Figure 4.14 shows the simplified ECM AC–DC–AC commutator of an AC–DC–AC inverter. The AC–DC rectifier is connected to the DC–AC inverter through a large series inductor. This inductor opposes any change in current and is of a sufficiently high inductance value that direct current is constrained to be almost constant.

As a result, the output produced is almost a square wave of current. A CS DC–AC inverter is used for large IEMD—about 150 kW—because of their simplicity, regeneration braking capabilities, reliability, and lower cost.

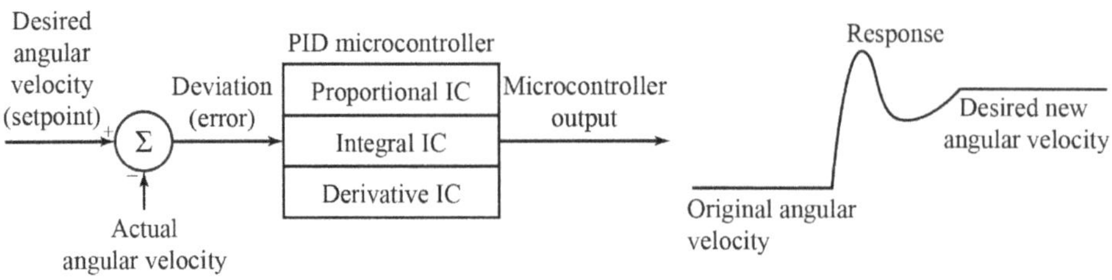

Figure 4.15. PID loop.

Since the CS DC–AC inverter monitors the actual EM motor angular velocity, they can be used to control only a single corresponding EM motor with characteristics that match the IEMD.

4.27 AC motor IEMD PID control

As a rule, AC motor IEMD applications require the AC–DC–AC commutator synchronous or asynchronous (induction) motor to run at a specific angular velocity as set by the keypad, angular-velocity potentiometer, or analogue input.

Certain AC motor IEMD provided an alternative option that allows precise process control through a set-point microcontroller or **proportional-integral-derivative** (PID) mode of operation. One AC motor IEMD come equipped with a built-in PID microcontroller. The PID loop is used to maintain a process variable, such as angular velocity, as illustrated in figure 4.15.

The desired angular velocity, or set-point, and the actual angular-velocity values are input to a summation point. These two signals are opposite in polarity and yield a zero error or deviation whenever the desired angular velocity equals the actual angular velocity. If the two signals differ in value, the error signal will have a positive or negative value, depending on whether the actual angular velocity is greater or less than the desired angular velocity. This error signal is input to the PID microcontroller. The terms proportional, integral, and derivative (PID) describe three basic mathematical functions then applied to the error signal. The PID output reacts to the error and outputs a frequency to try to reduce the error value to zero. The microcontroller's activity is to make the angular-velocity adjustments quickly, with minimum overshoot or oscillations. Tuning the PID microcontroller involves gain and time adjustments designed to improve performance and result in a fast response with minimum overshoot, allowing the EM motor to settle in quickly to the new angular velocity. As a rule, AC motor IEMD has a PID auto-tune function designed to ease the tuning process.

4.28 AC motor IEMD's parameter programming

The main AC motor IEMD program is contained in the microprocessor's firmware and not normally accessible to the IEMD user. A parameter is a variable associated with the operation of the IEMD that can be programmed or adjusted. Parameters provide a degree of adjustment so that the user can customise the IEMD to suit specific AC motor and driven equipment requirements. The number of parameters can range from 50, for a small basic IEMD, to over 200 for a larger, more complex IEMD. As a rule, an AC motor IEMD provides upload/download and parameter copy capability.

Common adjustable parameters include the following:

- pre-set values of the angular velocity;
- minimum and maximum values of the angular velocity;
- acceleration and deceleration rates;
- two- and three-wire remote control modes;
- stop modes: ramp, coast, DC injection;
- automatic torque boost;
- current limit;
- configurable input jog;
- V/F settings;
- carrier frequency;
- program password.

An AC motor IEMD comes with manufacturing works default settings for most parameters that are more conservative in nature. The default value settings simplify the start-up procedure. However, parameters for EM motor nameplate data are not factory-set (unless a matched IEMD and AC motor has been purchased) and must be entered in the field.

In general there are three types of parameter:

- ***Tuneable on the fly***—Parameters can be adjusted or changed while the AC motor IEMD is running or stopped. Angular velocity equals the actual angular velocity. If the two signals differ in value, the error signal will have a positive or negative value, depending on whether the actual angular velocity is greater or less than the desired angular velocity. This error signal is input to the PID microcontroller. The terms proportional, integral, and derivative describe three basic mathematical functions then applied to the error signal. The PID output reacts to the error and outputs a frequency to try to reduce the error value to zero. The PID microcontroller's activity is to make the angular-velocity adjustments quickly, with a minimum of overshoot or oscillations. Tuning the PID microcontroller involves gain and time adjustments designed to improve performance and result in a fast response with a minimum overshoot, allowing the DC motor to settle in quickly to the new angular velocity. Some DC motor IEMDs have a PID auto-tune function designed to ease the tuning process.
- ***Configurable***—Parameters can be adjusted or changed only while the DC motor IEMD is stopped.
- ***Read only***—Parameters cannot be adjusted.

An integral keypad with an LED display can be used to program and operate a small AC motor IEMD locally. The display shows either a parameter number or a parameter value. The AC motor IEMD's parameter menu outlines what the parameter number represents, and what numerical selections or options for the parameter are available. Parameter menu formats vary between make and model. For example, this AC motor IEMD has two kinds of parameters: program parameters (P-00 through P-64), which configure the IEMD operation, and display parameters (D-00 through D-64), which display information.

Examples of program parameters include:

P-00 minimum angular velocity—use this parameter to set the lowest frequency the IEMD will output. Default setting is 0.

P-01 maximum angular velocity—use this parameter to set the highest frequency the IEMD will output. Default setting is 50 or 60 Hz.

P-02 M-E motor overload current—set this parameter to the EM motor nameplate full-load ampere rating. Default setting is 100% of the rated IEMD current.

P-30 acceleration time—use this parameter to define the time it will take the IEMD to ramp-up from zero (0) Hz to maximum angular velocity. Default setting is 5.0 s.

Examples of display parameters include:

D-00 command frequency—this parameter represents the frequency that the IEMD is commanded to output.

D-01 output frequency—this parameter represents the output frequency at the EM motor terminals.

D-02 output current—this parameter represents the EM motor current.

D-03 bus voltage—this parameter represents the DC link voltage level.

4.29 AC motor IEMD diagnostics and troubleshooting

As a rule, an AC motor IEMD comes equipped with self-diagnostic controls to help trace the source of problems. Always observe the following precautions when troubleshooting the IEMD.

- Stop the IEMD.
- Disconnect, tag, and lock out AC power before operating on the IEMD.
- Verify that there is no voltage present at the AC input power terminals. It is important to remember that DC link capacitors retain hazardous voltages after input power has been disconnected. Therefore, wait 5 min for the DC link capacitors to discharge once power has been disconnected. Check the voltage with a voltmeter to ensure that the capacitors have discharged before touching any internal components.

Problem indicators may include the following:

LEDs provide a quick indication of problems. Normally, a steady glowing light means everything is running properly. Flashing yellow or red lights indicate a problem with the IEMD that should be checked. Consult the operator's manual for the specific IEMD to determine what a particular flashing light means.

Alarms indicate conditions that may affect IEMD operation or application performance. They are cleared automatically when the condition that caused the alarm is no longer present. Configurable alarms alert the operator to conditions that, if left untreated, may lead to an IEMD fault. The IEMD continues to operate during the alarm condition, and the alarms can be enabled or disabled by the programmer or operator. Non-configurable alarms alert the operator of conditions caused by improper programming and prevent the IEMD from starting until the problem is resolved. These alarms can never be disabled.

Fault parameters settings indicate conditions within the IEMD that require immediate attention. The IEMD responds to a fault by coasting to a stop and turning off output power to the EM motor. Auto-reset faults reset automatically if, after a pre-set time, the condition that caused the fault is no longer present. The IEMD then restarts.

Non-resettable faults may require EM drive or EM motor repair; the fault must be corrected before it can be cleared. User-configurable faults can be enabled and disabled to enunciate or ignore a fault condition.

Fault queues normally retain a history of faults. Typically, queues hold only a limited number of entries; therefore, when the queue is full, older faults are discarded when new faults occur. The system typically assigns a timestamp to the fault so that programmers or operators can determine when a fault occurred relative to the last drive power-up.

A complete listing of all the different types of faults and the appropriate corrective actions can typically be found in the operator's manual for a specific IEMD.

References

Barnes M 2003 *Practical Variable Speed Drives and Power Electronics, Automated Control Systems* (Perth, Australia: IDC Technologies)

Chan T-F and Shi K 2011 *Applied Intelligent Control of Induction Motor Drives* (New York: Wiley)

Geyer T 2017 *Model Predictive Control of High Power Converters and Industrial Drives* (New York: Wiley)

Holmes D G and Lipo T A 2003 *Pulse Width Modulation for Power Converters Principles and Practice* (Piscataway, NJ: Institute of Electrical and Electronics Engineers)

Hughes A and Drury W 2013 *Electric Motors and Drives Fundamentals, Types and Applications* 4th edn (Amsterdam: Elsevier)

Rashid M H 2018 *Power Electronics Handbook* 4th edn (Amsterdam: Elsevier)

Wach P 2011 *Dynamics and Control of Electrical Drives* (Berlin, Heidelberg: Springer)

Wilamowski B M and Irwin D J 2011 *The Industrial Electronics Handbook - Power Electronics and Motor Drives* (Boca Raton, FL: CRC Press)

Chapter 5

DC motor IEMD modus operandi

5.1 Introduction

DC motor IEMD technology is the oldest arrangement of mechatronic angular-velocity control. The angular velocity of a DC–AC/AC–DC commutator synchronous motor (in earlier times known as the DC motor) with the mechanical split-ring/flat commutators as well as the AC–DC–AC commutator synchronous or asynchronous (induction) motor with the macroelectronic commutators (macrocommutators) is the simplest to control, and can be varied over a very wide range. This DC motor IEMD is designed to deal with applications such as:

Crane/hoist—A DC motor IEMD offers several advantages in applications that operate at low values of the angular velocity, such as cranes and hoists. Advantages include low angular-velocity accuracy, short-time overload capacity, size, and torque providing control. Generated mechanical power from the DC–AC/AC–DC commutator synchronous motor/generator is used for braking and excess electrical power is fed back into the AC line. This electrical power helps reduce energy requirements and eliminates the necessity for heat-producing dynamic braking resistors. Peak current of at least 250% is available for short-term mechanical loads.

Winders/coilers—In DC motor IEMD winder operations, maintaining tension is very important. DC–AC commutator synchronous motors (commonly known as DC motors) are able to operate at rated current over a wide angular-velocity range, including low values of the angular velocity.

Mining/drilling—The DC motor IEMD is often preferred in the high mechanical-power applications required in the mining and drilling industry. For this arrangement of application, a DC motor IEMD offers advantages in size and cost. They are rugged, dependable, and industry proven.

doi:10.1088/2053-2563/aae7d7ch5

© IOP Publishing Ltd 2019

5.2 DC motor IEMD principles of operation

A DC motor IEMD varies the angular velocity of DC–AC commutator synchronous motors with greater efficiency and angular-velocity regulation than resistor control circuits. Since the angular velocity of a DC–AC commutator synchronous motor with the mechanical split-ring/flat or the macroelectronic commutator (macrocommutator) is directly proportional to armature voltage and inversely proportional to field current, either armature voltage or field current can be used to control angular velocity (Barnes 2003, Wilamowski and Irwin 2011, Chan and Shi 2011, Wach 2011, Hughes and Drury 2013, Holmes and Lipo 2003, Geyer 2017, Rashid 2018).

To change the sense of rotation direction of a DC–AC commutator synchronous motor, either the armature polarity can be reversed (figure 5.1), or the field polarity can be reversed.

The structural and functional diagram of a DC motor IEMD cyber-physical heterogeneous continuous or discrete complex dynamical hypersystem made up of a DC–AC commutator synchronous motor (DC motor) and a PID microcontroller is shown in figure 5.2.

The shunt-wound DC–AC synchronous motor is constructed with armature and field windings. A common classification of DC–AC synchronous motors is by the type of field excitation winding. Shunt-wound DC–AC synchronous motors are the most commonly used type for adjustable-velocity control. In most instances the shunt field winding is excited, as shown, with a constant-level voltage from the PID microcontroller.

The electrical valves, e.g. IGBTs, IGCTs or MOSFETs of the AC–DC rectifier section of the AC–DC–AC commutator converts the fixed-voltage **alternating current** (AC) of the power source to an adjustable-voltage, controlled **direct current** (DC) output which is applied to the armature of a DC–AC synchronous motor.

Angular-velocity control is achieved by regulating the armature voltage to the EM motor. EM motor angular velocity is directly proportional to the voltage applied to the armature. The main function of a DC motor IEMD is to rectify the fixed applied AC voltage into a variable rectified DC voltage. Electrical valves,

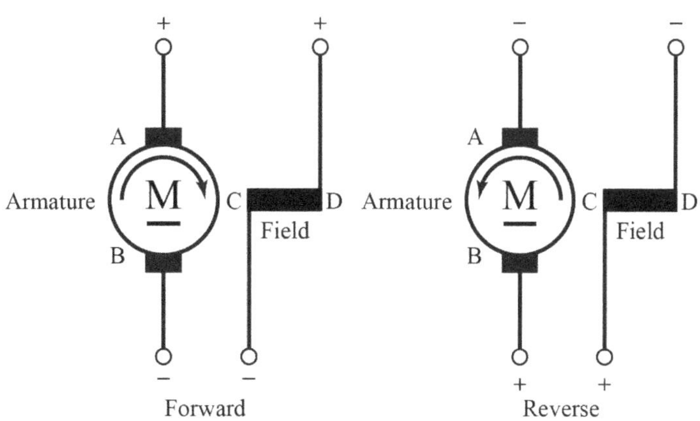

Figure 5.1. Changing the sense of rotation direction by reversing the polarity of the DC motor armature.

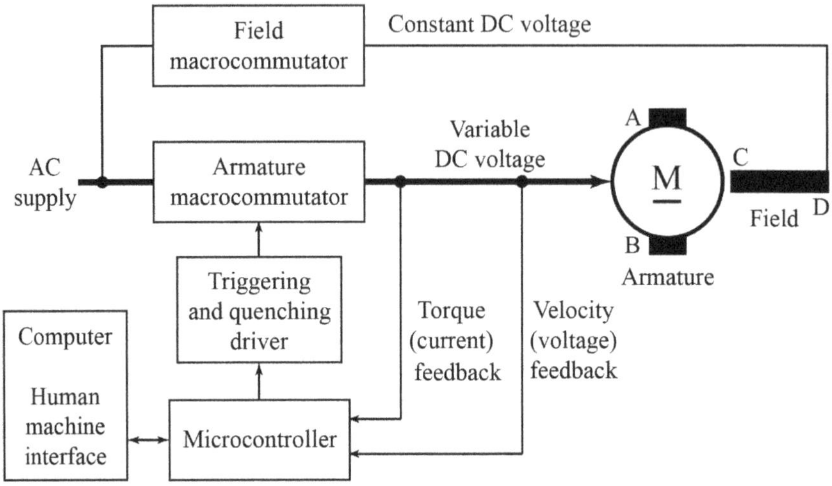

Figure 5.2. Structural and functional diagram of a DC motor IEMD cyber-physical heterogeneous continuous or discrete complex dynamical hypersystem.

e.g. IGBTs, IGCTs or MOSFETs macroelectronic commutators (macrocommutators) provide a convenient method of accomplishing this. They provide a controllable power output by phase angle control.

The triggering angle, or point in time where the electrical valve (IGBTs, IGCTs or MOSFETs) is triggered into conduction, is synchronised with the phase rotation of the AC power source, as illustrated in figure 5.3.

The amount of rectified DC voltage is controlled by timing the input pulse current to the gate. Applying gate current near the beginning of the sine wave cycle results in a higher average voltage applied to the EM motor armature. Gate current applied later in the cycle results in a lower average DC output voltage. The effect is similar to a very high-speed electrical valve (switch), capable of being turned on and off at an infinite number of points within each half-cycle. This occurs at a rate of 50 or 60 times a second on a 50 or 60 Hz line, to deliver a precise amount of power to the EM motor.

5.3 Single-phase AC supply input—DC motor IEMD

An armature voltage-controlled DC motor IEMD with DC–AC commutator synchronous motors (DC motors) with the mechanical split-ring/flat commutator, are a constant torque IEMD, capable of rated EM motor torque at any angular velocity up to rated EM motor base angular velocity. Fully controlled ECM AC–DC commutators, acting as AC–DC rectifiers, are built with electrical valves, e.g. IGBTs, IGCTs or MOSFETs.

Figure 5.4 shows an ECM AC–DC commutator acting as a fully controlled AC–DC rectifier powered by a single-phase AC supply.

The electrical valves, e.g. IGBTs, IGCTs, or MOSFETs rectify the supply voltage (changing the voltage from AC to DC) as well as controlling the output DC voltage level. In this ECM AC–DC commutator, electrical valves, e.g. IGBTs or MOSFETs

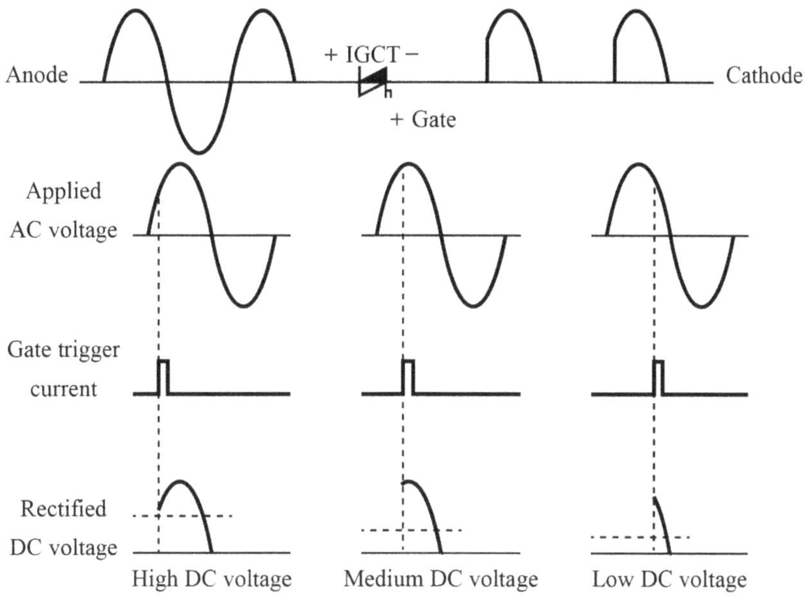

Figure 5.3. IGCT conversion from AC to variable DC.

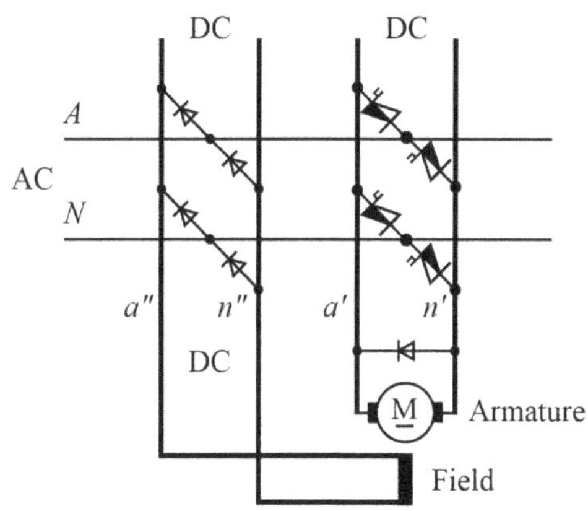

Figure 5.4. Single-phase IGCT ECM AC–DC commutator, acting as a fully controlled AC–DC rectifier powered by a single-phase AC supply.

V1 and V3 are triggered into conduction on the positive half of the input waveform and V2 and V4 on the negative half. Freewheeling diode D (also termed a suppressor diode) is connected across the armature to provide a path for release of energy stored in the armature when the applied voltage drops to zero. A separate diode AC–DC rectifier is used to convert the alternating current to a constant direct current required for the field circuit.

ECM AC–DC commutators, acting as single-phase controlled AC–DC rectifiers, are commonly used in the smaller mechanical power DC motor IEMD.

Features include:

- angular-velocity or torque control;
- tachometer input;
- fused input;
- angular velocity or current monitoring (0–10 V_{DC} or 4–20 mA).

5.4 DC motor IEMD three-phase AC supply input

ECM AC–DC commutators, acting as controlled AC–DC rectifiers, are not limited to single-phase designs. In most commercial and industrial control systems, AC power is available in three-phase configuration for maximum mechanical power and efficiency. Typically six electrical valves, e.g. IGBTs, IGCTs or MOSFETs are connected together, as shown in figure 5.5, to make an ECM AC–DC commutator acting as a three-phase fully controlled AC–DC rectifier for a DC–AC commutator's synchronous motor (DC motor) with the mechanical split-ring/flat and/or ECM macroelectronic commutator (macrocommutator). This ECM AC–DC commutator, acting as three-phase AC–DC rectifier has three legs, each phase connected to one of the three-phase voltages. It can be seen that the ECM AC–DC commutator has two halves, the positive half consisting of the electrical valves, e.g. IGBTs or MOSFETs V1, V3, and V5 and the negative half consisting of the electrical valves, e.g. IGBTs, IGCTs or MOSFETs V2, V4, and V6.

At any time when there is current flow, one electrical valve (IGBTs, IGCTs or MOSFETs) from each half conducts.

The variable DC output voltage from the ECM AC–DC commutator, acting AC–DC rectifier supplies voltage to the EM motor armature in order to run it at the desired angular velocity. The gate triggering angle of the electrical valves, e.g. IGBTs, IGCTs or MOSFETs in the ECM AC–DC commutator, acting as AC–DC

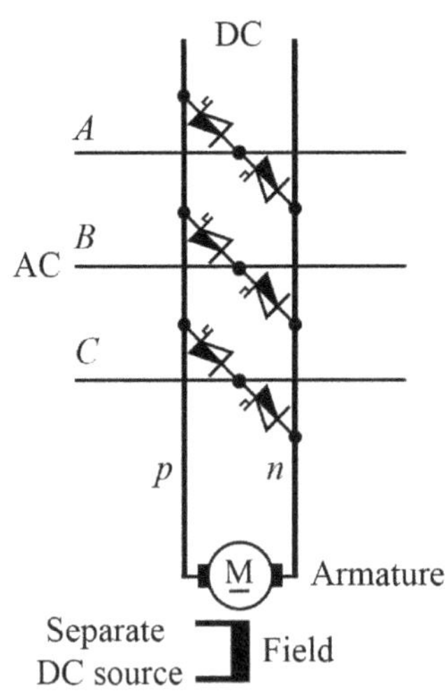

Figure 5.5. Three-phase ECM AC–DC commutator, acting as a fully controlled AC–DC rectifier powered by a three-phase AC supply.

rectifier, along with the maximum positive and negative values of the AC sine wave, determine the value of the EM motor armature voltage. The EM motor draws current from the three-phase AC power source in proportion to the amount of mechanical load applied to the EM motor shaft. Unlike a DC motor IEMD, bypassing the EMD to run the EM motor is not possible.

Larger mechanical-power three-phase DC motor IEMD panels often consist of a power module mounted on a chassis with line fuses and disconnect. This design simplifies mounting and makes connecting power cables easier as well. For example, a three-phase input DC motor IEMD has the following EMD power specifications:

- nominal line voltage for three-phase 230/400 V_{AC};
- voltage variation +15%, −10% of nominal;
- nominal line frequency 50 or 60 Hz;
- DC voltage rating 230 V AC line: Armature voltage 240 V_{DC}; field voltage 150 V_{DC};
- DC voltage; rating 460 V AC line: Armature voltage 500 V_{DC}; field voltage 300 V_{DC}.

5.5 Field voltage control of DC motor IEMD

To control the speed of a DC–AC commutator synchronous motor (DC motor) with the mechanical split-ring/flat commutator below its base angular velocity, the voltage applied to the armature of the EM motor is varied while the field voltage is held at its nominal value. To control the angular velocity above its base value, the armature is supplied with its rated voltage and the field is weakened. For this reason, an additional variable-voltage field regulator, as illustrated in figure 5.6, is necessary for a DC motor IEMD with field voltage control.

Field weakening is the act of reducing the current applied to a DC–AC commutator synchronous motor's shunt field. This action weakens the strength of

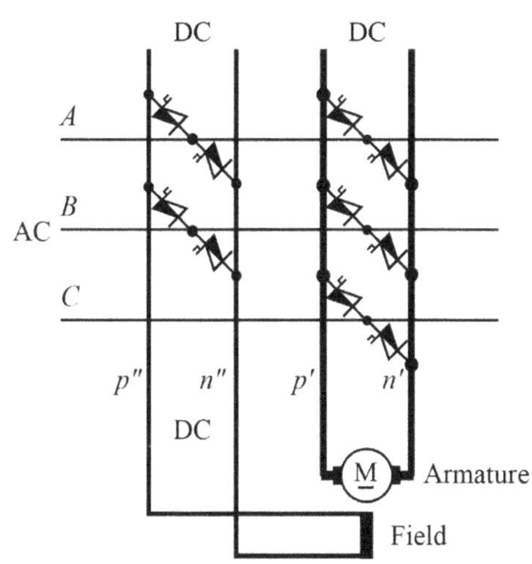

Figure 5.6. DC motor IEMD with an additional field ECM AC–DC commutator, acting as a fully controlled AC–DC rectifier that is a variable-voltage DC motor field regulator.

the magnetic field and thereby increases the EM motor angular velocity. The weakened field reduces the induced voltage of rotation (counter EMF) generated in the armature; therefore the armature current and the angular velocity increase. Field loss detection must be provided for a DC motor IEMD to protect against excessive EM motor angular velocity due to loss of EM motor field current.

A DC motor IEMD with EM motor field control provide coordinated automatic armature and field voltage control for extended angular-velocity range and constant mechanical power applications.

The DC–AC synchronous motor (DC motor) with the mechanical split-ring/flat commutator is armature-voltage-controlled for constant-torque, variable mechanical power operation to base angular velocity, where it is transferred to field control for constant mechanical power, variable-torque operation to EM motor maximum angular velocity.

5.6 DC motor IEMD non-regenerative and regenerative

Non-regenerative DC motor IEMD, also known as a single-quadrant DC motor IEMD, rotate in one sense of direction only and they have no inherent braking capabilities. Stopping the DC–AC commutator synchronous motor (DC motor) with the mechanical split-ring/flat commutator is done by removing voltage and allowing the EM motor to coast to a stop. Typically a non-regenerative DC motor IEMD operates high-friction loads such as mixers, where the load exerts a strong natural brake.

In applications where supplemental quick braking and/or EM motor reversing is required, dynamic braking and forward and reverse circuitry, such as shown in figures 5.7 and 5.8, may be provided by external means.

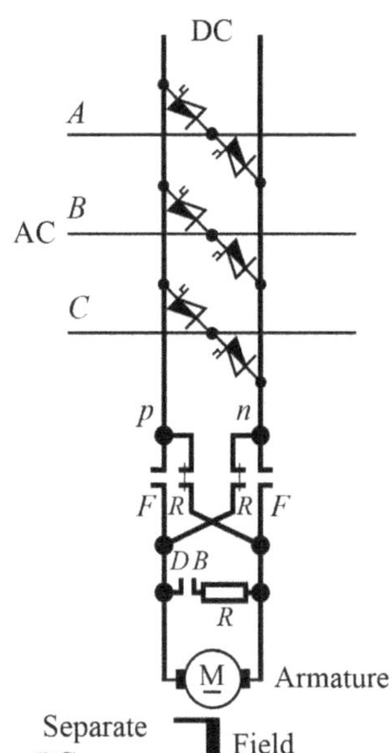

Figure 5.7. Non-regenerative DC motor IEMD with external dynamic braking and reversing contactors.

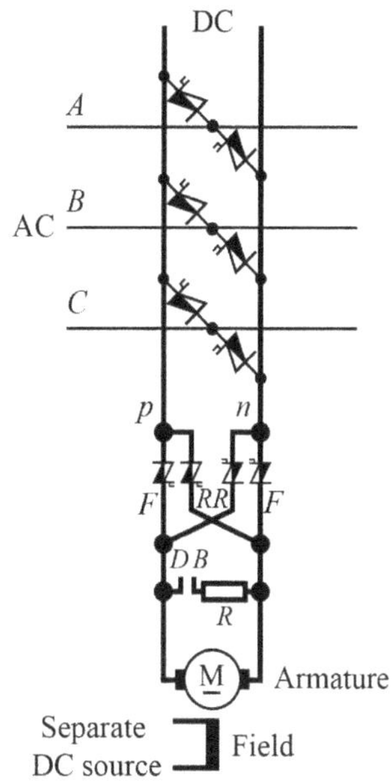

Figure 5.8. Non-regenerative DC motor IEMD with external dynamic braking and reversing IGCTs.

Dynamic braking (DB) requires the addition of a DB contactor and DB resistors that dissipate the braking energy as heat. The addition of an EM reversing contactor or IGCTs reverser or manual switch permits the reversing of the microcontroller polarity and therefore the sense of rotation direction of the EM motor armature. Field contactor reverse kits can also be installed to provide bilateral rotation by reversing the polarity of the shunt field.

All DC–AC commutator synchronous motors are AC–DC commutator synchronous generators as well. The term regenerative describes the ability of the IEMD under braking conditions to convert the generated energy of the EM motor into electrical energy, which is returned (or regenerated) to the AC power source.

A regenerative DC IEMD operates in all four quadrants purely electronically, without the use of EM switching contactors:

- ***Quadrant I***—an IEMD delivers forward torque, EM motor rotating forward (motoring mode of operation). This is the normal condition, providing power to a mechanical load similar to that of an EM motor starter.
- ***Quadrant II***—an IEMD delivers reverse torque, EM motor rotating forward (generating mode of operation). This is a regenerative condition, where the IEMD itself is absorbing power from a mechanical load, such as an overhauling load or deceleration.
- ***Quadrant III***—an IEMD delivers reverse torque, EM motor rotating reverse (motoring mode of operation). Basically the same as in quadrant I and similar to a reversing starter.
- ***Quadrant IV***—an IEMD delivers forward torque with EM motor rotating in reverse (generating mode of operation). This is the other regenerative

condition, where again, the IEMD is absorbing power from the mechanical load in order to bring the EM motor towards zero angular velocity.

A single-quadrant non-regenerative DC motor IEMD has one power ECM AC–DC commutator, acting as an AC–DC rectifier, with six controlled electrical valves, i.e. IGBTs or MOSFETs used to control the applied voltage level to the EM motor armature.

The non-regenerative IEMD can run in only motoring mode, and would require physically switching armature or field leads to reverse the sense of torque direction. A four-quadrant regenerative DC motor IEMD will have two complete sets of power ECM AC–DC commutators, with 12 controlled electrical valves, i.e. IGBTs, IGCTs or MOSFETs connected in inverse parallel as illustrated in figure 5.9.

One ECM AC–DC commutator controls forward torque, and the other controls reverse torque. During operation, only one set of ECM AC–DC commutators is active at a time.

For straight motoring in the forward direction, the forward ECM AC–DC commutator would be in control of the power to the EM motor. For straight motoring in the reverse direction, the reverse ECM AC–DC commutator is in control.

Cranes and hoists use a regenerative DC motor IEMD to hold back ‘overhauling loads’ such as a raised weight, or an electrical machine’s flywheel. Whenever the inertia of the EM motor load is greater than the EM motor rotor inertia, the mechanical load will be driving the EM motor and is termed an overhauling load. An overhauling load results in ME generator action within the EM motor, which will cause the ME generator to send current into the IEMD.

Regenerative braking is summarised as follows:

- During normal forward operation, the forward ECM AC–DC commutator acts as an AC–DC rectifier, supplying power to the EM motor. During this period the electrical valves’ gate pulses are withheld from reverse ECM AC–DC commutator so that it is inactive.

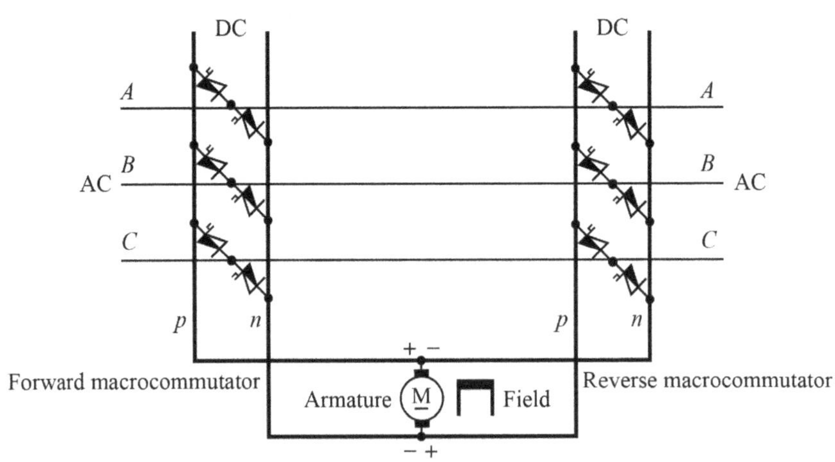

Figure 5.9. Four-quadrant regenerative DC motor IEMD.

- When EM motor angular velocity is reduced, the control circuit withholds the electrical valves' pulses to the forward ECM AC–DC commutator and simultaneously applies the electrical valves' pulses to reverse ECM AC–DC commutator.

During this period, the EM motor acts as a ME generator, and the reverse ECM AC–DC commutator conducts current through the armature in the reverse direction back to the AC line. This current reverses the torque, and the EM motor angular velocity decreases rapidly. Both regeneration and dynamic braking slow down a rotating DC–AC commutator synchronous motor (DC motor) with mechanical ring/disc commutator and its mechanical load. However, there are significant differences in stopping time and controllability during stopping, and safety issues depending on how one defines what should happen under emergency conditions. Regenerative braking will stop the mechanical load smoothly and faster than a dynamic brake for fast stop or emergency stop requirements. In addition, regenerative braking will regenerate power to the supply if the mechanical load is overhauling.

5.7 DC motor IEMD parameter programming

Programming parameters associated with DC are extensive and similar to those used in conjunction with a DC motor IEMD. An operator's panel is used for programming of control setup and operating parameters for a DC motor IEMD.

5.7.1 Angular velocity set-point

This signal is derived from a closely regulated fixed voltage source applied to a potentiometer. The potentiometer has the capability of accepting the fixed voltage and dividing it down to any value, for example, 10 to 0 V, depending on where it is set. A 10 V input to the IEMD from the angular velocity potentiometer corresponds to maximum EM motor angular velocity and 0 V corresponds to zero angular velocity. Similarly any angular velocity between zero and maximum can be obtained by adjusting the angular-velocity control to the appropriate setting.

5.7.2 Angular-velocity feedback information

In order to 'close the loop' and control EM motor angular velocity accurately, it is necessary to provide the control with a feedback signal related to EM motor angular velocity. The standard method of doing this in a simple control is by monitoring the armature voltage and feeding it back into the IEMD for comparison with the input set-point signal. The armature voltage feedback system is generally known as a voltage-regulated IEMD.

A second and more accurate method of obtaining the EM motor angular-velocity feedback information is from an EM motor-mounted tachometer. The output of this tachometer is directly related to the angular velocity of the EM motor.

When tachometer feedback is used, the IEMD is referred to as an angular-velocity-regulated IEMD. In some newer high-performance digital IEMDs, the feedback can come from an EM motor-mounted encoder that feeds back voltage pulses at a rate related to EM motor angular velocity. These pulses are counted and processed digitally and compared to the set-point, and an error signal is produced to regulate the armature voltage and speed.

5.7.3 Current feedback information

The second source of feedback information is obtained by monitoring the EM motor armature current. This is an accurate indication of the torque required by the mechanical load. The current feedback signal is used to eliminate the angular-velocity drop that normally would occur with increased torque load on the EM motor and to limit the current to a value that will protect the power electrical valves from damage. The current-limiting action of most controls is adjustable and is usually termed current limit or torque limit.

5.7.4 Minimum angular velocity

In most cases, when the microcontroller is initially installed the angular-velocity potentiometer can be turned down to its lowest point and the output voltage from the microcontroller will go to zero, causing the EM motor to stop. There are, however, situations where this is not desirable. For example, there are some applications that may need to be kept running at a minimum angular velocity and accelerated up to operating angular velocity as necessary. The typical minimum angular-velocity adjustment is from 0%–30% of EM motor base angular velocity.

5.7.5 Maximum angular velocity

The maximum angular-velocity adjustment sets the maximum angular velocity attainable. In some cases it is desirable to limit the EM motor angular velocity (and machine angular velocity) to something less than would be available at this maximum setting. The maximum adjustment allows this to be done.

5.7.6 *IR* compensation

Although a typical DC–AC commutator synchronous motor (DC motor) with the mechanical split-ring/flat commutator presents a mostly inductive load, there is always a small amount of fixed resistance in the armature circuit. IR compensation is a method used to adjust for the drop in an EM motor's angular velocity due to armature resistance. This helps stabilise the EM motor's angular velocity from a no-load to full-load condition. IR compensation should be applied only to voltage-regulated IEMD.

5.7.7 Acceleration time

As its name implies, the acceleration time adjustment will extend or shorten the amount of time for the EM motor to go from zero angular velocity up to the set angular velocity. It also regulates the time it takes to change angular velocities from one setting (e.g. 40%) to another setting (e.g. 80%).

5.7.8 Deceleration time

The deceleration time adjustment allows mechanical loads to be slowed down over an extended period of time.

For example, if power is removed from the EM motor and it stops in 3 s, then the deceleration time adjustment would allow the user to adjust this time typically within a 0.5–30 s range.

References

Barnes M 2003 *Practical Variable Speed Drives and Power Electronics, Automated Control Systems* (Perth, Australia: IDC Technologies)

Chan T-F and Shi K 2011 *Applied Intelligent Control of Induction Motor Drives* (New York: Wiley)

Geyer T 2017 *Model Predictive Control of High Power Converters and Industrial Drives* (New York: Wiley)

Holmes D G and Lipo T A 2003 *Pulse Width Modulation for Power Converters Principles and Practice* (Piscataway, NJ: Institute of Electrical and Electronics Engineers)

Hughes A and Drury W 2013 *Electric Motors and Drives Fundamentals, Types and Applications* 4th edn (Amsterdam: Elsevier)

Rashid M H 2018 *Power Electronics Handbook* 4th edn (Amsterdam: Elsevier)

Wach P 2011 *Dynamics and Control of Electrical Drives* (Berlin, Heidelberg: Springer)

Wilamowski B M and Irwin D J 2011 *The Industrial Electronics Handbook - Power Electronics and Motor Drives* (Boca Raton, FL: CRC Press)

IOP Publishing

The Integrated Electro-Mechanical Drive
A mechatronic approach
B T Fijalkowski and J Tutaj

Chapter 6

Integrated electro-mechanical drive (IEMD)

6.1 Introduction

A modern **integrated electro-mechanical drive** (IEMD) is expected to operate from a fixed voltage and fixed frequency grid (source) to feed the **electro-mechanical** (EM) motor with variable voltage and variable frequency to produce the desired angular velocity and torque for a given application.

In most applications, it is highly desirable to operate these EM motors in a wide range of angular-velocities and torques without loss of much energy in the electrical energy conversion unit (i.e. the ECM AC–AC or AC–DC–AC or DC–AC commutator).

The IEMD is expected to have minimum effect on the grid and EM motor connected to it, i.e. currents drawn from grid should be within specified limits and currents injecting in to electrical machine windings should not overheat the electrical machine windings to avoid insulation failure of the EM motor. It is also necessary that an IEMD should not disturb other loads connected to the **point of common coupling** (PCC). This IEMD should require minimal maintenance and service after the commissioning of the AC motor or DC motor IEMD cyber-physical heterogeneous dynamical hypersystem as it involves resources to allocate for this purpose (Barnes 2003, Wilamowski and Irwin 2011, Chan and Shi 2011, Wach 2011, Hughes and Drury 2013, Holmes and Lipo 2003, Geyer 2017, Rashid 2018).

The globally accepted and widely used prominent angular-velocity control technology—IEMD coupled with standard three-phase **alternating current** (AC) synchronous or asynchronous (induction) motors; replaced mostly solutions: mechanical, hydraulic as well as **direct current** (DC) motors.

The angular velocity of the rotating electrical field created by the synchronous or asynchronous (induction) motor stator windings is directly linked with the frequency of the supply applied to the stator windings. An electronically powered by the ECM AC–AC or AC–DC–AC or DC–AC commutator IEMD can produce variable

doi:10.1088/2053-2563/aae7d7ch6

© IOP Publishing Ltd 2019

frequency, variable voltage waveforms which can be utilised to control the EM motor angular velocity and torque at the shaft.

The adjustment of the EM motor angular velocity through the use of an IEMD can lead to better process control, less wear in the mechanical equipment, less acoustical noise, and importantly, significant energy savings. However, an IEMD can have some disadvantages such as **electromagnetic interference** (EMI) generation, current harmonics introduction into the supply and the possible reduction of efficiency and lifetime of EM motors. There are increasing demands for compactness and high-power density of **electrical commutation matrixer** (ECM) AC–AC, AC–DC–AC, DC–AC commutators used in industry. In industry and commercial applications, IEMDs are becoming very popular in textile drives, motive mining machinery, spindles of machine tools, hermetic pumps, and motion controls in the power range of fractional kW to 20 kW. Also, in current **hybrid-electric vehicles** (HEV) or **all-electric vehicles** (AEV) and more-electrical aircraft applications, the IEMD is an interesting technology for achieving compactness and reduced-mass design to meet more stringent requirements of on-board power and actuator control systems.

AC–AC, AC–DC–AC, DC–AC commutator synchronous or asynchronous (induction) motors with high torque density are required for many applications, including automotive and aerospace, and a high electric loading is often employed to increase the torque density.

For a particular application the designer can design the most suitable AC motor or DC motor IEMD cyber-physical heterogeneous dynamical hypersystem from all the possible choices.

Development of high-speed AC motor or DC motor IEMD cyber-physical heterogeneous dynamical hypersystems is needed for new emerging applications, such as generators/starters for micro gas turbines, turbo-compressor systems, drills for medical applications, and spindles for machining. Typically, the power ratings of these applications range from a few watts to kilowatts, and the rotational speeds from a few tens of thousands of rpm up to a million rpm. Recently, high-speed centrifugal turbo-compressors have been under intensive **research and development** (R&D). Since, compared to conventional compressors, high-speed centrifugal compressors have numerous qualities such as simple structure, light mass, small size, and high efficiency.

Integration of EM motor and of mechanical load has been already done and has resulted in energy efficient operation of overall IEMD cyber-physical heterogeneous dynamical hypersystem in HVAC compressors. An IEMD will be one step further and will improve overall performance of total IEMD cyber-physical heterogeneous dynamical hypersystem in these applications.

An IEMD offers many desirable features, such as high compactness, reduced material cost, reduced engineering time for installation or integration, lower IEMD cyber-physical heterogeneous dynamical hypersystem losses, more effective cooling arrangements and better protection against short circuiting and over-voltage due to *dv/dt* induced voltage reflection waves.

The major technological obstacle for the elevated temperature high-power density ECM AC–AC, AC–DC–AC, DC–AC commutator for an IEMD motor integrated is the unavailability of high temperature power components, such as high-power semiconductor electrical valves, large value DC-link capacitors, as well as control electronics devices and passive components.

Consequently, challenges presented to the designers for the next level of innovations include alternative ECM AC–AC, AC–DC–AC, DC–AC commutators, total optimisation of ECM topology and whole IEMD cyber-physical heterogeneous dynamical hypersystem (AC–AC, AC–DC–AC, DC–AC commutator synchronous or asynchronous motor), thermal management and cost-effective design.

6.2 Integrated electro-mechanical drive (IEMD) with sinusoidal pulse width modulation (SINPWM)

Consider the adjustable-velocity IEMD cyber-physical heterogeneous dynamical hypersystem shown in figure 6.1. Users distinguish between the (AC mains side) AC–DC commutator, acting as the AC–DC rectifier and the (EM motor windings side) DC–AC commutator, acting as the DC–AC inverter. The DC-link capacitor acts as a decoupling component.

The overall control task of the IEMD cyber-physical heterogeneous dynamical hypersystem is accordingly decomposed into the AC mains side microcontroller and the EM motor windings side microcontroller. The maintenance microcontroller maintains the DC-link voltage V_i at its reference value V_i^*using a cascaded control loop. The outer loop—the voltage microcontroller—regulates the DC-link voltage by adjusting the reference of the real power P^*. The reference for the reactive power Q^* is usually set to zero. The real and reactive power references are translated into the grid current reference I_g^{J*}, which is a **two-dimensional** (2-D) vector either in the stationary or in the rotating orthogonal coordinate system.

The inner loop, which is the current microcontroller, regulates the grid currents by manipulating the voltage applied by the AC–DC rectifier to the AC mains (grid).

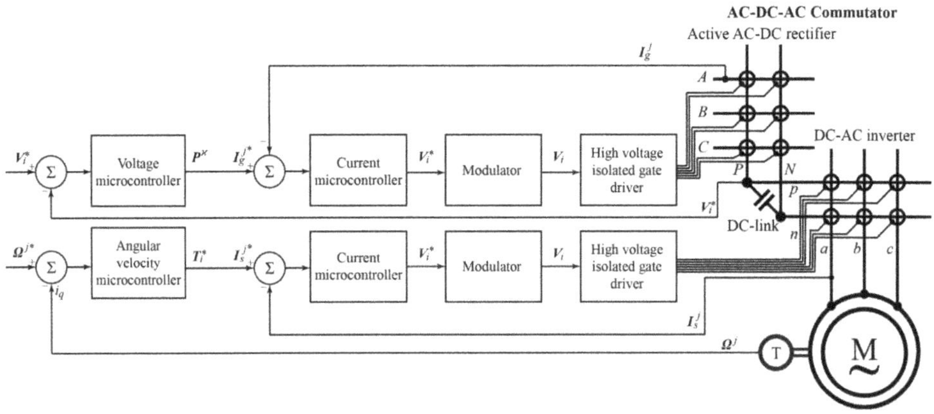

Figure 6.1. Structural and functional diagram of the AC–DC–AC commutator asynchronous (induction) motor IEMD cyber-physical heterogeneous dynamical hypersystem with cascaded control loops for the AC mains side and the EM motor windings side.

In most cases, the current microcontroller consists of two **proportional-integral** (PI) microcontrollers in a rotating orthogonal reference frame. The PI microcontrollers adjust the real-valued voltage reference V_i^*, which the PWM translates into the commutation (switching) signal V_i. Similarly, a cascaded control structure is used on the EM motor windings side. The outer angular-velocity control loop regulates the angular velocity of the rotor Ω along its reference Ω^*, by manipulating the reference of the electromagnetic torque T^*. Another outer control loop, which is omitted in figure 6.1, controls the rotor flux magnitude. The outputs of the outer control loops are translated into the stator current reference I_s^{j*}. The inner control loop regulates the stator current I_s^j along its reference, by manipulating the inverter and thus the stator voltage V_i^s. As on the AC mains side, the (inner) current microcontroller is typically based on two PI control loops with a subsequent PWM stage.

In summary, the AC mains side and the EM motor windings side are treated separately. Coupling between the two might be considered through a feedforward term, for example, a power feedforward term from the DC–AC inverter to the grid-side AC–DC rectifier. One or two outer control loops are used, which are **single-input single-output** (SISO) loops.

The inner current control loop is a **multiple-input multiple-output** (MIMO) control problem, which is often split into two orthogonal SISO loops. To mask the commutation characteristic of the AC–DC rectifier, a PWM is usually added to the inner control loop.

In this section, structural and functional diagrams of the AC–AC or AC–DC–AC or DC–AC commutator synchronous or asynchronous (induction) motor IEMD cyber-physical heterogeneous dynamical hypersystems with the **sinusoidal PWM** (SINPWM) control, commonly applied in industry EMD and traction propulsion are shown in figures 6.1–6.4.

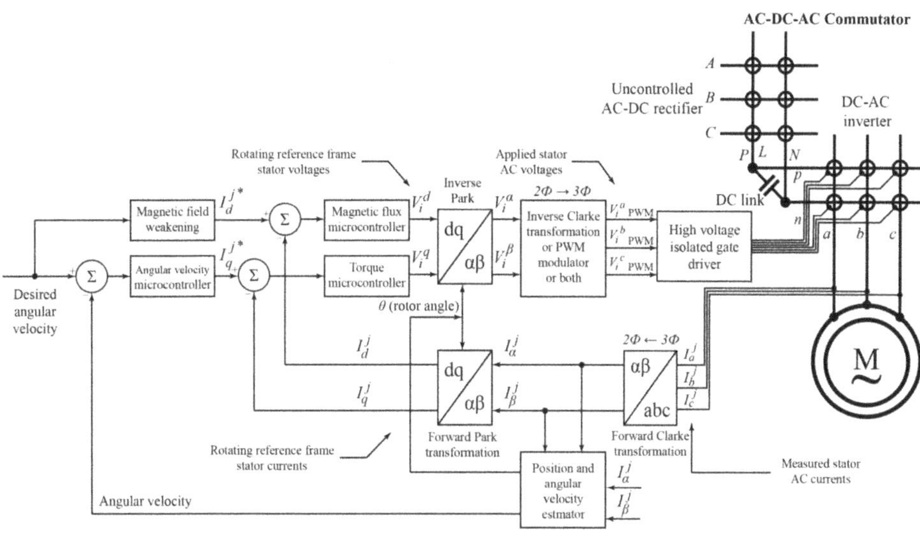

Figure 6.2. Structural and functional diagram of the AC–AC commutator asynchronous (induction) motor IEMD cyber-physical heterogeneous dynamical hypersystem with the sinusoidal PWN (SINPWM) control.

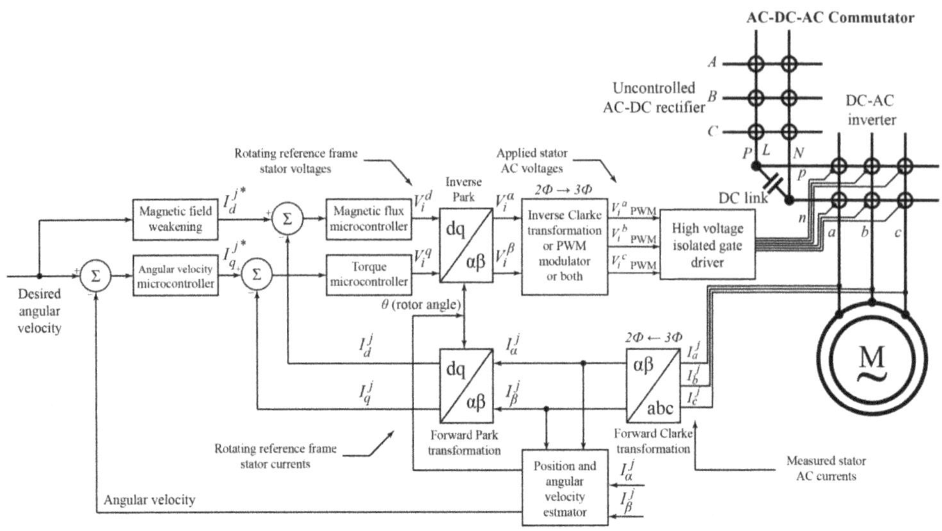

Figure 6.3. Structural and functional diagram of the AC–DC–AC commutator asynchronous (induction) motor IEMD cyber-physical heterogeneous dynamical hypersystem with the sinusoidal PWN (SINPWM) control.

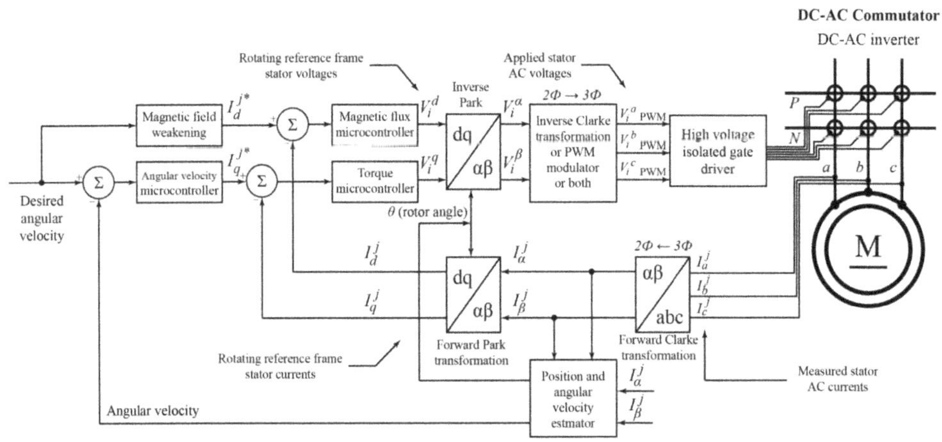

Figure 6.4. Structural and functional diagram of the DC–AC commutator asynchronous (induction) motor IEMD cyber-physical heterogeneous dynamical hypersystem with the sinusoidal PWN (SINPWM) control.

6.3 Integrated electro-mechanical drive (IEMD) with vector pulse width modulation (VECPWM)

6.3.1 Introduction

It is essential that an IEMD rotates in both senses of direction with adjustable-velocity of wide range. The AC–DC–AC or DC–AC commutator synchronous or asynchronous (induction) motor is applied since it has high-power density, and vector control technique can further improve the efficiency and reduce the noise. The total cost of the IEMD cyber-physical heterogeneous dynamical hypersystem can be reduced and reliability can be improved using sensorless technology. Here are some key requirements of an IEMD cyber-physical heterogeneous dynamical hypersystem:

- The AC–DC–AC or DC–AC commutator synchronous or asynchronous (induction) motor must be able to operate at four quadrants with adjustable velocity.
- Time of commutation from **counter clockwise** (CCW) maximal angular velocity to **clockwise** (CW) maximal angular velocity and vice versa must be as short as possible.
- High efficiency and low noise are desired. Since a fundamental part of the EM motor's **back electromotive force** (BEMF) is much more dominating as compared to harmonics, $I_d = 0$ vector control strategy can be applied which leads to high efficiency, and sinusoidal phase currents contribute to smooth torque, and hence, low noise.
- IEMD can be started even if the EM motor's rotor is still rotating freely due to inertia.
- IEMD cannot stop but must be able to operate properly when AC input voltage is down for a couple of seconds.
- An IEMD cyber-physical dynamical system must have protection against **over-voltage** (OV), **under-voltage** (UV), and **over-current** (OC). The IEMD cyber-physical heterogeneous dynamical hypersystem must have the 'lock of rotor detection' feature which means that the system has a mechanism coping with the situation where motor's rotor is locked for any reason.

6.3.2 IEMD—sensorless sinusoidal vector control

The following section demonstrates the implementation of an IEMD cyber-physical heterogeneous dynamical hypersystem with sensorless vector control.

Hardware block—An EMC AC–DC–AC or DC–AC commutator, acting as a DC–AC inverter is used to drive the AC–DC–AC or DC–AC commutator synchronous or asynchronous (induction) motor. The main power for the whole system comes from the outlet of 380/220 V 50/60 Hz, where a voltage doubler may be necessary for 230 V input. This AC input voltage is rectified through an ECM AC–DC commutator, acting as an AC–DC rectifier, and DC-link voltage of the ECM DC–AC commutator, acting as the DC–AC inverter comes from the output of this ECM AC–DC commutator, acting as an AC–DC rectifier which is in the range of 280—320 V. A voltage of 15 V is derived from the DC-link through a DC–DC chopper, which is used for the DC–AC inverter's gate drivers, and a voltage of 3.3 V is derived from this 15 V through a regulator, which is used for MC56F8006 and analog amplifier circuits.

Figure 6.5 shows the hardware block of the IEMD cyber-physical heterogeneous dynamical hypersystem with the **vector pulse width modulation** (VECPWM). Three phase currents must be known to perform vector control. Shunt resistors on the ECM DC–AC commutator (acting as a DC–AC inverter) lower ECM columns and related operational amplifiers are used to get phase currents. Besides, a DC-link voltage is also essential for EM motor control; it is sampled using only divider

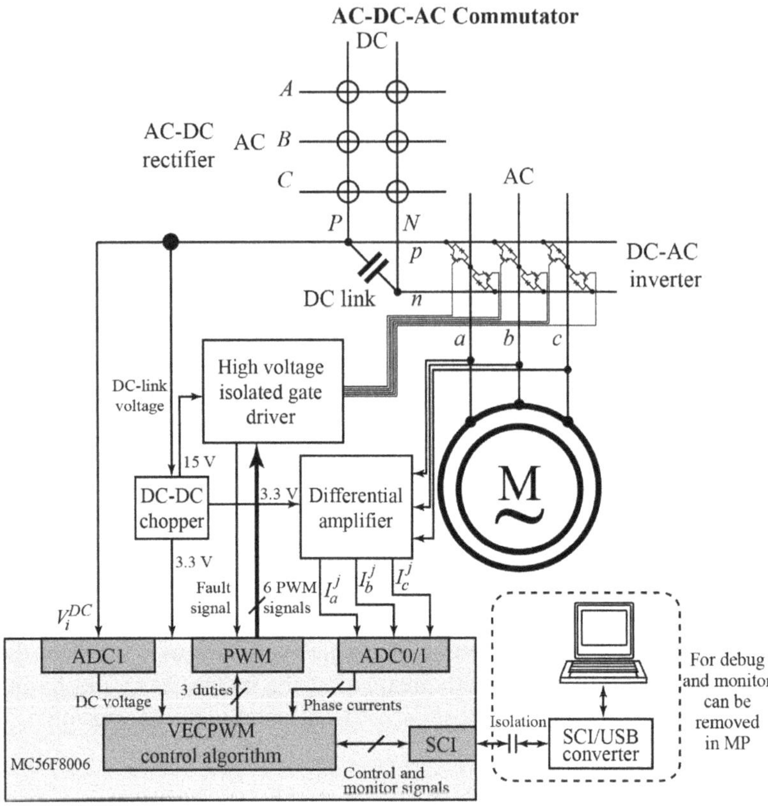

Figure 6.5. Hardware block of the IEMD cyber-physical heterogeneous dynamical hypersystem with the VECPWM control.

resistors. The whole IEMD cyber-physical heterogeneous dynamical hypersystem can be controlled and monitored using FreeMASTER through SCI.

EM motor control strategy

The whole control strategy is depicted in figure 6.5 and can be explained as follows. Three phase currents and DC-link voltage are sampled to realise VECPWM control, or **magnetic-field-oriented control** (MFOC). Three phase currents are converted into direct and quadrature currents through Clarke and Park transformations.

- Two inner current loops calculate the direct and quadrature stator voltages required to create the desired magnetic-flux and torque currents. Direct and quadrature voltages are transformed into voltages in $\alpha - \beta$ stationary coordinate, where these voltages must be compensated per DC-link voltage.
- Three PWM duties are generated utilising compensated $\alpha - \beta$ voltages and a VECPWM module.
- The required direct current is set to 0 since the **interior permanent magnet** (IPM) of the rotor provides magnetic-flux.

- The outer angular-velocity loop adjusts the reference quadrature current which is proportional to the torque. The required angular velocity can be set through a FreeMASTER interface.

FreeMASTER is a user-friendly real-time debug monitor and data visualisation tool that users can use for application development and information management. It supports non-intrusive monitoring of variables on a running system.

Users can display multiple variables changing over time on an oscilloscope-like display, or view the data in text form. FreeMASTER also supports additional capabilities and targets with an on-target driver for transmitting data from the target to the host computer.

Almost all the blocks in figure 6.5 can be found, for example, in **Freescale Embedded Software Libraries** (FSLESL) including all the transformations, PI microcontroller, Observer, VECPWM, DC-link ripple elimination, etc.

Sensorless control—It is essential to know the angle between the *d*-axis of the rotor and α-axis of the stator winding to realise the transformation between two-phase stationary coordinate and two-phase rotating coordinate in every current loop. EM motor angular-velocity information is also necessary to realise angular-velocity loop. Since there is no sensor on the EM motor due to cost problems, the position and angular velocity of the rotor are estimated through an observer. The observer is based on the EM motor mathematical model, thus EM motor parameters must be available. It is a Luenberger state observer which estimates **back electromotive force** (BEMF) in a quasi-synchronous rotating coordinate γ–δ using angle of $\theta_{\gamma\delta}$ (angle between the γ-axis and the α-axis) for Park transformation. The BEMF in coordinate γ–δ reflects the angle error (θ_{error}) between the γ-axis and the *d*-axis of the rotor. Figure 6.6 shows the two coordinates γ–δ and *d*–*q*.

A Luenberger state observer is used to get BEMF information in coordinate γ–δ as shown in figure 6.7.

Then, a tracking observer is used to make θ_{error} approach to 0, so that $\theta_{\gamma\delta}$ will be the real rotor position in the end. Figure 6.8 shows the whole block of position and speed estimation.

Start-up—Since the observer actually utilises BEMF to get the estimated rotor position and angular velocity, there will be no reliable position or angular-velocity

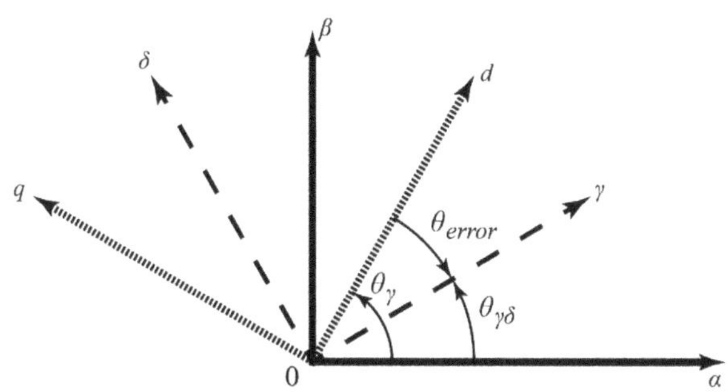

Figure 6.6. Quasi-synchronous rotating coordinate γ–δ and real synchronous rotating coordinate *d*–*q*.

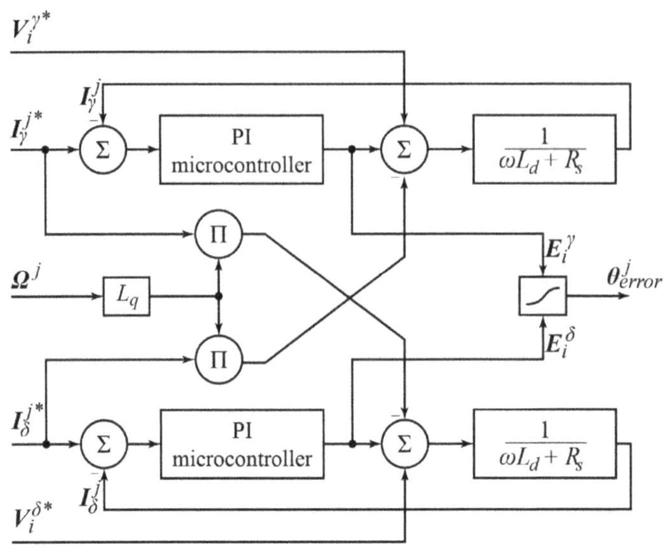

Figure 6.7. State observer for position error.

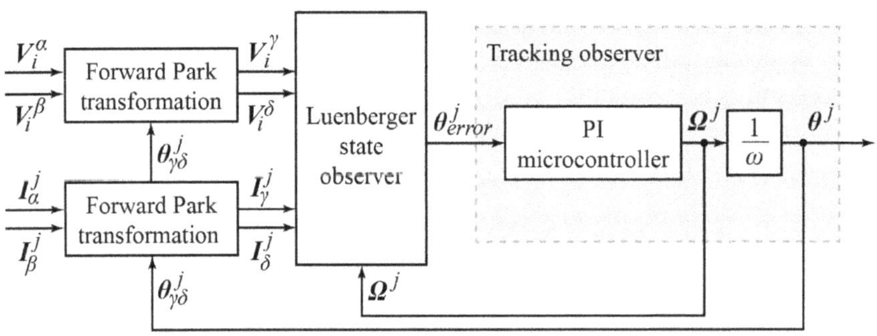

Figure 6.8. Angular velocity and position estimation.

information when the motor is stopped or angular velocity is too low because the BEMF is not large enough to be accurately observed. In this case, the EM motor is started by generating a slowly rotating stator current vector so that the rotor is pulled to synchronise with this rotating stator current vector; this phase is termed the angular-velocity open loop phase. During open loop running, the observer operates in parallel. When BEMF is large enough to be accurately observed, the estimated rotor position is used for forward Park and inverse Park transformations and the estimated angular velocity is used for angular-velocity loop control; this phase is termed the angular-velocity closed loop phase. In both the phases, stator currents are controlled to be sinusoidal, so that smooth electromagnetic torque is generated, which minimises acoustic noise and mechanical vibration.

Initial rotor position detection—The start torque could be very small or even not in the desired sense of direction if the initial position of the rotor is not known. In this case, there might be little vibration during start-up. If the initial position of the rotor is known, maximal start torque can be achieved, the time duration of the angular-velocity open loop phase can be shortened and the EM motor can get into the closed loop very soon. A popular method is used to estimate the initial rotor position at standstill, utilising the saturation effect of the stator iron core due to the permanent magnet. A stator inductance varies with the rotor position due to the saturation caused by the rotor magnetic field, as does the rate of change of the current in stator winding when a constant voltage is applied to the windings.

Six basic voltage vectors with the same length and lasting time which are used in the VECPWM module are applied to the windings one by one, so that corresponding six current pulses are generated and rotor position of $\pi/6$ electrical radian resolution can be estimated according to the differences between the maximal values of these current pulses.

An appropriate current vector can be generated according to this estimated initial rotor position to achieve the maximal start-up torque. In this application, the time duration of open loop lasts about five electrical rotating periods and reaches 30 rpm (four magnetic-pole pairs) before entering the angular-velocity closed loop control.

Strategy of start-up when IEMD rotates freely—The IEMD may rotate freely when it is to be started, under the following circumstances:

- the IEMD is installed in an open environment, and it is rotating due to the breeze;
- separate to the IEMD, and as a rule, another IEMD is installed in the same hall and is started ahead of the others, the IEMDs started later may rotate due to the airflow generated by the one started first;
- the IEMD is rotating and then turned *OFF* manually. It will continue to rotate due to inertia.

Each time an IEMD is started, its rotating status must be checked to know whether it is rotating freely or not. Initial rotor position detection and start-up will fail if it is rotating. A simple way to cope with this is: turn *ON* the three lower electrical valves of the ECM DC–AC commutator, acting as the DC–AC inverter (turn *OFF* the upper electrical valves) and sample the phase currents before starting the IEMD. Forward Clarke transformation is applied to get the $\alpha - \beta$ currents and then the length of the current vector can be calculated. The current vector length can be used to tell whether the IEMD is at a standstill or not; the IEMD is rotating if the current length is larger than a certain value and vice versa. Keep turning *ON* the lower three electrical valves of the ECM DC–AC commutator, acting as the DC–AC inverter if the IEMD is freely rotating, so that the EM motor operates in M-E generator mode and the mechanical energy will turn into heat consumed by stator winding resistance. The generated current vector length will decrease gradually until it gets smaller than a certain value, and the IEMD is assumed to be stopped at this point. Outer force is assumed to be big if the current vector length cannot decrease to a certain value in a certain period of time, like 20 s. The IEMD cannot be started in this case, and an error flag will be set.

The structural and functional diagram IEMD cyber-physical heterogeneous dynamical hypersystem with the VECPWM for different, integrated AC–AC,

AC–DC–AC or DC–AC commutator asynchronous (induction) motors using direct vector control method is shown in figures 6.9–6.11.

For the operation of the integrated AC–AC, AC–DC–AC or DC–AC commutator asynchronous (induction) motor in constant torque region requires magnetic-flux command to be maintained at its rated value.

The operation in a constant power region is referred to as magnetic-field weakening mode where the magnetic-flux command is reduced. The desired angular velocity and actual angular velocity measured from the EM motor windings are compared and the angular-velocity loop generates I_q^{j*}.

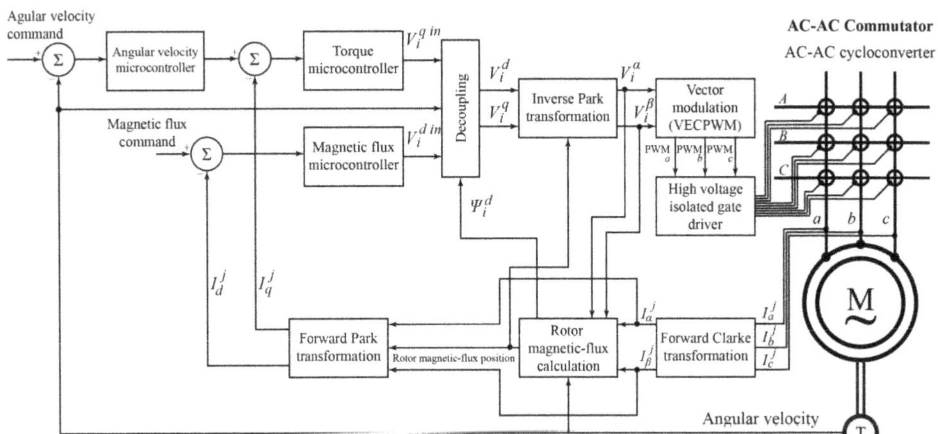

Figure 6.9. Structural and functional diagram of the IEMD cyber-physical heterogeneous dynamical hypersystem with the VECPWM control for an integrated AC–AC commutator asynchronous (induction) motor.

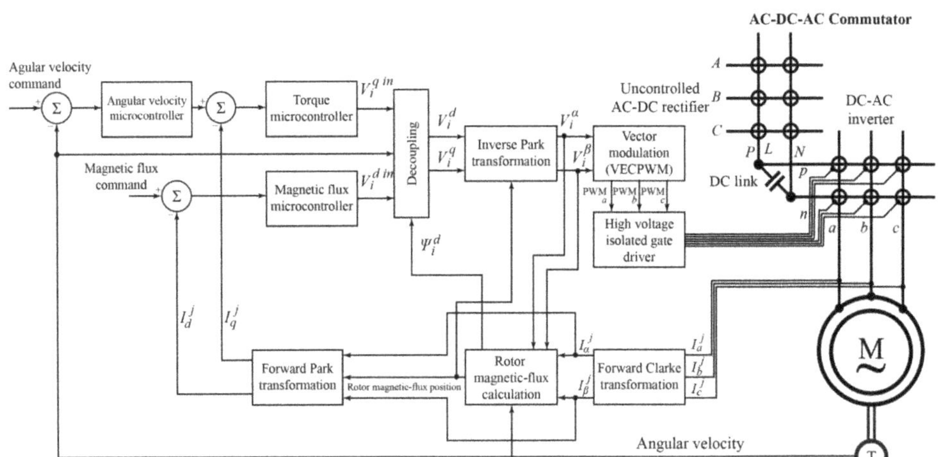

Figure 6.10. Structural and functional diagram of the IEMD cyber-physical heterogeneous dynamical hypersystem with the VECPWM control for an integrated AC–DC–AC commutator asynchronous (induction) motor.

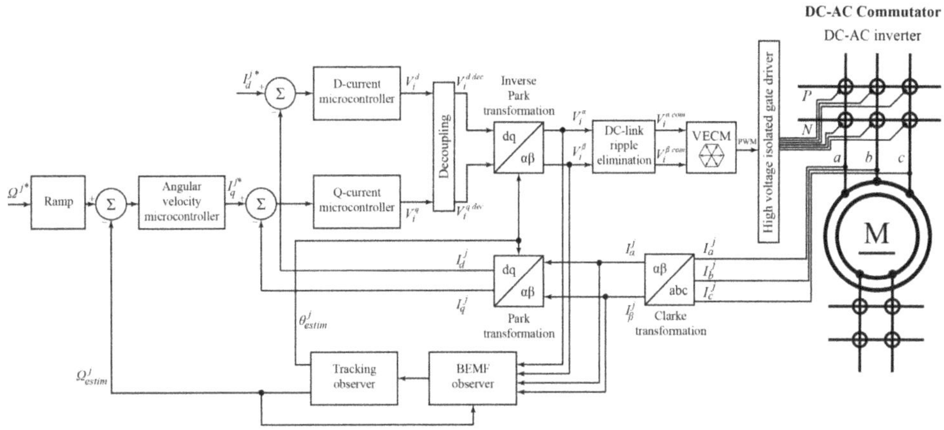

Figure 6.11. Structural and functional diagram of the IEMD cyber-physical heterogeneous dynamical hypersystem with the VECPWM control for an integrated DC–AC commutator asynchronous (induction) motor.

The desired magnetic-flux and actual magnetic-flux estimated from voltage model based magnetic-flux vector estimator are compared in the current loop to generate I_d^{j*}. Inverse transformations are then carried out to produce the required values of stator phase currents.

In every real IEMD cyber-physical heterogeneous dynamical hypersystem there is a limit beyond which microcontroller cannot act on the growth of its dynamics. When driving EM motors, the maximum value must be limited voltage and currents. If a saturation-type non-linearity is inserted behind the microcontroller output, it limits the output of the microcontroller, but if the microcontroller used comprises an integration component or a summing component in the form of discrete microprocessors, the value of these parts will continue to increase in the case of saturation. This fact is very unfavourable to the quality of the transition, because when changed the value of the set-point must be 'unintegrated' on the integrator. The solution is offered using the 'anti-windup' algorithm.

The algorithm designed in this textbook solves the given problem even though users do not work with the real values of voltage and current, but their vector components in *d–q* coordinates. Each of the four microcontrollers acts with a single voltage component (PI) and current component (PID).

Magnetic-flux microprocessors and V_i^dcomponents use a standard 'anti-windup' algorithm. If the output from the microcontroller is higher than after the saturation, it is this difference that is read by the time constant from the integrator.

In this subsection, structural and functional diagrams of the AC–AC or AC–DC–AC or DC–AC commutator synchronous or asynchronous (induction) motor IEMD cyber-physical heterogeneous dynamical hypersystems with the VECPWM control, commonly applied in industry EMD and traction propulsion are shown in figures 6.9–6.17.

Angular velocity microcontrollers and V_i^q folders use an algorithm of the same principle, but with the appearance modification vector components of voltage and

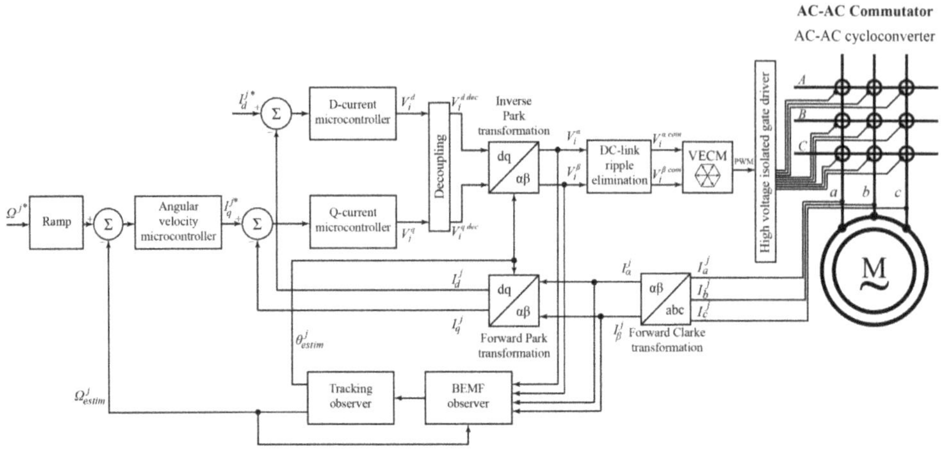

Figure 6.12. Structural and functional diagram of the IEMD cyber-physical heterogeneous dynamical hypersystem with the VECPWM control for an integrated AC–AC commutator asynchronous (induction) motor.

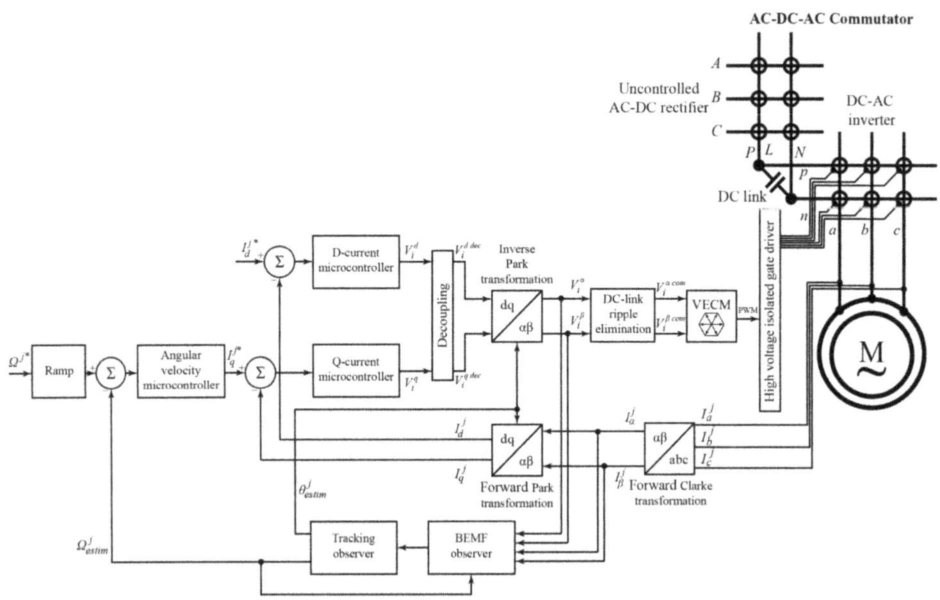

Figure 6.13. Structural and functional diagram of the IEMD cyber-physical heterogeneous dynamical hypersystem with the VECPWM control for an integrated AC–DC–AC commutator asynchronous (induction) motor.

current. To find the actual current value voltage, users have to calculate the absolute value of the vector components, e.g. for the voltage:

$$\overline{V}_i^r = \sqrt{(V_i^d)^2 + (V_i^q)^2};$$

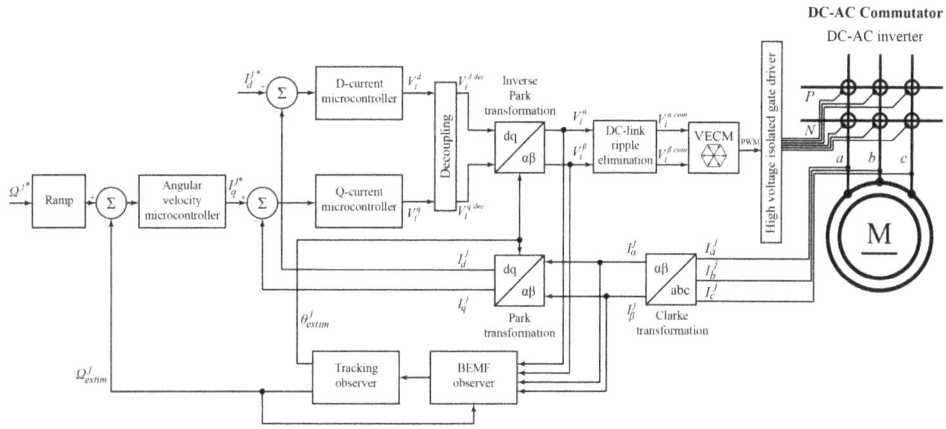

Figure 6.14. Structural and functional diagram of the IEMD cyber-physical heterogeneous dynamical hypersystem with the VECPWM control for an integrated DC–AC commutator asynchronous (induction) motor.

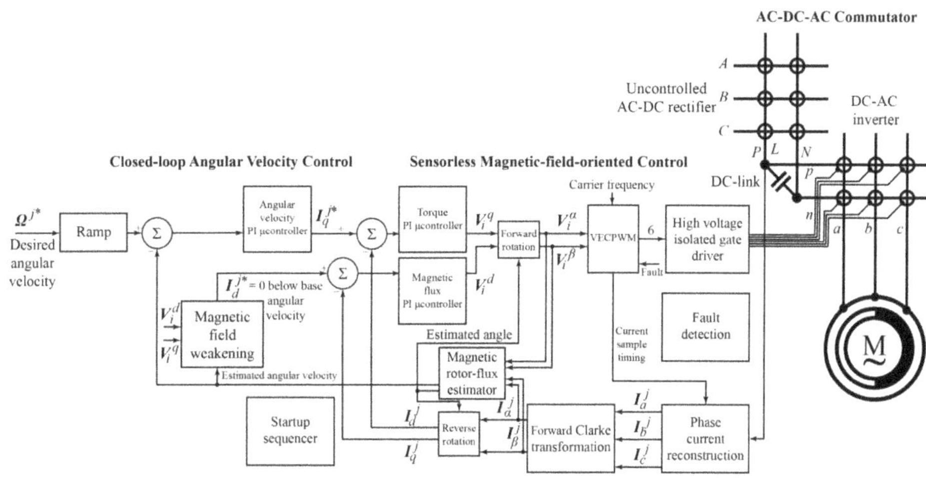

Figure 6.15. Structural and functional diagram of the IEMD cyber-physical heterogeneous dynamical hypersystem with the VECPWM control for an integrated AC–DC–AC commutator IPM synchronous motor.

For this reason, the angular-velocity microcontroller and V_i^q microcontroller are included component instead of a standard saturation modified version which, from knowledge of the maximum allowed current, voltage and knowledge of the preferred vector component I_d^j and V_i^d calculates the maximum allowed value of the remaining components q and V_i^d. Cross-linked components are added to the controlled voltage using the 'decoupling' block. This modification enables efficient use of voltage and current constraints together with the 'anti-windup' algorithm in vector control.

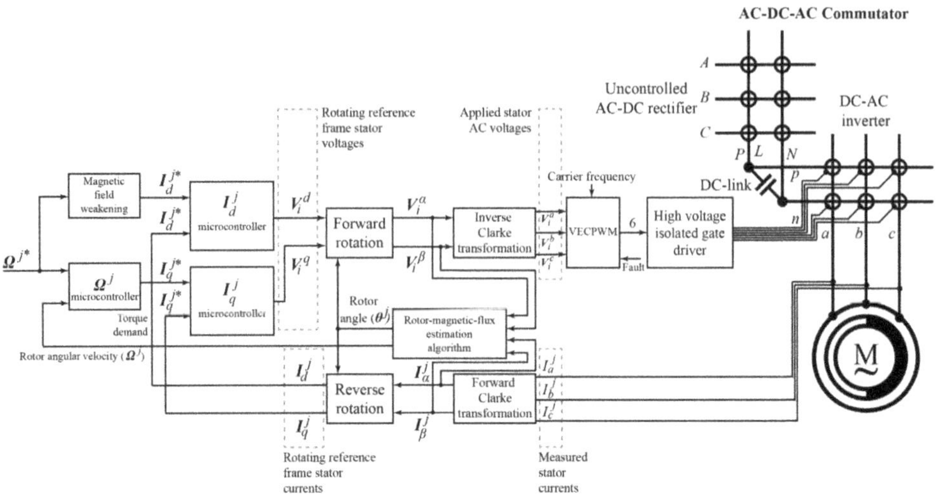

Figure 6.16. Structural and functional diagram of the IEMD cyber-physical heterogeneous dynamical hypersystem with the VECPWM control for an integrated AC–DC–AC commutator IPM synchronous motor.

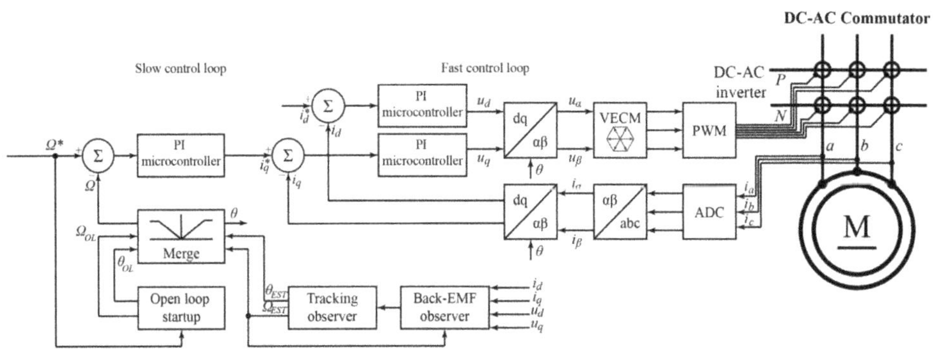

Figure 6.17. Structural and functional diagram of the IEMD cyber-physical heterogeneous dynamical hypersystem with the VECPWM control for an integrated DC–AC commutator asynchronous (induction) motor.

Figures 6.9–6.14 show the basic structure of the vector control of the AC–AC or AC–DC–AC or DC–AC commutator asynchronous (induction) motor.

To perform vector control, follow these steps:

- Measure the EM motor quantities (phase voltages and currents).
- Transform them to the two-phase system (α–β) using a forward Clarke transformation.
- Calculate the rotor magnetic-flux space vector magnitude and position angle.
- Transform stator currents to the d–q coordinate system using a forward Park transformation.
- The stator current torque- (I_q^j) and flux- (I_d^j) producing components are separately controlled.

- The output stator voltage space vector is calculated using the decoupling block.
- An inverse Park transformation transforms the stator voltage space vector back from the *d–q* coordinate system to the two-phase system fixed with the stator.
- Using the VECPWM, the output three-phase voltage is generated.

6.4 Integrated electro-mechanical drive (IEMD) with the holor pulse width modulation (HOLPWM)

In this textbook on DC and AC motor IEMD cyber-physical heterogeneous dynamical hypersystems with the **holor pulse width modulation** (HOLPWM), the authors broke down the major dynamical systems, hyposystems and components in a 'generic' **adjustable-velocity** (AV) IEMD and discussed some of the requirements in those hypersystems (figure 6.18).

Next, let us look at some of the variations in a real-world IEMD in terms of architectures and topologies. As noted earlier, these are the essential components comprising an AV IEMD's power conversion section:

- Input–output voltage ratings.
- DC-link topology.
- Power semiconductor electrical valves (electronic switches) found in the DC–AC inverter subsection.
- DC–AC inverter subsection technology.

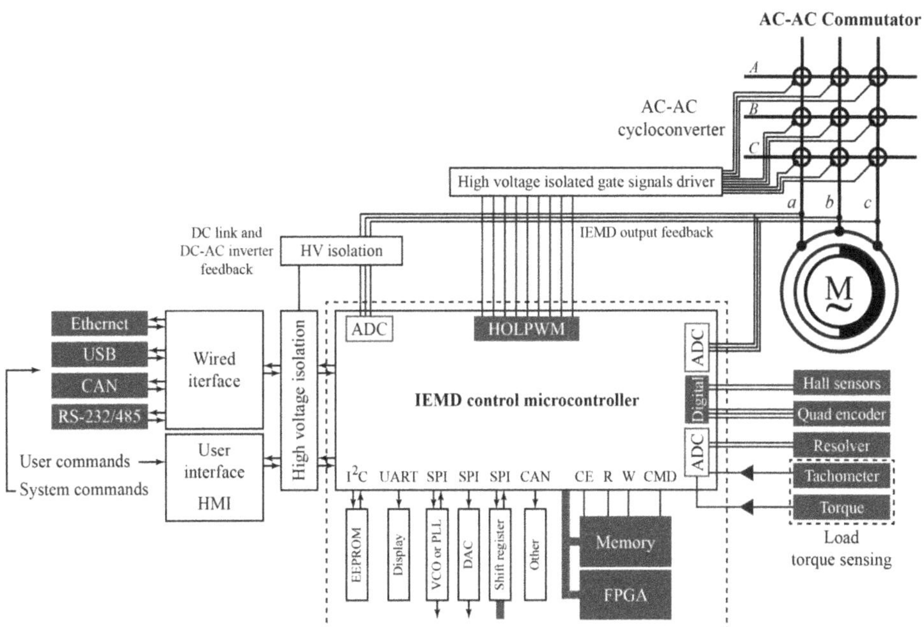

Figure 6.18. Structural and functional diagram of the AC–AC commutator IPM synchronous motor IEMD cyber-physical heterogeneous dynamical hypersystem with the holor PWM (HOLPWM) control.

Meanwhile, in the IEMD control and software section, one finds:

- High-voltage isolation gate-drive PWM techniques.
- DC and AC motor IEMD control architecture and algorithms.

The input to the IEMD can be either single- or three-phase AC, or DC (storage battery). The output is nearly always three-phase, even with a single-phase input. The function of the DC link is to store energy for input to the DC–AC inverter subsection. The DC link must store enough energy so that the bus is *'stiff'* and does not allow appreciable voltage sag when under load.

Ripple on the DC link could mean an inadequate mains supply, poor DC–AC inverter design, or issues with DC–AC inverter operation. It is common to monitor the DC-link ripple for correlation with other IEMD behaviours.

There are three main topologies for the DC link:

- Voltage-sourced (VS) DC–AC inverter.
- Current-sourced (CS) DC–AC inverter.
- Load-commutated (LC) DC–AC inverter.

There are the following difference between CS DC–AC inverter and VS DC–AC inverter:

- The input voltage in the VS DC–AC inverter is constant whereas the input current may be constant or variable. The input current in the DC–AC inverter is constant.
- The input supply source may be short circuited due to misfiring of the commutation electrical valve in the VS DC–AC inverter whereas there is no difficulty due to misfiring of electrical valves to the input supply current constant in the CS DC–AC inverter.
- The maximum current passes through the electrical valve in the VS DC–AC inverter depending upon the ECM condition whereas the input current on the CS DC–AC inverter is constant therefore the maximum current which passes through the electrical valve is limited.
- The ECM of the electrical valves on the CS DC–AC inverter is simpler than that of the VS DC–AC inverter.
- There is no need for a freewheeling diode to be is used in the VS DC–AC inverter for reactive power or regenerative load.
- The large inductor is connected at the input side in order to keep a constant current at the input side of the CS DC–AC inverter.
- There is a need for a controlled AC–DC rectifier in order to input voltage control in the VS DC–AC inverter, similarly a controlled AC–DC rectifier is necessary to control input current in the CS DC–AC inverter.

In a VS DC–AC inverter, the DC link stores energy as voltage in a capacitor in parallel with the DC–AC inverter subsection. A series inductor provides harmonics filtering. The VS DC–AC inverter topology is the most common in IEMD DC links, and provide faster peak current delivery for dynamic control of EM motor torque. CS DC–AC inverters store current in an inductor. Response times of a CS DC–AC inverter are slower than that of a VS DC–AC inverter owing to a delay in current delivery from the inductor to the DC–AC inverter subsection's input.

An LC DC–AC inverter uses thyristors and cannot use PWM signalling for the output voltage. This has fallen largely out of use with the advent of IGBTs of higher power and voltage. The AC–AC cycloconverter is a fourth type that dispenses with the DC link and uses a direct connection between the AC-rectified output and the inverter input. This topology is used in high-voltage power line commutation (switching) or frequency conversion. EM motor windings DC–AC inverter subsections use various kinds of power semiconductor electrical valves, generally either MOSFETs, IGBTs or IGCTs. MOSFETs are typically found in 240 V_{AC} and a lower input IEMD while IGBTs are used in an input IEMD of 380–600 V_{AC}.

Electrical valves using wide-bandgap materials such as **silicon carbide** (SiC) or **gallium nitride** (GaN) have faster rise times and are seeing more use in an IEMD for efficiency's sake. Analysis of commutation (switching) and conduction losses of these electrical valves calls for higher-bandwidth voltage and current measurements. When measuring IEMD outputs, lower bandwidth is usually sufficient, as the highest speed wide-bandgap materials are not typically driven at their highest speed due to reliability and/or **electromagnetic interference/radio-frequency interference** (EMI/RFI) emissions concerns.

We show in figures 6.18–6.25 AC and DC motor IEMD cyber-physical heterogeneous dynamical hypersystems with the HOLPWM for different, integrated AC–

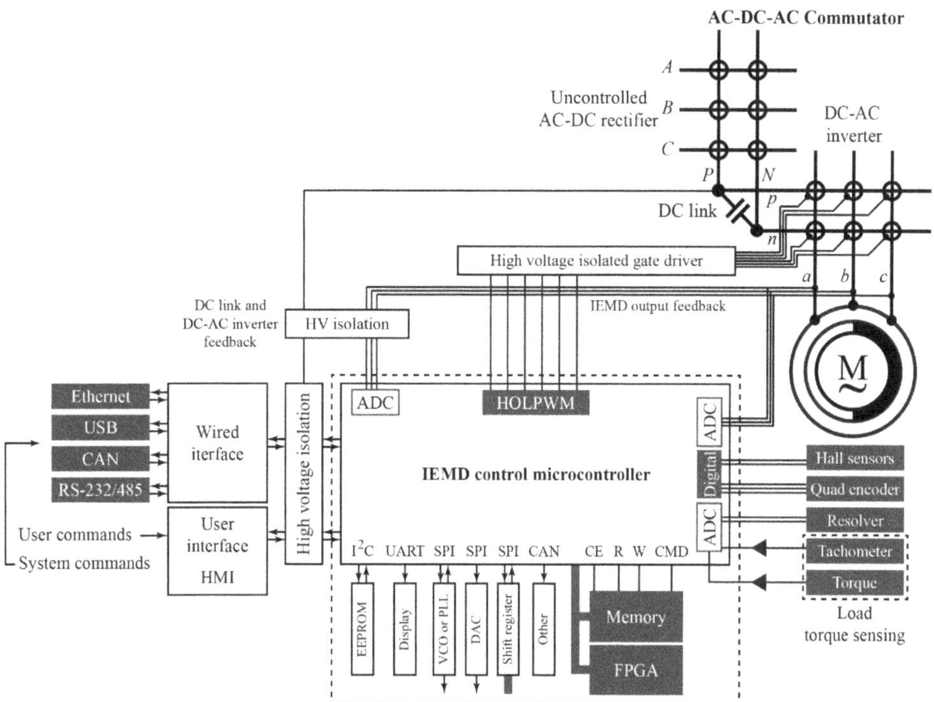

Figure 6.19. Structural and functional diagram of the AC–DC–AC commutator IPM synchronous motor IEMD cyber-physical heterogeneous dynamical hypersystem with the holor PWM (HOLPWM) control.

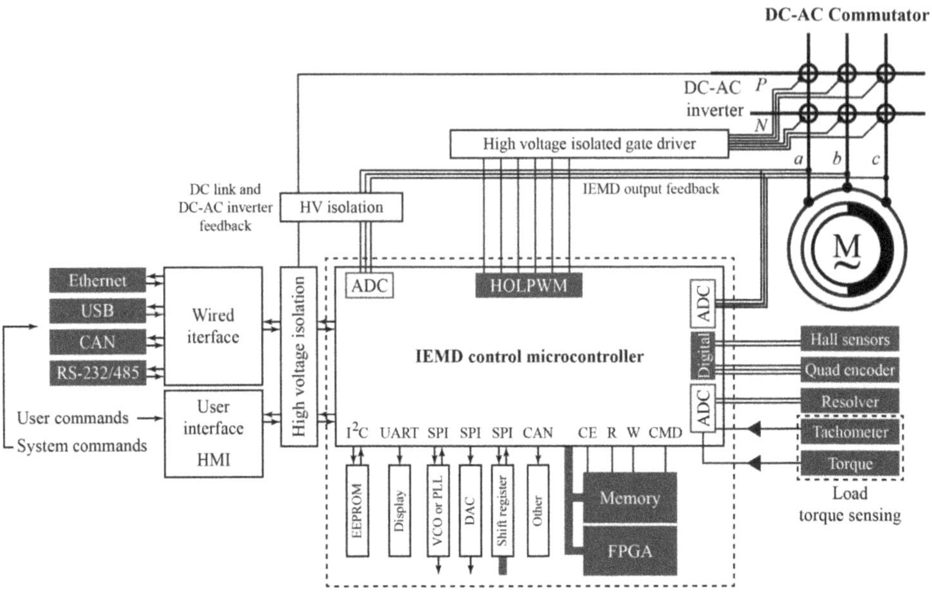

Figure 6.20. Structural and functional diagram of the DC–AC commutator IPM synchronous motor IEMD cyber-physical heterogeneous dynamical hypersystem with the holor PWM (HOLPWM) control.

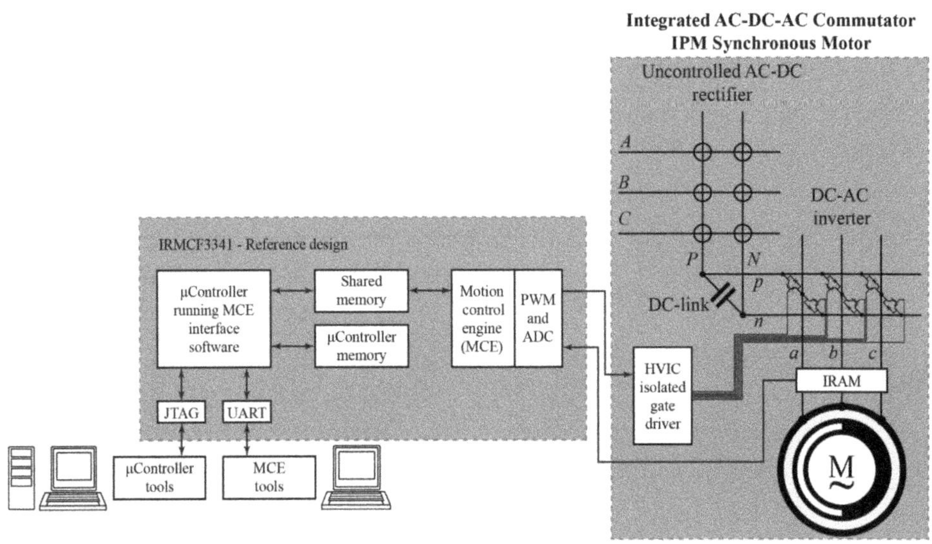

Figure 6.21. Structural and functional diagram of the AC–DC–AC commutator IPM synchronous motor IEMD cyber-physical heterogeneous dynamical hypersystem with the holor PWM (HOLPWM) control.

AC, AC–DC–AC or DC–AC commutator asynchronous (induction) motors using IEMD control microcontroller with the **field programmable gate array** (FPGA) with an AC–AC, AC–DC–AC or DC–AC commutator, acting as a AC–AC cyclo-converter, AC–DC–AC frequency changer or DC–AC inverter, respectively. The DC–AC inverter using high-frequency electrical valves, i.e. MOSFETs may achieve

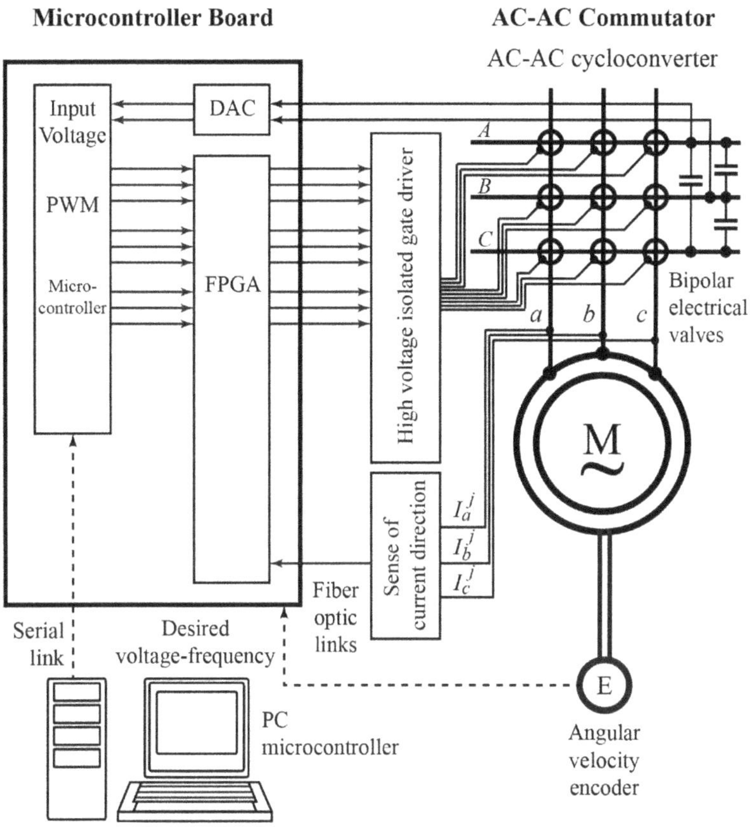

Figure 6.22. Structural and functional diagram of the AC–AC commutator IPM asynchronous (induction) motor IEMD cyber-physical heterogeneous dynamical hypersystem with the holor PWM (HOLPWM) control.

high-frequency operation up to 100 kHz with a commutation HOLPWM technique. High response IEMD cyber-physical heterogeneous dynamical hypersystem is achieved by using MOSFETs so that the EM motor can recover to a steady state value as soon as possible. It requires only 0.02 s to reach the steady state value. Because the processing speed gets faster, software control principally has a calculation time limitation. FPGA is used for angular velocity and torque control because high-speed calculation is obtained using the ability of the hardware processing (Hanamoto *et al* 2008).

The IEMD DC–AC inverter subsection topologies break down like this: a single-phase IEMD uses ECM topologies, while a three-phase IEMD also uses cascaded ECM topologies. Both result in two-level output signals.

Other topologies produce more than two levels of output. The IEMD in the 600 V class and lower EM motor applications are using two-level, cascaded ECM topologies.

In this subsection, structural and functional diagrams of the AC–AC or AC–DC–AC or DC–AC commutator synchronous or asynchronous (induction) motor

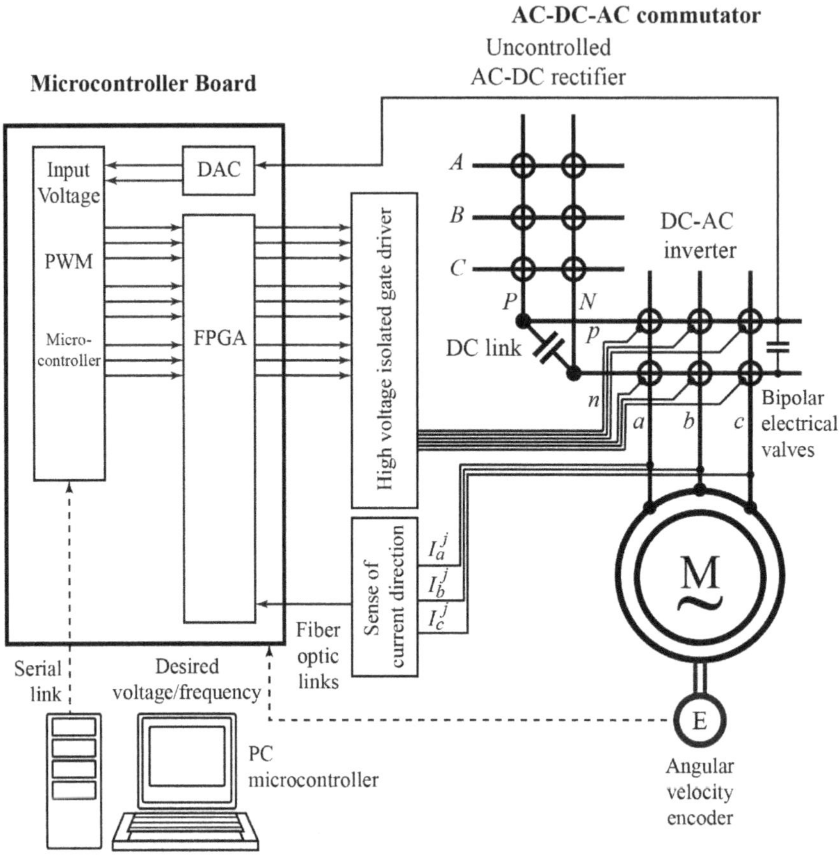

Figure 6.23. Structural and functional diagram of the AC–DC–AC commutator IPM asynchronous (induction) motor IEMD cyber-physical heterogeneous dynamical hypersystem with the holor PWM (HOLPWM) control.

IEMD cyber-physical heterogeneous dynamical hypersystems with the HOLPWM control, commonly applied in industry EMD and traction propulsion are shown in figures 6.18–6.25.

6.5 Conclusion

The most efficient method of controlling output voltage is to incorporate **pulse width modulation** (PWM) control within ECM DC–AC commutators, acting as DC–AC inverters. In this method, a fixed DC voltage is supplied to the ECM DC–AC commutator, acting as a DC–AC inverter and a controlled AC output voltage is obtained by adjusting the *ON–OFF* period of DC–AC inverter electrical valves. A PWM **adjustable-velocity** (AV) IEMD is increasingly applied in many new industrial applications that require superior performance. ECM commutators are used in industrial applications to convert and deliver their required energy to the EM motor windings or load because of advances in solid-state electrical valves and micro-processors. There are so many techniques which are used for controlling AC or DC

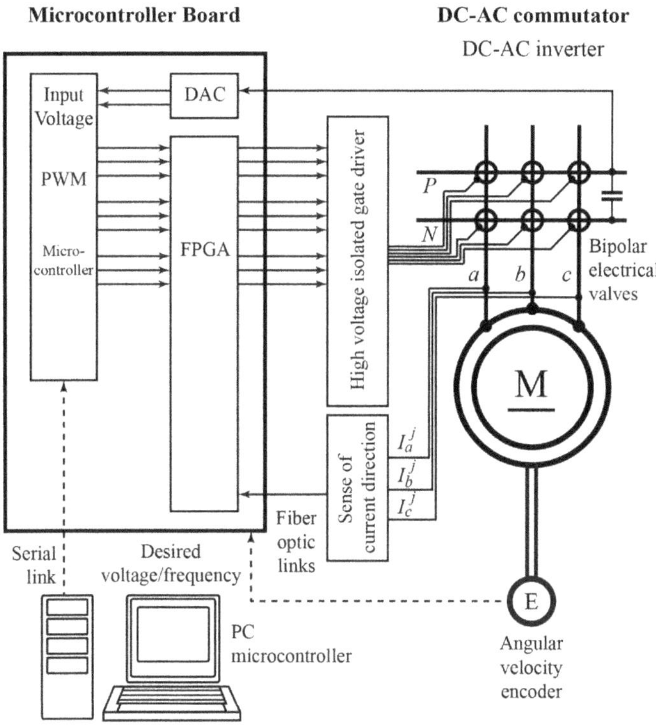

Figure 6.24. Structural and functional diagram of the DC–AC commutator IPM asynchronous (induction) motor IEMD cyber-physical heterogeneous dynamical hypersystem with the holor PWM (HOLPWM) control.

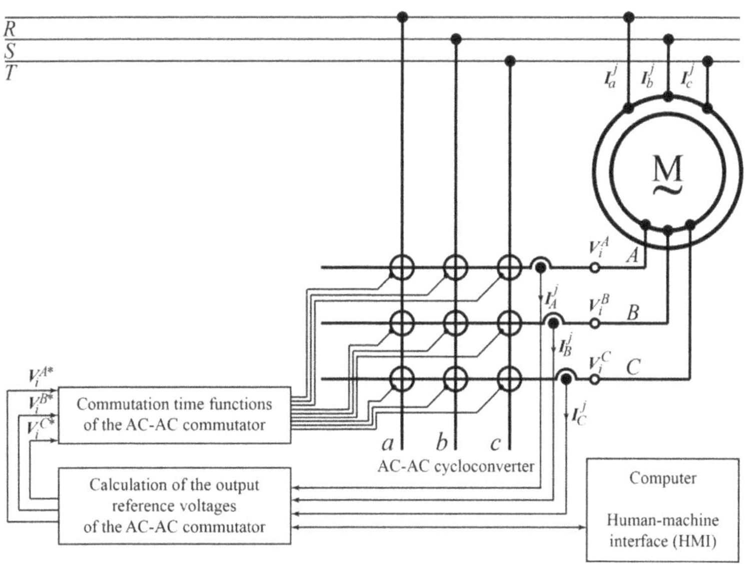

Figure 6.25. Structural and functional diagram of the AC–AC commutator split-rings or double fed asynchronous motor IEMD cyber-physical heterogeneous dynamical hypersystem with the holor PWM (HOLPWM) control.

motor IEMDs and HOLPWM and VECPWM techniques improve the quality of the current and reduce the torque ripple in AC motor or DC motor IEMD efficiently while maintaining the other performance characteristics of the AC motor or DC motor IEMD cyber-physical heterogeneous continuous or discrete dynamical hypersystem.

References

Barnes M 2003 *Practical Variable Speed Drives and Power Electronics, Automated Control Systems* (Perth, Australia: IDC Technologies)

Chan T-F and Shi K 2011 *Applied Intelligent Control of Induction Motor Drives* (New York: Wiley)

Geyer T 2017 *Model Predictive Control of High Power Converters and Industrial Drives* (New York: Wiley)

Hanamoto T, Deriha M, Ikeda H and Tsuji T 2008 Digital hardware circuit using FPGA for speed control system of permanent magnet synchronous motor *Proc. of the ICEM* 1–5

Holmes D G and Lipo T A 2003 *Pulse Width Modulation for Power Converters Principles and Practice* (Piscataway, NJ: Institute of Electrical and Electronics Engineers)

Hughes A and Drury W 2013 *Electric Motors and Drives Fundamentals, Types and Applications* 4th edn (Amsterdam: Elsevier)

Rashid M H 2018 *Power Electronics Handbook* 4th edn (Amsterdam: Elsevier)

Wach P 2011 *Dynamics and Control of Electrical Drives* (Berlin, Heidelberg: Springer)

Wilamowski B M and Irwin D J 2011 *The Industrial Electronics Handbook - Power Electronics and Motor Drives* (Boca Raton, FL: CRC Press)

Chapter 7

Physical and mathematical models of AC–AC or AC–DC–AC or DC–AC commutator synchronous or asynchronous motors

7.1 Introduction

A mathematical model of an integrated EM motor is a set of analyses about its properties expressed in mathematical form. Thus, mathematical modelling is nothing but the course of creating a formal description of these properties, limited to the set of properties exhibited by the physical model, using mathematical relationships. This means that the physical model determines the form of the mathematical model. The mathematical model should be unambiguously equivalent to the physical model.

An analysis of dynamic phenomena during an operation of integrated EM motors is necessary at the stage of design efforts as well as during usage of an IEMD. Mathematical models of the integrated EM motors should be developed in order to carry out this analysis. Depending on requirements, the AC and DC motor MMD electrical machines' mathematical models are applied for mathematical description of EM motors and the IEMD, as well.

The mathematical models (based on lumped parameters) are the basis for the mathematical description of an IEMD, including an automated IEMD with AC–AC, AC–DC–AC or DC–AC commutator synchronous or asynchronous (induction) motors, used in the IEMD for numerous technological processes.

The mathematical models of an IEMD that contains AC–AC, AC–DC–AC or DC–AC commutator synchronous or asynchronous (induction) motors with macro-commutators and microcontrollers, are commonly used despite the intensive development of the field models of MMD electrical machines and methods of their simulation.

In the case of mathematical modelling, the problem is to solve two fundamental tasks:

doi:10.1088/2053-2563/aae7d7ch7

© IOP Publishing Ltd 2019

- formulating a set of the second-order Euler–Lagrange differential equations of dynamics in a matrix notation;
- determining values of a coefficient's (parameters) of the abovementioned equations.

A well carried out estimation of mathematical model parameters of the chosen MMD electrical machine's structure is the key to its diagnosis and simulation. If the mathematical model and estimation methods were properly selected for the given MMD electrical machine's structure, then this solution may be copied, providing satisfactory results of simulation and diagnosis of the considered MMD electrical machine's structure.

In order to formulate a mathematical model of the AC and DC MMD electrical machines for the electrically-powered and mechatronically-controlled IEMD, analogous to the mathematical model of the MCM DC–AC/AC–DC commutator electromagnetically- or magnetoelectrically-excited motor/generator with a rotating MCM DC–AC/AC–DC commutator (in which phenomenon of current commutation (switching) takes place as a function of the angular displacement of the rotor θ), taking into consideration equations of unholonomic constraints of the ECM AC–AC, AC–DC–AC or DC–AC/AC–DC macrocommutator, a set of the second-order Euler–Lagrange differential equations of dynamics in a matrix notation for the AC synchronous or asynchronous (induction) motor/generator can be written.

The dynamic behaviour of the AC synchronous or asynchronous (induction) motor/generator is considered, the activity and state system being formed from a set of the second-order Euler–Lagrange differential equations of dynamics (see appendix C).

When an activity and state system is capable of dissipating kinetic co-energy is considered, then a set of the second-order Euler–Lagrange differential equations of dynamics in high-level system matrix notation may be expressed (White and Woodson 1959, Szklarski et al. 1963, Fijalkowski 1987a, 1987b, 2016, Tutaj 1996, 2012).

$$\left\{\frac{d}{dt}\left(\frac{\partial T^*}{\partial \dot{\boldsymbol{q}}}\right) - \frac{\partial T^*}{\partial \boldsymbol{q}} + \frac{\partial U}{\partial \boldsymbol{q}} + \frac{d}{dt}\left(\frac{\partial T_R^*}{\partial \dot{\boldsymbol{q}}}\right) - \boldsymbol{f}^T\right\}\delta\boldsymbol{q} = 0, \tag{7.1a}$$

or

$$\left\{\frac{d}{dt}\left(\frac{\partial \mathcal{T}^*}{\partial v}\right) - \frac{\partial \mathcal{T}^*}{\partial \boldsymbol{q}} + \frac{\partial \mathcal{U}}{\partial \boldsymbol{q}} + \frac{d}{dt}\left(\frac{\partial \mathcal{T}_R^*}{\partial v}\right) - \boldsymbol{f}^T\right\}\delta\boldsymbol{q} = 0, \tag{7.1b}$$

where:

$$\mathcal{T}^* \triangleq \frac{1}{2}\,\dot{\boldsymbol{q}}^T\boldsymbol{K}(\boldsymbol{q},\, \dot{\boldsymbol{q}})\dot{\boldsymbol{q}} \triangleq \frac{1}{2}\,v^T\boldsymbol{K}(\boldsymbol{q},\, v)v \tag{7.1c}$$

$$\mathcal{U} \triangleq \frac{1}{2}\,\boldsymbol{q}^T\boldsymbol{P}(\boldsymbol{q},\, \dot{\boldsymbol{q}})\boldsymbol{q} \triangleq \frac{1}{2}\,\boldsymbol{q}^T\boldsymbol{P}\,(\boldsymbol{q},\, v)\boldsymbol{q} \tag{7.1d}$$

$$\mathcal{T}_R^* \triangleq \int_0^t \mathcal{R}dt \triangleq \int_0^t (\mathcal{R}_V + \mathcal{R}_C)dt \triangleq \int_0^t \left(\frac{1}{2} v^T \dot{\boldsymbol{D}} v + \frac{1}{2} (\Delta \boldsymbol{f} \mathrm{sgn} v)\, v \right) dt \qquad (7.1e)$$

$$\boldsymbol{f}^T = (\mathrm{grad}\, \mathcal{U}_F)^T = \frac{\partial \mathcal{U}_F}{\partial \boldsymbol{q}} \qquad (7.1f)$$

$$\mathcal{R}_V \triangleq \frac{1}{2} \dot{\boldsymbol{q}}^T \boldsymbol{D}(q, \dot{q}) \dot{\boldsymbol{q}} \triangleq \frac{1}{2} v^T \boldsymbol{D}(\dot{q}, v) v \qquad (7.1g)$$

$$\mathcal{R}_C \triangleq \frac{1}{2} (\Delta \boldsymbol{f} \mathrm{sgn} \dot{\boldsymbol{q}})\, \dot{\boldsymbol{q}} \triangleq \frac{1}{2} (\Delta \boldsymbol{f} \mathrm{sgn} v)\, v \qquad (7.1h)$$

$$\mathrm{sgn}\, \dot{\boldsymbol{q}} = \left[\mathrm{sgn} \|\dot{q}_1\|^T \ldots \mathrm{sgn} \|\dot{q}_\vartheta\|^T \right]^T;$$

$$\mathrm{sgn}\, \dot{\boldsymbol{q}} \triangleq \mathrm{sgn} v \triangleq \begin{cases} +1 & v > 0; \\ -1 & v < 0. \end{cases}$$

Taking into account the above, the mathematical model of the AC synchronous or asynchronous (induction) motor may be obtained by differentiating, respectively, the conservative kinetic co-energy and conservative potential energy as well as dissipative kinetic co-energy so obtained with respect to generalised coordinates and velocities, respectively, as well as time to form a modified set of the second-order Euler–Lagrange differential equations of dynamics in high-level system matrix notation may be rewritten in the form:

$$\begin{Bmatrix} \frac{d}{dt}(v^T \boldsymbol{K}(\boldsymbol{q}, v)) - \frac{1}{2} v^T \frac{\partial}{\partial \boldsymbol{q}} \boldsymbol{K}(\boldsymbol{q}, v) v + \boldsymbol{q}^T \boldsymbol{P}(\boldsymbol{q}, v) \\ + v^T \boldsymbol{D}(\boldsymbol{q}, v) + \mathrm{sgn} v^T \boldsymbol{\Delta f} - \boldsymbol{f}^T \end{Bmatrix} \delta \boldsymbol{q} = 0 \qquad (7.2)$$

where: $\boldsymbol{K}(\boldsymbol{q}, v)$— a matrix of the conservative kinetic co-energy coefficients;
$\boldsymbol{P}(\boldsymbol{q}, v)$ —a matrix of the conservative potential energy coefficients;
$\boldsymbol{D}(\boldsymbol{q}, v)$ —a matrix of the dissipative kinetic co-energy coefficients;
$\Delta \boldsymbol{f}$—a matrix of the generalized force drops;
$\boldsymbol{q}$—a vector of the generalised coordinates;
$\boldsymbol{v}$—a vector of the generalised velocities;
$\boldsymbol{f}$—a vector of the generalised forces.

In a modified set of the second-order Euler–Lagrange differential equations of dynamics the regarded MMD electrical machine as the cyber-physical heterogeneous continuous dynamical hypersystem, i.e. the activity and state dynamical hypersystem decomposed on three activity and state physical homogeneous continuous dynamical hyposystems: wounded stator electrical dynamical hyposystem, wounded rotor electrical dynamical hyposystem and rotor mechanical dynamical system, the particular high-level dynamical system's vectors and matrices incorporating three low-level dynamical hyposystem's hypovectors and hypomatrices:

$$\boldsymbol{q} = \left[\left\|q_e^j\right\|^T \ \left\|q_m^j\right\|^T\right]^T = \left[\left\|\langle q_s^j\rangle^T \ \ \langle q_r^j\rangle^T\right\|^T \ \left\|q_m^j\right\|^T\right]^T;$$

Introducing to equation (7.2) the abovementioned physical quantities as follows:

$$\begin{aligned}
&\left\{\frac{d}{\mathrm{dt}}\left(\begin{bmatrix}\|v_e^j\| \\ \|v_m^j\|\end{bmatrix}^T \begin{bmatrix}\|K_e(q,v)\| & \|0\| \\ \|0\| & \|K_m(q,v)\|\end{bmatrix}\right)\right. \\
&-\frac{1}{2}\begin{bmatrix}\|v_e^j\| \\ \|v_m^j\|\end{bmatrix}^T \frac{\partial}{\partial\begin{bmatrix}\left\|q_e^j\right\| \\ \left\|q_m^j\right\|\end{bmatrix}}\begin{bmatrix}\|K_e(q,v)\| & \|0\| \\ \|0\| & \|K_m(q,v)\|\end{bmatrix}\begin{bmatrix}\|v_e^j\| \\ \|v_m^j\|\end{bmatrix} \\
&+\begin{bmatrix}\|v_e^j\| \\ \|v_m^j\|\end{bmatrix}^T \begin{bmatrix}\|P_e(q,v)\| & \|0\| \\ \|0\| & \|P_m(q,v)\|\end{bmatrix} \\
&+\begin{bmatrix}\|v_e^j\| \\ \|v_m^j\|\end{bmatrix}^T \begin{bmatrix}\|D_e(q,v)\| & \|0\| \\ \|0\| & \|D_m(q,v)\|\end{bmatrix} \\
&\left.+\operatorname{sgn}\begin{bmatrix}\|v_e^j\| \\ \|v_m^j\|\end{bmatrix}^T \begin{bmatrix}\left\|\Delta f_i^e\right\| & \|0\| \\ \|0\| & \left\|\Delta f_i^m\right\|\end{bmatrix} - \begin{bmatrix}\left\|f_i^e\right\| \\ \left\|f_i^m\right\|\end{bmatrix}^T\right\}\delta\begin{bmatrix}\left\|q_e^j\right\| \\ \left\|q_m^j\right\|\end{bmatrix} = 0
\end{aligned} \tag{7.3}$$

or in detail

$$\begin{aligned}
&\left\{\frac{d}{dt}\left(\begin{bmatrix}\left\|\begin{matrix}\langle v_s^j\rangle \\ \langle v_r^j\rangle\end{matrix}\right\| \\ \|v_m^j\|\end{bmatrix}^T \begin{bmatrix}\left\|\begin{matrix}\langle K_s(q,v)\rangle & \langle K_{\mathrm{sr}}(q,v)\rangle \\ \langle K_{\mathrm{rs}}(q,v)\rangle & \langle K_r(q,v)\rangle\end{matrix}\right\| & \|0\| \\ \|0\| & \|K_m(q,v)\|\end{bmatrix}\right)\right. \\
&-\frac{1}{2}\begin{bmatrix}\left\|\begin{matrix}\langle v_s^j\rangle \\ \langle v_r^j\rangle\end{matrix}\right\| \\ \|v_m^j\|\end{bmatrix}^T \frac{\partial}{\partial\begin{bmatrix}\left\|\begin{matrix}\langle q_s^j\rangle \\ \langle q_r^j\rangle\end{matrix}\right\| \\ \left\|q_m^j\right\|\end{bmatrix}}\begin{bmatrix}\left\|\begin{matrix}\langle K_s(q,v)\rangle & \langle K_{\mathrm{sr}}(q,v)\rangle \\ \langle K_{\mathrm{rs}}(q,v)\rangle & \langle K_r(q,v)\rangle\end{matrix}\right\| & \|0\| \\ \|0\| & \|K_m(q,v)\|\end{bmatrix}\begin{bmatrix}\left\|\begin{matrix}\langle v_s^j\rangle \\ \langle v_r^j\rangle\end{matrix}\right\| \\ \|v_m^j\|\end{bmatrix} \\
&+\begin{bmatrix}\left\|\begin{matrix}\langle q_s^j\rangle \\ \langle q_r^j\rangle\end{matrix}\right\| \\ \left\|q_m^j\right\|\end{bmatrix}^T \begin{bmatrix}\left\|\begin{matrix}\langle P_s(q,v)\rangle & \langle 0\rangle \\ \langle 0\rangle & \langle P_r(q,v)\rangle\end{matrix}\right\| & \|0\| \\ \|0\| & \|P_m(q,v)\|\end{bmatrix} \\
&+\begin{bmatrix}\left\|\begin{matrix}\langle v_s^j\rangle \\ \langle v_r^j\rangle\end{matrix}\right\| \\ \|v_m^j\|\end{bmatrix}^T \begin{bmatrix}\left\|\begin{matrix}\langle D_s(q,v)\rangle & \langle 0\rangle \\ \langle 0\rangle & \langle D_r(q,v)\rangle\end{matrix}\right\| & \|0\| \\ \|0\| & \|D_m(q,v)\|\end{bmatrix} \\
&\left.+\operatorname{sgn}\begin{bmatrix}\left\|\begin{matrix}\langle v_s^j\rangle \\ \langle v_r^j\rangle\end{matrix}\right\| \\ \|v_m^j\|\end{bmatrix}^T \begin{bmatrix}\left\|\begin{matrix}\langle \Delta f_i^s\rangle & \langle 0\rangle \\ \langle 0\rangle & \langle \Delta f_i^r\rangle\end{matrix}\right\| & \|0\| \\ \|0\| & \left\|\Delta f_i^m\right\|\end{bmatrix} - \begin{bmatrix}\left\|\begin{matrix}\langle f_i^s\rangle \\ \langle f_i^r\rangle\end{matrix}\right\| \\ \left\|f_i^m\right\|\end{bmatrix}^T\right\}\begin{bmatrix}\left\|\begin{matrix}\langle q_s^j\rangle \\ \langle q_r^j\rangle\end{matrix}\right\| \\ \left\|q_m^j\right\|\end{bmatrix} = 0
\end{aligned} \tag{7.4}$$

However, the assumption was made that

$$\langle \boldsymbol{P}(\boldsymbol{q}, v)\rangle = 0.$$

Thus, a mathematical model of the generalised MMD electrical machine may be expressed:

$$
\left\{\frac{d}{dt}\left(\begin{bmatrix}\left\|\begin{matrix}\langle i_s^j\rangle\\ \langle i_r^j\rangle\end{matrix}\right\|\\ \|\Omega^j\|\end{bmatrix}^T\begin{bmatrix}\left\|\begin{matrix}\langle L_s(\theta, i_s)\rangle & \langle L_{\mathrm{sr}}(\theta, i_s, i_r)\rangle\\ \langle L_{\mathrm{rs}}(\theta, i_r, i_s)\rangle & \langle L_r(\theta, i_r)\rangle\end{matrix}\right\| & \|0\|\\ \|0\| & \|J(\theta, \Omega)\|\end{bmatrix}\right)\right.
$$

$$
-\frac{1}{2}\begin{bmatrix}\left\|\begin{matrix}\langle i_s^j\rangle\\ \langle i_r^j\rangle\end{matrix}\right\|\\ \|\Omega^j\|\end{bmatrix}^T\frac{\partial}{\partial\begin{bmatrix}\left\|\begin{matrix}\langle q_s^j\rangle\\ \langle q_r^j\rangle\end{matrix}\right\|\\ \|\theta^j\|\end{bmatrix}}\begin{bmatrix}\left\|\begin{matrix}\langle L_s(\theta, i_s)\rangle & \langle L_{\mathrm{sr}}(\theta, i_s, i_r)\rangle\\ \langle L_{\mathrm{rs}}(\theta, i_r, i_s)\rangle & \langle L_r(\theta, i_r)\rangle\end{matrix}\right\| & \|0\|\\ \|0\| & \|J(\theta, \Omega)\|\end{bmatrix}\begin{bmatrix}\left\|\begin{matrix}\langle i_s^j\rangle\\ \langle i_r^j\rangle\end{matrix}\right\|\\ \|\Omega^j\|\end{bmatrix} \tag{7.5}
$$

$$
+\begin{bmatrix}\left\|\begin{matrix}\langle i_s^j\rangle\\ \langle i_r^j\rangle\end{matrix}\right\|\\ \|\Omega^j\|\end{bmatrix}^T\begin{bmatrix}\left\|\begin{matrix}\langle R_s(i_s)\rangle & \langle 0\rangle\\ \langle 0\rangle & \langle R_r(i_r)\rangle\end{matrix}\right\| & \|0\|\\ \|0\| & \|D(\Omega)\|\end{bmatrix}
$$

$$
\left.+\,\mathrm{sgn}\begin{bmatrix}\left\|\begin{matrix}\langle i_s^j\rangle\\ \langle i_r^j\rangle\end{matrix}\right\|\\ \|\Omega^j\|\end{bmatrix}^T\begin{bmatrix}\left\|\begin{matrix}\langle \Delta V_i^s\rangle & \langle 0\rangle\\ \langle 0\rangle & \langle \Delta V_i^r\rangle\end{matrix}\right\| & \|0\|\\ \|0\| & \|\Delta T_i\|\end{bmatrix}-\begin{bmatrix}\left\|\begin{matrix}\langle V_i^s\rangle\\ \langle V_i^r\rangle\end{matrix}\right\|\\ \|T_i\|\end{bmatrix}^T\right\}\delta\begin{bmatrix}\left\|\begin{matrix}\langle q_s^j\rangle\\ \langle q_r^j\rangle\end{matrix}\right\|\\ \|\theta^j\|\end{bmatrix}=0
$$

Taking into account magnetic saturation in stator and rotor circuits, the mathematical model may be obtained:

$$
\left\{\frac{d}{dt}\left(\begin{bmatrix}\left\|\begin{matrix}\langle i_s^j\rangle\\ \langle i_r^j\rangle\end{matrix}\right\|\\ \|\Omega^j\|\end{bmatrix}^T\begin{bmatrix}\left\|\begin{matrix}\langle L_s(\theta, i_s)\rangle & \langle L_{\mathrm{sr}}(\theta, i_s, i_r)\rangle\\ \langle L_{rs}(\theta, i_r, i_s)\rangle & \langle L_r(\theta, i_r)\rangle\end{matrix}\right\| & \|0\|\\ \|0\| & \|J(\theta, \Omega^j)\|\end{bmatrix}\right)\right.
$$

$$
-\frac{1}{2}\begin{bmatrix}\left\|\begin{matrix}\langle i_s^j\rangle\\ \langle i_r^j\rangle\end{matrix}\right\|\\ \|\Omega^j\|\end{bmatrix}^T\frac{\partial}{\partial[\|\theta^j\|]}\begin{bmatrix}\left\|\begin{matrix}\langle L_s(\theta, i_s)\rangle & \langle L_{sr}(\theta, i_s, i_r)\rangle\\ \langle L_{rs}(\theta, i_r, i_s)\rangle & \langle L_r(\theta, i_r)\rangle\end{matrix}\right\| & \|0\|\\ \|0\| & \|J(\theta, \Omega^j)\|\end{bmatrix}\begin{bmatrix}\left\|\begin{matrix}\langle i_s^j\rangle\\ \langle i_r^j\rangle\end{matrix}\right\|\\ \|\Omega^j\|\end{bmatrix} \tag{7.6}
$$

$$
+\begin{bmatrix}\left\|\begin{matrix}\langle i_s^j\rangle\\ \langle i_r^j\rangle\end{matrix}\right\|\\ \|\Omega^j\|\end{bmatrix}^T\begin{bmatrix}\left\|\begin{matrix}\langle R_s\rangle & \langle 0\rangle\\ \langle 0\rangle & \langle R_r\rangle\end{matrix}\right\| & \|0\|\\ \|0\| & \|D\|\end{bmatrix}
$$

$$
\left.+\,\mathrm{sgn}\begin{bmatrix}\left\|\begin{matrix}\langle i_s^j\rangle\\ \langle i_r^j\rangle\end{matrix}\right\|\\ \|\Omega^j\|\end{bmatrix}^T\begin{bmatrix}\left\|\begin{matrix}\langle \Delta V_i^s\rangle & \langle 0\rangle\\ \langle 0\rangle & \langle \Delta V_i^r\rangle\end{matrix}\right\| & \|0\|\\ \|0\| & \|\Delta T_i\|\end{bmatrix}-\begin{bmatrix}\left\|\begin{matrix}\langle V_i^s\rangle\\ \langle V_i^r\rangle\end{matrix}\right\|\\ \|T_i\|\end{bmatrix}^T\right\}\delta\begin{bmatrix}\left\|\begin{matrix}\langle q_s^j\rangle\\ \langle q_r^j\rangle\end{matrix}\right\|\\ \|\theta^j\|\end{bmatrix}=0
$$

Differentiating the first part of the equation (7.6) one will get:

$$
\begin{aligned}
&\left\{\left(\frac{d}{dt}\begin{bmatrix}\left\|\begin{matrix}\langle i_s^j\rangle\\ \langle i_r^j\rangle\end{matrix}\right\|\\ \|\Omega^j\|\end{bmatrix}^T\right)\begin{bmatrix}\left\|\begin{matrix}\langle L_s(\theta, i_s)\rangle & \langle L_{sr}(\theta, i_s, i_r)\rangle\\ \langle L_{rs}(\theta, i_r, i_s)\rangle & \langle L_r(\theta, i_r)\rangle\end{matrix}\right\| & \|0\|\\ \|0\| & \|J(\theta, \Omega^j)\|\end{bmatrix}\right.\\
&+\begin{bmatrix}\left\|\begin{matrix}\langle i_s^j\rangle\\ \langle i_r^j\rangle\end{matrix}\right\|\\ \|\Omega^j\|\end{bmatrix}^T\frac{\partial}{\partial[\|\theta^j\|]}\begin{bmatrix}\left\|\begin{matrix}\langle L_s(\theta, i_s)\rangle & \langle L_{sr}(\theta, i_s, i_r)\rangle\\ \langle L_{rs}(\theta, i_r, i_s)\rangle & \langle L_r(\theta, i_r)\rangle\end{matrix}\right\| & \|0\|\\ \|0\| & \|J(\theta, \Omega^j)\|\end{bmatrix}\Omega^j\\
&+\begin{bmatrix}\left\|\begin{matrix}\langle i_s^j\rangle\\ \langle i_r^j\rangle\end{matrix}\right\|\\ \|\Omega^j\|\end{bmatrix}^T\frac{\partial}{\partial\left[\left\|\langle i_s^j\rangle\right\|\right]}\begin{bmatrix}\left\|\begin{matrix}\langle L_s(\theta, i_s)\rangle & \langle L_{sr}(\theta, i_s, i_r)\rangle\\ \langle L_{rs}(\theta, i_r, i_s)\rangle & \langle L_r(\theta, i_r)\rangle\end{matrix}\right\| & \|0\|\\ \|0\| & \|0\|\end{bmatrix}\frac{d}{dt}\begin{bmatrix}\left\|\begin{matrix}\langle i_s^j\rangle\\ \langle i_r^j\rangle\end{matrix}\right\|\\ \|\Omega^j\|\end{bmatrix}\\
&-\frac{1}{2}\begin{bmatrix}\left\|\begin{matrix}\langle i_s^j\rangle\\ \langle i_r^j\rangle\end{matrix}\right\|\\ \|\Omega^j\|\end{bmatrix}^T\frac{\partial}{\partial[\|\theta^j\|]}\begin{bmatrix}\left\|\begin{matrix}\langle L_s(\theta, i_s)\rangle & \langle L_{sr}(\theta, i_s, i_r)\rangle\\ \langle L_{rs}(\theta, i_r, i_s)\rangle & \langle L_r(\theta, i_r)\rangle\end{matrix}\right\| & \|0\|\\ \|0\| & \|J(\theta, \Omega^j)\|\end{bmatrix}\begin{bmatrix}\left\|\begin{matrix}\langle i_s^j\rangle\\ \langle i_r^j\rangle\end{matrix}\right\|\\ \|\Omega^j\|\end{bmatrix}\\
&+\begin{bmatrix}\left\|\begin{matrix}\langle i_s^j\rangle\\ \langle i_r^j\rangle\end{matrix}\right\|\\ \|\Omega^j\|\end{bmatrix}^T\begin{bmatrix}\left\|\begin{matrix}\langle R_s\rangle & \langle 0\rangle\\ \langle 0\rangle & \langle R_r\rangle\end{matrix}\right\| & \|0\|\\ \|0\| & \|D\|\end{bmatrix}\\
&\left.+\operatorname{sgn}\begin{bmatrix}\left\|\begin{matrix}\langle i_s^j\rangle\\ \langle i_r^j\rangle\end{matrix}\right\|\\ \|\Omega^j\|\end{bmatrix}^T\begin{bmatrix}\left\|\begin{matrix}\langle \Delta V_i^s\rangle & \langle 0\rangle\\ \langle 0\rangle & \langle \Delta V_i^r\rangle\end{matrix}\right\| & \|0\|\\ \|0\| & \|\Delta T_i\|\end{bmatrix}-\begin{bmatrix}\left\|\begin{matrix}\langle V_i^s\rangle\\ \langle V_i^r\rangle\end{matrix}\right\|\\ \|T_i\|\end{bmatrix}^T\right\}\delta\begin{bmatrix}\left\|\begin{matrix}\langle q_s^j\rangle\\ \langle q_r^j\rangle\end{matrix}\right\|\\ \|\theta^j\|\end{bmatrix}=0
\end{aligned}
\tag{7.7}
$$

or in detail

$$
\begin{aligned}
&\left\{\left(\frac{d}{dt}\begin{bmatrix}\left\|\begin{matrix}\langle i_s^j\rangle\\ \langle i_r^j\rangle\end{matrix}\right\|\\ \|\Omega^j\|\end{bmatrix}^T\right)\begin{bmatrix}\left\|\begin{matrix}\langle L_s(\theta, i_s)\rangle & \langle L_{sr}(\theta, i_s, i_r)\rangle\\ \langle L_{rs}(\theta, i_r, i_s)\rangle & \langle L_r(\theta, i_r)\rangle\end{matrix}\right\| & \|0\|\\ \|0\| & \|J(\theta, \Omega^j)\|\end{bmatrix}\right.\\
&+\begin{bmatrix}\left\|\begin{matrix}\langle i_s^j\rangle\\ \langle i_r^j\rangle\end{matrix}\right\|\\ \|\Omega^j\|\end{bmatrix}^T\begin{bmatrix}\left\|\begin{matrix}\langle L_s^*(\theta, i_s)\rangle & \langle L_{sr}^*(\theta, i_s, i_r)\rangle\\ \langle L_{rs}^*(\theta, i_s, i_r)\rangle & \langle L_r^*(\theta, i_r)\rangle\end{matrix}\right\| & \|0\|\\ \|0\| & \|J*(\theta, \Omega^j)\|\end{bmatrix}\Omega^j\\
&+\begin{bmatrix}\left\|\begin{matrix}\langle i_s^j\rangle\\ \langle i_r^j\rangle\end{matrix}\right\|\\ \|\Omega^j\|\end{bmatrix}^T\begin{bmatrix}\left\|\begin{matrix}\langle L_s^{**}(\theta, i_s)\rangle & \langle L_{sr}^{**}(\theta, i_s, i_r)\rangle\\ \langle L_{rs}^{**}(\theta, i_s)\rangle & \langle L_r^{**}(\theta, i_r)\rangle\end{matrix}\right\| & \|0\|\\ \|0\| & \|0\|\end{bmatrix}\frac{d}{dt}\begin{bmatrix}\left\|\begin{matrix}\langle i_s^j\rangle\\ \langle i_r^j\rangle\end{matrix}\right\|\\ \|\Omega^j\|\end{bmatrix}\\
&-\frac{1}{2}\begin{bmatrix}\left\|\begin{matrix}\langle i_s^j\rangle\\ \langle i_r^j\rangle\end{matrix}\right\|\\ \|\Omega^j\|\end{bmatrix}^T\begin{bmatrix}\left\|\begin{matrix}\langle L_s^*(\theta, i_s)\rangle & \langle L_{sr}^*(\theta, i_s, i_r)\rangle\\ \langle L_{rs}^*(\theta, i_s, i_r)\rangle & \langle L_r^*(\theta, i_r)\rangle\end{matrix}\right\| & \|0\|\\ \|0\| & \|J*(\theta, \Omega^j)\|\end{bmatrix}\begin{bmatrix}\left\|\begin{matrix}\langle i_s^j\rangle\\ \langle i_r^j\rangle\end{matrix}\right\|\\ \|\Omega^j\|\end{bmatrix}\\
&+\begin{bmatrix}\left\|\begin{matrix}\langle i_s^j\rangle\\ \langle i_r^j\rangle\end{matrix}\right\|\\ \|\Omega^j\|\end{bmatrix}^T\begin{bmatrix}\left\|\begin{matrix}\langle R_s\rangle & \langle 0\rangle\\ \langle 0\rangle & \langle R_r\rangle\end{matrix}\right\| & \|0\|\\ \|0\| & \|D\|\end{bmatrix}\\
&\left.+\operatorname{sgn}\begin{bmatrix}\left\|\begin{matrix}\langle i_s^j\rangle\\ \langle i_r^j\rangle\end{matrix}\right\|\\ \|\Omega^j\|\end{bmatrix}^T\begin{bmatrix}\left\|\begin{matrix}\langle \Delta V_i^s\rangle & \langle 0\rangle\\ \langle 0\rangle & \langle \Delta V_i^r\rangle\end{matrix}\right\| & \|0\|\\ \|0\| & \|\Delta T_i\|\end{bmatrix}-\begin{bmatrix}\left\|\begin{matrix}\langle V_i^s\rangle\\ \langle V_i^r\rangle\end{matrix}\right\|\\ \|T_i\|\end{bmatrix}^T\right\}\delta\begin{bmatrix}\left\|\begin{matrix}\langle q_s^j\rangle\\ \langle q_r^j\rangle\end{matrix}\right\|\\ \|\theta^j\|\end{bmatrix}=0
\end{aligned}
\tag{7.8}
$$

Each MMD electrical machine, consisting of stator winding and rotor winding, is the physical heterogeneous dynamical hypersystem of six magnetically-coupled circuits: three immovable circuits of stator and three circuits of rotor (factual or equivalent one for squirrel-cage rotor).

The following assumptions are usually adopted in order to simplify the mathematical model of MMD electrical machine:

- lumped phase windings of MMD electrical machine are taken into account;
- phase windings of MMD electrical machine are balanced;
- magnetic circuits are linear and the equivalent inductances of MMD electrical machine are independent of currents;
- the regularity of airgap of MMD electrical machine is assumed;
- the influence of phase winding capacitances of MMD electrical machine are omitted;
- the influence of MMD electrical machine's anisotropy is omitted;
- hysteresis loss, eddy-current loss and anomalous loss of MMD electrical machine are omitted;
- harmonics of spatial distribution of magnetic field in airgap of MMD electrical machine are omitted;
- the moment of inertia has a constant value and is not a function of generalized coordinate and speed (generalised angular position and angular-velocity).

Thus, a mathematical model of the generalised AC synchronous or asynchronous (induction) motor may expressed as follows:

$$\begin{aligned}
&\left\{\left(\frac{d}{dt}\begin{bmatrix}\left\|\begin{matrix}\langle i_s^j\rangle\\ \langle i_r^j\rangle\end{matrix}\right\|\\ \Omega^j\end{bmatrix}^T\right)\begin{bmatrix}\left\|\begin{matrix}\langle L_s(\theta)\rangle & \langle L_{sr}(\theta)\rangle\\ \langle L_{rs}(\theta)\rangle & \langle L_r(\theta)\rangle\end{matrix}\right\| & \|0\|\\ \|0\| & J\end{bmatrix}\right.\\
&+\begin{bmatrix}\left\|\begin{matrix}\langle i_s^j\rangle\\ \langle i_r^j\rangle\end{matrix}\right\|\\ \Omega^j\end{bmatrix}^T\frac{\partial}{\partial\theta^j}\begin{bmatrix}\left\|\begin{matrix}\langle L_s(\theta)\rangle & \langle L_{sr}(\theta)\rangle\\ \langle L_{rs}(\theta)\rangle & \langle L_r(\theta)\rangle\end{matrix}\right\| & \|0\|\\ \|0\| & J\end{bmatrix}\Omega^j\\
&-\frac{1}{2}\begin{bmatrix}\left\|\begin{matrix}\langle i_s^j\rangle\\ \langle i_r^j\rangle\end{matrix}\right\|\\ \Omega^j\end{bmatrix}^T\frac{\partial}{\partial\theta^j}\begin{bmatrix}\left\|\begin{matrix}\langle L_s(\theta)\rangle & \langle L_{sr}(\theta)\rangle\\ \langle L_{rs}(\theta)\rangle & \langle L_r(\theta)\rangle\end{matrix}\right\| & \|0\|\\ \|0\| & J\end{bmatrix}\begin{bmatrix}\left\|\begin{matrix}\langle i_s^j\rangle\\ \langle i_r^j\rangle\end{matrix}\right\|\\ \Omega^j\end{bmatrix}\\
&+\begin{bmatrix}\left\|\begin{matrix}\langle i_s^j\rangle\\ \langle i_r^j\rangle\end{matrix}\right\|\\ \Omega^j\end{bmatrix}^T\begin{bmatrix}\left\|\begin{matrix}\langle R_s\rangle & \langle 0\rangle\\ \langle 0\rangle & \langle R_r\rangle\end{matrix}\right\| & \|0\|\\ \|0\| & D\end{bmatrix}\\
&\left.+\operatorname{sgn}\begin{bmatrix}\left\|\begin{matrix}\langle i_s^j\rangle\\ \langle i_r^j\rangle\end{matrix}\right\|\\ \Omega^j\end{bmatrix}^T\begin{bmatrix}\left\|\begin{matrix}\langle \Delta V_i^s\rangle & \langle 0\rangle\\ \langle 0\rangle & \langle \Delta V_i^r\rangle\end{matrix}\right\| & \|0\|\\ \|0\| & \Delta T_i\end{bmatrix}-\begin{bmatrix}\left\|\begin{matrix}\langle V_i^s\rangle\\ \langle V_i^r\rangle\end{matrix}\right\|\\ T_i\end{bmatrix}^T\right\}\delta\begin{bmatrix}\left\|\begin{matrix}\langle q_s^j\rangle\\ \langle q_r^j\rangle\end{matrix}\right\|\\ \theta^j\end{bmatrix}=0
\end{aligned}\tag{7.9}$$

or in detail

$$\left\{\left(\frac{d}{dt}\begin{bmatrix} \left\| \begin{matrix} \langle i_s^j \rangle \\ \langle i_r^j \rangle \end{matrix} \right\| \\ \Omega^j \end{bmatrix}^T\right) \begin{bmatrix} \left\| \begin{matrix} \langle L_s(\theta) \rangle & \langle L_{sr}(\theta) \rangle \\ \langle L_{rs}(\theta) \rangle & \langle L_r(\theta) \rangle \end{matrix} \right\| & \|0\| \\ \|0\| & J \end{bmatrix}\right.$$

$$+\begin{bmatrix} \left\| \begin{matrix} \langle i_s^j \rangle \\ \langle i_r^j \rangle \end{matrix} \right\| \\ \Omega^j \end{bmatrix}^T \begin{bmatrix} \left\| \begin{matrix} \langle L_s^*(\theta) \rangle & \langle L_{sr}^*(\theta) \rangle \\ \langle L_{rs}^*(\theta) \rangle & \langle L_r^*(\theta) \rangle \end{matrix} \right\| & \|0\| \\ \|0\| & 0 \end{bmatrix} \Omega^j +$$

$$-\frac{1}{2}\begin{bmatrix} \left\| \begin{matrix} \langle i_s^j \rangle \\ \langle i_r^j \rangle \end{matrix} \right\| \\ \Omega^j \end{bmatrix}^T \begin{bmatrix} \left\| \begin{matrix} \langle L_s^*(\theta) \rangle & \langle L_{sr}^*(\theta) \rangle \\ \langle L_{rs}^*(\theta) \rangle & \langle L_r^*(\theta) \rangle \end{matrix} \right\| & \|0\| \\ \|0\| & 0 \end{bmatrix} \begin{bmatrix} \left\| \begin{matrix} \langle i_s^j \rangle \\ \langle i_r^j \rangle \end{matrix} \right\| \\ \Omega^j \end{bmatrix} \qquad (7.10)$$

$$+\begin{bmatrix} \left\| \begin{matrix} \langle i_s^j \rangle \\ \langle i_r^j \rangle \end{matrix} \right\| \\ \Omega^j \end{bmatrix}^T \begin{bmatrix} \left\| \begin{matrix} \langle R_s \rangle & \langle 0 \rangle \\ \langle 0 \rangle & \langle R_r \rangle \end{matrix} \right\| & \|0\| \\ \|0\| & D \end{bmatrix}$$

$$\left. + \operatorname{sgn}\begin{bmatrix} \left\| \begin{matrix} \langle i_s^j \rangle \\ \langle i_r^j \rangle \end{matrix} \right\| \\ \Omega^j \end{bmatrix}^T \begin{bmatrix} \left\| \begin{matrix} \langle \Delta V_i^s \rangle & \langle 0 \rangle \\ \langle 0 \rangle & \langle \Delta V_i^r \rangle \end{matrix} \right\| & \|0\| \\ \|0\| & \Delta T_i \end{bmatrix} - \begin{bmatrix} \left\| \begin{matrix} \langle V_i^s \rangle \\ \langle V_i^r \rangle \end{matrix} \right\| \\ T_i \end{bmatrix}^T \right\} \delta \begin{bmatrix} \left\| \begin{matrix} \langle q_s^j \rangle \\ \langle q_r^j \rangle \end{matrix} \right\| \\ \theta^j \end{bmatrix} = 0$$

In a modified set of the second-order Euler–Lagrange differential equations of dynamics the regarded AC synchronous or asynchronous (induction) motor, i.e. the activity state system decomposed on three activity and state hyposystems: rotor mechanical hyposystem, interior-permanent-magnet (IPM) magnetoelectrical or wounded electromagnetical exciter electrical hyposystem and wounded armature electrical hyposystems, the particular high-level system vectors and matrices incorporating three low-level system hypovectors and hypomatrices:

$$\begin{aligned} \boldsymbol{q} &= \left[\|q_e^j\|^T \ \ \|q_m^j\|^T \right]^T \\ &= \left[\left\| \langle q_s^j \rangle^T \ \ \langle q_r^j \rangle^T \right\|^T \ \ \|q_m^j\|^T \right]^T \\ &= \left[\langle q_a^j \ \ q_b^j \ \ q_c^j \rangle^T \ \ \langle q_u^j \ \ q_v^j \ \ q_w^j \rangle^T \ \ \langle \theta^j \rangle^T \right]^T \\ &= \left[q_a^j \ \ q_b^j \ \ q_c^j \ q_u^j \ q_v^j \ q_w^j \ \theta^j \right]^T; \end{aligned} \qquad (7.11)$$

where

$$\left\|q_e^j\right\|^T = \left\|\langle q_s^j\rangle^T \quad \langle q_r^j\rangle^T\right\|^T = .\left\|\langle q_a^j \quad q_b^j \quad q_c^j\rangle^T \quad \langle q_u^j \quad q_v^j \quad q_w^j\rangle^T\right\|^T$$

$$= \left\|q_a^j\, q_b^j\, q_c^j \quad q_u^j\, q_v^j\, q_w^j\right\|^T;$$

$$\left\|q_m^j\right\|^T = \|\langle \theta^j\rangle\|^T = \theta^j$$

$$\begin{aligned}
\acute{\boldsymbol{q}} &= \left[\left\|\acute{q}_e^j\right\|^T \quad \left\|\acute{q}_m^j\right\|^T\right]^T = \left[\left\|\langle \acute{q}_s^j\rangle^T \quad \langle \acute{q}_r^j\rangle^T\right\|^T \quad \left\|\acute{q}_m^j\right\|\right]^T = \left[\acute{q}_s^j \quad \acute{q}_r^j \quad \acute{q}_m^j\right]^T \\
&= \boldsymbol{v} = \left[\|v_e^j\|^T \quad \|v_m^j\|^T\right]^T = \left[\left\|\langle v_s^j\rangle^T \quad \langle v_r^j\rangle^T\right\|^T \quad \|v_m^j\|^T\right]^T \\
&= \left[\left\|\langle i_s^j\rangle^T \quad \langle i_r^j\rangle^T\right\|^T \quad \|\Omega^j\|^T\right]^T = \left[\langle i_a^j \quad i_b^j \quad i_c^j\rangle^T \quad \langle i_u^j \quad i_v^j \quad i_w^j\rangle^T \quad \langle \Omega^j\rangle^T\right]^T \\
&= \left[i_a^j \quad i_b^j \quad i_c^j\, i_u^j\, i_v^j\, i_w^j\, \Omega^j\right]^T;
\end{aligned} \tag{7.12}$$

where

$$\left\|\acute{q}_e^j\right\|^T = \left\|\langle \acute{q}_s^j\rangle^T \quad \langle \acute{q}_r^j\rangle^T\right\|^T;$$

or

$$\|v_e^j\|^T = \left\|\langle v_s^j\rangle^T \quad \langle v_r^j\rangle^T\right\|^T.$$

$$\left\|\acute{q}_m^j\right\|^T = \|\Omega^j\|^T = \Omega^j$$

$$\begin{aligned}
\boldsymbol{f} &= \left[\left\|f_i^e\right\|^T \quad \left\|f_i^m\right\|^T\right]^T = \left[\left\|\langle f_i^s\rangle^T \quad \langle f_i^r\rangle^T\right\|^T \quad \left\|f_i^m\right\|^T\right]^T = \left[f_i^s \quad f_i^r \quad f_i^m\right]^T \\
&= \left[\left\|\langle V_i^s\rangle^T \quad \langle V_i^r\rangle^T\right\|^T \quad \|T_i\|^T\right]^T \\
&= \left[\langle V_i^a \quad V_i^b \quad V_i^c\rangle^T \quad \langle V_i^u \quad V_i^v \quad V_i^w\rangle^T \quad \langle T_i\rangle^T\right]^T \\
&= \left[V_i^a \quad V_i^b \quad V_i^c\, V_i^u\, V_i^v\, V_i^w\, T_i\right]^T.
\end{aligned} \tag{7.13}$$

$$\Delta \boldsymbol{f} = \begin{bmatrix} \|\Delta f_i^e\| & \|0\| \\ \|0\| & \|\Delta f_i^m\| \end{bmatrix} = \begin{bmatrix} \left\| \begin{matrix} \langle \Delta f_i^s \rangle & \langle 0 \rangle \\ \langle 0 \rangle & \langle \Delta f_i^r \rangle \end{matrix} \right\| & \|0\| \\ \|0\| & \|\Delta f_i^m\| \end{bmatrix} = \begin{bmatrix} \begin{matrix} \langle \Delta f_i^s \rangle & \langle 0 \rangle \\ \langle 0 \rangle & \langle \Delta f_i^r \rangle \end{matrix} & \langle 0 \rangle \\ \langle 0 \rangle & \Delta f_i^m \end{bmatrix}$$

$$= \begin{bmatrix} \begin{pmatrix} \Delta V_i^a & 0 & 0 \\ 0 & \Delta V_i^b & 0 \\ 0 & 0 & \Delta V_i^c \end{pmatrix} & \langle 0 \rangle & \\ \langle 0 \rangle & \begin{pmatrix} \Delta V_i^u & 0 & 0 \\ 0 & \Delta V_i^v & 0 \\ 0 & 0 & \Delta V_i^w \end{pmatrix} & \langle 0 \rangle \\ & \langle 0 \rangle & \Delta T_i \end{bmatrix} \tag{7.14}$$

$$= \begin{bmatrix} \Delta V_i^a & 0 & 0 & & & & \\ 0 & \Delta V_i^b & 0 & & 0 & & \\ 0 & 0 & \Delta V_i^c & & & & 0 \\ & & & \Delta V_i^u & 0 & 0 & \\ & 0 & & 0 & \Delta V_i^v & 0 & \\ & & & 0 & 0 & \Delta V_i^w & \\ & & & 0 & & & \Delta T_i \end{bmatrix}$$

$$= \operatorname{diag}\left[\Delta V_i^a \;\; \Delta V_i^b \;\; \Delta V_i^c \;\; \Delta V_i^u \;\; \Delta V_i^v \;\; \Delta V_i^w \;\; \Delta T_i\right].$$

$$\boldsymbol{K}(q, v) = \begin{bmatrix} \|K_e(q, v)\| & \|0\| \\ \|0\| & \|K_m(q, v)\| \end{bmatrix} = \begin{bmatrix} \left\| \begin{matrix} \langle K_s(q, v) \rangle & \langle K_{sr}(q, v) \rangle \\ \langle K_{rs}(q, v) \rangle & \langle K_r(q, v) \rangle \end{matrix} \right\| & \|0\| \\ \|0\| & \|K_m(q, v)\| \end{bmatrix}$$

$$= \begin{bmatrix} \begin{pmatrix} L_a(\theta) & L_{ab}(\theta) & L_{ac}(\theta) \\ L_{ba}(\theta) & L_b(\theta) & L_{bc}(\theta) \\ L_{ca}(\theta) & L_{cb}(\theta) & L_c(\theta) \end{pmatrix} & \begin{pmatrix} L_{au}(\theta) & L_{av}(\theta) & L_{aw}(\theta) \\ L_{bu}(\theta) & L_{bv}(\theta) & L_{bw}(\theta) \\ L_{cu}(\theta) & L_{cv}(\theta) & L_{cw}(\theta) \end{pmatrix} & \langle 0 \rangle \\ \begin{pmatrix} L_{ua}(\theta) & L_{ub}(\theta) & L_{uc}(\theta) \\ L_{va}(\theta) & L_{vb}(\theta) & L_{vc}(\theta) \\ L_{wa}(\theta) & L_{wb}(\theta) & L_{wc}(\theta) \end{pmatrix} & \begin{pmatrix} L_u(\theta) & L_{uv}(\theta) & L_{uw}(\theta) \\ L_{vu}(\theta) & L_v(\theta) & L_{vw}(\theta) \\ L_{wu}(\theta) & L_{wv}(\theta) & L_w(\theta) \end{pmatrix} & \langle 0 \rangle \\ \langle 0 \rangle & \langle 0 \rangle & J \end{bmatrix} \tag{7.15}$$

$$\langle L_s^*(\theta, i_s)\rangle = \frac{\partial}{\partial\theta}\langle L_s(\theta, i_s)\rangle, \; \langle L_{sr}^*(\theta, i_s)\rangle = \frac{\partial}{\partial\theta}\langle L_{sr}(\theta, i_s, i_r)\rangle = \langle L_{rs}^*(\theta, i_s, i_r)\rangle,$$

$$\langle L_r^*(\theta, i_r)\rangle = \frac{\partial}{\partial\theta}\langle L_r(\theta, i_r)\rangle$$

$$\langle L_s^{**}(\theta, i_s)\rangle = \frac{\partial}{\partial i_s}\langle L_s(\theta, i_s)\rangle, \; \langle L_r^*(\theta, i_r)\rangle = \frac{\partial}{\partial\theta}\langle L_r(\theta, i_r)\rangle,$$

$$\langle L_{sr}^{**}(\theta, i_s, i_r)\rangle = \frac{\partial}{\partial i_s}\langle L_{sr}(\theta, i_s, i_r)\rangle + \frac{\partial}{\partial i_r}\langle L_{sr}(\theta, i_s, i_r)\rangle = \langle L_{rs}^{**}(\theta, i_s, i_r)\rangle$$

$$\begin{aligned} \boldsymbol{P}(q, v) &= \operatorname{diag}\left[\|P_e(q, v)\| \;\; \|P_m(q, v)\|\right] \\ &= \operatorname{diag}\left[\left\|\langle P_s(q, v)\rangle \;\; \langle P_r(q, v)\rangle\right\| \;\; \|P_m(q, v)\|\right] = [0]. \end{aligned} \tag{7.16}$$

$$\begin{aligned} \boldsymbol{D}(q, v) &= \begin{bmatrix} \|D_e(q, v)\| & \|0\| \\ \|0\| & \|D_m(q, v)\| \end{bmatrix} \\ &= \begin{bmatrix} \left\|\begin{matrix} \langle D_s(q, v)\rangle & 0 \\ 0 & \langle D_r(q, v)\rangle \end{matrix}\right\| & 0 \\ 0 & \|D_m(q, v)\| \end{bmatrix} \\ &= \begin{bmatrix} \langle D_s(q, v)\rangle & 0 & \\ 0 & \langle D_r(q, v)\rangle & 0 \\ & 0 & \|D_m(q, v)\| \end{bmatrix} = \begin{bmatrix} \|R_e\| & 0 \\ 0 & \|D\| \end{bmatrix} \\ &= \begin{bmatrix} \left\|\begin{matrix} \langle R_s\rangle & 0 \\ 0 & \langle R_r\rangle \end{matrix}\right\| & 0 \\ 0 & \|D\| \end{bmatrix} = \begin{bmatrix} R_{as} & 0 & 0 & & & & \\ 0 & R_{bs} & 0 & & 0 & & \\ 0 & 0 & R_{cs} & & & & 0 \\ & & & R_{ar} & 0 & 0 & \\ & 0 & & 0 & R_{br} & 0 & \\ & & & 0 & 0 & R_{cr} & \\ & & 0 & & & & D \end{bmatrix} \\ &= \operatorname{diag}\left[R_{as} \;\; R_{bs} \;\; R_{cs} \; R_{ar} \; R_{br} \; R_{cr} \; D\right]. \end{aligned} \tag{7.17}$$

where

$$\|D_e(q, v)\| = \begin{Vmatrix} \langle R_s \rangle & 0 \\ 0 & \langle R_r \rangle \end{Vmatrix} = \begin{Vmatrix} \begin{matrix} R_{as} & 0 & 0 \\ 0 & R_{bs} & 0 \\ 0 & 0 & R_{cs} \end{matrix} & 0 \\ 0 & \begin{matrix} R_{ar} & 0 & 0 \\ 0 & R_{br} & 0 \\ 0 & 0 & R_{cr} \end{matrix} \end{Vmatrix}$$

$$= \mathrm{diag}\begin{Vmatrix} R_{as} & R_{bs} & R_{cs} & R_{ar} & R_{br} & R_{cr} \end{Vmatrix};$$

$$\langle R_s \rangle = \begin{pmatrix} R_{as} & 0 & 0 \\ 0 & R_{bs} & 0 \\ 0 & 0 & R_{cs} \end{pmatrix};$$

$$\langle R_r \rangle = \begin{pmatrix} R_{ar} & 0 & 0 \\ 0 & R_{\mathrm{br}} & 0 \\ 0 & 0 & R_{cr} \end{pmatrix};$$

$$\|D_m(q, v)\| = D.$$

7.2 DC–AC commutator synchronous motor with the mechanical split-ring or flat commutator and electromagnetical exciter (DC motor)

There are a number of variations of the DC–AC commutator synchronous motor with the mechanical split-ring or flat commutator and electromagnetical exciter (known as the 'DC motor') and each has different characteristics and require different means of control. The DC motor has two major magnetic components, the 'exciter' and the 'armature'. The DC motor also has a means of converting DC–AC to create the rotating magnetic field. This EM motor may comprise a set of connections arranged around the shaft of the EM motor and connected to discrete windings (coils) on the armature with carbon brushes to connect the DC voltage to the armature. A traditional DC–AC commutator synchronous motor with the mechanical split-ring or flat commutator and electromagnetical exciter (DC motor) comprises a wound stator known as the 'exciter', and a wound-rotor known as the 'armature'. DC voltage is applied to the armature by means of a mechanical split-ring or flat commutator with a set of carbon brushes. The armature consists of a number of discrete windings which are terminated on the mechanical split-ring or flat commutator which is in effect a rotary switch. The DC motor can be configured as a series connection, shunt connection or a compound connection.

A physical model of the DC–AC commutator synchronous motor with the mechanical split-ring or flat commutator and electromagnetical exciter (DC motor) is shown in figure 7.1.

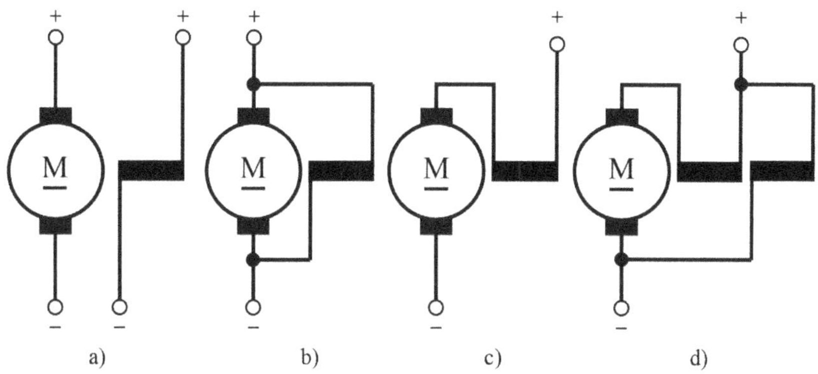

Figure 7.1. Physical model of the DC–AC commutator synchronous motor with the mechanical split-ring or flat commutator and electromagnetical exciter (DC motor): (a) separate field; (b) shunt field; (c) series field; (d) compound field.

Taking into account equation (7.10) one will get:

$$
\begin{aligned}
&\Bigg\{\left(\frac{d}{dt}\begin{bmatrix}\left\|\begin{matrix}\langle i_r^j\rangle\\ \langle i_s^j\rangle\end{matrix}\right\|\\ \Omega^j\end{bmatrix}^T\right)\begin{bmatrix}\left\|\begin{matrix}\langle L_r(\theta)\rangle & \langle L_{rs}(\theta)\rangle\\ \langle L_{sr}(\theta)\rangle & \langle L_s\rangle\end{matrix}\right\| & \|0\|\\ \|0\| & J\end{bmatrix}\\
&+\begin{bmatrix}\left\|\begin{matrix}\langle i_r^j\rangle\\ \langle i_s^j\rangle\end{matrix}\right\|\\ \Omega^j\end{bmatrix}^T\begin{bmatrix}\left\|\begin{matrix}\langle L_r^*(\theta)\rangle & \langle L_{rs}^*(\theta)\rangle\\ \langle L_{sr}^*(\theta)\rangle & \langle 0\rangle\end{matrix}\right\| & \|0\|\\ \|0\| & \|0\|\end{bmatrix}\Omega^j+\\
&-\frac{1}{2}\begin{bmatrix}\left\|\begin{matrix}\langle i_r^j\rangle\\ \langle i_s^j\rangle\end{matrix}\right\|\\ \Omega^j\end{bmatrix}^T\begin{bmatrix}\left\|\begin{matrix}\langle L_r^*(\theta)\rangle & \langle L_{rs}^*(\theta)\rangle\\ \langle L_{sr}^*(\theta)\rangle & \langle 0\rangle\end{matrix}\right\| & \|0\|\\ \|0\| & 0\end{bmatrix}\\
&+\begin{bmatrix}\left\|\begin{matrix}\langle i_r^j\rangle\\ \langle i_s^j\rangle\end{matrix}\right\|\\ \Omega^j\end{bmatrix}^T\begin{bmatrix}\left\|\begin{matrix}\langle R_r\rangle & \langle 0\rangle\\ \langle 0\rangle & \langle R_s\rangle\end{matrix}\right\| & \|0\|\\ \|0\| & D\end{bmatrix}\\
&+\operatorname{sgn}\begin{bmatrix}\left\|\begin{matrix}\langle i_r^j\rangle\\ \langle i_s^j\rangle\end{matrix}\right\|\\ \Omega^j\end{bmatrix}^T\begin{bmatrix}\left\|\begin{matrix}\langle \Delta V_i^r\rangle & \langle 0\rangle\\ \langle 0\rangle & \langle 0\rangle\end{matrix}\right\| & \|0\|\\ \|0\| & T_i\end{bmatrix}-\begin{bmatrix}\left\|\begin{matrix}\langle V_i^r\rangle\\ \langle V_i^s\rangle\end{matrix}\right\|\\ T_i\end{bmatrix}^T\Bigg\}\,\delta\begin{bmatrix}\left\|\begin{matrix}\langle q_r^j\rangle\\ \langle q_s^j\rangle\end{matrix}\right\|\\ \theta^j\end{bmatrix}=0
\end{aligned}
$$

or in detail

$$\begin{aligned}
&\Bigg\{ \left(\frac{d}{dt} \left\| \begin{matrix} \langle i^{j}_{r(\mathrm{AC})} \rangle \\ i^{j}_{s(\mathrm{DC})} \\ \Omega^{j} \end{matrix} \right\|^{T} \right) \left\| \begin{matrix} \langle L_{r}(\theta) \rangle & \langle L_{rs}(\theta) \rangle & 0 \\ \langle L_{sr}(\theta) \rangle & L_{s} & 0 \\ 0 & 0 & J \end{matrix} \right\| \\
&+ \left\| \begin{matrix} \langle i^{j}_{r(\mathrm{AC})} \rangle \\ i^{j}_{s(\mathrm{DC})} \\ \Omega^{j} \end{matrix} \right\|^{T} \left\| \begin{matrix} \langle L^{*}_{r}(\theta) \rangle & \langle L^{*}_{rs}(\theta) \rangle & 0 \\ \langle L^{*}_{sr}(\theta) \rangle & 0 & 0 \\ 0 & 0 & 0 \end{matrix} \right\| \Omega^{j} \\
&- \frac{1}{2} \left\| \begin{matrix} \langle i^{j}_{r(\mathrm{AC})} \rangle \\ i^{j}_{s(\mathrm{DC})} \\ \Omega^{j} \end{matrix} \right\|^{T} \left\| \begin{matrix} \langle L^{*}_{r}(\theta) \rangle & \langle L^{*}_{rs}(\theta) \rangle & 0 \\ \langle L^{*}_{sr}(\theta) \rangle & 0 & 0 \\ 0 & 0 & 0 \end{matrix} \right\| \left\| \begin{matrix} \langle i^{j}_{r(\mathrm{AC})} \rangle \\ i^{j}_{s(\mathrm{DC})} \\ \Omega^{j} \end{matrix} \right\| + \left\| \begin{matrix} \langle i^{j}_{r(\mathrm{AC})} \rangle \\ i^{j}_{s(\mathrm{DC})} \\ \Omega^{j} \end{matrix} \right\|^{T} \left\| \begin{matrix} \langle R_{s} \rangle & \langle 0 \rangle & 0 \\ \langle 0 \rangle & R_{r} & 0 \\ 0 & 0 & D \end{matrix} \right\| \\
&+ \operatorname{sgn} \left\| \begin{matrix} \langle i^{j}_{r(\mathrm{AC})} \rangle \\ i^{j}_{s(\mathrm{DC})} \\ \Omega^{j} \end{matrix} \right\|^{T} \left\| \begin{matrix} \langle \Delta V^{r(\mathrm{AC})}_{i} \rangle & \langle 0 \rangle & 0 \\ \langle 0 \rangle & 0 & 0 \\ 0 & 0 & \Delta T_{i} \end{matrix} \right\| - \left\| \begin{matrix} \langle V^{r(\mathrm{AC})}_{i} \rangle \\ V^{s(\mathrm{DC})}_{i} \\ T_{i} \end{matrix} \right\|^{T} \Bigg\} \delta \left\| \begin{matrix} \langle q^{j}_{r(\mathrm{AC})} \rangle \\ q^{j}_{s(\mathrm{DC})} \\ \theta^{j} \end{matrix} \right\| = 0
\end{aligned}$$

Introducing to the armature circuit a mechanical split-ring/flat commutator one will get:

$$\begin{aligned}
&\Bigg\{ \left(\frac{d}{dt} \left\| \begin{matrix} \langle C^{ji}_{rr} \rangle \langle i^{r(\mathrm{DC})}_{i} \rangle \\ i^{j}_{s(\mathrm{DC})} \\ \Omega^{j} \end{matrix} \right\|^{T} \right) \left\| \begin{matrix} \langle L_{rr}(\theta) \rangle & \langle L_{rs}(\theta) \rangle & 0 \\ \langle L_{sr}(\theta) \rangle & L_{s} & 0 \\ 0 & 0 & J \end{matrix} \right\| \\
&+ \left\| \begin{matrix} \langle C^{ji}_{rr} \rangle \langle i^{r(\mathrm{DC})}_{i} \rangle \\ i^{j}_{s(\mathrm{DC})} \\ \Omega^{j} \end{matrix} \right\|^{T} \left\| \begin{matrix} \langle L^{*}_{r}(\theta) \rangle & \langle L^{*}_{rs}(\theta) \rangle & 0 \\ \langle L^{*}_{sr}(\theta) \rangle & 0 & 0 \\ 0 & 0 & 0 \end{matrix} \right\| \Omega^{j} \\
&- \frac{1}{2} \left\| \begin{matrix} \langle C^{ji}_{rr} \rangle \langle i^{r(\mathrm{DC})}_{i} \rangle \\ i^{j}_{s(\mathrm{DC})} \\ \Omega^{j} \end{matrix} \right\|^{T} \left\| \begin{matrix} \langle L^{*}_{r}(\theta) \rangle & \langle L^{*}_{rs}(\theta) \rangle & 0 \\ \langle L^{*}_{sr}(\theta) \rangle & 0 & 0 \\ 0 & 0 & 0 \end{matrix} \right\| \left\| \begin{matrix} \langle C^{ji}_{rr} \rangle \langle i^{r(\mathrm{DC})}_{i} \rangle \\ i^{j}_{s(\mathrm{DC})} \\ \Omega^{j} \end{matrix} \right\| \\
&+ \left\| \begin{matrix} \langle C^{ji}_{rr} \rangle \langle i^{r(\mathrm{DC})}_{i} \rangle \\ i^{j}_{s(\mathrm{DC})} \\ \Omega^{j} \end{matrix} \right\|^{T} \left\| \begin{matrix} \langle R_{s} \rangle & \langle 0 \rangle & 0 \\ \langle 0 \rangle & R_{r} & 0 \\ 0 & 0 & D \end{matrix} \right\| + \operatorname{sgn} \left\| \begin{matrix} \langle C^{rr}_{i} \rangle \langle V^{j}_{r(\mathrm{DC})} \rangle & \langle 0 \rangle & 0 \\ \langle 0 \rangle & 0 & 0 \\ 0 & 0 & \Delta T_{i} \end{matrix} \right\| \\
&- \left\| \begin{matrix} \langle C^{rr}_{ij} \rangle \langle V^{j}_{r(\mathrm{DC})} \rangle \\ V^{s(\mathrm{DC})}_{i} \\ T_{i} \end{matrix} \right\|^{T} \Bigg\} \delta \left\| \begin{matrix} \langle C^{ji}_{rr} \rangle \langle q^{r(\mathrm{DC})}_{i} \rangle \\ q^{j}_{s(\mathrm{DC})} \\ \theta^{j} \end{matrix} \right\| = 0
\end{aligned}$$

where in:

$$\langle V_i^{r(\mathrm{AC})} \rangle = \langle C_{ij}^{rr} \rangle \langle V_{r(\mathrm{DC})}^{j} \rangle$$

$$\begin{pmatrix} V_i^{a(\mathrm{AC})} \\ V_i^{b(\mathrm{AC})} \\ V_i^{c(\mathrm{AC})} \end{pmatrix} = \begin{pmatrix} C_{ij}^{\mathrm{Pa}} & C_{ij}^{\mathrm{Na}} \\ C_{ij}^{\mathrm{Pb}} & C_{ij}^{\mathrm{Nb}} \\ C_{ij}^{\mathrm{Pc}} & C_{ij}^{\mathrm{Nc}} \end{pmatrix} \begin{pmatrix} V_{P(\mathrm{DC})}^{j} \\ V_{N(\mathrm{DC})}^{j} \end{pmatrix}$$

or

$$\langle i_i^{r(\mathrm{DC})} \rangle = \langle C_{rr}^{ji} \rangle \langle i_{r(\mathrm{AC})}^{j} \rangle$$

$$\begin{pmatrix} i_i^{P(\mathrm{DC})} \\ i_i^{N(\mathrm{DC})} \end{pmatrix} = \begin{pmatrix} C_{\mathrm{Pa}}^{ji} & C_{\mathrm{Na}}^{ji} \\ C_{\mathrm{Pb}}^{ji} & C_{\mathrm{Nb}}^{ji} \\ C_{\mathrm{Pc}}^{ji} & C_{\mathrm{Nc}}^{ji} \end{pmatrix}^{T} \begin{pmatrix} i_{a(\mathrm{AC})}^{j} \\ i_{b(\mathrm{AC})}^{j} \\ i_{c(\mathrm{AC})}^{j} \end{pmatrix}$$

$$\begin{pmatrix} C_{ij}^{\mathrm{Pa}} & C_{ij}^{\mathrm{Na}} \\ C_{ij}^{\mathrm{Pb}} & C_{ij}^{\mathrm{Nb}} \\ C_{ij}^{\mathrm{Pc}} & C_{ij}^{\mathrm{Nc}} \end{pmatrix} = \begin{pmatrix} C_{\mathrm{Pa}}^{ji} & C_{\mathrm{Na}}^{ji} \\ C_{\mathrm{Pb}}^{ji} & C_{\mathrm{Nb}}^{ji} \\ C_{\mathrm{Pc}}^{ji} & C_{\mathrm{Nc}}^{ji} \end{pmatrix}$$

7.3 DC–AC commutator synchronous motor with the mechanical split-ring or flat commutator and magnetoelectrical exciter (DC motor)

A traditional DC–AC commutator synchronous motor with the mechanical split-ring or flat commutator and magnetoelectrical exciter (DC motor) comprises a wound stator known as the 'exciter', and a wound-rotor known as the 'armature'. DC voltage is applied to the armature by means of a mechanical split-ring or flat commutator with a set of carbon brushes.

The armature consists of a number of discrete windings which are terminated on the mechanical split-ring or flat commutator which is in effect a set of rotary electrical valves (mechanical switches). The DC–AC commutator synchronous motor with the mechanical split-ring or flat commutator and magnetoelectrical exciter (DC motor) can be configured as a series connection, shunt connection or a compound connection.

In the expression for conservative co-energy of a generalised MMD electrical machine with **interior permanent magnets** (IPM) considered as an electrical dynamical system containing IPMs, the words taking into account the interaction of electric currents of the dynamical system with the current i_o and the own energy of IPMs are taken into account $1/2L_o i_o^2$ cannot be taken into account, because with the assumption $i_o = \text{const}$ it has no effect on the form of Euler–Lagrange differential equations of the second-order.

In this case, it is not necessary to derive the differential equation of the dynamics of the exciter of the MMD electrical machine corresponding to the coordinate of the electric charge

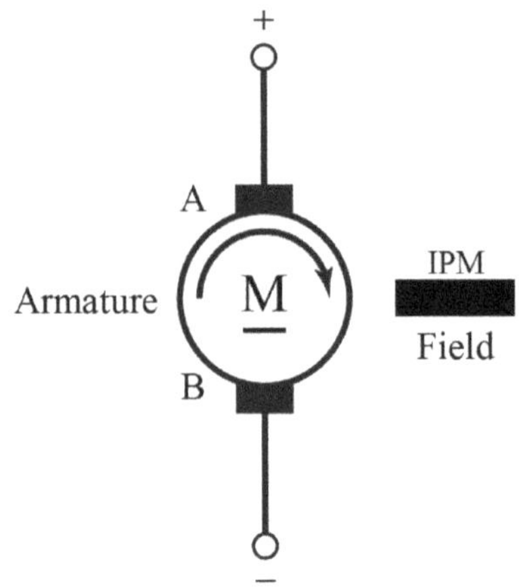

Figure 7.2. Physical model of the DC–AC commutator synchronous motor with the mechanical split-ring or flat commutator and magnetoelectrical exciter (DC motor).

$$q_0 = \int_0^t i_0 \, dt$$

representing the current of the exciter i_0 because such equation assuming $i_o = \text{const}$ would give a condition for the source voltage applied to the electromagnet, maintaining a constant value of current i_0.

A physical model of the DC–AC commutator synchronous motor with the mechanical split-ring or flat commutator and magnetoelectrical exciter (DC motor) is shown in figure 7.2.

Taking into account equation (7.10) one will get:

$$\begin{aligned}
&\left\{ \left(\frac{d}{dt} \begin{bmatrix} \left\| \begin{matrix} \langle i_s^j \rangle \\ \langle i_r^j \rangle \end{matrix} \right\| \\ \Omega^j \end{bmatrix}^T \right) \begin{bmatrix} \left\| \begin{matrix} \langle L_s(\theta) \rangle & \langle L_{sr}(\theta) \rangle \\ \langle L_{rs}(\theta) \rangle & \langle L_r \rangle \end{matrix} \right\| & \|0\| \\ \|0\| & J \end{bmatrix} \right. \\
&+ \begin{bmatrix} \left\| \begin{matrix} \langle i_s^j \rangle \\ \langle i_r^j \rangle \end{matrix} \right\| \\ \Omega^j \end{bmatrix}^T \begin{bmatrix} \left\| \begin{matrix} \langle L_s^*(\theta) \rangle & \langle L_{sr}^*(\theta) \rangle \\ \langle L_{rs}^*(\theta) \rangle & \langle 0 \rangle \end{matrix} \right\| & \|0\| \\ \|0\| & 0 \end{bmatrix} \Omega^j \\
&- \frac{1}{2} \begin{bmatrix} \left\| \begin{matrix} \langle i_s^j \rangle \\ \langle i_r^j \rangle \end{matrix} \right\| \\ \Omega^j \end{bmatrix}^T \begin{bmatrix} \left\| \begin{matrix} \langle L_s^*(\theta) \rangle & \langle L_{sr}^*(\theta) \rangle \\ \langle L_{rs}^*(\theta) \rangle & \langle 0 \rangle \end{matrix} \right\| & \|0\| \\ \|0\| & 0 \end{bmatrix} \begin{bmatrix} \left\| \begin{matrix} \langle i_s^j \rangle \\ \langle i_r^j \rangle \end{matrix} \right\| \\ \Omega^j \end{bmatrix} \\
&+ \begin{bmatrix} \left\| \begin{matrix} \langle i_s^j \rangle \\ \langle i_r^j \rangle \end{matrix} \right\| \\ \Omega^j \end{bmatrix}^T \begin{bmatrix} \left\| \begin{matrix} \langle R_s \rangle & \langle 0 \rangle \\ \langle 0 \rangle & \langle R_r \rangle \end{matrix} \right\| & \|0\| \\ \|0\| & D \end{bmatrix} \\
&\left. + \operatorname{sgn} \begin{bmatrix} \left\| \begin{matrix} \langle i_s^j \rangle \\ \langle i_r^j \rangle \end{matrix} \right\| \\ \Omega^j \end{bmatrix}^T \begin{bmatrix} \left\| \begin{matrix} \langle \Delta V_i^s \rangle & \langle 0 \rangle \\ \langle 0 \rangle & \langle 0 \rangle \end{matrix} \right\| & \|0\| \\ \|0\| & \Delta T_i \end{bmatrix} - \begin{bmatrix} \left\| \begin{matrix} \langle V_i^s \rangle \\ \langle 0 \rangle \end{matrix} \right\| \\ T_i \end{bmatrix}^T \right\} \delta \begin{bmatrix} \left\| \begin{matrix} \langle q_s^j \rangle \\ \langle q_r^j \rangle \end{matrix} \right\| \\ \theta^j \end{bmatrix} = 0
\end{aligned}$$

or in detail

$$
\left\{\left(\frac{d}{dt}\begin{Vmatrix}\langle i^j_{r(AC)}\rangle\\ i^j_{s(DC)}\\ \Omega^j\end{Vmatrix}^T\right)\begin{Vmatrix}\langle L_r(\theta)\rangle & \langle L_{rs}(\theta)\rangle & 0\\ \langle L_{sr}(\theta)\rangle & L_s & 0\\ 0 & 0 & J\end{Vmatrix}\right.
$$

$$
+\begin{Vmatrix}\langle i^j_{r(AC)}\rangle\\ i^j_{s(DC)}\\ \Omega^j\end{Vmatrix}^T\begin{Vmatrix}\langle L^*_r(\theta)\rangle & \langle L^*_{rs}(\theta)\rangle & 0\\ \langle L^*_{sr}(\theta)\rangle & 0 & 0\\ 0 & 0 & 0\end{Vmatrix}\Omega^j
$$

$$
-\frac{1}{2}\begin{Vmatrix}\langle i^j_{r(AC)}\rangle\\ i^j_{s(DC)}\\ \Omega^j\end{Vmatrix}^T\begin{Vmatrix}\langle L^*_r(\theta)\rangle & \langle L^*_{rs}(\theta)\rangle & 0\\ \langle L^*_{sr}(\theta)\rangle & 0 & 0\\ 0 & 0 & 0\end{Vmatrix}+\begin{Vmatrix}\langle i^j_{r(AC)}\rangle\\ i^j_{s(DC)}\\ \Omega^j\end{Vmatrix}^T\begin{Vmatrix}\langle R_r\rangle & \langle 0\rangle & 0\\ \langle 0\rangle & R_s & 0\\ 0 & 0 & D\end{Vmatrix}
$$

$$
\left.+\operatorname{sgn}\begin{Vmatrix}\langle i^j_{r(AC)}\rangle\\ i^j_{s(DC)}\\ \Omega^j\end{Vmatrix}^T\begin{Vmatrix}\langle \Delta V_i^{r(AC)}\rangle & \langle 0\rangle & 0\\ \langle 0\rangle & 0 & 0\\ 0 & 0 & \Delta T_i\end{Vmatrix}-\begin{Vmatrix}\langle V_i^{r(AC)}\rangle\\ 0\\ T_i\end{Vmatrix}^T\right\}\delta\begin{Vmatrix}\langle q^j_{r(AC)}\rangle\\ q^j_{s(DC)}\\ \theta^j\end{Vmatrix}=0
$$

Introducing to the armature circuit a mechanical split-ring/flat commutator one will get:

$$
\left\{\left(\frac{d}{dt}\begin{Vmatrix}\langle C^{ji}_{rr}\rangle\langle i_i^{r(DC)}\rangle\\ i_i^{s(DC)}\\ \Omega^j\end{Vmatrix}^T\right)\begin{Vmatrix}\langle L_r(\theta)\rangle & \langle L_{sr}(\theta)\rangle & 0\\ \langle L_{sr}(\theta)\rangle & L_s & 0\\ 0 & 0 & J\end{Vmatrix}+\right.
$$

$$
\begin{Vmatrix}\langle C^{ji}_{rr}\rangle\langle i_i^{r(DC)}\rangle\\ i_i^{s(DC)}\\ \Omega^j\end{Vmatrix}^T\begin{Vmatrix}\langle L^*_r(\theta)\rangle & \langle L^*_{rs}(\theta)\rangle & 0\\ \langle L^*_{sr}(\theta)\rangle & 0 & 0\\ 0 & 0 & 0\end{Vmatrix}\Omega^j
$$

$$
-\frac{1}{2}\begin{Vmatrix}\langle C^{ji}_{rr}\rangle\langle i_i^{r(DC)}\rangle\\ i_i^{s(DC)}\\ \Omega^j\end{Vmatrix}^T\begin{Vmatrix}\langle L^*_r(\theta)\rangle & \langle L^*_{rs}(\theta)\rangle & 0\\ \langle L^*_{sr}(\theta)\rangle & 0 & 0\\ 0 & 0 & 0\end{Vmatrix}\begin{Vmatrix}\langle C^{ji}_{rr}\rangle\langle i_i^{r(DC)}\rangle\\ i_i^{s(DC)}\\ \Omega^j\end{Vmatrix}
$$

$$
+\begin{Vmatrix}\langle C^{ji}_{rr}\rangle\langle i_i^{r(DC)}\rangle\\ i_i^{s(DC)}\\ \Omega^j\end{Vmatrix}^T\begin{Vmatrix}\langle R_r\rangle & 0 & 0\\ 0 & 0 & 0\\ 0 & 0 & D\end{Vmatrix}
$$

$$
+\operatorname{sgn}\begin{Vmatrix}\langle C^{ji}_{rr}\rangle\langle i_i^{r(DC)}\rangle\\ i_i^{s(DC)}\\ \Omega^j\end{Vmatrix}^T\begin{Vmatrix}\langle \Delta V_i^{r(AC)}\rangle & \langle 0\rangle & 0\\ \langle 0\rangle & 0 & 0\\ 0 & 0 & \Delta T_i\end{Vmatrix}
$$

$$
\left.-\begin{Vmatrix}\langle C^{rr}_{ij}\rangle\langle V^j_{r(DC)}\rangle\\ 0\\ T\end{Vmatrix}^T\right\}\delta\begin{Vmatrix}\langle C^{ji}_{rr}\rangle\langle q_i^{r(DC)}\rangle\\ q_i^{s(DC)}\\ \theta^j\end{Vmatrix}=0
$$

where in:

$$\langle V_i^{r(AC)} \rangle = \langle C_{ij}^{rr} \rangle \langle V_{r(DC)}^{j} \rangle$$

$$\begin{pmatrix} V_i^{a(AC)} \\ V_i^{b(AC)} \\ V_i^{c(AC)} \end{pmatrix} = \begin{pmatrix} C_{ij}^{Pa} & C_{ij}^{Na} \\ C_{ij}^{Pb} & C_{ij}^{Nb} \\ C_{ij}^{Pc} & C_{ij}^{Nc} \end{pmatrix} \begin{pmatrix} V_{P(DC)}^{j} \\ V_{N(DC)}^{j} \end{pmatrix}$$

$$\langle i_i^{r(DC)} \rangle = \langle C_{rr}^{ji} \rangle \langle i_{r(AC)}^{j} \rangle$$

$$\begin{pmatrix} i_i^{P(DC)} \\ i_i^{N(DC)} \end{pmatrix} = \begin{pmatrix} C_{Pa}^{ji} & C_{Na}^{ji} \\ C_{Pb}^{ji} & C_{Nb}^{ji} \\ C_{Pc}^{ji} & C_{Nc}^{ji} \end{pmatrix}^T \begin{pmatrix} i_{a(AC)}^{j} \\ i_{b(AC)}^{j} \\ i_{c(AC)}^{j} \end{pmatrix}$$

$$\begin{pmatrix} C_{Pa}^{ji} & C_{Na}^{ji} \\ C_{Pb}^{ji} & C_{Nb}^{ji} \\ C_{Pc}^{ji} & C_{Nc}^{ji} \end{pmatrix} = \begin{pmatrix} C_{ij}^{Pa} & C_{ij}^{Na} \\ C_{ij}^{Pb} & C_{ij}^{Nb} \\ C_{ij}^{Pc} & C_{ij}^{Nc} \end{pmatrix}.$$

7.4 DC–AC commutator synchronous motor with the macroelectronic commutator (macrocommutator) and electromagnetical exciter

DC–AC commutator synchronous motors with a macroelectronic commutator (macrocommutator) and electromagnetical exciter is a rotating MMD electrical machine that converts electrical energy into mechanical energy. It operates on DC electricity. Similar to a generic DC motor, a DC–AC commutator synchronous motors with the macrocommutator has both a rotor and a stator. In a traditional DC motor the armature conductors will rotate and the magnetic-field exciter comprising the stator remains physically static. But in a DC–AC commutator synchronous motors with the macrocommutator, the roles of the armature conductors and the magnetic-field exciter are reversed. Here the armature conductors remain stationary and the magnetic-field exciter rotates. Due to this feature a DC–AC commutator synchronous motors with the macrocommutator is equivalent to a reversed DC commutator motor. In this the armature remains static and the magnets rotate.

DC–AC commutator synchronous motors with a macrocommutator and electromagnetical exciter have a macrocommutator, acting as a DC–AC inverter to create the AC voltage required to establish the rotating magnetic field. The DC–AC commutator synchronous motor with the macrocommutator and electromagnetical exciter does not have carbon brushes and so requires far less maintenance than a traditional DC motor. The speed of this DC motor is determined by the

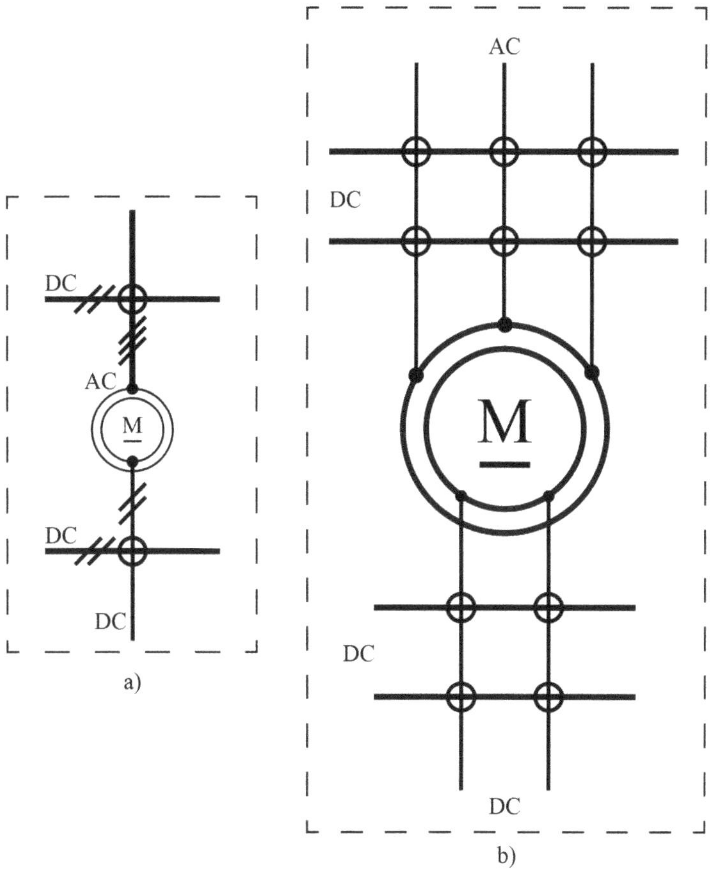

Figure 7.3. Physical model of the DC–AC commutator synchronous motor with the macrocommutator and electromagnetical exciter (DC motor).

macroelectronics and so better speed accuracy can be achieved than in DC–AC commutator synchronous motors with the mechanical split-ring or flat commutator and electromagnetical exciter (traditional DC motors).

A physical model of the DC–AC commutator synchronous motor with the macroelectronic commutator (macrocommutator) and electromagnetical exciter (DC motor) is shown in figure 7.3.

Taking into account equation (7.10) we get:

$$\left\{\frac{d}{dt}\left(\begin{bmatrix}\left\|\begin{matrix}\langle i_s^j\rangle\\ \langle i_r^j\rangle\end{matrix}\right\|\\ \|\Omega^j\|\end{bmatrix}^T\right)\begin{bmatrix}\left\|\begin{matrix}\langle L_s(\theta)\rangle & \langle L_{sr}(\theta)\rangle\\ \langle L_{rs}(\theta)\rangle & \langle L_r(\theta)\rangle\end{matrix}\right\| & \|0\|\\ \|0\| & \|J(\theta,\Omega)\|\end{bmatrix}\right.$$

$$+\begin{bmatrix}\left\|\begin{matrix}\langle i_s^j\rangle\\ \langle i_r^j\rangle\end{matrix}\right\|\\ \|\Omega^j\|\end{bmatrix}^T\begin{bmatrix}\left\|\begin{matrix}\langle L_s^*(\theta)\rangle & \langle L_{sr}^*(\theta)\rangle\\ \langle L_{rs}^*(\theta)\rangle & \langle L_r^*(\theta)\rangle\end{matrix}\right\| & \|0\|\\ \|0\| & \|J*(\theta,\Omega)\|\end{bmatrix}\big[\|\Omega^j\|\big]$$

$$-\frac{1}{2}\begin{bmatrix}\left\|\begin{matrix}\langle i_s^j\rangle\\ \langle i_r^j\rangle\end{matrix}\right\|\\ \|\Omega^j\|\end{bmatrix}^T\begin{bmatrix}\left\|\begin{matrix}\langle L_s^*(\theta)\rangle & \langle L_{sr}^*(\theta)\rangle\\ \langle L_{rs}^*(\theta)\rangle & \langle L_r^*(\theta)\rangle\end{matrix}\right\| & \|0\|\\ \|0\| & \|J*(\theta,\Omega)\|\end{bmatrix}\begin{bmatrix}\left\|\begin{matrix}\langle i_s^j\rangle\\ \langle i_r^j\rangle\end{matrix}\right\|\\ \|\Omega^j\|\end{bmatrix}$$

$$+\begin{bmatrix}\left\|\begin{matrix}\langle i_s^j\rangle\\ \langle i_r^j\rangle\end{matrix}\right\|\\ \|\Omega^j\|\end{bmatrix}^T\begin{bmatrix}\left\|\begin{matrix}\langle R_s\rangle & \langle 0\rangle\\ \langle 0\rangle & \langle R_r\rangle\end{matrix}\right\| & \|0\|\\ \|0\| & \|D\|\end{bmatrix}$$

$$\left.+\,\mathrm{sgn}\begin{bmatrix}\left\|\begin{matrix}\langle i_s^j\rangle\\ \langle i_r^j\rangle\end{matrix}\right\|\\ \|\Omega^j\|\end{bmatrix}^T\begin{bmatrix}\left\|\begin{matrix}\langle \Delta V_i^s\rangle & \langle 0\rangle\\ \langle 0\rangle & \langle \Delta V_i^r\rangle\end{matrix}\right\| & \|0\|\\ \|0\| & \|\Delta T_i\|\end{bmatrix}-\begin{bmatrix}\left\|\begin{matrix}\langle V_i^s\rangle\\ \langle V_i^r\rangle\end{matrix}\right\|\\ \|T_i\|\end{bmatrix}^T\right\}\delta\begin{bmatrix}\left\|\begin{matrix}\langle q_s^j\rangle\\ \langle q_r^j\rangle\end{matrix}\right\|\\ \|\theta^j\|\end{bmatrix}=0$$

or in detail

$$\left\{\left(\frac{d}{dt}\left\|\begin{matrix}\langle i_{s(AC)}^j\rangle\\ i_{r(DC)}^j\\ \Omega^j\end{matrix}\right\|^T\right)\left\|\begin{matrix}\langle L_s(\theta)\rangle & \langle L_{sr}(\theta)\rangle & 0\\ \langle L_{rs}(\theta)\rangle & L_r & 0\\ 0 & 0 & J\end{matrix}\right\|+\left\|\begin{matrix}\langle i_{s(AC)}^j\rangle\\ i_{r(DC)}^j\\ \Omega^j\end{matrix}\right\|^T\left\|\begin{matrix}\langle L_s^*(\theta)\rangle & \langle L_{sr}^*(\theta)\rangle & 0\\ \langle L_{rs}^*(\theta)\rangle & 0 & 0\\ 0 & 0 & 0\end{matrix}\right\|\Omega^j\right.$$

$$-\frac{1}{2}\left\|\begin{matrix}\langle i_{s(AC)}^j\rangle\\ i_{r(DC)}^j\\ \Omega^j\end{matrix}\right\|^T\left\|\begin{matrix}\langle L_s^*(\theta)\rangle & \langle L_{sr}^*(\theta)\rangle & 0\\ \langle L_{rs}^*(\theta)\rangle & 0 & 0\\ 0 & 0 & 0\end{matrix}\right\|\left\|\begin{matrix}\langle i_{s(AC)}^j\rangle\\ i_{r(DC)}^j\\ \Omega^j\end{matrix}\right\|$$

$$+\left\|\begin{matrix}\langle i_{s(AC)}^j\rangle\\ i_{r(DC)}^j\\ \Omega^j\end{matrix}\right\|^T\left\|\begin{matrix}\langle R_s\rangle & \langle 0\rangle & 0\\ \langle 0\rangle & R_r & 0\\ 0 & 0 & D\end{matrix}\right\|+\mathrm{sgn}\left\|\begin{matrix}\langle i_{s(AC)}^j\rangle\\ i_{r(DC)}^j\\ \Omega^j\end{matrix}\right\|^T\left\|\begin{matrix}\langle \Delta V_i^{s(AC)}\rangle & \langle 0\rangle & 0\\ \langle 0\rangle & \Delta V_i^{r(DC)} & 0\\ 0 & 0 & \Delta T_i\end{matrix}\right\|$$

$$\left.-\left\|\begin{matrix}\langle V_i^{s(AC)}\rangle\\ V_i^{r(DC)}\\ T_i\end{matrix}\right\|^T\right\}\delta\left\|\begin{matrix}\langle q_{s(AC)}^j\rangle\\ q_{r(DC)}^j\\ \theta^j\end{matrix}\right\|=0$$

Introducing to the armature circuit a macroelectronic commutator (macrocommutator) we get:

$$\left\{\left(\frac{d}{dt}\left\|\begin{matrix}\langle C_{ss}^{ji}\rangle\langle i_i^{s(DC)}\rangle\\ i_i^{r(DC)}\\ \Omega^j\end{matrix}\right\|^T\right)\left\|\begin{matrix}\langle L_s(\theta)\rangle & \langle L_{sr}(\theta)\rangle & 0\\ \langle L_{rs}(\theta)\rangle & L_r & 0\\ 0 & 0 & J\end{matrix}\right\|\right.$$

$$+\left\|\begin{matrix}\langle C_{ss}^{ji}\rangle\langle i_i^{s(DC)}\rangle\\ i_i^{r(DC)}\\ \Omega^j\end{matrix}\right\|^T\left\|\begin{matrix}\langle L_s^*(\theta)\rangle & \langle L_{sr}^*(\theta)\rangle & 0\\ \langle L_{rs}^*(\theta)\rangle & 0 & 0\\ 0 & 0 & 0\end{matrix}\right\|\Omega^j$$

$$-\frac{1}{2}\left\|\begin{matrix}\langle C_{ss}^{ji}\rangle\langle i_i^{s(DC)}\rangle\\ i_i^{r(DC)}\\ \Omega^j\end{matrix}\right\|^T\left\|\begin{matrix}\langle L_s^*(\theta)\rangle & \langle L_{sr}^*(\theta)\rangle & 0\\ \langle L_{rs}^*(\theta)\rangle & 0 & 0\\ 0 & 0 & 0\end{matrix}\right\|\left\|\begin{matrix}\langle C_{ss}^{ji}\rangle\langle i_i^{s(DC)}\rangle\\ i_i^{r(DC)}\\ \Omega^j\end{matrix}\right\|$$

$$+\left\|\begin{matrix}\langle C_{ss}^{ji}\rangle\langle i_i^{s(DC)}\rangle\\ i_i^{r(DC)}\\ \Omega^j\end{matrix}\right\|^T\left\|\begin{matrix}\langle R_s\rangle & 0 & 0\\ 0 & R_r & 0\\ 0 & 0 & D\end{matrix}\right\|$$

$$+\,\mathrm{sgn}\left\|\begin{matrix}\langle C_{ss}^{ji}\rangle\langle i_i^{s(DC)}\rangle\\ i_i^{r(DC)}\\ \Omega^j\end{matrix}\right\|^T\left\|\begin{matrix}\langle C_{ij}^{ss}\rangle\langle \Delta V_{s(DC)}^j\rangle & 0 & 0\\ 0 & 0 & 0\\ 0 & 0 & \Delta T\end{matrix}\right\|$$

$$\left.-\left\|\begin{matrix}\langle C_{ij}^{ss}\rangle\langle V_{s(DC)}^j\rangle\\ V_i^{r(DC)}\\ T_i\end{matrix}\right\|^T\right\}\delta\left\|\begin{matrix}\langle C_{ss}^{ji}\rangle\langle q_i^{s(DC)}\rangle\\ q_i^{r(DC)}\\ \theta^j\end{matrix}\right\|=0$$

where in:

$$\langle V_i^{s(AC)}\rangle=\langle C_{ij}^{ss}\rangle\langle V_{s(DC)}^j\rangle$$

$$\begin{pmatrix}V_i^{a(AC)}\\ V_i^{b(AC)}\\ V_i^{c(AC)}\end{pmatrix}=\begin{pmatrix}C_{ij}^{Pa} & C_{ij}^{Na}\\ C_{ij}^{Pb} & C_{ij}^{Nb}\\ C_{ij}^{Pc} & C_{ij}^{Nc}\end{pmatrix}\begin{pmatrix}V_{P(DC)}^j\\ V_{N(DC)}^j\end{pmatrix}$$

$$\langle i_i^{s(DC)}\rangle=\langle C_{ij}^{ss}\rangle^T\langle i_{s(AC)}^j\rangle$$

$$\begin{pmatrix}i_i^{P(DC)}\\ i_i^{N(DC)}\end{pmatrix}=\begin{pmatrix}C_{ij}^{Pa} & C_{ij}^{Pb} & C_{ij}^{Pc}\\ C_{ij}^{Na} & C_{ij}^{Nb} & C_{ij}^{Nc}\end{pmatrix}\begin{pmatrix}i_{a(AC)}^j\\ i_{b(AC)}^j\\ i_{c(AC)}^j\end{pmatrix}$$

or

$$\langle i^{j}_{s(AC)} \rangle = \langle C^{ji}_{ss} \rangle \langle i^{s(DC)}_{i} \rangle$$

$$\begin{pmatrix} i^{j}_{a(AC)} \\ i^{j}_{b(AC)} \\ i^{j}_{c(AC)} \end{pmatrix} = \begin{pmatrix} C^{ji}_{aP} & C^{ji}_{aN} \\ C^{ji}_{bP} & C^{ji}_{bN} \\ C^{ji}_{cP} & C^{ji}_{cN} \end{pmatrix} \begin{pmatrix} i_{i}^{P(DC)} \\ i_{i}^{N(DC)} \end{pmatrix}$$

$$\begin{pmatrix} C^{ji}_{aP} & C^{ji}_{aN} \\ C^{ji}_{bP} & C^{ji}_{bN} \\ C^{ji}_{cP} & C^{ji}_{cN} \end{pmatrix} = \begin{pmatrix} C_{ij}^{Pa} & C_{ij}^{Pb} & C_{ij}^{Pc} \\ C_{ij}^{Na} & C_{ij}^{Nb} & C_{ij}^{Nc} \end{pmatrix}^{T}$$

7.5 DC–AC commutator synchronous motor with the macroelectronic commutator (macrocommutator) and magnetoelectrical exciter

DC–AC commutator synchronous motors with the macroelectronic commutator (macrocommutator) and electromagnetical exciter is a rotating MMD electrical machine that converts electrical energy into mechanical energy. It operates on DC electricity. Similar to a generic DC motor, a DC–AC commutator synchronous motor with a macrocommutator has both a rotor and a stator. In a traditional DC motor the armature conductors will rotate and the magnetic-field exciter comprising the stator remains physically static. But in a DC–AC commutator synchronous motor with a macrocommutator, the roles of the armature conductors and the magnetic-field exciter are reversed. Here the armature conductors remain stationary and the magnetic-field exciter rotates. Due to this feature a DC–AC commutator synchronous motors with a macrocommutator is equivalent to a reversed DC commutator motor. In this the armature remains static and the magnets rotate. DC–AC commutator synchronous motors with a macroelectronic commutator (macrocommutator) and electromagnetical exciter do not have a wound armature in the rotor, rather they use an IPM rotor and macroelectronics to create a rotating magnetic field in the stator. Commonly, the magnetic-field exciter is provided by IPMs on the rotor and the armature is provided by a stator winding. Hall effect sensors are used to determine the position of the rotor and this synchronises the AC voltage applied to the stator. The maximum size of this DC motor is limited by the size and strength of the IPMs that can be used.

In the expression for conservative co-energy of a generalised MMD electrical machine with IPMs considered as an electrical dynamical system containing IPMs, the words taking into account the interaction of electric currents of the dynamical system with the current i_o and the own energy of IPMs are taken into account $1/2L_o i_o^2$ cannot be taken into account, because with the assumption i_o = const it has no effect on the form of Euler–Lagrange differential equations of the second-order.

In this case, it is not necessary to derive the differential equation of the dynamics of the exciter of the MMD electrical machine corresponding to the coordinate of the electric charge

$$q_0 = \int_0^t i_0 dt$$

representing the current of the exciter i_0 because such an equation assuming i_0 = const would give a condition for the source voltage applied to the electromagnet, maintaining a constant value of current i_0.

A physical model of the DC–AC commutator synchronous motor with the macroelectronic commutator (macrocommutator) and magnetoelectrical exciter (DC motor) is shown in figure 7.4.

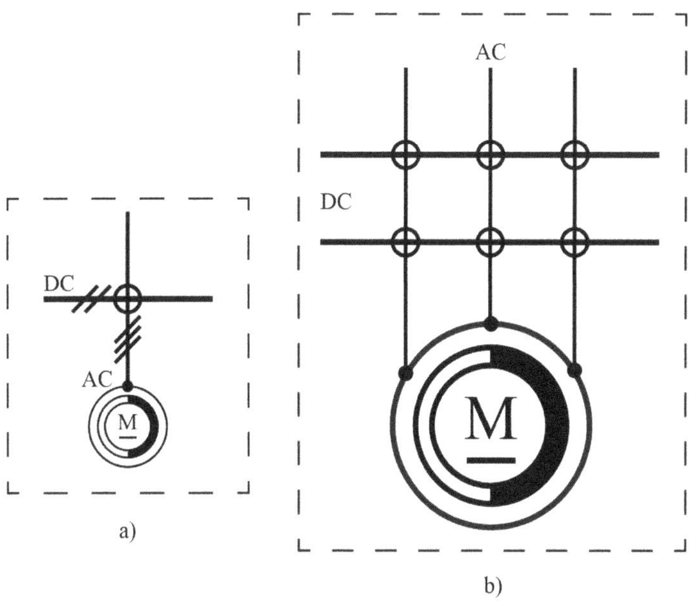

Figure 7.4. Physical model of the DC–AC commutator synchronous motor with macroelectronic commutator (macrocommutator) and magnetoelectrical exciter (DC motor).

Taking into account equation (7.10) one will get:

$$\left\{\left(\frac{d}{dt}\begin{bmatrix}\left\|\begin{matrix}\langle i_s^j\rangle\\ \langle i_r^j\rangle\end{matrix}\right\|\\ \Omega^j\end{bmatrix}^T\right)\begin{bmatrix}\left\|\begin{matrix}\langle L_s(\theta)\rangle & \langle L_{sr}(\theta)\rangle\\ \langle L_{rs}(\theta)\rangle & \langle L_r(\theta)\rangle\end{matrix}\right\| & \|0\|\\ \|0\| & J\end{bmatrix}\right.$$

$$+\begin{bmatrix}\left\|\begin{matrix}\langle i_s^j\rangle\\ \langle i_r^j\rangle\end{matrix}\right\|\\ \Omega^j\end{bmatrix}^T\begin{bmatrix}\left\|\begin{matrix}\langle L_s^*(\theta)\rangle & \langle L_{sr}^*(\theta)\rangle\\ \langle L_{rs}^*(\theta)\rangle & \langle L_r^*(\theta)\rangle\end{matrix}\right\| & \|0\|\\ \|0\| & 0\end{bmatrix}\Omega^j$$

$$-\frac{1}{2}\begin{bmatrix}\left\|\begin{matrix}\langle i_s^j\rangle\\ \langle i_r^j\rangle\end{matrix}\right\|\\ \Omega^j\end{bmatrix}^T\begin{bmatrix}\left\|\begin{matrix}\langle L_s^*(\theta)\rangle & \langle L_{sr}^*(\theta)\rangle\\ \langle L_{rs}^*(\theta)\rangle & \langle L_r^*(\theta)\rangle\end{matrix}\right\| & \|0\|\\ \|0\| & 0\end{bmatrix}\begin{bmatrix}\left\|\begin{matrix}\langle i_s^j\rangle\\ \langle i_r^j\rangle\end{matrix}\right\|\\ \Omega^j\end{bmatrix}$$

$$+\begin{bmatrix}\left\|\begin{matrix}\langle i_s^j\rangle\\ \langle i_r^j\rangle\end{matrix}\right\|\\ \Omega^j\end{bmatrix}^T\begin{bmatrix}\left\|\begin{matrix}\langle R_s\rangle & \langle 0\rangle\\ \langle 0\rangle & \langle R_r\rangle\end{matrix}\right\| & \|0\|\\ \|0\| & D\end{bmatrix}$$

$$\left.+\,\mathrm{sgn}\begin{bmatrix}\left\|\begin{matrix}\langle i_s^j\rangle\\ \langle i_r^j\rangle\end{matrix}\right\|\\ \Omega^j\end{bmatrix}^T\begin{bmatrix}\left\|\begin{matrix}\langle \Delta V_i^s\rangle & \langle 0\rangle\\ \langle 0\rangle & \langle \Delta V_i^r\rangle\end{matrix}\right\| & \|0\|\\ \|0\| & \Delta T_i\end{bmatrix}-\begin{bmatrix}\left\|\begin{matrix}\langle V_i^s\rangle\\ \langle V_i^r\rangle\end{matrix}\right\|\\ T_i\end{bmatrix}^T\right\}\delta\begin{bmatrix}\left\|\begin{matrix}\langle q_s^j\rangle\\ \langle q_r^j\rangle\end{matrix}\right\|\\ \theta^j\end{bmatrix}=0$$

or in detail

$$\left\{\left(\frac{d}{dt}\left\|\begin{matrix}\langle i_{s(AC)}^j\rangle\\ i_{r(DC)}^j\\ \Omega^j\end{matrix}\right\|^T\right)\left\|\begin{matrix}\langle L_s(\theta)\rangle & \langle L_{sr}(\theta)\rangle & 0\\ \langle L_{rs}(\theta)\rangle & L_r & 0\\ 0 & 0 & J\end{matrix}\right\|+\left\|\begin{matrix}\langle i_{s(AC)}^j\rangle\\ i_{r(DC)}^j\\ \Omega^j\end{matrix}\right\|^T\left\|\begin{matrix}\langle L_s^*(\theta)\rangle & \langle L_{sr}^*(\theta)\rangle & 0\\ \langle L_{rs}^*(\theta)\rangle & 0 & 0\\ 0 & 0 & 0\end{matrix}\right\|\Omega^j\right.$$

$$-\frac{1}{2}\left\|\begin{matrix}\langle i_{s(AC)}^j\rangle\\ i_{r(DC)}^j\\ \Omega^j\end{matrix}\right\|^T\left\|\begin{matrix}\langle L_s^*(\theta)\rangle & \langle L_{sr}^*(\theta)\rangle & 0\\ \langle L_{rs}^*(\theta)\rangle & 0 & 0\\ 0 & 0 & 0\end{matrix}\right\|\left\|\begin{matrix}\langle i_{s(AC)}^j\rangle\\ i_{r(DC)}^j\\ \Omega^j\end{matrix}\right\|$$

$$+\left\|\begin{matrix}\langle i_{s(AC)}^j\rangle\\ i_{r(DC)}^j\\ \Omega^j\end{matrix}\right\|^T\left\|\begin{matrix}\langle R_s\rangle & \langle 0\rangle & 0\\ \langle 0\rangle & 0 & 0\\ 0 & 0 & D\end{matrix}\right\|+\,\mathrm{sgn}\left\|\begin{matrix}\langle i_{s(AC)}^j\rangle\\ i_{r(DC)}^j\\ \Omega^j\end{matrix}\right\|^T\left\|\begin{matrix}\langle \Delta V_i^{s(AC)}\rangle & \langle 0\rangle & 0\\ \langle 0\rangle & 0 & 0\\ 0 & 0 & \Delta T_i\end{matrix}\right\|$$

$$\left.-\left\|\begin{matrix}\langle V_i^{s(AC)}\rangle\\ 0\\ T_i\end{matrix}\right\|^T\right\}\delta\left\|\begin{matrix}\langle q_{s(AC)}^j\rangle\\ q_{r(DC)}^j\\ \theta^j\end{matrix}\right\|=0$$

Introducing to the armature circuit a macroelectronic commutator (macrocommutator) one will get:

$$\left\{\left(\frac{d}{dt}\begin{Vmatrix}\langle C_{ss}^{ji}\rangle\langle i_i^{s(DC)}\rangle\\ i_i^{r(DC)}\\ \Omega^j\end{Vmatrix}^T\right)\begin{Vmatrix}\langle L_s(\theta)\rangle & \langle L_{sr}(\theta)\rangle & 0\\ \langle L_{rs}(\theta)\rangle & L_r & 0\\ 0 & 0 & J\end{Vmatrix}\right.$$

$$+\begin{Vmatrix}\langle C_{ss}^{ji}\rangle\langle i_i^{s(DC)}\rangle\\ i_i^{r(DC)}\\ \Omega^j\end{Vmatrix}^T\begin{Vmatrix}\langle L_s^*(\theta)\rangle & \langle L_{sr}^*(\theta)\rangle & 0\\ \langle L_{rs}^*(\theta)\rangle & 0 & 0\\ 0 & 0 & 0\end{Vmatrix}\Omega^j$$

$$-\frac{1}{2}\begin{Vmatrix}\langle C_{ss}^{ji}\rangle\langle i_i^{s(DC)}\rangle\\ i_i^{r(DC)}\\ \Omega^j\end{Vmatrix}^T\begin{Vmatrix}\langle L_s^*(\theta)\rangle & \langle L_{sr}^*(\theta)\rangle & 0\\ \langle L_{rs}^*(\theta)\rangle & 0 & 0\\ 0 & 0 & 0\end{Vmatrix}\begin{Vmatrix}\langle C_{ss}^{ji}\rangle\langle i_i^{s(DC)}\rangle\\ i_i^{r(DC)}\\ \Omega^j\end{Vmatrix}$$

$$+\begin{Vmatrix}\langle C_{ss}^{ji}\rangle\langle i_i^{s(DC)}\rangle\\ i_i^{r(DC)}\\ \Omega^j\end{Vmatrix}^T\begin{Vmatrix}\langle R_s\rangle & 0 & 0\\ 0 & 0 & 0\\ 0 & 0 & D\end{Vmatrix}$$

$$+\operatorname{sgn}\begin{Vmatrix}\langle C_{ss}^{ji}\rangle\langle i_i^{s(DC)}\rangle\\ i_i^{r(DC)}\\ \Omega^j\end{Vmatrix}^T\begin{Vmatrix}\langle C_{ij}^{ss}\rangle\langle \Delta V_{s(DC)}^j\rangle & 0 & 0\\ 0 & 0 & 0\\ 0 & 0 & \Delta T\end{Vmatrix}$$

$$\left.-\begin{Vmatrix}\langle C_{ij}^{ss}\rangle\langle V_{s(DC)}^j\rangle\\ 0\\ T_i\end{Vmatrix}^T\right\}\delta\begin{Vmatrix}\langle C_{ss}^{ji}\rangle\langle q_i^{s(DC)}\rangle\\ q_i^{r(DC)}\\ \theta^j\end{Vmatrix}=0$$

where in:

$$\langle V_i^{s(AC)}\rangle=\langle C_{ij}^{ss}\rangle\langle V_{s(DC)}^j\rangle$$

$$\begin{pmatrix}V_i^{a(AC)}\\ V_i^{b(AC)}\\ V_i^{c(AC)}\end{pmatrix}=\begin{pmatrix}C_{ij}^{Pa} & C_{ij}^{Na}\\ C_{ij}^{Pb} & C_{ij}^{Nb}\\ C_{ij}^{Pc} & C_{ij}^{Nc}\end{pmatrix}\begin{pmatrix}V_{P(DC)}^j\\ V_{N(DC)}^j\end{pmatrix}$$

$$\langle i_i^{s(DC)}\rangle=\langle C_{ij}^{ss}\rangle\langle i_{s(AC)}^j\rangle$$

$$\begin{pmatrix}i_i^{P(DC)}\\ i_i^{N(DC)}\end{pmatrix}=\begin{pmatrix}C_{ij}^{Pa} & C_{ij}^{Pb} & C_{ij}^{Pc}\\ C_{ij}^{Na} & C_{ij}^{Nb} & C_{ij}^{Nc}\end{pmatrix}\begin{pmatrix}i_{a(AC)}^j\\ i_{b(AC)}^j\\ i_{c(AC)}^j\end{pmatrix}$$

or

$$\langle i^{j}_{s(AC)} \rangle = \langle C^{ji}_{ss} \rangle \langle i^{s(DC)}_{i} \rangle$$

$$\begin{pmatrix} i^{j}_{a(AC)} \\ i^{j}_{b(AC)} \\ i^{j}_{c(AC)} \end{pmatrix} = \begin{pmatrix} C^{ji}_{aP} & C^{ji}_{aN} \\ C^{ji}_{bP} & C^{ji}_{bN} \\ C^{ji}_{cP} & C^{ji}_{cN} \end{pmatrix} \begin{pmatrix} i^{P(DC)}_{i} \\ i^{N(DC)}_{i} \end{pmatrix}$$

$$\begin{pmatrix} C^{ji}_{aP} & C^{ji}_{aN} \\ C^{ji}_{bP} & C^{ji}_{bN} \\ C^{ji}_{cP} & C^{ji}_{cN} \end{pmatrix} = \begin{pmatrix} C^{Pa}_{ij} & C^{Pb}_{ij} & C^{Pc}_{ij} \\ C^{Na}_{ij} & C^{Nb}_{ij} & C^{Nc}_{ij} \end{pmatrix}^{T}$$

7.6 DC–AC commutator variable-reluctance synchronous motor with the macroelectronic commutator (macrocommutator)

This section describes a mathematical model of the DC–AC commutator variable-reluctance synchronous motor with a macroelectronic commutator (macrocommutator). The mathematical model of the DC–AC commutator reluctance synchronous motor is nonparametric and can only be established with experimental data, instead of an analytical representation. Because the reluctance varies with rotor position and magnetic saturation is part of the normal operation of DC–AC commutator variable-reluctance synchronous motors, there is no simple analytical expression for the magnetic field produced by the phase windings. The shape of phase current before commutation is of interest because it varies widely depending on when the phase winding is excited and what the rotor speed is. To illustrate this effect, two step response simulations may be done in Matlab/Simulink.

The DC–AC commutator variable-reluctance synchronous motor model used in these two simulations is a 6/4 linear magnetics model. For the first simulation, a step voltage is fed into phase *a* and the initial rotor position is set to be 1° instead of 0° so that the rotor will move in the positive sense of direction. The results show that the rotor stops at 45° after some oscillation which is the aligned position of this phase *a*.

For the second simulation, a step voltage is fed into phase *c*. The initial position is 0°. According to this, the rotor will move towards the aligned position of phase *c*, i.e. 15°.

Multi-phase DC–AC commutator variable-reluctance synchronous motor:

- take into account the internal and mutual inductances of an m-phase armature winding as functions of angular position and current;
  - the influence of auxiliary circuits of the macroelectronic commutator (macrocommutator) on the course of commutation is neglected;
  - losses on eddy currents and hysteresis in the magnetic circuit of the rotor and armature are omitted;
  - the rotor has no electrical circuits.

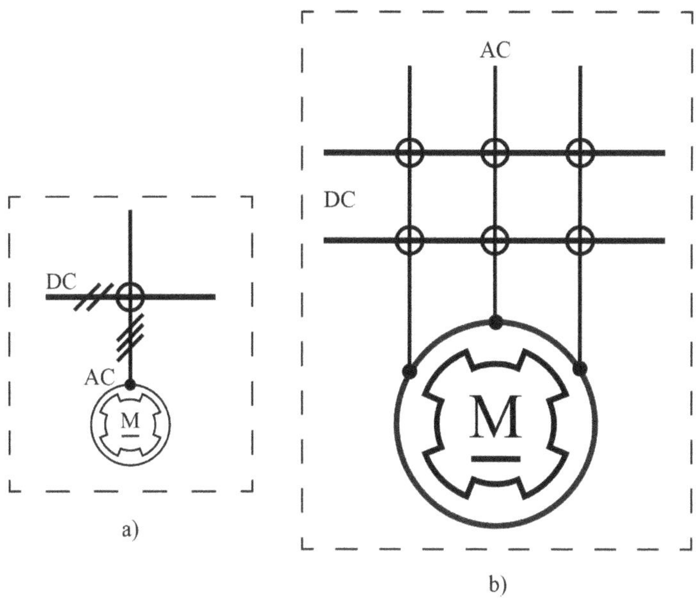

Figure 7.5. Physical model of the DC–AC commutator variable-reluctance synchronous motor with macro-electronic commutator (macrocommutator).

A physical model of the DC–AC commutator variable-reluctance synchronous motor with the macroelectronic commutator (macrocommutator) is shown in figure 7.5.

Taking into account equation (7.10) one will get:

$$\left\{ \left(\frac{d}{dt} \begin{bmatrix} \left\| \begin{matrix} \langle i_s^j \rangle \\ \langle i_r^j \rangle \end{matrix} \right\| \\ \Omega^j \end{bmatrix}^T \right) \begin{bmatrix} \left\| \begin{matrix} \langle L_s(\theta) \rangle & \langle L_{sr}(\theta) \rangle \\ \langle L_{rs}(\theta) \rangle & \langle L_r(\theta) \rangle \end{matrix} \right\| & \|0\| \\ \|0\| & J \end{bmatrix} + \begin{bmatrix} \left\| \begin{matrix} \langle i_s^j \rangle \\ \langle i_r^j \rangle \end{matrix} \right\| \\ \Omega^j \end{bmatrix}^T \begin{bmatrix} \left\| \begin{matrix} \langle L_s^*(\theta) \rangle & \langle L_{sr}^*(\theta) \rangle \\ \langle L_{rs}^*(\theta) \rangle & \langle L_r^*(\theta) \rangle \end{matrix} \right\| & \|0\| \\ \|0\| & 0 \end{bmatrix} \Omega^j \right.$$

$$- \frac{1}{2} \begin{bmatrix} \left\| \begin{matrix} \langle i_s^j \rangle \\ \langle i_r^j \rangle \end{matrix} \right\| \\ \Omega^j \end{bmatrix}^T \begin{bmatrix} \left\| \begin{matrix} \langle L_s^*(\theta) \rangle & \langle L_{sr}^*(\theta) \rangle \\ \langle L_{rs}^*(\theta) \rangle & \langle L_r^*(\theta) \rangle \end{matrix} \right\| & \|0\| \\ \|0\| & 0 \end{bmatrix} \begin{bmatrix} \left\| \begin{matrix} \langle i_s^j \rangle \\ \langle i_r^j \rangle \end{matrix} \right\| \\ \Omega^j \end{bmatrix} + \begin{bmatrix} \left\| \begin{matrix} \langle i_s^j \rangle \\ \langle i_r^j \rangle \end{matrix} \right\| \\ \Omega^j \end{bmatrix}^T \begin{bmatrix} \left\| \begin{matrix} \langle R_s \rangle & \langle 0 \rangle \\ \langle 0 \rangle & \langle R_r \rangle \end{matrix} \right\| & \|0\| \\ \|0\| & D \end{bmatrix}$$

$$\left. + \operatorname{sgn} \begin{bmatrix} \left\| \begin{matrix} \langle i_s^j \rangle \\ \langle i_r^j \rangle \end{matrix} \right\| \\ \Omega^j \end{bmatrix}^T \begin{bmatrix} \left\| \begin{matrix} \langle \Delta V_i^s \rangle & \langle 0 \rangle \\ \langle 0 \rangle & \langle \Delta V_i^r \rangle \end{matrix} \right\| & \|0\| \\ \|0\| & \Delta T_i \end{bmatrix} - \begin{bmatrix} \left\| \begin{matrix} \langle V_i^s \rangle \\ \langle V_i^r \rangle \end{matrix} \right\| \\ T_i \end{bmatrix}^T \right\} \delta \begin{bmatrix} \left\| \begin{matrix} \langle q_s^j \rangle \\ \langle q_r^j \rangle \end{matrix} \right\| \\ \theta^j \end{bmatrix} = 0$$

or in detail:

$$
\begin{aligned}
&\Bigg\{ \left(\frac{d}{dt} \begin{Vmatrix} \langle i_i^{s(AC)} \rangle \\ \Omega^j \end{Vmatrix}^T \right) \begin{Vmatrix} \langle L_s(\theta, i) \rangle & 0 \\ 0 & J \end{Vmatrix} + \begin{Vmatrix} \langle i_i^{s(AC)} \rangle \\ \Omega^j \end{Vmatrix}^T \begin{Vmatrix} \langle L_s^*(\theta, i) \rangle & 0 \\ 0 & 0 \end{Vmatrix} \Omega^j \\
&+ \left(\begin{Vmatrix} \langle i_i^{s(AC)} \rangle \\ \Omega^j \end{Vmatrix}^T \begin{Vmatrix} \langle L_s^{**}(\theta, i) \rangle & 0 \\ 0 & 0 \end{Vmatrix} \right) \frac{d}{dt} \begin{Vmatrix} \langle i_i^{s(AC)} \rangle \\ \Omega^j \end{Vmatrix} \\
&- \frac{1}{2} \begin{Vmatrix} \langle i_i^{s(AC)} \rangle \\ \Omega^j \end{Vmatrix}^T \begin{Vmatrix} \langle L_s^*(\theta, i) \rangle & 0 \\ 0 & 0 \end{Vmatrix} \begin{Vmatrix} \langle i_i^{s(AC)} \rangle \\ \Omega^j \end{Vmatrix} + \begin{Vmatrix} \langle i_i^{s(AC)} \rangle \\ \Omega^j \end{Vmatrix}^T \begin{Vmatrix} \langle R_s \rangle & 0 \\ 0 & D \end{Vmatrix} \\
&+ \operatorname{sgn} \begin{Vmatrix} \langle i_i^{s(AC)} \rangle \\ \Omega^j \end{Vmatrix}^T \begin{Vmatrix} \langle V_i^{s(AC)} \rangle & 0 \\ 0 & \Delta T_i \end{Vmatrix} - \begin{Vmatrix} \langle V_i^{s(AC)} \rangle \\ T_i \end{Vmatrix}^T \Bigg\} \delta \begin{Vmatrix} \langle q_i^{s(AC)} \rangle \\ \theta^j \end{Vmatrix} = 0
\end{aligned}
$$

Introducing to the armature circuit a macroelectronic commutator (macrocommutator) one will get:

$$
\begin{aligned}
&\Bigg\{ \left(\frac{d}{dt} \begin{Vmatrix} \langle C_{ss}^{ji} \rangle \langle i_i^{s(DC)} \rangle \\ \Omega^j \end{Vmatrix}^T \right) \begin{Vmatrix} \langle L_s(\theta, i) \rangle & 0 \\ 0 & J \end{Vmatrix} \\
&+ \begin{Vmatrix} \langle C_{ss}^{ji} \rangle \langle i_i^{s(DC)} \rangle \\ \Omega^j \end{Vmatrix}^T \begin{Vmatrix} \langle L_s^*(\theta, i) \rangle & 0 \\ 0 & 0 \end{Vmatrix} \Omega^j \\
&+ \left(\begin{Vmatrix} \langle C_{ss}^{ji} \rangle \langle i_i^{s(DC)} \rangle \\ \Omega^j \end{Vmatrix}^T \begin{Vmatrix} \langle L_s^{**}(\theta, i) \rangle & 0 \\ 0 & 0 \end{Vmatrix} \right) \frac{d}{dt} \begin{Vmatrix} \langle C_{ss}^{ji} \rangle \langle i_i^{s(DC)} \rangle \\ \Omega^j \end{Vmatrix} \\
&- \frac{1}{2} \begin{Vmatrix} \langle C_{ss}^{ji} \rangle \langle i_i^{s(DC)} \rangle \\ \Omega^j \end{Vmatrix}^T \begin{Vmatrix} \langle L_s^*(\theta, i) \rangle & 0 \\ 0 & 0 \end{Vmatrix} \begin{Vmatrix} \langle C_{ss}^{ji} \rangle \langle i_i^{s(DC)} \rangle \\ \Omega^j \end{Vmatrix} \\
&+ \begin{Vmatrix} \langle C_{ss}^{ji} \rangle \langle i_i^{s(DC)} \rangle \\ \Omega^j \end{Vmatrix}^T \begin{Vmatrix} \langle R_s \rangle & 0 \\ 0 & D \end{Vmatrix} \\
&+ \operatorname{sgn} \begin{Vmatrix} \langle C_{ss}^{ji} \rangle \langle i_i^{s(DC)} \rangle \\ \Omega^j \end{Vmatrix}^T \begin{Vmatrix} \langle C_{ij}^{ss} \rangle \langle \Delta V_{s(DC)}^j \rangle & 0 \\ 0 & \Delta T_i \end{Vmatrix} \\
&- \begin{Vmatrix} \langle C_{ij}^{ss} \rangle \langle V_i^{s(DC)} \rangle \\ T_i \end{Vmatrix}^T \Bigg\} \delta \begin{Vmatrix} \langle C_{ss}^{ji} \rangle \langle q_i^{s(DC)} \rangle \\ \theta^j \end{Vmatrix} = 0
\end{aligned}
$$

where in:

$$\langle V_i^{s(AC)} \rangle = \langle C_{ij}^{ss} \rangle \langle V_{s(DC)}^{j} \rangle$$

$$\begin{pmatrix} V_i^{a(AC)} \\ V_i^{b(AC)} \\ V_i^{c(AC)} \end{pmatrix} = \begin{pmatrix} C_{ij}^{Pa} & C_{ij}^{Na} \\ C_{ij}^{Pb} & C_{ij}^{Nb} \\ C_{ij}^{Pc} & C_{ij}^{Nc} \end{pmatrix} \begin{pmatrix} V_{P(DC)}^{j} \\ V_{N(DC)}^{j} \end{pmatrix}$$

$$\langle i_i^{s(DC)} \rangle = \langle C_{ij}^{ss} \rangle^T \langle i_{s(AC)}^{j} \rangle$$

$$\begin{pmatrix} i_i^{P(DC)} \\ i_i^{N(DC)} \end{pmatrix} = \begin{pmatrix} C_{ij}^{Pa} & C_{ij}^{Pb} & C_{ij}^{Pc} \\ C_{ij}^{Na} & C_{ij}^{Nb} & C_{ij}^{Nc} \end{pmatrix} \begin{pmatrix} i_{a(AC)}^{j} \\ i_{b(AC)}^{j} \\ i_{c(AC)}^{j} \end{pmatrix}$$

or

$$\langle i_{s(AC)}^{j} \rangle = \langle C_{ss}^{ji} \rangle \langle i_i^{s(DC)} \rangle$$

$$\begin{pmatrix} i_{a(AC)}^{j} \\ i_{b(AC)}^{j} \\ i_{c(AC)}^{j} \end{pmatrix} = \begin{pmatrix} C_{aP}^{ji} & C_{aN}^{ji} \\ C_{bP}^{ji} & C_{bN}^{ji} \\ C_{cP}^{ji} & C_{cN}^{ji} \end{pmatrix} \begin{pmatrix} i_i^{P(DC)} \\ i_i^{N(DC)} \end{pmatrix}$$

$$\begin{pmatrix} C_{aP}^{ji} & C_{aN}^{ji} \\ C_{bP}^{ji} & C_{bN}^{ji} \\ C_{cP}^{ji} & C_{cN}^{ji} \end{pmatrix} = \begin{pmatrix} C_{ij}^{Pa} & C_{ij}^{Pb} & C_{ij}^{Pc} \\ C_{ij}^{Na} & C_{ij}^{Nb} & C_{ij}^{Nc} \end{pmatrix}^T$$

7.7 DC–AC commutator squirrel-cage-rotor asynchronous (induction) motor with the macroelectronic commutator (macrocommutator)

The stator is the outer body of the DC–AC commutator squirrel-cage rotor asynchronous (induction) motor with the macroelectronic commutator (macrocommutator) motor which houses the driven windings on an iron core. In a single-speed three-phase EM motor design, the standard stator has three windings, while a single-phase EM motor typically has two windings. The number of magnetic poles determines the speed of the DC–AC commutator squirrel-cage-rotor asynchronous (induction) motor. A two magnetic-pole EM motor has a synchronous speed of 3000 rpm at 50 Hz and 3600 rpm 60 Hz. A four magnetic-pole EM motor runs at half this speed, a six magnetic-pole EM motor at one third speed and an eight magnetic pole EM motor at one quarter speed. The winding configuration, slot configuration and lamination steel all have an effect on the performance of the EM motor. The voltage rating of the EM motor is determined by the number of turns on the stator and the power rating of the EM motor is determined by the losses which comprise copper loss and iron loss, and the ability of the EM motor to dissipate the heat generated by these losses. The stator design determines the rated speed of the EM motor and most of the full load, full speed characteristics.

The rotor comprises a cylinder made up of round laminations pressed onto the EM motor shaft, and a number of short-circuited windings. The rotor windings are made up of rotor bars passed through the rotor, from one end to the other, around the surface of the rotor. The bars protrude beyond the rotor and are connected together by a shorting ring at each end. The bars are usually made of aluminium or copper, but are sometimes made of brass. The position relative to the surface of the rotor, shape, cross-sectional area and material of the bars determine the rotor characteristics.

Essentially, the rotor windings exhibit inductance and resistance, and these characteristics can effectively be dependent on the frequency of the current flowing in the rotor.

A bar with a large cross-sectional area will exhibit a low resistance, while a bar of a small cross-sectional area will exhibit a high resistance. Likewise a copper bar will have a low resistance compared to a brass bar of equal proportions. Positioning the bar deeper into the rotor, increases the amount of iron around the bar, and consequently increases the inductance exhibited by the rotor.

The impedance of the bar is made up of both resistance and inductance, and so two bars of equal dimensions will exhibit a different AC impedance depending on their position relative to the surface of the rotor.

A thin bar which is inserted radially into the rotor, with one edge near the surface of the rotor and the other edge towards the shaft, will effectively change in resistance as the frequency of the current changes. This is because the AC impedance of the outer portion of the bar is lower than the inner impedance at high frequencies lifting the effective impedance of the bar relative to the impedance of the bar at low frequencies where the impedance of both edges of the bar will be lower and almost equal. The rotor design determines the starting characteristics.

Additionally, the full voltage starting current of a particular EM motor is voltage and speed dependant, but not load dependant. The magnetising current varies depending on the design of the DC motor. For small DC motors, the magnetising current may be as high as 60%, but for large two magnetic-pole EM motors, the magnetising current is more typically 20%–25%. At the design voltage, the iron is typically near saturation, so the iron loss and magnetising current do not vary linearly with voltage with small increases in voltage resulting in a high increase in magnetising current and iron loss.

The speed of DC–AC commutator squirrel-cage-rotor asynchronous (induction) motors can be controlled by variation of the frequency of the voltage applied to the EM motor. Due to magnetic-flux saturation problems with AC induction motors, the voltage applied to the EM motor must alter with the frequency. The AC induction motor is an AC pseudo-synchronous electrical machine and so behaves as a speed source. The running speed is set by the frequency applied to it and is independent of load torque provided the EM motor is not overloaded.

A modern adjustable-velocity IEMD come in two major formats, **voltage-to-frequency** (*V*/*F*) and holor or vector. The *V*/*F* IEMD is an EMD where the voltage applied to the EM motor is directly related to the frequency. In the ideal EM motor, the magnetic circuit would be purely inductive and keeping a constant *V*/*F* ratio would maintain a constant magnetic-flux in the iron.

The real EM motor has resistance in series with the magnetising inductance. This has no bearing on the operation at line frequency, however, as the frequency of the EMD is reduced, the resistance begins to become significant relative to the inductive reactance. This causes the magnetic flux to reduce at very low frequencies and so it is difficult to get sufficient torque at low speeds. For many applications, this low torque is not a problem, but there are some that do need a high torque from a low speed. Early EMDs were designed with a voltage boost to provide a measure of torque increase at low speed.

A holor or vector IEMD have a mathematical model of the EMD in software and by measuring the current holors or vectors in relation to the applied voltage, they are able to maintain a constant magnetic field at all frequencies below the line frequency. This IEMD needs to be tuned to the EM motor and typically includes a self-tuning algorithm that is enabled at commissioning to determine the component values for the mathematical model. If the EM motor is replaced, the EMD needs to be returned to learn the characteristics of the new EM motors.

A holor or vector IEMD comes in three major formats, closed loop, open loop and **direct torque control** (DTC). The closed loop microcontrollers were the first holor or vector microcontrollers and are still the best option for accurate control at zero speed. The open loop holor or vector and DTC are suitable for applications requiring good control above 3–5 Hz. Quite a number of a modern IEMDs can operate as *V*/*F*, open loop holor or vector or closed loop holor or vector just by changing a parameter—closed loop requires a shaft encoder to give accurate speed feedback.

The major differentiation between a modern IEMD is the enclosure, auxiliary functionality, programming and user interface. A low cost IEMD is often very poorly filtered and can create major RFI (EMC) issues. As a rule, an IEMD includes no filtering and must be installed with external filters, and others include all the filtering required. DC and AC reactors help to reduce the noise generated by the EMD, and to improve the distortion power factor of the IEMD. Because the IEMD rectifies the incoming supply, the current waveform is exceedingly distorted and so the harmonics are high. A low cost IEMD without the reactors has a very poor power factor. NB: Most IEMD suppliers quote the power factor as better than 0.95 implying a high power factor.

While the displacement power factor is high, the distortion power factor can be less than 0.7. Distortion power factor cannot be corrected with capacitors, but can be improved with expensive filters. There are 'active front end' or 'regenerative' IEMDs that have an AC–DC–AC commutator's inverter stage on the input as well as the output and these can draw sinusoidal current from the supply resulting in a high power factor. It is possible that this technology may become a mandatory requirement at some time in the future. An IEMD is typically used in some form of automation process and so they are now including additional functionality and controls to simplify the automation process. There are a number of programmable inputs and outputs and relays and an IEMD also includes a PID loop and a motorised potentiometer is also common. A holor or vector and certain *V*/*F* IEMD can be set-up for angular-velocity control or torque control. Torque control is used in tensioning applications such as paper machines where the master controls a winding drum and the diameter increases as the drum fills up. This requires other IEMDs feeding the paper to run at different speeds. Traditionally, this was achieved by conventional DC electrical machines as they naturally operate in torque mode.

A DC–AC commutator squirrel-cage-rotor asynchronous (induction) motor with the macroelectronic commutator (macrocommutator) and a compact frame has the following properties:

- three-phase winding on the stator and rotor;
- own inductances and mutual windings of the stator are not functions of the generalised mechanical coordinate (angle of the shaft position), similarly to the inductance of the rotor;
- stator–rotor mutual inductances are functions of a generalised mechanical coordinate (angle of the shaft position);
- the rotor phase circuits are shorted;
- no additional voltage drops are taken into account at the phase connections of the rotor winding ($\Delta V_i^{r(AC)}$).

A physical model of the DC–AC commutator squirrel-cage-rotor asynchronous (induction) motor with the macroelectronic commutator (macrocommutator) is shown in figure 7.6.

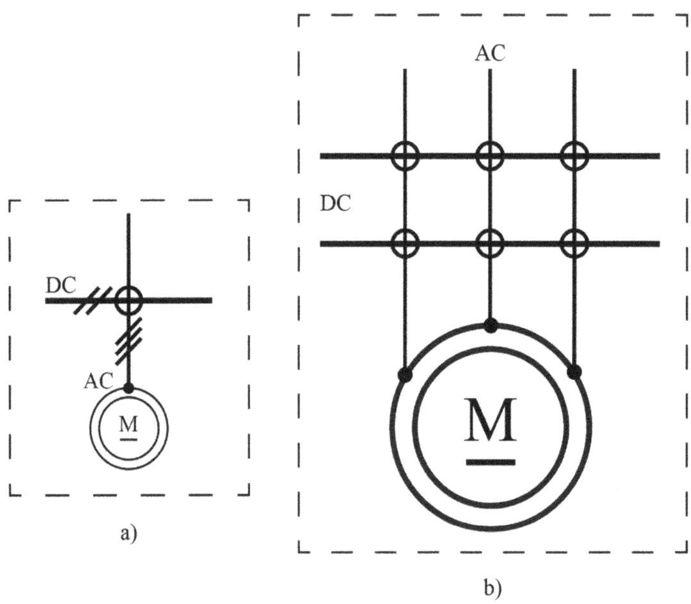

Figure 7.6. Physical model of the DC–AC commutator squirrel-cage-rotor asynchronous (induction) motor with the macroelectronic commutator (macrocommutator).

Taking into account equation (7.10) one will get:

$$
\begin{aligned}
&\left\{\left(\frac{d}{dt}\begin{bmatrix}\left\|\begin{matrix}\langle i_s^j\rangle\\ \langle i_r^j\rangle\end{matrix}\right\|\\ \Omega^j\end{bmatrix}^T\right)\begin{bmatrix}\left\|\begin{matrix}\langle L_s(\theta)\rangle & \langle L_{sr}(\theta)\rangle\\ \langle L_{rs}(\theta)\rangle & \langle L_r(\theta)\rangle\end{matrix}\right\| & \|0\|\\ \|0\| & J\end{bmatrix} + \begin{bmatrix}\left\|\begin{matrix}\langle i_s^j\rangle\\ \langle i_r^j\rangle\end{matrix}\right\|\\ \Omega^j\end{bmatrix}^T\begin{bmatrix}\left\|\begin{matrix}\langle L_s^*(\theta)\rangle & \langle L_{sr}^*(\theta)\rangle\\ \langle L_{rs}^*(\theta)\rangle & \langle L_r^*(\theta)\rangle\end{matrix}\right\| & \|0\|\\ \|0\| & 0\end{bmatrix}\Omega^j\right.\\
&-\frac{1}{2}\begin{bmatrix}\left\|\begin{matrix}\langle i_s^j\rangle\\ \langle i_r^j\rangle\end{matrix}\right\|\\ \Omega^j\end{bmatrix}^T\begin{bmatrix}\left\|\begin{matrix}\langle L_s^*(\theta)\rangle & \langle L_{sr}^*(\theta)\rangle\\ \langle L_{rs}^*(\theta)\rangle & \langle L_r^*(\theta)\rangle\end{matrix}\right\| & \|0\|\\ \|0\| & \Omega^j\end{bmatrix}\begin{bmatrix}\left\|\begin{matrix}\langle i_s^j\rangle\\ \langle i_r^j\rangle\end{matrix}\right\|\\ \Omega^j\end{bmatrix} + \begin{bmatrix}\left\|\begin{matrix}\langle i_s^j\rangle\\ \langle i_r^j\rangle\end{matrix}\right\|\\ \Omega^j\end{bmatrix}^T\begin{bmatrix}\left\|\begin{matrix}\langle R_s\rangle & \langle 0\rangle\\ \langle 0\rangle & \langle R_r\rangle\end{matrix}\right\| & \|0\|\\ \|0\| & D\end{bmatrix}\\
&\left.+\,\mathrm{sgn}\begin{bmatrix}\left\|\begin{matrix}\langle i_s^j\rangle\\ \langle i_r^j\rangle\end{matrix}\right\|\\ \Omega^j\end{bmatrix}^T\begin{bmatrix}\left\|\begin{matrix}\langle \Delta V_i^s\rangle & \langle 0\rangle\\ \langle 0\rangle & 0\end{matrix}\right\| & \|0\|\\ \|0\| & \Delta T_i\end{bmatrix} - \begin{bmatrix}\left\|\begin{matrix}\langle V_i^s\rangle\\ \langle V_i^r\rangle\end{matrix}\right\|\\ T_i\end{bmatrix}^T\right\}\delta\begin{bmatrix}\left\|\begin{matrix}\langle q_s^j\rangle\\ \langle q_r^j\rangle\end{matrix}\right\|\\ \theta^j\end{bmatrix} = 0
\end{aligned}
$$

Taking into account the above assumptions, the equations of the AC asynchronous machine have the form:

$$
\begin{aligned}
&\left\{\left(\frac{d}{dt}\left\|\begin{matrix}\langle i_{s(AC)}^j\rangle\\ \langle i_{r(AC)}^j\rangle\\ \Omega^j\end{matrix}\right\|^T\right)\left\|\begin{matrix}\langle L_s\rangle & \langle L_{sr}(\theta)\rangle & 0\\ \langle L_{rs}(\theta)\rangle & \langle L_r\rangle & 0\\ 0 & 0 & J\end{matrix}\right\| + \left\|\begin{matrix}\langle i_{s(AC)}^j\rangle\\ \langle i_{r(AC)}^j\rangle\\ \Omega^j\end{matrix}\right\|^T\left\|\begin{matrix}\langle 0\rangle & \langle L_{sr}^*(\theta)\rangle & 0\\ \langle L_{rs}^*(\theta)\rangle & \langle 0\rangle & 0\\ 0 & 0 & 0\end{matrix}\right\|\Omega^j\right.\\
&-\frac{1}{2}\left\|\begin{matrix}\langle i_{s(AC)}^j\rangle\\ \langle i_{r(AC)}^j\rangle\\ \Omega^j\end{matrix}\right\|^T\left\|\begin{matrix}\langle 0\rangle & \langle L_{sr}^*(\theta)\rangle & 0\\ \langle L_{rs}^*(\theta)\rangle & \langle 0\rangle & 0\\ 0 & 0 & 0\end{matrix}\right\|\left\|\begin{matrix}\langle i_{s(AC)}^j\rangle\\ \langle i_{r(AC)}^j\rangle\\ \Omega^j\end{matrix}\right\| + \left\|\begin{matrix}\langle i_{s(AC)}^j\rangle\\ \langle i_{r(AC)}^j\rangle\\ \Omega^j\end{matrix}\right\|^T\left\|\begin{matrix}\langle R_s\rangle & \langle 0\rangle & 0\\ \langle 0\rangle & \langle R_r\rangle & 0\\ 0 & 0 & D\end{matrix}\right\|\\
&\left.+\,\mathrm{sgn}\left\|\begin{matrix}\langle i_{s(AC)}^j\rangle\\ \langle i_{r(AC)}^j\rangle\\ \Omega^j\end{matrix}\right\|^T\left\|\begin{matrix}\langle \Delta V_i^{s(AC)}\rangle & \langle 0\rangle & 0\\ \langle 0\rangle & \langle 0\rangle & 0\\ 0 & 0 & \Delta T_i\end{matrix}\right\| - \left\|\begin{matrix}\langle V_i^{s(AC)}\rangle\\ \langle 0\rangle\\ T_i\end{matrix}\right\|^T\right\}\delta\left\|\begin{matrix}\langle q_{s(AC)}^j\rangle\\ \langle q_{r(AC)}^j\rangle\\ \theta^j\end{matrix}\right\| = 0
\end{aligned}
$$

Introducing to the armature circuit a macroelectronic commutator (macrocommutator) one will get:

$$\left\{\left(\frac{d}{dt}\left\|\begin{matrix}\langle C_{ss}^{ji}\rangle\langle i_i^{s(DC)}\rangle\\ \langle i_{r(AC)}^{j}\rangle\\ \Omega^j\end{matrix}\right\|^T\right)\left\|\begin{matrix}\langle L_s\rangle & \langle L_{sr}(\theta)\rangle & 0\\ \langle L_{rs}(\theta)\rangle & \langle L_r\rangle & 0\\ 0 & 0 & J\end{matrix}\right\|\right.$$

$$+\left\|\begin{matrix}\langle C_{ss}^{ji}\rangle\langle i_i^{s(DC)}\rangle\\ \langle i_{r(AC)}^{j}\rangle\\ \Omega^j\end{matrix}\right\|^T\left\|\begin{matrix}\langle 0\rangle & \langle L_{sr}^{*}(\theta)\rangle & 0\\ \langle L_{rs}^{*}(\theta)\rangle & \langle 0\rangle & 0\\ 0 & 0 & 0\end{matrix}\right\|\Omega^j$$

$$-\frac{1}{2}\left\|\begin{matrix}\langle C_{ss}^{ji}\rangle\langle i_i^{s(DC)}\rangle\\ \langle i_{r(AC)}^{j}\rangle\\ \Omega^j\end{matrix}\right\|^T\left\|\begin{matrix}\langle 0\rangle & \langle L_{sr}^{*}(\theta)\rangle & 0\\ \langle L_{rs}^{*}(\theta)\rangle & \langle 0\rangle & 0\\ 0 & 0 & 0\end{matrix}\right\|\left\|\begin{matrix}\langle C_{ss}^{ji}\rangle\langle i_i^{s(DC)}\rangle\\ \langle i_{r(AC)}^{j}\rangle\\ \Omega^j\end{matrix}\right\|$$

$$+\left\|\begin{matrix}\langle C_{ss}^{ji}\rangle\langle i_i^{s(DC)}\rangle\\ \langle i_{r(AC)}^{j}\rangle\\ \Omega^j\end{matrix}\right\|^T\left\|\begin{matrix}\langle R_s\rangle & \langle 0\rangle & 0\\ \langle 0\rangle & \langle R_r\rangle & 0\\ 0 & 0 & D\end{matrix}\right\|$$

$$+\,\mathrm{sgn}\left\|\begin{matrix}\langle C_{ss}^{ji}\rangle\langle i_i^{s(DC)}\rangle\\ \langle i_{r(AC)}^{j}\rangle\\ \Omega^j\end{matrix}\right\|^T\left\|\begin{matrix}\langle C_{ij}^{ss}\rangle\langle \Delta V_{s(DC)}^{j}\rangle & \langle 0\rangle & 0\\ \langle 0\rangle & \langle 0\rangle & 0\\ 0 & 0 & \Delta T_i\end{matrix}\right\|$$

$$\left.-\left\|\begin{matrix}\langle C_{ij}^{ss}\rangle\langle V_{s(DC)}^{j}\rangle\\ \langle 0\rangle\\ T_i\end{matrix}\right\|^T\right\}\delta\left\|\begin{matrix}\langle C_{ss}^{ji}\rangle\langle q_i^{s(DC)}\rangle\\ \langle q_{r(AC)}^{j}\rangle\\ \theta^j\end{matrix}\right\|=0$$

where in:

$$\langle V_i^{s(AC)}\rangle=\langle C_{ij}^{ss}\rangle\langle V_{P(DC)}^{j}\rangle$$

$$\begin{pmatrix}V_i^{a(AC)}\\ V_i^{b(AC)}\\ V_i^{c(AC)}\end{pmatrix}=\begin{pmatrix}C_{ij}^{Pa} & C_{ij}^{Na}\\ C_{ij}^{Pb} & C_{ij}^{Nb}\\ C_{ij}^{Pc} & C_{ij}^{Nc}\end{pmatrix}\begin{pmatrix}V_{P(DC)}^{j}\\ V_{N(DC)}^{j}\end{pmatrix}$$

$$\langle i_i^{P(DC)}\rangle=\langle C_{ij}^{ss}\rangle^T\langle i_{s(AC)}^{j}\rangle$$

$$\begin{pmatrix}i_i^{P(DC)}\\ i_i^{N(DC)}\end{pmatrix}=\begin{pmatrix}C_{ij}^{Pa} & C_{ij}^{Pb} & C_{ij}^{Pc}\\ C_{ij}^{Na} & C_{ij}^{Nb} & C_{ij}^{Nc}\end{pmatrix}\begin{pmatrix}i_{a(AC)}^{j}\\ i_{b(AC)}^{j}\\ i_{c(AC)}^{j}\end{pmatrix}$$

or

$$\langle i^j_{s(AC)} \rangle = \langle C^{ji}_{ss} \rangle \langle i_i^{P(DC)} \rangle$$

$$\begin{pmatrix} i^j_{a(AC)} \\ i^j_{b(AC)} \\ i^j_{c(AC)} \end{pmatrix} = \begin{pmatrix} C^{ji}_{aP} & C^{ji}_{aN} \\ C^{ji}_{bP} & C^{ji}_{bN} \\ C^{ji}_{cP} & C^{ji}_{cN} \end{pmatrix} \begin{pmatrix} i_i^{P(DC)} \\ i_i^{N(DC)} \end{pmatrix}$$

$$\begin{pmatrix} C^{ji}_{aP} & C^{ji}_{aN} \\ C^{ji}_{bP} & C^{ji}_{bN} \\ C^{ji}_{cP} & C^{ji}_{cN} \end{pmatrix} = \begin{pmatrix} C_{ij}^{Pa} & C_{ij}^{Pb} & C_{ij}^{Pc} \\ C_{ij}^{Na} & C_{ij}^{Nb} & C_{ij}^{Nc} \end{pmatrix}^T$$

7.8 AC–AC or AC–DC–AC commutator synchronous motor with the macroelectronic commutator (macrocommutator) and electromagnetic exciter

The speed of AC–AC or AC–DC–AC commutator synchronous motors with the macroelectronic commutator (macrocommutator) and electromagnetical exciter can be controlled by variation of the frequency of the voltage applied to the EM motor. Due to magnetic-flux saturation problems with EM motors, the voltage applied to the EM motor must alter with the frequency. The running speed is set by the frequency applied to it and is independent of load torque provided the EM motor is not overloaded.

A modern adjustable-velocity IEMD comes in two major formats, *V/F* and space holor or vector. The *V/F* IEMD is one where the voltage applied to the EM motor is directly related to the frequency. In the ideal EM motor, the magnetic circuit would be purely inductive and keeping a constant *V*/Hz ratio would maintain a constant magnetic flux in the iron.

The real EM motor has resistance in series with the magnetising inductance. This has no bearing on the operation at line frequency, however, as the frequency of the IEMD is reduced, the resistance begins to become significant relative to the inductive reactance. This causes the magnetic flux to reduce at very low frequencies and so it is difficult to get sufficient torque at low speeds. For many applications, this low torque is not a problem, but there are some that do need a high torque from a low speed. Early EMDs were designed with a voltage boost to provide a measure of torque increase at low speed.

A holor or vector IEMD has a mathematical model of the IEMD in software and by measuring the current holors or vectors in relation to the applied voltage, they are able to maintain a constant magnetic field at all frequencies below the line frequency. This IEMD needs to be tuned to the EM motor and typically includes a self-tuning algorithm that is enabled at commissioning to determine the component values for the mathematical model. If the EM motor is replaced, the IEMD needs to be returned to learn the characteristics of the new EM motors.

A holor or vector IEMD comes in three major formats, closed loop, open-loop control and **direct torque control** (DTC). The closed loop microcontrollers were the first holor or vector microcontrollers and are still the best option for accurate control at zero speed. The open loop holor or vector and DTC are suitable for applications requiring good control above 3–5 Hz.

Quite a number of a modern IEMD can operate as *V*/*F*, open-loop holor or vector and/or closed loop holor or vector just by changing a parameter—closed loop requires a shaft encoder to give accurate speed feedback. The major differentiation between a modern IEMD is the enclosure, auxiliary functionality, programming and user interface. A low cost IEMD is often very poorly filtered and can create major RFI (EMC) issues. As a rule, an IEMD includes no filtering and must be installed with external filters, and others include all the filtering required. AC and DC reactors help to reduce the noise generated by the IEMD, and to improve the distortion power factor of the IEMD. Because the IEMD rectifies the incoming supply, the current waveform is much distorted and so the harmonics are high. A low cost IEMD without the reactors has a very poor power factor. NB Most IEMD manufacturers quote the $\cos\varphi$ as better than 0.95 implying a high power factor. While the displacement power factor is high, the distortion power factor can be less than 0.7. Distortion power factor cannot be corrected with capacitors, but can be improved with expensive filters. There is 'active front end' IEMD or 'regenerative' IEMD that has an ECM DC–AC inverter stage on the input as well as the output and this can draw sinusoidal current from the supply resulting in a high power factor. It is possible that this technology may become a mandatory requirement at some time in the future. An IEMD is typically used in some form of automation process and so it now includes additional functionality and controls to simplify the automation process. There are a number of programmable inputs and outputs and relays and as a rule, an IEMD also includes a PID loop and a motorised potentiometer is also common. A holor or vector IEMD and certain *V*/Hz IEMD can be set up for angular-velocity control or torque control. Torque control is used in tensioning applications such as paper machines where the master controls a winding drum and the diameter increases as the drum fills up. This requires other IEMD feeding the paper to run at different speeds. Traditionally, this was achieved by DC motors as they naturally operate in torque mode.

The design of the IEMD power sections comprises an AC–DC commutator, acting as an AC–DC rectifier to convert the incoming power from AC to DC. This is followed by a power DC filter which comprises a number of high voltage high current DC capacitors commonly in a series parallel arrangement. The DC filter will commonly include one or two DC chokes in series with the rectified DC. After the DC filter, comes the output DC–AC commutator, acting as the DC–AC inverter stage which is made up of a series of solid-state electrical valves (electronic switches). There are three ECM columns for a three-phase output with two electrical valves on each column. One electrical valve connects the positive DC link to the output of that phase, and the other electrical valve connects the negative DC link to the output on that phase. Control of the output electrical valves produces a PWM output waveform designed to cause a sinusoidal current to flow into the EM motor windings.

There are a number of structural and functional diagrams and algorithms for the generation of the output waveforms, one common algorithm is the space holor or vector modulation technique. The waveform generation is usually done in firmware or in a special function chip.

AC–DC commutator—The AC–DC commutator is a full wave ECM AC–DC rectifier, single-phase or three-phase depending on the input requirements. The AC–DC rectifier can be controlled using a combination of transistors or thyristors and diodes, or more commonly uncontrolled using diodes only. Because the output of the AC–DC rectifier is connected to a large capacitive filter, there must be a means of providing the initial charge to the capacitors without damaging the AC–DC rectifier. The initial charging current for discharged capacitors connected to the full rectified voltage is very high and would cause AC–DC rectifier failure. The initial charge current is commonly limited by a series resistance in one of the DC outputs (see figure 7.7). This soft charge resistance is shorted out as soon as the capacitors are fully charged. The shorting device can be a relay or contactor, or it can be a transistor or thyristor. The alternative means of limiting the charge current is to use a controlled ECM and slowly increase the output voltage applied to the filter.

DC filter—The DC filter provides smoothing of the DC link applied to the output DC–DC commutator, acting as the DC–AC inverter. There must be sufficient capacitance to provide the smoothing required for the output current required. The capacitors must have sufficient ripple current rating to avoid excess heating and life shortening and voltage rating to withstand the maximum expected input voltages. There are two types of DC filter used, a capacitive input filter and an inductive input filter as shown in figure 7.8.

The capacitive input filter comprises a capacitor bank and an inductive input filter has an inductor in series with at least one of the DC inputs to the capacitive filter. With the capacitive input filter, current will flow from the supply, through the AC–DC rectifiers into the capacitors only when the supply voltage is higher than the DC voltage.

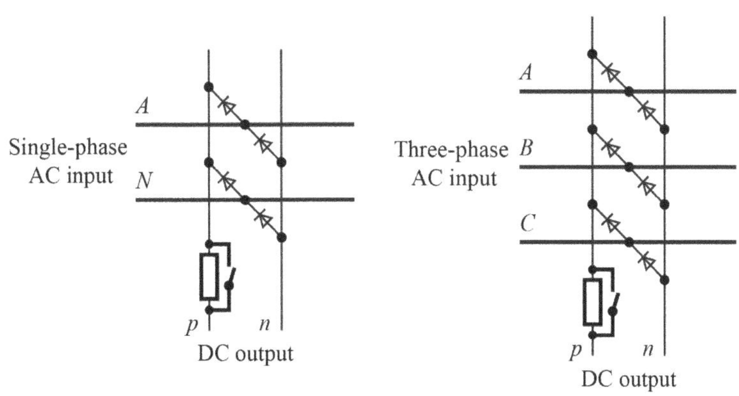

Figure 7.7. Limitation of the initial charge current by a series resistance in one of the DC outputs.

The result of this is that a very high current flows for a short time at the crest of the waveform only. This results in a very low distortion power factor, lots of harmonics and excessive heating of the AC–DC rectifier and capacitors. The reason for the addition of the DC link choke(s) is that a lower current flows for longer in each half cycle reducing the harmonics and increasing the distortion power factor. Another advantage of the DC link choke is that it helps to decrease the amount of commutation noise that leaks back onto the supply, reducing EMC radiation. The filter values are very different for single-phase inputs and three-phase inputs due to the magnitudes and frequency of the ripple currents. For a single-phase input, the ripple frequency is twice the line frequency and for a three-phase input, the ripple frequency is six times the line frequency.

Output DC–AC inverter—The output DC–AC inverter (figure 7.9) for a three-phase output stage comprises six solid-state electrical valves. In a small low voltage and low current IEMD, the output stages will typically be MOSFETs and in a larger IEMD, they are typically IGBTs. The output electrical valves operate at a high frequency, typically between 3 kHz and 16 kHz, and are controlled to produce a PWM output waveform which causes a sinusoidal current to flow in the EM motor. There are many different PWM structural and functional diagrams, and algorithms with different advantages. One common waveform generator structural and functional diagram is the space vector modulation (VECM) algorithm. VECM is covered here. The output voltage must provide both variable voltage and variable frequency control.

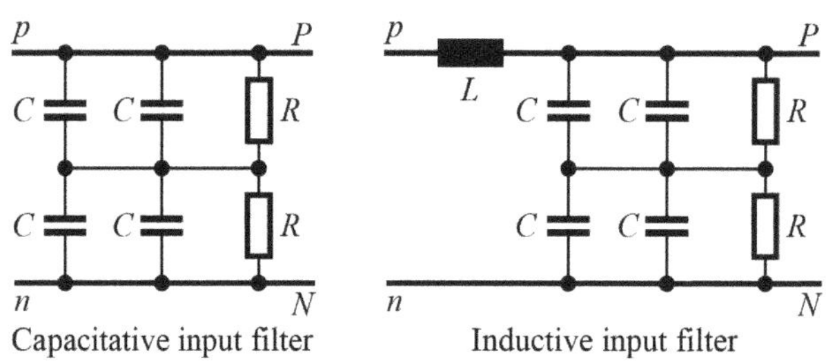

Figure 7.8. A capacitive input filter and an inductive input filter.

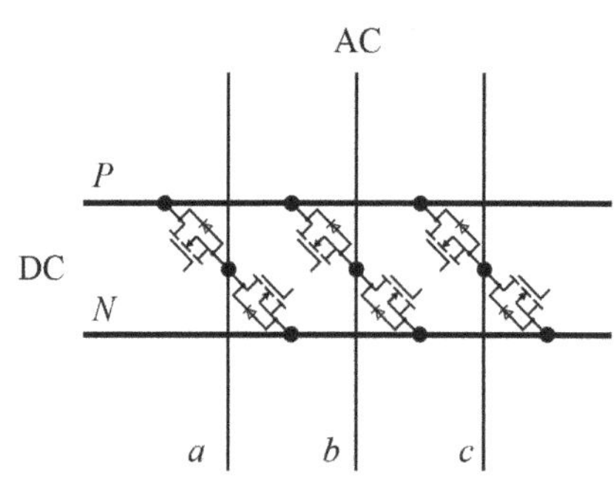

Figure 7.9. Output DC–AC inverter.

Each electrical valve needs to have a high voltage isolation gate driver that is isolated from the control electronics and is able to provide sufficient energy to fully control the electrical valves. In some cases, this would mean three isolated supplies to run the three upper electrical valves, and one isolated supply to run the lower electrical valves. The high voltage isolation gate driver must be capable of withstanding very high rates of change of voltage with minimum delays. Care must be taken to prevent the upper and lower electrical valves on one phase being on at the same time, this includes through the commutation stage. This requires an interlock delay between one electrical valve turning *OFF* and the other electrical valves turning *ON*.

Braking—Rapid slowing of the load can require energy to be removed from the load. This energy goes back into the IEMD and will result in an increasing DC link voltage. If the DC link voltage goes too high, the IEMD will be damaged. The excess energy can be dumped out into large resistors provided that the IEMD is fitted with

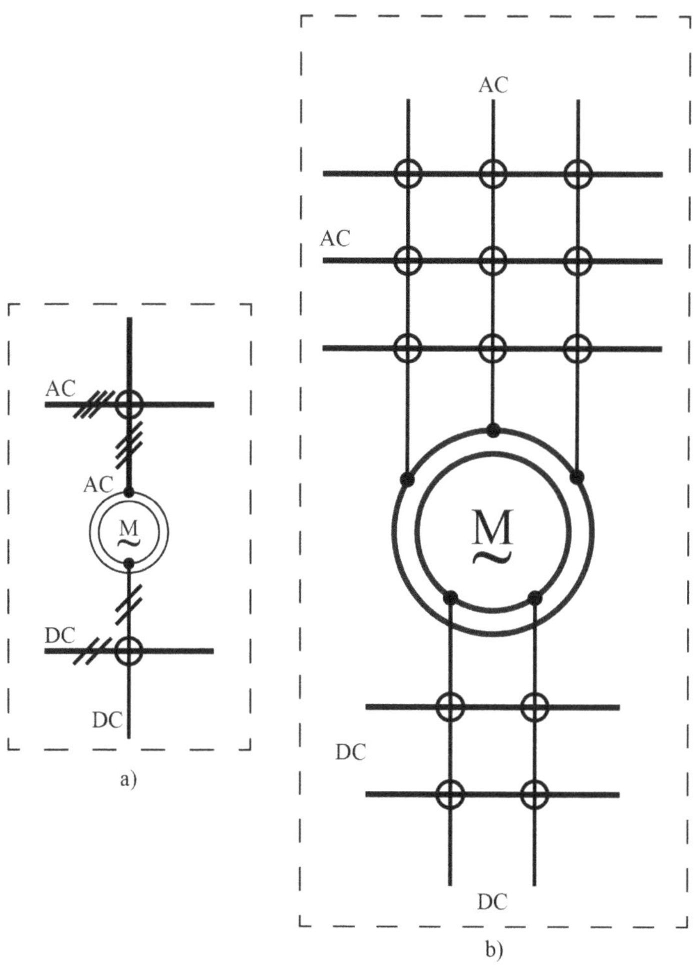

Figure 7.10. Physical model of the AC–AC commutator synchronous motor with the macroelectronic commutator (macrocommutator) and electromagnetical exciter.

a braking module, or can be fed back into the supply if the IEMD has an active front end. If there are multiple IEMDs in operation but with different duty cycles, it is possible to join all the DC-links and the excess energy can then go into driving other EM motors. The braking resistors need to be sized to suit the IEMD (resistance) and to suit the mechanical load (brake mechanical energy). A physical model of the AC–AC or AC–DC–AC commutator synchronous motor with the macroelectronic commutator (macrocommutator) and electromagnetical exciter is shown in figures 7.10 and 7.11.

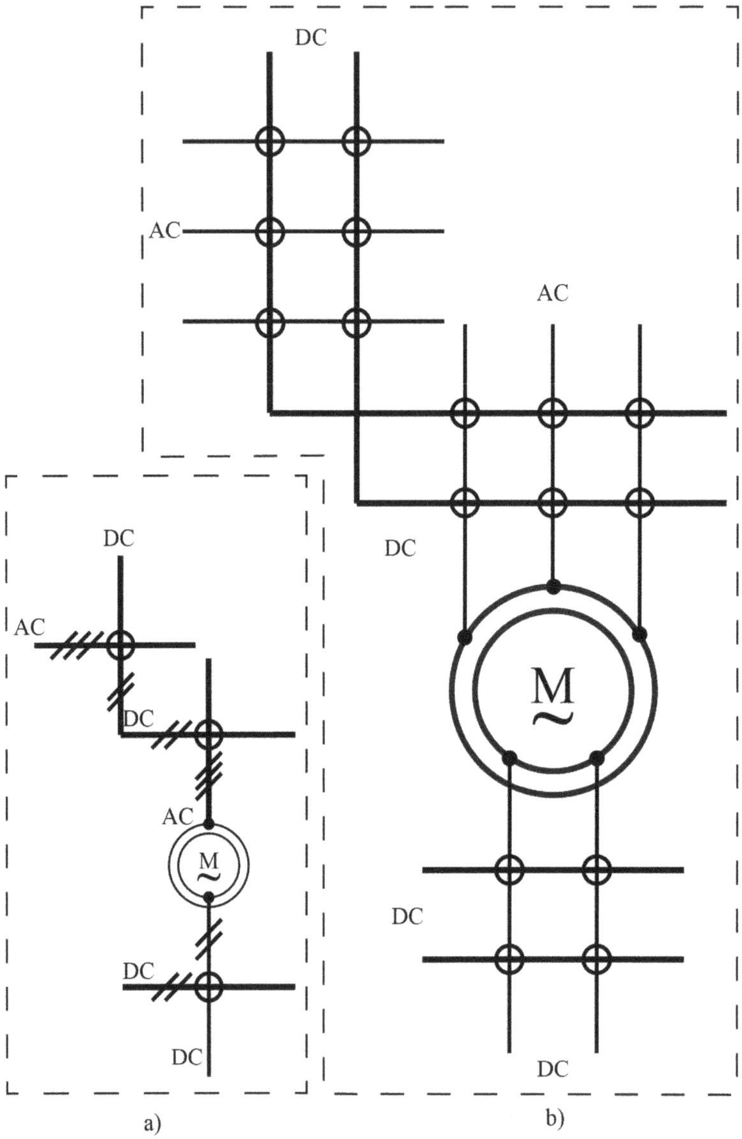

Figure 7.11. Physical model of the AC–DC–AC commutator synchronous motor with the macroelectronic commutator (macrocommutator) and electromagnetical exciter.

Taking into account equation (7.10) one will get:

$$\left\{\left(\frac{d}{dt}\begin{bmatrix}\left\|\begin{matrix}\langle i_s^j\rangle\\ \langle i_r^j\rangle\end{matrix}\right\|\\ \Omega^j\end{bmatrix}^T\right)\begin{bmatrix}\left\|\begin{matrix}\langle L_s(\theta)\rangle & \langle L_{sr}(\theta)\rangle\\ \langle L_{rs}(\theta)\rangle & \langle L_r(\theta)\rangle\end{matrix}\right\| & \|0\|\\ \|0\| & J\end{bmatrix}+\begin{bmatrix}\left\|\begin{matrix}\langle i_s^j\rangle\\ \langle i_r^j\rangle\end{matrix}\right\|\\ \Omega^j\end{bmatrix}^T\begin{bmatrix}\left\|\begin{matrix}\langle L_s^*(\theta)\rangle & \langle L_{sr}^*(\theta)\rangle\\ \langle L_{rs}^*(\theta)\rangle & \langle L_r^*(\theta)\rangle\end{matrix}\right\| & \|0\|\\ \|0\| & 0\end{bmatrix}\Omega^j\right.$$

$$-\frac{1}{2}\begin{bmatrix}\left\|\begin{matrix}\langle i_s^j\rangle\\ \langle i_r^j\rangle\end{matrix}\right\|\\ \Omega^j\end{bmatrix}^T\begin{bmatrix}\left\|\begin{matrix}\langle L_s^*(\theta)\rangle & \langle L_{sr}^*(\theta)\rangle\\ \langle L_{rs}^*(\theta)\rangle & \langle L_r^*(\theta)\rangle\end{matrix}\right\| & \|0\|\\ \|0\| & 0\end{bmatrix}\begin{bmatrix}\left\|\begin{matrix}\langle i_s^j\rangle\\ \langle i_r^j\rangle\end{matrix}\right\|\\ \Omega^j\end{bmatrix}$$

$$+\begin{bmatrix}\left\|\begin{matrix}\langle i_s^j\rangle\\ \langle i_r^j\rangle\end{matrix}\right\|\\ \Omega^j\end{bmatrix}^T\begin{bmatrix}\left\|\begin{matrix}\langle R_s\rangle & \langle 0\rangle\\ \langle 0\rangle & \langle R_r\rangle\end{matrix}\right\| & \|0\|\\ \|0\| & D\end{bmatrix}+\operatorname{sgn}\begin{bmatrix}\left\|\begin{matrix}\langle i_s^j\rangle\\ \langle i_r^j\rangle\end{matrix}\right\|\\ \Omega^j\end{bmatrix}^T\begin{bmatrix}\left\|\begin{matrix}\langle \Delta V_i^s\rangle & \langle 0\rangle\\ \langle 0\rangle & 0\end{matrix}\right\| & \|0\|\\ \|0\| & \Delta T_i\end{bmatrix}$$

$$\left.-\begin{bmatrix}\left\|\begin{matrix}\langle V_i^s\rangle\\ \langle 0\rangle\end{matrix}\right\|\\ T_i\end{bmatrix}^T\right\}\delta\begin{bmatrix}\left\|\begin{matrix}\langle q_s^j\rangle\\ \langle q_r^j\rangle\end{matrix}\right\|\\ \theta^j\end{bmatrix}=0$$

or in detail:

$$\left\{\left(\frac{d}{dt}\left\|\begin{matrix}\langle i_{s(AC)}^j\rangle\\ i_{r(DC)}^j\\ \Omega^j\end{matrix}\right\|^T\right)\left\|\begin{matrix}\langle L_s(\theta)\rangle & \langle L_{sr}(\theta)\rangle & 0\\ \langle L_{rs}(\theta)\rangle & L_r & 0\\ 0 & 0 & J\end{matrix}\right\|+\left\|\begin{matrix}\langle i_{s(AC)}^j\rangle\\ i_{r(DC)}^j\\ \Omega^j\end{matrix}\right\|^T\left\|\begin{matrix}\langle L_s^*(\theta)\rangle & \langle L_{sr}^*(\theta)\rangle & 0\\ \langle L_{rs}^*(\theta)\rangle & 0 & 0\\ 0 & 0 & 0\end{matrix}\right\|\Omega^j\right.$$

$$-\frac{1}{2}\left\|\begin{matrix}\langle i_{s(AC)}^j\rangle\\ i_{r(DC)}^j\\ \Omega^j\end{matrix}\right\|^T\left\|\begin{matrix}\langle L_s^*(\theta)\rangle & \langle L_{sr}^*(\theta)\rangle & 0\\ \langle L_{rs}^*(\theta)\rangle & 0 & 0\\ 0 & 0 & 0\end{matrix}\right\|\left\|\begin{matrix}\langle i_{s(AC)}^j\rangle\\ i_{r(DC)}^j\\ \Omega^j\end{matrix}\right\|$$

$$+\left\|\begin{matrix}\langle i_{s(AC)}^j\rangle\\ i_{r(DC)}^j\\ \Omega^j\end{matrix}\right\|^T\left\|\begin{matrix}\langle R_s\rangle & \langle 0\rangle & 0\\ \langle 0\rangle & R_r & 0\\ 0 & 0 & D\end{matrix}\right\|+\operatorname{sgn}\left\|\begin{matrix}\langle i_{s(AC)}^j\rangle\\ i_{r(DC)}^j\\ \Omega^j\end{matrix}\right\|^T\left\|\begin{matrix}\langle \Delta V_i^{s(AC)}\rangle & \langle 0\rangle & 0\\ \langle 0\rangle & \Delta V_i^{r(DC)} & 0\\ 0 & 0 & \Delta T_i\end{matrix}\right\|$$

$$\left.-\left\|\begin{matrix}\langle V_i^{s(AC)}\rangle\\ V_i^{r(DC)}\\ T_i\end{matrix}\right\|^T\right\}\delta\left\|\begin{matrix}\langle q_{s(AC)}^j\rangle\\ q_{r(DC)}^j\\ \theta^j\end{matrix}\right\|=0$$

Introducing to the armature circuit a macroelectronic commutator (macrocommutator) one will get:

$$\left\{\left(\frac{d}{dt}\left\|\begin{matrix}\langle C_{ss}^{ji}\rangle\langle i_i^{s(AC)}\rangle \\ i_i^{r(DC)} \\ \Omega^j\end{matrix}\right\|^T\right)\left\|\begin{matrix}\langle L_s(\theta)\rangle & \langle L_{sr}(\theta)\rangle & 0 \\ \langle L_{rs}(\theta)\rangle & L_r & 0 \\ 0 & 0 & J\end{matrix}\right\|\right.$$

$$+\left\|\begin{matrix}\langle C_{ss}^{ji}\rangle\langle i_i^{s(AC)}\rangle \\ i_i^{r(DC)} \\ \Omega^j\end{matrix}\right\|^T\left\|\begin{matrix}\langle L_s^*(\theta)\rangle & \langle L_{sr}^*(\theta)\rangle & 0 \\ \langle L_{rs}^*(\theta)\rangle & 0 & 0 \\ 0 & 0 & 0\end{matrix}\right\|\Omega^j$$

$$-\frac{1}{2}\left\|\begin{matrix}\langle C_{ss}^{ji}\rangle\langle i_i^{s(AC)}\rangle \\ i_i^{r(DC)} \\ \Omega^j\end{matrix}\right\|^T\left\|\begin{matrix}\langle L_s^*(\theta)\rangle & \langle L_{sr}^*(\theta)\rangle & 0 \\ \langle L_{rs}^*(\theta)\rangle & 0 & 0 \\ 0 & 0 & 0\end{matrix}\right\|\left\|\begin{matrix}\langle C_{ss}^{ji}\rangle\langle i_i^{s(AC)}\rangle \\ i_i^{r(DC)} \\ \Omega^j\end{matrix}\right\|$$

$$+\left\|\begin{matrix}\langle C_{ss}^{ji}\rangle\langle i_i^{s(AC)}\rangle \\ i_i^{r(DC)} \\ \Omega^j\end{matrix}\right\|^T\left\|\begin{matrix}\langle R_s\rangle & \langle 0\rangle & 0 \\ \langle 0\rangle & R_r & 0 \\ 0 & 0 & D\end{matrix}\right\|$$

$$+\operatorname{sgn}\left\|\begin{matrix}\langle C_{ss}^{ji}\rangle\langle i_i^{s(AC)}\rangle \\ i_i^{r(DC)} \\ \Omega^j\end{matrix}\right\|^T\left\|\begin{matrix}\langle C_{ij}^{ss}\rangle\langle \Delta V_{s(AC)}^j\rangle & \langle 0\rangle & 0 \\ \langle 0\rangle & \Delta V_r^{DC} & 0 \\ 0 & 0 & \Delta T\end{matrix}\right\|$$

$$\left.-\left\|\begin{matrix}\langle C_{ij}^{ss}\rangle\langle V_{s(AC)}^j\rangle \\ V_{r(DC)}^j \\ T\end{matrix}\right\|^T\right\}\delta\left\|\begin{matrix}\langle C_{ss}^{ji}\rangle\langle q_i^{s(AC)}\rangle \\ q_i^{r(AC)} \\ \theta^j\end{matrix}\right\| = 0$$

where in:

$$\langle V_i^{s(AC)}\rangle = \langle C_{ij}^{ss}\rangle\langle V_{s(AC)}^j\rangle$$

$$\begin{pmatrix} V_i^{a(AC)} \\ V_i^{b(AC)} \\ V_i^{c(AC)}\end{pmatrix} = \begin{pmatrix} C_{ij}^{Aa} & C_{ij}^{Ba} & C_{ij}^{Ca} \\ C_{ij}^{Ab} & C_{ij}^{Bb} & C_{ij}^{Cb} \\ C_{ij}^{Ac} & C_{ij}^{Bc} & C_{ij}^{Cc}\end{pmatrix}\begin{pmatrix} V_{A(AC)}^j \\ V_{B(AC)}^j \\ V_{C(AC)}^j\end{pmatrix}$$

$$\langle i_{s(AC)}^j\rangle = \langle C_{ij}^{ss}\rangle^T\langle i_i^{s(AC)}\rangle$$

$$\begin{pmatrix} i_i^{A(AC)} \\ i_i^{B(AC)} \\ i_i^{C(AC)}\end{pmatrix} = \begin{pmatrix} C_{ij}^{Aa} & C_{ij}^{Ba} & C_{ij}^{Ca} \\ C_{ij}^{Ab} & C_{ij}^{Bb} & C_{ij}^{Cb} \\ C_{ij}^{Ac} & C_{ij}^{Bc} & C_{ij}^{Cc}\end{pmatrix}^T\begin{pmatrix} i_{a(AC)}^j \\ i_{b(AC)}^j \\ i_{c(AC)}^j\end{pmatrix}$$

or

$$\langle i_s^j \rangle = \langle C_{ss}^{ji} \rangle \langle i_i^{s(AC)} \rangle$$

$$\begin{pmatrix} i_{a(AC)}^j \\ i_{b(AC)}^j \\ i_{c(AC)}^j \end{pmatrix} = \begin{pmatrix} C_{aA}^{ji} & C_{aB}^{ji} & C_{aC}^{ji} \\ C_{bA}^{ji} & C_{bB}^{ji} & C_{bC}^{ji} \\ C_{cA}^{ji} & C_{cB}^{ji} & C_{cC}^{ji} \end{pmatrix} \begin{pmatrix} i_i^{A(AC)} \\ i_i^{B(AC)} \\ i_i^{C(AC)} \end{pmatrix}$$

$$\begin{pmatrix} C_{aA}^{ji} & C_{aB}^{ji} & C_{aC}^{ji} \\ C_{bA}^{ji} & C_{bB}^{ji} & C_{bC}^{ji} \\ C_{cA}^{ji} & C_{cB}^{ji} & C_{cC}^{ji} \end{pmatrix} = \begin{pmatrix} C_{ij}^{Aa} & C_{ij}^{Ba} & C_{ij}^{Ca} \\ C_{ij}^{Ab} & C_{ij}^{Bb} & C_{ij}^{Cb} \\ C_{ij}^{Ac} & C_{ij}^{Bc} & C_{ij}^{Cc} \end{pmatrix}^T$$

7.9 AC–AC or AC–DC–AC commutator synchronous motor with the macroelectronic commutator (macrocommutator) and magnetoelectrical exciter

In the expression for conservative co-energy of a generalised MMD electrical machine with interior permanent magnets (IPM) considered as an electrical dynamical system containing IPMs, the words taking into account the interaction of electric currents of the dynamical system with the current i_o and the own energy of IPMs are taken into account $1/2L_o i_o^2$ cannot be taken into account, because with the assumption i_o = const it has no effect on the form of Euler–Lagrange differential equations of the second-order. In this case, it is not necessary to derive the differential equation of the dynamics of the exciter of the MMD electrical machine corresponding to the coordinate of the electric charge

$$q_0 = \int_0^t i_0 dt$$

representing the current of the exciter i_0 because such an equation assuming i_o = const would give a condition for the source voltage applied to the electromagnet, maintaining a constant value of current i_0.

A physical model of the AC–AC or AC–DC–AC commutator synchronous motor with the macroelectronic commutator (macrocommutator) and magnetoelectrical exciter is shown in figures 7.12 and 7.13.

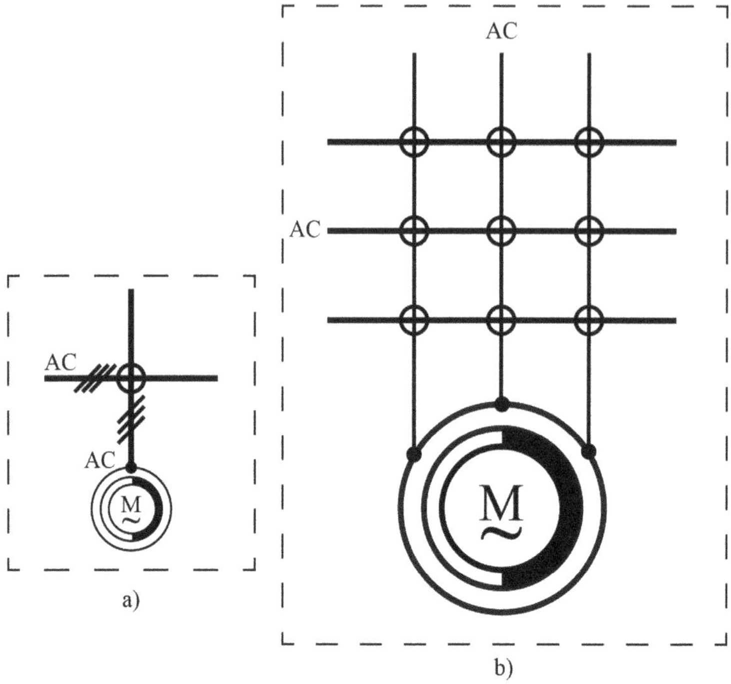

Figure 7.12. Physical model of the AC–AC commutator synchronous motor with the macroelectronic commutator (macrocommutator) and magnetoelectrical exciter.

Taking into account equation (7.10) one will get:

$$
\begin{aligned}
&\left\{\left(\frac{d}{dt}\begin{bmatrix}\left\|\begin{matrix}\langle i_s^j\rangle\\ \langle i_r^j\rangle\end{matrix}\right\|\\ \Omega^j\end{bmatrix}^T\right)\begin{bmatrix}\left\|\begin{matrix}\langle L_s(\theta)\rangle & \langle L_{sr}(\theta)\rangle\\ \langle L_{rs}(\theta)\rangle & \langle L_r(\theta)\rangle\end{matrix}\right\| & \|0\|\\ \|0\| & J\end{bmatrix}+\begin{bmatrix}\left\|\begin{matrix}\langle i_s^j\rangle\\ \langle i_r^j\rangle\end{matrix}\right\|\\ \Omega^j\end{bmatrix}^T\begin{bmatrix}\left\|\begin{matrix}\langle L_s^*(\theta)\rangle & \langle L_{sr}^*(\theta)\rangle\\ \langle L_{rs}^*(\theta)\rangle & \langle L_r^*(\theta)\rangle\end{matrix}\right\| & \|0\|\\ \|0\| & 0\end{bmatrix}\Omega^j\right.\\
&-\frac{1}{2}\begin{bmatrix}\left\|\begin{matrix}\langle i_s^j\rangle\\ \langle i_r^j\rangle\end{matrix}\right\|\\ \Omega^j\end{bmatrix}^T\begin{bmatrix}\left\|\begin{matrix}\langle L_s^*(\theta)\rangle & \langle L_{sr}^*(\theta)\rangle\\ \langle L_{rs}^*(\theta)\rangle & \langle L_r^*(\theta)\rangle\end{matrix}\right\| & \|0\|\\ \|0\| & 0\end{bmatrix}\begin{bmatrix}\left\|\begin{matrix}\langle i_s^j\rangle\\ \langle i_r^j\rangle\end{matrix}\right\|\\ \Omega^j\end{bmatrix}\\
&+\begin{bmatrix}\left\|\begin{matrix}\langle i_s^j\rangle\\ \langle i_r^j\rangle\end{matrix}\right\|\\ \Omega^j\end{bmatrix}^T\begin{bmatrix}\left\|\begin{matrix}\langle R_s\rangle & \langle 0\rangle\\ \langle 0\rangle & \langle R_r\rangle\end{matrix}\right\| & \|0\|\\ \|0\| & D\end{bmatrix}+\operatorname{sgn}\begin{bmatrix}\left\|\begin{matrix}\langle i_s^j\rangle\\ \langle i_r^j\rangle\end{matrix}\right\|\\ \Omega^j\end{bmatrix}^T\begin{bmatrix}\left\|\begin{matrix}\langle \Delta V_i^s\rangle & \langle 0\rangle\\ \langle 0\rangle & 0\end{matrix}\right\| & \|0\|\\ \|0\| & \Delta T_i\end{bmatrix}\\
&\left.-\begin{bmatrix}\left\|\begin{matrix}\langle V_i^s\rangle\\ \langle 0\rangle\end{matrix}\right\|\\ T_i\end{bmatrix}^T\right\}\delta\begin{bmatrix}\left\|\begin{matrix}\langle q_s^j\rangle\\ \langle q_r^j\rangle\end{matrix}\right\|\\ \theta^j\end{bmatrix}=0
\end{aligned}
$$

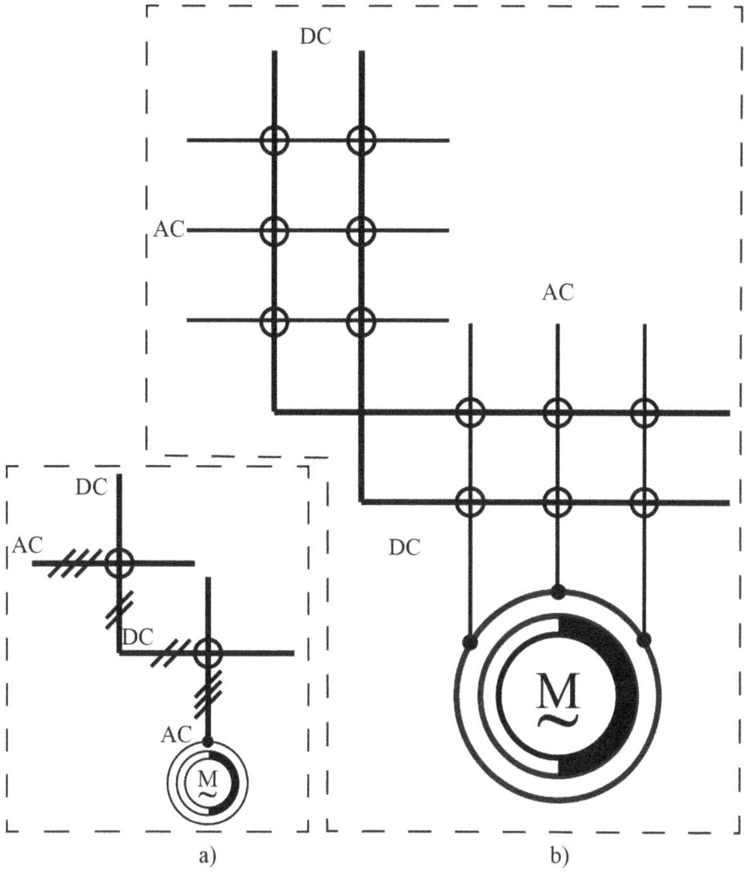

Figure 7.13. Physical model of the AC–DC–AC commutator synchronous motor with the macroelectronic commutator (macrocommutator) and magnetoelectrical exciter.

or in detail:

$$\left\{\left(\frac{d}{dt}\left\|\begin{matrix}\langle i^j_{s(AC)}\rangle\\ i^j_{r(DC)}\\ \Omega^j\end{matrix}\right\|^T\right)\left\|\begin{matrix}\langle L_s(\theta)\rangle & \langle L_{sr}(\theta)\rangle & 0\\ \langle L_{rs}(\theta)\rangle & L_r & 0\\ 0 & 0 & J\end{matrix}\right\| + \left\|\begin{matrix}\langle i^j_{s(AC)}\rangle\\ i^j_{r(DC)}\\ \Omega^j\end{matrix}\right\|^T \left\|\begin{matrix}\langle L^*_s(\theta)\rangle & \langle L^*_{sr}(\theta)\rangle & 0\\ \langle L^*_{rs}(\theta)\rangle & 0 & 0\\ 0 & 0 & 0\end{matrix}\right\|\Omega^j\right.$$

$$-\frac{1}{2}\left\|\begin{matrix}\langle i^j_{s(AC)}\rangle\\ i^j_{r(DC)}\\ \Omega^j\end{matrix}\right\|^T\left\|\begin{matrix}\langle L^*_s(\theta)\rangle & \langle L^*_{sr}(\theta)\rangle & 0\\ \langle L^*_{rs}(\theta)\rangle & 0 & 0\\ 0 & 0 & 0\end{matrix}\right\|\left\|\begin{matrix}\langle i^j_{s(AC)}\rangle\\ i^j_{r(DC)}\\ \Omega^j\end{matrix}\right\| + \left\|\begin{matrix}\langle i^j_{s(AC)}\rangle\\ i^j_{r(DC)}\\ \Omega^j\end{matrix}\right\|^T\left\|\begin{matrix}\langle R_s\rangle & \langle 0\rangle & 0\\ \langle 0\rangle & 0 & 0\\ 0 & 0 & D\end{matrix}\right\|$$

$$\left.+\,\mathrm{sgn}\left\|\begin{matrix}\langle i^j_{s(AC)}\rangle\\ i^j_{r(DC)}\\ \Omega^j\end{matrix}\right\|^T\left\|\begin{matrix}\langle \Delta V_i^{s(AC)}\rangle & \langle 0\rangle & 0\\ \langle 0\rangle & 0 & 0\\ 0 & 0 & \Delta T_i\end{matrix}\right\| - \left\|\begin{matrix}\langle V_i^{s(AC)}\rangle\\ 0\\ T_i\end{matrix}\right\|^T\right\}\delta\left\|\begin{matrix}\langle q^j_{s(AC)}\rangle\\ q^j_{r(DC)}\\ \theta^j\end{matrix}\right\| = 0$$

Introducing to the armature circuit a macroelectronic commutator (macrocommutator) one will get:

$$\left\{\left(\frac{d}{dt}\left\|\begin{matrix}\langle C_{ss}^{ji}\rangle\langle i_i^{s(AC)}\rangle\\ i_i^{r(DC)}\\ \Omega^j\end{matrix}\right\|^T\right)\left\|\begin{matrix}\langle L_s(\theta)\rangle & \langle L_{sr}(\theta)\rangle & 0\\ \langle L_{rs}(\theta)\rangle & L_r & 0\\ 0 & 0 & J\end{matrix}\right\|\right.$$

$$+\left\|\begin{matrix}\langle C_{ss}^{ji}\rangle\langle i_i^{s(AC)}\rangle\\ i_i^{r(DC)}\\ \Omega^j\end{matrix}\right\|^T\left\|\begin{matrix}\langle L_s^*(\theta)\rangle & \langle L_{sr}^*(\theta)\rangle & 0\\ \langle L_{rs}^*(\theta)\rangle & 0 & 0\\ 0 & 0 & 0\end{matrix}\right\|\Omega^j$$

$$-\frac{1}{2}\left\|\begin{matrix}\langle C_{ss}^{ji}\rangle\langle i_i^{s(AC)}\rangle\\ i_i^{r(DC)}\\ \Omega^j\end{matrix}\right\|^T\left\|\begin{matrix}\langle L_s^*(\theta)\rangle & \langle L_{sr}^*(\theta)\rangle & 0\\ \langle L_{rs}^*(\theta)\rangle & 0 & 0\\ 0 & 0 & 0\end{matrix}\right\|\left\|\begin{matrix}\langle C_{ss}^{ji}\rangle\langle i_i^{s(AC)}\rangle\\ i_i^{r(DC)}\\ \Omega^j\end{matrix}\right\|$$

$$+\left\|\begin{matrix}\langle C_{ss}^{ji}\rangle\langle i_i^{s(AC)}\rangle\\ i_i^{r(DC)}\\ \Omega^j\end{matrix}\right\|^T\left\|\begin{matrix}\langle R_s\rangle & \langle 0\rangle & 0\\ \langle 0\rangle & 0 & 0\\ 0 & 0 & D\end{matrix}\right\|$$

$$+\operatorname{sgn}\left\|\begin{matrix}\langle C_{ss}^{ji}\rangle\langle i_i^{s(AC)}\rangle\\ i_i^{r(DC)}\\ \Omega^j\end{matrix}\right\|^T\left\|\begin{matrix}\langle C_{ij}^{ss}\rangle\langle \Delta V_{s(AC)}^j\rangle & \langle 0\rangle & 0\\ \langle 0\rangle & 0 & 0\\ 0 & 0 & \Delta T\end{matrix}\right\|$$

$$\left.-\left\|\begin{matrix}\langle C_{ij}^{ss}\rangle\langle V_{s(AC)}^j\rangle\\ 0\\ T\end{matrix}\right\|^T\right\}\delta\left\|\begin{matrix}\langle C_{ss}^{ji}\rangle\langle q_i^{s(AC)}\rangle\\ q_i^{r(AC)}\\ \theta^j\end{matrix}\right\|=0$$

where in:

$$\langle V_i^{s(AC)}\rangle=\langle C_{ij}^{ss}\rangle\langle V_{s(AC)}^j\rangle$$

$$\begin{pmatrix}V_i^{a(AC)}\\ V_i^{b(AC)}\\ V_i^{c(AC)}\end{pmatrix}=\begin{pmatrix}C_{ij}^{Aa} & C_{ij}^{Ba} & C_{ij}^{Ca}\\ C_{ij}^{Ab} & C_{ij}^{Bb} & C_{ij}^{Cb}\\ C_{ij}^{Ac} & C_{ij}^{Bc} & C_{ij}^{Cc}\end{pmatrix}\begin{pmatrix}V_{A(AC)}^j\\ V_{B(AC)}^j\\ V_{C(AC)}^j\end{pmatrix}$$

$$\langle i_i^{s(AC)}\rangle=\langle C_{ij}^{ss}\rangle^T\langle i_{s(AC)}^j\rangle$$

$$\begin{pmatrix}i_i^{A(AC)}\\ i_i^{B(AC)}\\ i_i^{C(AC)}\end{pmatrix}=\begin{pmatrix}C_{ij}^{Aa} & C_{ij}^{Ba} & C_{ij}^{Ca}\\ C_{ij}^{Ab} & C_{ij}^{Bb} & C_{ij}^{Cb}\\ C_{ij}^{Ac} & C_{ij}^{Bc} & C_{ij}^{Cc}\end{pmatrix}\begin{pmatrix}i_{a(AC)}^j\\ i_{b(AC)}^j\\ i_{c(AC)}^j\end{pmatrix}$$

or

$$\langle i_s^j \rangle = \langle C_{ss}^{ji} \rangle \langle i_i^{s(AC)} \rangle$$

$$\begin{pmatrix} i_{a(AC)}^j \\ i_{b(AC)}^j \\ i_{c(AC)}^j \end{pmatrix} = \begin{pmatrix} C_{aA}^{ji} & C_{aB}^{ji} & C_{aC}^{ji} \\ C_{bA}^{ji} & C_{bB}^{ji} & C_{bC}^{ji} \\ C_{cA}^{ji} & C_{cB}^{ji} & C_{cC}^{ji} \end{pmatrix} \begin{pmatrix} i_i^{A(AC)} \\ i_i^{B(AC)} \\ i_i^{C(AC)} \end{pmatrix}$$

$$\begin{pmatrix} C_{aA}^{ji} & C_{aB}^{ji} & C_{aC}^{ji} \\ C_{bA}^{ji} & C_{bB}^{ji} & C_{bC}^{ji} \\ C_{cA}^{ji} & C_{cB}^{ji} & C_{cC}^{ji} \end{pmatrix} = \begin{pmatrix} C_{ij}^{Aa} & C_{ij}^{Ba} & C_{ij}^{Ca} \\ C_{ij}^{Ab} & C_{ij}^{Bb} & C_{ij}^{Cb} \\ C_{ij}^{Ac} & C_{ij}^{Bc} & C_{ij}^{Cc} \end{pmatrix}^T$$

7.10 AC–AC or AC–DC–AC commutator split-ring or wound-rotor asynchronous (induction) doubly-fed motor with the macroelectronic commutator (macrocommutator)

The AC–AC or AC–DC–AC commutator split-ring or wound-rotor asynchronous (induction) doubly-fed motor has two distinctly separate parts, the stator and the rotor. The stator circuit is rated as with an AC–AC or AC–DC–AC commutator squirrel-cage-rotor asynchronous (induction) motor and the rotor is rated in frame voltage and short circuit current. The frame voltage is the open circuit voltage when the rotor is not rotating and gives a measure of the turns ratio between the rotor and the stator. The short circuit current is the current flowing when the AC–AC or AC–DC–AC commutator split-ring or wound-rotor asynchronous (induction) doubly-fed motor is operating at full speed with the slip-rings rotor shorted and full load is applied to the EM motor shaft.

The AC–AC or AC–DC–AC commutator split-ring or wound-rotor asynchronous (induction) doubly-fed motor with the macroelectronic commutator (macrocommutator) is an AC induction machine where the rotor comprises a set of discrete windings (coils) that are terminated in slip-rings to which external AC–AC or AC–DC–AC commutator can be connected.

The stator is the same as is used with an AC–AC or AC–DC–AC commutator squirrel-cage-rotor asynchronous (induction) motor with the macrocommutator.

AC–AC or AC–DC–AC commutator split-ring or wound-rotor asynchronous (induction) doubly-fed motors with the macrocommutator are a variation on the AC–AC or AC–DC–AC commutator squirrel-cage-rotor asynchronous (induction) motors with the macrocommutator.

The AC–AC or AC–DC–AC commutator split-ring or wound-rotor asynchronous (induction) doubly-fed motor with the macrocommutator has a set of windings on the rotor which are not short-circuited, but are terminated to a set of slip-rings for connection to external resistors and contactors or AC–AC or AC–DC–AC commutators.

The AC–AC or AC–DC–AC commutator split-ring or wound-rotor asynchronous (induction) doubly-fed motor with the macrocommutator enables the starting characteristics of the AC motor to be totally controlled and modified to suit the load.

The down-side of the AC–AC or AC–DC–AC commutator split-ring or wound-rotor asynchronous (induction) doubly-fed motor with the macrocommutator is that the slip-rings and carbon-brush assemblies need regular maintenance which is a cost not applicable to the AC–AC or AC–DC–AC commutator squirrel-cage-rotor asynchronous (induction) motors with the macrocommutator.

If the rotor windings are shorted and a start is attempted, i.e. the AC motor is converted to an AC–AC or AC–DC–AC commutator squirrel-cage-rotor asynchronous (induction) motor with the macrocommutator, it will exhibit an extremely high locked rotor current, typically as high as 1400% and a very low locked rotor torque, perhaps as low as 60%.

In most applications, this is not an option. Another use of the AC–AC or AC–DC–AC commutator split-ring or wound-rotor asynchronous (induction) doubly-fed motor with the macrocommutator is as a means of adjustable-velocity control.

By modifying the angular-velocity versus torque curve, by controlling the AC–AC or AC–DC–AC commutator, the angular-velocity at which the AC motor will drive a particular mechanical load can be altered.

This section reports adjustable-velocity and torque control of an AC–AC or AC–DC–AC commutator split-ring or wound-rotor asynchronous (induction) doubly-fed motor with an ECM AC–AC or AC–DC–AC commutator, acting as an AC–AC cycloconverter or AC–DC–AC frequency changer, respectively, or using the slip energy recovery principle.

The proposed IEMD cyber-physical heterogeneous dynamical hypersystem uses an ECM AC–AC commutator, acting as an AC–AC cycloconverter or an ECM AC–DC–AC commutator, acting as AC–DC–AC frequency changer to extract the slip energy from the rotor into the mains instead of using the line commutated AC–AC cycloconverter or back-to-back PWM converter as a static frequency changer.

The AC motor IEMD cyber-physical heterogeneous dynamical hypersystem enables operations at both hyposynchronous and hypersynchronous speed regions.

The well-known simulation results for various operating conditions show the good control performance of the IEMD cyber-physical heterogeneous dynamical hypersystem.

A physical model of the AC–AC or AC–DC–AC commutator split-ring or wound-rotor asynchronous doubly-fed motor with the macroelectronic commutator (macrocommutator) is shown in figures 7.14 and 7.15.

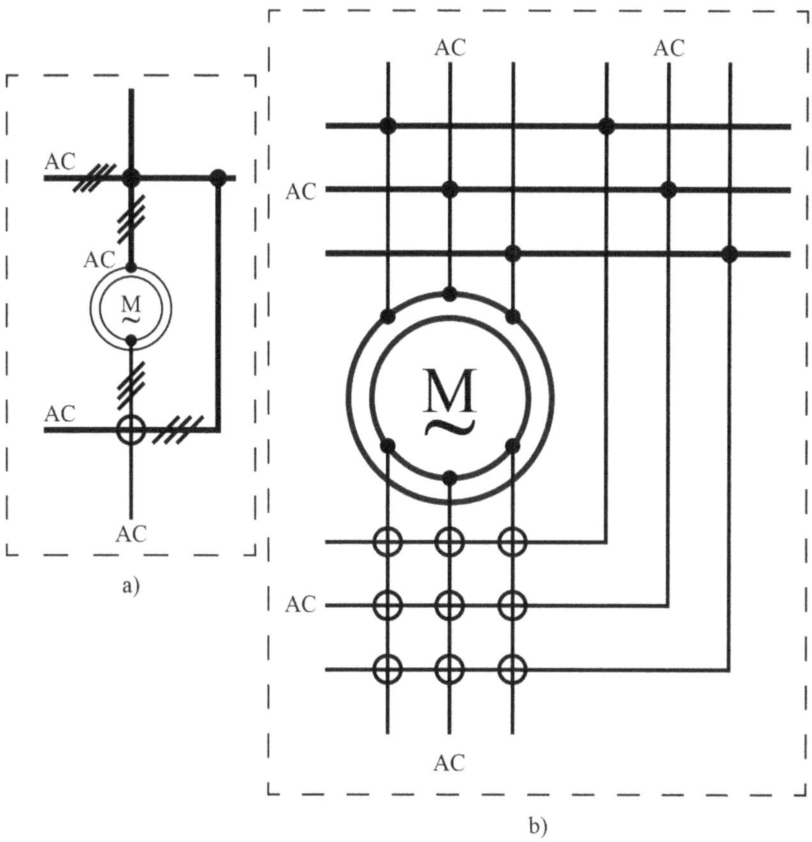

Figure 7.14. Physical model of the AC–AC commutator split-ring or wound-rotor asynchronous doubly-fed motor with the macroelectronic commutator (macrocommutator).

Taking into account equation (7.10) one will get:

$$
\begin{aligned}
&\left\{\left(\frac{d}{dt}\begin{bmatrix}\left\|\begin{matrix}\langle i_s^j\rangle\\ \langle i_r^j\rangle\end{matrix}\right\|\\ \Omega^j\end{bmatrix}^T\right)\begin{bmatrix}\left\|\begin{matrix}\langle L_s(\theta)\rangle & \langle L_{sr}(\theta)\rangle\\ \langle L_{rs}(\theta)\rangle & \langle L_r(\theta)\rangle\end{matrix}\right\| & \|0\|\\ \|0\| & J\end{bmatrix} + \begin{bmatrix}\left\|\begin{matrix}\langle i_s^j\rangle\\ \langle i_r^j\rangle\end{matrix}\right\|\\ \Omega^j\end{bmatrix}^T\begin{bmatrix}\left\|\begin{matrix}\langle L_s^*(\theta)\rangle & \langle L_{sr}^*(\theta)\rangle\\ \langle L_{rs}^*(\theta)\rangle & \langle L_r^*(\theta)\rangle\end{matrix}\right\| & \|0\|\\ \|0\| & 0\end{bmatrix}\Omega^j\right.\\
&-\frac{1}{2}\begin{bmatrix}\left\|\begin{matrix}\langle i_s^j\rangle\\ \langle i_r^j\rangle\end{matrix}\right\|\\ \Omega^j\end{bmatrix}^T\begin{bmatrix}\left\|\begin{matrix}\langle L_s^*(\theta)\rangle & \langle L_{sr}^*(\theta)\rangle\\ \langle L_{rs}^*(\theta)\rangle & \langle L_r^*(\theta)\rangle\end{matrix}\right\| & \|0\|\\ \|0\| & 0\end{bmatrix}\begin{bmatrix}\left\|\begin{matrix}\langle i_s^j\rangle\\ \langle i_r^j\rangle\end{matrix}\right\|\\ \Omega^j\end{bmatrix} + \begin{bmatrix}\left\|\begin{matrix}\langle i_s^j\rangle\\ \langle i_r^j\rangle\end{matrix}\right\|\\ \Omega^j\end{bmatrix}^T\begin{bmatrix}\left\|\begin{matrix}\langle R_s\rangle & \langle 0\rangle\\ \langle 0\rangle & \langle R_r\rangle\end{matrix}\right\| & \|0\|\\ \|0\| & D\end{bmatrix}\\
&\left.+\,\mathrm{sgn}\begin{bmatrix}\left\|\begin{matrix}\langle i_s^j\rangle\\ \langle i_r^j\rangle\end{matrix}\right\|\\ \Omega^j\end{bmatrix}^T\begin{bmatrix}\left\|\begin{matrix}\langle \Delta V_i^s\rangle & \langle 0\rangle\\ \langle 0\rangle & \langle \Delta V_i^r\rangle\end{matrix}\right\| & \|0\|\\ \|0\| & \Delta T_i\end{bmatrix} - \begin{bmatrix}\left\|\begin{matrix}\langle V_i^s\rangle\\ \langle V_i^r\rangle\end{matrix}\right\|\\ T_i\end{bmatrix}^T\right\}\delta\begin{bmatrix}\left\|\begin{matrix}\langle q_s^j\rangle\\ \langle q_r^j\rangle\end{matrix}\right\|\\ \theta^j\end{bmatrix} = 0
\end{aligned}
$$

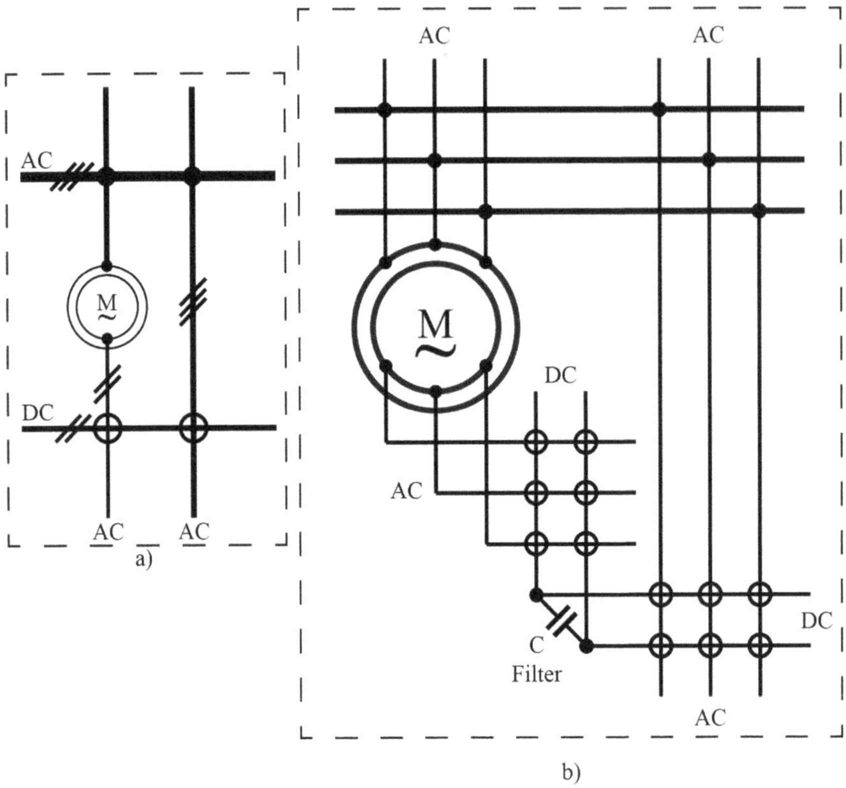

Figure 7.15. Physical model of the AC–DC–AC commutator split-ring or wound-rotor asynchronous doubly-fed motor with the macroelectronic commutator (macrocommutator).

An AC–AC or AC–DC–AC commutator split-ring or wound-rotor asynchronous doubly-fed motor with the macroelectronic commutator (macrocommutator) has the following properties:

- three-phase winding on the stator and rotor;
- own inductances and mutual windings of the stator are not functions of the generalised mechanical coordinate (angle of the shaft position), similarly to the inductance of the rotor;
- stator–rotor mutual inductances are functions of a generalised mechanical coordinate (angle of the shaft position);
- the rotor phase circuits are shorted;
- no additional voltage drops are taken into account at the phase connections of the rotor winding ($\Delta V_i^{r(AC)}$).

Taking into account the above assumptions, the equations of the AC asynchronous machine have the form:

$$
\left\{\left(\frac{d}{dt}\begin{Vmatrix}\langle i^j_{s(AC)}\rangle\\ \langle i^j_{r(AC)}\rangle\\ \Omega^j\end{Vmatrix}^T\right)\begin{Vmatrix}\langle L_s\rangle & \langle L_{sr}(\theta)\rangle & 0\\ \langle L_{rs}(\theta)\rangle & \langle L_r\rangle & 0\\ 0 & 0 & J\end{Vmatrix} + \begin{Vmatrix}\langle i^j_{s(AC)}\rangle\\ \langle i^j_{r(AC)}\rangle\\ \Omega^j\end{Vmatrix}^T \begin{Vmatrix}\langle 0\rangle & \langle L^*_{sr}(\theta)\rangle & 0\\ \langle L^*_{rs}(\theta)\rangle & \langle 0\rangle & 0\\ 0 & 0 & 0\end{Vmatrix}\Omega^j\right.
$$

$$
-\frac{1}{2}\begin{Vmatrix}\langle i^j_{s(AC)}\rangle\\ \langle i^j_{r(AC)}\rangle\\ \Omega^j\end{Vmatrix}^T\begin{Vmatrix}\langle 0\rangle & \langle L^*_{sr}(\theta)\rangle & 0\\ \langle L^*_{rs}(\theta)\rangle & \langle 0\rangle & 0\\ 0 & 0 & 0\end{Vmatrix}\begin{Vmatrix}\langle i^j_{s(AC)}\rangle\\ \langle i^j_{r(AC)}\rangle\\ \Omega^j\end{Vmatrix} + \begin{Vmatrix}\langle i^j_{s(AC)}\rangle\\ \langle i^j_{r(AC)}\rangle\\ \Omega^j\end{Vmatrix}^T\begin{Vmatrix}\langle R_s\rangle & \langle 0\rangle & 0\\ \langle 0\rangle & \langle R_r\rangle & 0\\ 0 & 0 & D\end{Vmatrix}
$$

$$
\left. + \operatorname{sgn}\begin{Vmatrix}\langle i^j_{s(AC)}\rangle\\ \langle i^j_{r(AC)}\rangle\\ \Omega^j\end{Vmatrix}^T\begin{Vmatrix}\langle \Delta V_i^{s(AC)}\rangle & \langle 0\rangle & 0\\ \langle 0\rangle & \langle \Delta V_i^{r(AC)}\rangle & 0\\ 0 & 0 & \Delta T_i\end{Vmatrix} - \begin{Vmatrix}\langle V_i^{s(AC)}\rangle\\ \langle V_i^{r(AC)}\rangle\\ T_i\end{Vmatrix}^T\right\}\delta\begin{Vmatrix}\langle q^j_{s(AC)}\rangle\\ \langle q^j_{r(AC)}\rangle\\ \theta^j\end{Vmatrix} = 0
$$

Introducing to the armature circuit a macroelectronic commutator (macrocommutator) one will get:

$$
\left\{\left(\frac{d}{dt}\begin{Vmatrix}\langle i^j_{s(AC)}\rangle\\ \langle C^{ji}_{rr}\rangle\langle i_i^{r(AC)}\rangle\\ \Omega^j\end{Vmatrix}^T\right)\begin{Vmatrix}\langle L_s\rangle & \langle L_{sr}(\theta)\rangle & 0\\ \langle L_{rs}(\theta)\rangle & \langle L_r\rangle & 0\\ 0 & 0 & J\end{Vmatrix}\right.
$$

$$
+\begin{Vmatrix}\langle i^j_{s(AC)}\rangle\\ \langle C^{ji}_{rr}\rangle\langle i_i^{r(AC)}\rangle\\ \Omega^j\end{Vmatrix}^T\begin{Vmatrix}\langle 0\rangle & \langle L^*_{sr}(\theta)\rangle & 0\\ \langle L^*_{rs}(\theta)\rangle & \langle 0\rangle & 0\\ 0 & 0 & 0\end{Vmatrix}\Omega^j
$$

$$
-\frac{1}{2}\begin{Vmatrix}\langle i^j_{s(AC)}\rangle\\ \langle C^{ji}_{rr}\rangle\langle i_i^{r(AC)}\rangle\\ \Omega^j\end{Vmatrix}^T\begin{Vmatrix}\langle 0\rangle & \langle L^*_{sr}(\theta)\rangle & 0\\ \langle L^*_{rs}(\theta)\rangle & \langle 0\rangle & 0\\ 0 & 0 & 0\end{Vmatrix}\begin{Vmatrix}\langle i^j_{s(AC)}\rangle\\ \langle C^{ji}_{rr}\rangle\langle i_i^{r(AC)}\rangle\\ \Omega^j\end{Vmatrix}
$$

$$
+\begin{Vmatrix}\langle i^j_{s(AC)}\rangle\\ \langle C^{ji}_{rr}\rangle\langle i_i^{r(AC)}\rangle\\ \Omega^j\end{Vmatrix}^T\begin{Vmatrix}\langle R_s\rangle & \langle 0\rangle & 0\\ \langle 0\rangle & \langle R_r\rangle & 0\\ 0 & 0 & D\end{Vmatrix}
$$

$$
+\operatorname{sgn}\begin{Vmatrix}\langle i^j_{s(AC)}\rangle\\ \langle C^{ji}_{rr}\rangle\langle i_i^{r(AC)}\rangle\\ \Omega^j\end{Vmatrix}^T\begin{Vmatrix}\langle \Delta V_i^{s(AC)}\rangle & \langle 0\rangle & 0\\ \langle 0\rangle & \langle C^{rr}_{ij}\rangle\langle V^j_{r(AC)}\rangle & 0\\ 0 & 0 & \Delta T_i\end{Vmatrix}
$$

$$
\left.-\begin{Vmatrix}\langle \Delta V_i^{s(AC)}\rangle\\ \langle C^{rr}_{ij}\rangle\langle V^j_{r(AC)}\rangle\\ T_i\end{Vmatrix}^T\right\}\delta\begin{Vmatrix}\langle q^j_{s(AC)}\rangle\\ \langle C^{ji}_{rr}\rangle\langle q_i^{r(AC)}\rangle\\ \theta^j\end{Vmatrix} = 0
$$

where in:

$$\langle V_i^{r(AC)} \rangle = \langle C_{ij}^{rr} \rangle \langle V_{r(AC)}^j \rangle$$

$$\begin{pmatrix} V_i^{a(AC)} \\ V_i^{b(AC)} \\ V_i^{c(AC)} \end{pmatrix} = \begin{pmatrix} C_{ij}^{Aa} & C_{ij}^{Ba} & C_{ij}^{Ca} \\ C_{ij}^{Ab} & C_{ij}^{Bb} & C_{ij}^{Cb} \\ C_{ij}^{Ac} & C_{ij}^{Bc} & C_{ij}^{Cc} \end{pmatrix} \begin{pmatrix} V_{A(AC)}^j \\ V_{B(AC)}^j \\ V_{C(AC)}^j \end{pmatrix}$$

$$\langle i_i^{r(AC)} \rangle = \langle C_{ij}^{rr} \rangle^T \langle i_{r(AC)}^j \rangle$$

$$\begin{pmatrix} i_i^{A(AC)} \\ i_i^{B(AC)} \\ i_i^{C(AC)} \end{pmatrix} = \begin{pmatrix} C_{ij}^{Aa} & C_{ij}^{Ba} & C_{ij}^{Ca} \\ C_{ij}^{Ab} & C_{ij}^{Bb} & C_{ij}^{Cb} \\ C_{ij}^{Ac} & C_{ij}^{Bc} & C_{ij}^{Cc} \end{pmatrix} \begin{pmatrix} i_{a(AC)}^j \\ i_{b(AC)}^j \\ i_{c(AC)}^j \end{pmatrix}$$

or

$$\langle i_r^j \rangle = \langle C_{rr}^{ji} \rangle \langle i_i^{r(AC)} \rangle$$

$$\begin{pmatrix} i_{a(AC)}^j \\ i_{b(AC)}^j \\ i_{c(AC)}^j \end{pmatrix} = \begin{pmatrix} C_{aA}^{ji} & C_{aB}^{ji} & C_{aC}^{ji} \\ C_{bA}^{ji} & C_{bB}^{ji} & C_{bC}^{ji} \\ C_{cA}^{ji} & C_{cB}^{ji} & C_{cC}^{ji} \end{pmatrix} \begin{pmatrix} i_i^{A(AC)} \\ i_i^{B(AC)} \\ i_i^{C(AC)} \end{pmatrix}$$

$$\begin{pmatrix} C_{aA}^{ji} & C_{aB}^{ji} & C_{aC}^{ji} \\ C_{bA}^{ji} & C_{bB}^{ji} & C_{bC}^{ji} \\ C_{cA}^{ji} & C_{cB}^{ji} & C_{cC}^{ji} \end{pmatrix} = \begin{pmatrix} C_{ij}^{Aa} & C_{ij}^{Ba} & C_{ij}^{Ca} \\ C_{ij}^{Ab} & C_{ij}^{Bb} & C_{ij}^{Cb} \\ C_{ij}^{Ac} & C_{ij}^{Bc} & C_{ij}^{Cc} \end{pmatrix}^T$$

7.11 AC–AC or AC–DC–AC commutator squirrel-cage-rotor asynchronous (induction) motor with the macroelectronic commutator (macrocommutator)

The stator is the outer body of the AC–AC or AC–DC–AC commutator squirrel-cage-rotor asynchronous (induction) motor with the macroelectronic commutator (macrocommutator) motor which houses the driven windings on an iron core. In a single-speed three-phase AC motor design, the standard stator has three windings, while a single-phase AC motor typically has two windings. The number of magnetic poles determines the speed of the AC–AC or AC–DC–AC commutator split-ring-rotor asynchronous (induction) motor. The winding configuration, slot configuration and lamination steel all have an effect on the performance of the AC motor.

The voltage rating of the AC motor is determined by the number of turns on the stator and the power rating of the AC motor is determined by the losses which comprise copper loss and iron loss, and the ability of the AC motor to dissipate the heat generated by these losses. The stator design determines the rated speed of the AC motor and most of the full load, full speed characteristics.

The rotor comprises a cylinder made up of round laminations pressed onto the AC motor shaft, and a number of short-circuited windings. The rotor windings are made up of rotor bars passed through the rotor, from one end to the other, around the surface of the rotor. The bars protrude beyond the rotor and are connected together by a shorting ring at each end. The bars are usually made of aluminium or copper, but are sometimes made of brass. The position relative to the surface of the rotor, shape, cross-sectional area and material of the bars determine the rotor characteristics. Essentially, the rotor windings exhibit inductance and resistance, and these characteristics can effectively be dependent on the frequency of the current flowing in the rotor.

A bar with a large cross-sectional area will exhibit a low resistance, while a bar of a small cross-sectional area will exhibit a high resistance. Likewise a copper bar will have a low resistance compared to a brass bar of equal proportions. Positioning the bar deeper into the rotor increases the amount of iron around the bar, and consequently increases the inductance exhibited by the rotor.

A physical model of the AC–AC or AC–DC–AC commutator squirrel-cage-rotor asynchronous (induction) motor with the macroelectronic commutator (macrocommutator) is shown in figures 7.16 and 7.17.

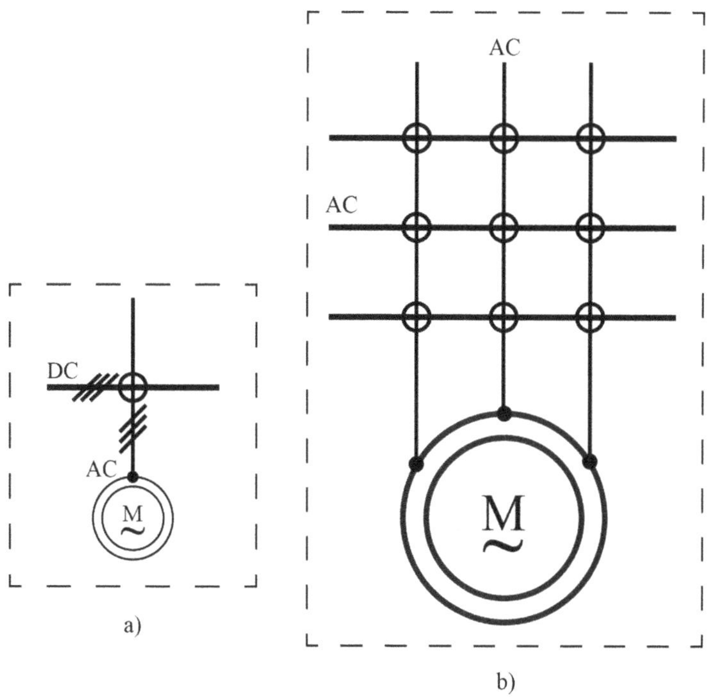

Figure 7.16. Physical model of the AC–AC commutator squirrel-cage-rotor asynchronous (induction) motor with the macroelectronic commutator (macrocommutator).

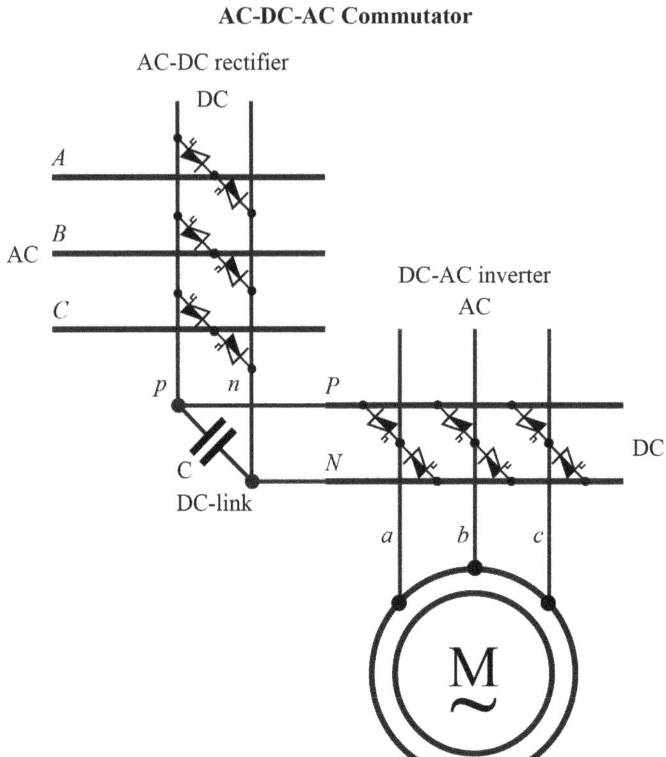

Figure 7.17. Physical model of the AC–DC–AC commutator squirrel-cage-rotor asynchronous (induction) motor with the macroelectronic commutator (macrocommutator).

Taking into account the equation (7.10) one will get:

$$\begin{aligned}
&\Bigg\{\left(\frac{d}{dt}\begin{bmatrix}\left\|\begin{matrix}\langle i_s^j\rangle\\ \langle i_r^j\rangle\end{matrix}\right\|\\ \Omega^j\end{bmatrix}^T\right)\begin{bmatrix}\left\|\begin{matrix}\langle L_s(\theta)\rangle & \langle L_{sr}(\theta)\rangle\\ \langle L_{rs}(\theta)\rangle & \langle L_r(\theta)\rangle\end{matrix}\right\| & \|0\|\\ \|0\| & J\end{bmatrix}+\begin{bmatrix}\left\|\begin{matrix}\langle i_s^j\rangle\\ \langle i_r^j\rangle\end{matrix}\right\|\\ \Omega^j\end{bmatrix}^T\begin{bmatrix}\left\|\begin{matrix}\langle L_s^*(\theta)\rangle & \langle L_{sr}^*(\theta)\rangle\\ \langle L_{rs}^*(\theta)\rangle & \langle L_r^*(\theta)\rangle\end{matrix}\right\| & \|0\|\\ \|0\| & 0\end{bmatrix}\Omega^j\\
&-\frac{1}{2}\begin{bmatrix}\left\|\begin{matrix}\langle i_s^j\rangle\\ \langle i_r^j\rangle\end{matrix}\right\|\\ \Omega^j\end{bmatrix}^T\begin{bmatrix}\left\|\begin{matrix}\langle L_s^*(\theta)\rangle & \langle L_{sr}^*(\theta)\rangle\\ \langle L_{rs}^*(\theta)\rangle & \langle L_r^*(\theta)\rangle\end{matrix}\right\| & \|0\|\\ \|0\| & \Omega^j\end{bmatrix}\begin{bmatrix}\left\|\begin{matrix}\langle i_s^j\rangle\\ \langle i_r^j\rangle\end{matrix}\right\|\\ \Omega^j\end{bmatrix}+\begin{bmatrix}\left\|\begin{matrix}\langle i_s^j\rangle\\ \langle i_r^j\rangle\end{matrix}\right\|\\ \Omega^j\end{bmatrix}^T\begin{bmatrix}\left\|\begin{matrix}\langle R_s\rangle & \langle 0\rangle\\ \langle 0\rangle & \langle R_r\rangle\end{matrix}\right\| & \|0\|\\ \|0\| & D\end{bmatrix}\\
&+\operatorname{sgn}\begin{bmatrix}\left\|\begin{matrix}\langle i_s^j\rangle\\ \langle i_r^j\rangle\end{matrix}\right\|\\ \Omega^j\end{bmatrix}^T\begin{bmatrix}\left\|\begin{matrix}\langle \Delta V_i^s\rangle & \langle 0\rangle\\ \langle 0\rangle & 0\end{matrix}\right\| & \|0\|\\ \|0\| & \Delta T_i\end{bmatrix}-\begin{bmatrix}\left\|\begin{matrix}\langle V_i^s\rangle\\ \langle V_i^r\rangle\end{matrix}\right\|\\ T_i\end{bmatrix}^T\Bigg\}\delta\begin{bmatrix}\left\|\begin{matrix}\langle q_s^j\rangle\\ \langle q_r^j\rangle\end{matrix}\right\|\\ \theta^j\end{bmatrix}=0
\end{aligned}$$

An AC–AC or AC–DC–AC commutator squirrel-cage-rotor asynchronous motor with the macroelectronic commutator (macrocommutator) has the following properties:

- three-phase winding on the stator and rotor;
- own inductances and mutual windings of the stator are not functions of the generalised mechanical coordinate (angle of the shaft position), similarly to the inductance of the rotor;
- stator–rotor mutual inductances are functions of a generalised mechanical coordinate (angle of the shaft position);
- the rotor phase circuits are shorted;
- no additional voltage drops are taken into account at the phase connections of the rotor winding ($\Delta V_i^{r(AC)}$).

Taking into account the above assumptions, the equations of the AC asynchronous machine have the form:

$$\left\{\left(\frac{d}{dt}\left\|\begin{matrix}\langle i^j_{s(AC)}\rangle\\ \langle i^j_{r(AC)}\rangle\\ \Omega^j\end{matrix}\right\|^T\right)\left\|\begin{matrix}\langle L_s\rangle & \langle L_{sr}(\theta)\rangle & 0\\ \langle L_{rs}(\theta)\rangle & \langle L_r\rangle & 0\\ 0 & 0 & J\end{matrix}\right\|\right.$$

$$+\left\|\begin{matrix}\langle i^j_{s(AC)}\rangle\\ \langle i^j_{r(AC)}\rangle\\ \Omega^j\end{matrix}\right\|^T\left\|\begin{matrix}\langle 0\rangle & \langle L^*_{sr}(\theta)\rangle & 0\\ \langle L^*_{rs}(\theta)\rangle & \langle 0\rangle & 0\\ 0 & 0 & 0\end{matrix}\right\|\Omega^j$$

$$-\frac{1}{2}\left\|\begin{matrix}\langle i^j_{s(AC)}\rangle\\ \langle i^j_{r(AC)}\rangle\\ \Omega^j\end{matrix}\right\|^T\left\|\begin{matrix}\langle 0\rangle & \langle L^*_{sr}(\theta)\rangle & 0\\ \langle L^*_{rs}(\theta)\rangle & \langle 0\rangle & 0\\ 0 & 0 & 0\end{matrix}\right\|\left\|\begin{matrix}\langle i^j_{s(AC)}\rangle\\ \langle i^j_{r(AC)}\rangle\\ \Omega^j\end{matrix}\right\|$$

$$+\left\|\begin{matrix}\langle i^j_{s(AC)}\rangle\\ \langle i^j_{r(AC)}\rangle\\ \Omega^j\end{matrix}\right\|^T\left\|\begin{matrix}\langle R_s\rangle & \langle 0\rangle & 0\\ \langle 0\rangle & \langle R_r\rangle & 0\\ 0 & 0 & D\end{matrix}\right\|$$

$$+\,\mathrm{sgn}\left\|\begin{matrix}\langle i^j_{s(AC)}\rangle\\ \langle i^j_{r(AC)}\rangle\\ \Omega^j\end{matrix}\right\|^T\left\|\begin{matrix}\langle \Delta V_i^{s(AC)}\rangle & \langle 0\rangle & 0\\ \langle 0\rangle & \langle 0\rangle & 0\\ 0 & 0 & \Delta T_i\end{matrix}\right\|$$

$$\left.-\left\|\begin{matrix}\langle V_i^{s(AC)}\rangle\\ \langle 0\rangle\\ T_i\end{matrix}\right\|^T\right\}\delta\left\|\begin{matrix}\langle q^j_{s(AC)}\rangle\\ \langle q^j_{r(AC)}\rangle\\ \theta^j\end{matrix}\right\| = 0$$

Introducing to the armature circuit a macroelectronic commutator (macrocommutator) one will get:

$$\left\{\left(\frac{d}{dt}\left\|\begin{matrix}\langle C_{ss}^{ji}\rangle\langle i_i^{s(DC)}\rangle\\ \langle i_{r(AC)}^{j}\rangle\\ \Omega^j\end{matrix}\right\|^T\right)\left\|\begin{matrix}\langle L_s\rangle & \langle L_{sr}(\theta)\rangle & 0\\ \langle L_{rs}(\theta)\rangle & \langle L_r\rangle & 0\\ 0 & 0 & J\end{matrix}\right\|\right.$$

$$+\left\|\begin{matrix}\langle C_{ss}^{ji}\rangle\langle i_i^{s(DC)}\rangle\\ \langle i_{r(AC)}^{j}\rangle\\ \Omega^j\end{matrix}\right\|^T\left\|\begin{matrix}\langle 0\rangle & \langle L_{sr}^{*}(\theta)\rangle & 0\\ \langle L_{rs}^{*}(\theta)\rangle & \langle 0\rangle & 0\\ 0 & 0 & 0\end{matrix}\right\|\Omega^j$$

$$-\frac{1}{2}\left\|\begin{matrix}\langle C_{ss}^{ji}\rangle\langle i_i^{s(DC)}\rangle\\ \langle i_{r(AC)}^{j}\rangle\\ \Omega^j\end{matrix}\right\|^T\left\|\begin{matrix}\langle 0\rangle & \langle L_{sr}^{*}(\theta)\rangle & 0\\ \langle L_{rs}^{*}(\theta)\rangle & \langle 0\rangle & 0\\ 0 & 0 & 0\end{matrix}\right\|\left\|\begin{matrix}\langle C_{ss}^{ji}\rangle\langle i_i^{s(DC)}\rangle\\ \langle i_{r(AC)}^{j}\rangle\\ \Omega^j\end{matrix}\right\|$$

$$+\left\|\begin{matrix}\langle C_{ss}^{ji}\rangle\langle i_i^{s(DC)}\rangle\\ \langle i_{r(AC)}^{j}\rangle\\ \Omega^j\end{matrix}\right\|^T\left\|\begin{matrix}\langle R_s\rangle & \langle 0\rangle & 0\\ \langle 0\rangle & \langle R_r\rangle & 0\\ 0 & 0 & D\end{matrix}\right\|$$

$$+\operatorname{sgn}\left\|\begin{matrix}\langle C_{ss}^{ji}\rangle\langle i_i^{s(DC)}\rangle\\ \langle i_{r(AC)}^{j}\rangle\\ \Omega^j\end{matrix}\right\|^T\left\|\begin{matrix}\langle C_{ij}^{ss}\rangle\langle \Delta V_{s(DC)}^{j}\rangle & \langle 0\rangle & 0\\ \langle 0\rangle & \langle 0\rangle & 0\\ 0 & 0 & \Delta T_i\end{matrix}\right\|$$

$$\left.-\left\|\begin{matrix}\langle C_{ij}^{ss}\rangle\langle V_{s(DC)}^{j}\rangle\\ \langle 0\rangle\\ T_i\end{matrix}\right\|^T\right\}\delta\left\|\begin{matrix}\langle C_{ss}^{ji}\rangle\langle q_i^{s(DC)}\rangle\\ \langle q_{r(AC)}^{j}\rangle\\ \theta^j\end{matrix}\right\|=0$$

where in:

$$\langle V_i^{s(AC)}\rangle=\langle C_{ij}^{ss}\rangle\langle V_{s(AC)}^{j}\rangle$$

$$\begin{pmatrix}V_i^{a(AC)}\\ V_i^{b(AC)}\\ V_i^{c(AC)}\end{pmatrix}=\begin{pmatrix}C_{ij}^{Aa} & C_{ij}^{Ba} & C_{ij}^{Ca}\\ C_{ij}^{Ab} & C_{ij}^{Bb} & C_{ij}^{Cb}\\ C_{ij}^{Ac} & C_{ij}^{Bc} & C_{ij}^{Cc}\end{pmatrix}\begin{pmatrix}V_{A(AC)}^{j}\\ V_{B(AC)}^{j}\\ V_{C(AC)}^{j}\end{pmatrix}$$

$$\langle i_i^{s(AC)}\rangle=\langle C_{ij}^{ss}\rangle^T\langle i_{s(AC)}^{j}\rangle$$

$$\begin{pmatrix}i_i^{A(AC)}\\ i_i^{B(AC)}\\ i_i^{C(AC)}\end{pmatrix}=\begin{pmatrix}C_{ij}^{Aa} & C_{ij}^{Ba} & C_{ij}^{Ca}\\ C_{ij}^{Ab} & C_{ij}^{Bb} & C_{ij}^{Cb}\\ C_{ij}^{Ac} & C_{ij}^{Bc} & C_{ij}^{Cc}\end{pmatrix}\begin{pmatrix}i_{a(AC)}^{j}\\ i_{b(AC)}^{j}\\ i_{c(AC)}^{j}\end{pmatrix}$$

or

$$\langle i_s^j \rangle = \langle C_{ss}^{ji} \rangle \langle i_i^{s(AC)} \rangle$$

$$\begin{pmatrix} i_{a(AC)}^j \\ i_{b(AC)}^j \\ i_{c(AC)}^j \end{pmatrix} = \begin{pmatrix} C_{aA}^{ji} & C_{aB}^{ji} & C_{aC}^{ji} \\ C_{bA}^{ji} & C_{bB}^{ji} & C_{bC}^{ji} \\ C_{cA}^{ji} & C_{cB}^{ji} & C_{cC}^{ji} \end{pmatrix} \begin{pmatrix} i_i^{A(AC)} \\ i_i^{B(AC)} \\ i_i^{C(AC)} \end{pmatrix}$$

$$\begin{pmatrix} C_{aA}^{ji} & C_{aB}^{ji} & C_{aC}^{ji} \\ C_{bA}^{ji} & C_{bB}^{ji} & C_{bC}^{ji} \\ C_{cA}^{ji} & C_{cB}^{ji} & C_{cC}^{ji} \end{pmatrix} = \begin{pmatrix} C_{ij}^{Aa} & C_{ij}^{Ba} & C_{ij}^{Ca} \\ C_{ij}^{Ab} & C_{ij}^{Bb} & C_{ij}^{Cb} \\ C_{ij}^{Ac} & C_{ij}^{Bc} & C_{ij}^{Cc} \end{pmatrix}^T$$

7.12 AC–AC or AC–DC–AC commutator variable-reluctance synchronous motor with the macroelectronic commutator (macrocommutator)

This section describes a mathematical model of the DC–AC commutator variable-reluctance synchronous motor with the macroelectronic commutator (macrocommutator). The mathematical model of the DC–AC commutator variable-reluctance synchronous motor is nonparametric and can only be established with experimental data, instead of an analytical representation. Because the reluctance varies with rotor position and magnetic saturation is part of the normal operation of DC–AC commutator variable-reluctance synchronous motors, there is no simple analytical expression for the magnetic field produced by the phase windings. The shape of phase current before commutation is of interest because it varies widely depending on when the phase winding is excited and what the rotor speed is. To illustrate this effect, two step response simulations may be done in Matlab/Simulink.

The DC–AC commutator variable-reluctance synchronous motor model used in these two simulations is a 6/4 linear magnetics model. For the first simulation, a step voltage is fed into phase *a* and the initial rotor position is set to be 1° instead of 0° so that the rotor will move in the positive sense of direction. The results show that the rotor stops at 45° after some oscillation which is the aligned position of this phase *a*. For the second simulation, a step voltage is fed into phase *c*. The initial position is 0°. According to this, the rotor will move towards the aligned position of phase *c*, i.e. 15°.

A physical model of the AC–AC or AC–DC–AC commutator variable-reluctance synchronous motor with the macroelectronic commutator (macrocommutator) is shown in figures 7.18 and 7.19.

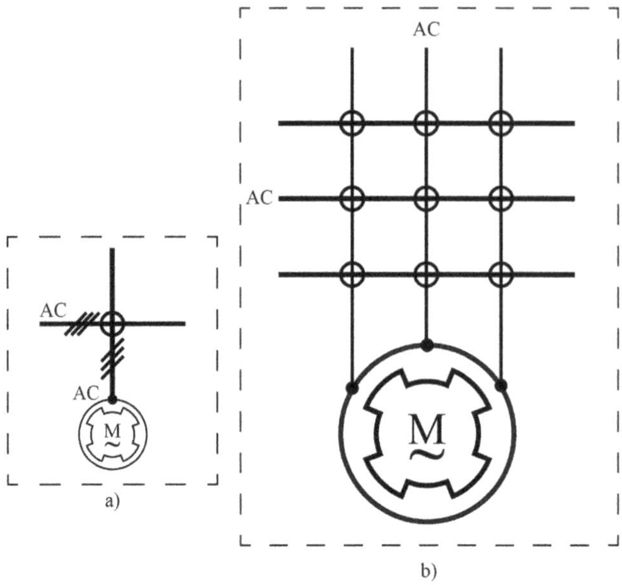

Figure 7.18. Physical model of the AC–AC commutator variable-reluctance synchronous motor with the macroelectronic commutator (macrocommutator).

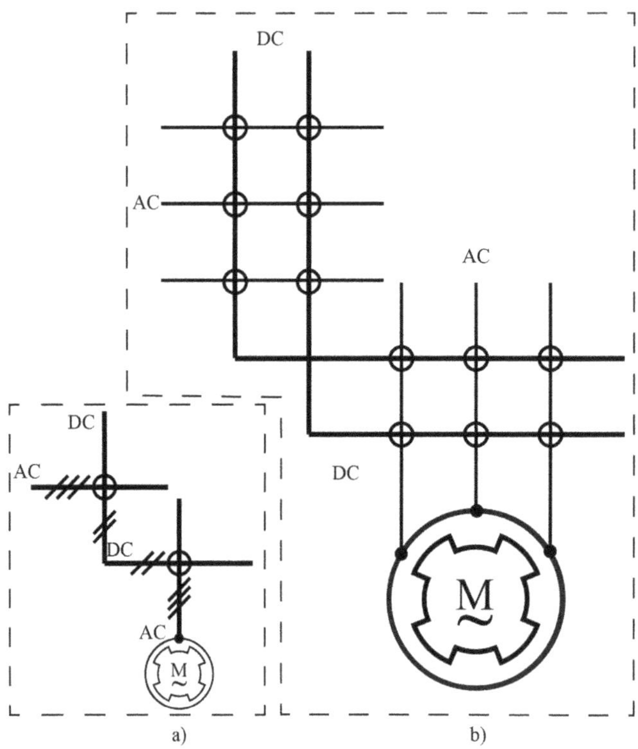

Figure 7.19. Physical model of the AC–DC–AC commutator variable-reluctance synchronous motor with the macroelectronic commutator (macrocommutator).

Multi-phase DC–AC commutator variable-reluctance synchronous motor:

- take into account the internal and mutual inductances of a m-phase armature winding as functions of angular position and current;
  - the influence of auxiliary circuits of the macroelectronic commutator (macrocommutator) on the course of commutation is neglected;
  - losses on eddy currents and hysteresis in the magnetic circuit of the rotor and armature are omitted;
  - the rotor has no electrical circuits.

Taking into account equation (7.10) one will get:

$$\left\{\left(\frac{d}{dt}\begin{bmatrix}\left\|\begin{matrix}\langle i_s^j\rangle\\ \langle i_r^j\rangle\end{matrix}\right\|\\ \Omega^j\end{bmatrix}^T\right)\begin{bmatrix}\left\|\begin{matrix}\langle L_s(\theta)\rangle & \langle L_{sr}(\theta)\rangle\\ \langle L_{rs}(\theta)\rangle & \langle L_r(\theta)\rangle\end{matrix}\right\| & \|0\|\\ \|0\| & J\end{bmatrix}\right.$$

$$+\begin{bmatrix}\left\|\begin{matrix}\langle i_s^j\rangle\\ \langle i_r^j\rangle\end{matrix}\right\|\\ \Omega^j\end{bmatrix}^T\begin{bmatrix}\left\|\begin{matrix}\langle L_s^*(\theta)\rangle & \langle L_{sr}^*(\theta)\rangle\\ \langle L_{rs}^*(\theta)\rangle & \langle L_r^*(\theta)\rangle\end{matrix}\right\| & \|0\|\\ \|0\| & 0\end{bmatrix}\Omega^j$$

$$-\frac{1}{2}\begin{bmatrix}\left\|\begin{matrix}\langle i_s^j\rangle\\ \langle i_r^j\rangle\end{matrix}\right\|\\ \Omega^j\end{bmatrix}^T\begin{bmatrix}\left\|\begin{matrix}\langle L_s^*(\theta)\rangle & \langle L_{sr}^*(\theta)\rangle\\ \langle L_{rs}^*(\theta)\rangle & \langle L_r^*(\theta)\rangle\end{matrix}\right\| & \|0\|\\ \|0\| & \Omega^j\end{bmatrix}\begin{bmatrix}\left\|\begin{matrix}\langle i_s^j\rangle\\ \langle i_r^j\rangle\end{matrix}\right\|\\ \Omega^j\end{bmatrix}$$

$$+\begin{bmatrix}\left\|\begin{matrix}\langle i_s^j\rangle\\ \langle i_r^j\rangle\end{matrix}\right\|\\ \Omega^j\end{bmatrix}^T\begin{bmatrix}\left\|\begin{matrix}\langle R_s\rangle & \langle 0\rangle\\ \langle 0\rangle & \langle R_r\rangle\end{matrix}\right\| & \|0\|\\ \|0\| & D\end{bmatrix}$$

$$+\,\mathrm{sgn}\begin{bmatrix}\left\|\begin{matrix}\langle i_s^j\rangle\\ \langle i_r^j\rangle\end{matrix}\right\|\\ \Omega^j\end{bmatrix}^T\begin{bmatrix}\left\|\begin{matrix}\langle \Delta V_i^s\rangle & \langle 0\rangle\\ \langle 0\rangle & 0\end{matrix}\right\| & \|0\|\\ \|0\| & \Delta T_i\end{bmatrix}$$

$$\left.-\begin{bmatrix}\left\|\begin{matrix}\langle V_i^s\rangle\\ \langle V_i^r\rangle\end{matrix}\right\|\\ T_i\end{bmatrix}^T\right\}\delta\begin{bmatrix}\left\|\begin{matrix}\langle q_s^j\rangle\\ \langle q_r^j\rangle\end{matrix}\right\|\\ \theta^j\end{bmatrix}=0$$

or in detail:

$$
\begin{aligned}
&\left\{\left(\frac{d}{dt}\left\|\begin{matrix}\langle i_i^{s(AC)}\rangle\\ \Omega^j\end{matrix}\right\|^T\right)\left\|\begin{matrix}\langle L_s(\theta,\, i)\rangle & 0\\ 0 & J\end{matrix}\right\| + \left\|\begin{matrix}\langle i_i^{s(AC)}\rangle\\ \Omega^j\end{matrix}\right\|^T\left\|\begin{matrix}\langle L_s^{*}(\theta,\, i)\rangle & 0\\ 0 & 0\end{matrix}\right\|\Omega^j\right.\\
&+\left(\left\|\begin{matrix}\langle i_i^{s(AC)}\rangle\\ \Omega^j\end{matrix}\right\|^T\left\|\begin{matrix}\langle L_s^{**}(\theta,\, i)\rangle & 0\\ 0 & 0\end{matrix}\right\|\right)\frac{d}{dt}\left\|\begin{matrix}\langle i_i^{s(AC)}\rangle\\ \Omega^j\end{matrix}\right\|\\
&-\frac{1}{2}\left\|\begin{matrix}\langle i_i^{s(AC)}\rangle\\ \Omega^j\end{matrix}\right\|^T\left\|\begin{matrix}\langle L_s^{*}(\theta,\, i)\rangle & 0\\ 0 & 0\end{matrix}\right\|\left\|\begin{matrix}\langle i_i^{s(AC)}\rangle\\ \Omega^j\end{matrix}\right\|\\
&+\left\|\begin{matrix}\langle i_i^{s(AC)}\rangle\\ \Omega^j\end{matrix}\right\|^T\left\|\begin{matrix}\langle R_s\rangle & 0\\ 0 & D\end{matrix}\right\|\\
&+\operatorname{sgn}\left\|\begin{matrix}\langle i_i^{s(AC)}\rangle\\ \Omega^j\end{matrix}\right\|^T\left\|\begin{matrix}\langle V_i^{s(AC)}\rangle & 0\\ 0 & \Delta T_i\end{matrix}\right\|\\
&\left.-\left\|\begin{matrix}\langle V_i^{s(AC)}\rangle\\ T_i\end{matrix}\right\|^T\right\}\delta\left\|\begin{matrix}\langle q_i^{s(AC)}\rangle\\ \theta^j\end{matrix}\right\| = 0
\end{aligned}
$$

Introducing to the armature circuit a macroelectronic commutator (macrocommutator) one will get:

$$
\begin{aligned}
&\left\{\left(\frac{d}{dt}\left\|\begin{matrix}\langle C_{ss}^{ji}\rangle\langle i_i^{s(AC)}\rangle\\ \Omega^j\end{matrix}\right\|^T\right)\left\|\begin{matrix}\langle L_s(\theta,\, i)\rangle & 0\\ 0 & J\end{matrix}\right\|\right.\\
&+\left\|\begin{matrix}\langle C_{ss}^{ji}\rangle\langle i_i^{s(AC)}\rangle\\ \Omega^j\end{matrix}\right\|^T\left\|\begin{matrix}\langle L_s^{*}(\theta,\, i)\rangle & 0\\ 0 & 0\end{matrix}\right\|\Omega^j\\
&+\left(\left\|\begin{matrix}\langle C_{ss}^{ji}\rangle\langle i_i^{s(AC)}\rangle\\ \Omega^j\end{matrix}\right\|^T\left\|\begin{matrix}\langle L_s^{**}(\theta,\, i)\rangle & 0\\ 0 & 0\end{matrix}\right\|\right)\frac{d}{dt}\left\|\begin{matrix}\langle C_{ss}^{ji}\rangle\langle i_i^{s(AC)}\rangle\\ \Omega^j\end{matrix}\right\|\\
&-\frac{1}{2}\left\|\begin{matrix}\langle C_{ss}^{ji}\rangle\langle i_i^{s(AC)}\rangle\\ \Omega^j\end{matrix}\right\|^T\left\|\begin{matrix}\langle L_s^{*}(\theta,\, i)\rangle & 0\\ 0 & 0\end{matrix}\right\|\left\|\begin{matrix}\langle C_{ss}^{ji}\rangle\langle i_i^{s(AC)}\rangle\\ \Omega^j\end{matrix}\right\|\\
&+\left\|\begin{matrix}\langle C_{ss}^{ji}\rangle\langle i_i^{s(AC)}\rangle\\ \Omega^j\end{matrix}\right\|^T\left\|\begin{matrix}\langle R_s\rangle & 0\\ 0 & D\end{matrix}\right\|\\
&+\operatorname{sgn}\left\|\begin{matrix}\langle C_{ss}^{ji}\rangle\langle i_i^{s(AC)}\rangle\\ \Omega^j\end{matrix}\right\|^T\left\|\begin{matrix}\langle C_{ij}^{ss}\rangle\langle \Delta V_{s(AC)}^{j}\rangle & 0\\ 0 & \Delta T_i\end{matrix}\right\|\\
&\left.-\left\|\begin{matrix}\langle C_{ij}^{ss}\rangle\langle V_i^{s(AC)}\rangle\\ T_i\end{matrix}\right\|^T\right\}\delta\left\|\begin{matrix}\langle C_{ss}^{ji}\rangle\langle q_i^{s(AC)}\rangle\\ \theta^j\end{matrix}\right\| = 0
\end{aligned}
$$

where in:

$$\langle V_i^{s(AC)}\rangle = \langle C_{ij}^{ss}\rangle\langle V_{s(AC)}^{j}\rangle$$

$$\begin{pmatrix} V_i^{a(AC)} \\ V_i^{b(AC)} \\ V_i^{c(AC)} \end{pmatrix} = \begin{pmatrix} C_{ij}^{Aa} & C_{ij}^{Ba} & C_{ij}^{Ca} \\ C_{ij}^{Ab} & C_{ij}^{Bb} & C_{ij}^{Cb} \\ C_{ij}^{Ac} & C_{ij}^{Bc} & C_{ij}^{Cc} \end{pmatrix} \begin{pmatrix} V_{A(AC)}^{j} \\ V_{B(AC)}^{j} \\ V_{C(AC)}^{j} \end{pmatrix}$$

$$\langle i_i^{s(AC)}\rangle = \langle C_{ij}^{ss}\rangle^T \langle i_{s(AC)}^{j}\rangle$$

$$\begin{pmatrix} i_i^{A(AC)} \\ i_i^{B(AC)} \\ i_i^{C(AC)} \end{pmatrix} = \begin{pmatrix} C_{ij}^{Aa} & C_{ij}^{Ba} & C_{ij}^{Ca} \\ C_{ij}^{Ab} & C_{ij}^{Bb} & C_{ij}^{Cb} \\ C_{ij}^{Ac} & C_{ij}^{Bc} & C_{ij}^{Cc} \end{pmatrix} \begin{pmatrix} i_{a(AC)}^{j} \\ i_{b(AC)}^{j} \\ i_{c(AC)}^{j} \end{pmatrix}$$

or

$$\langle i_s^{j}\rangle = \langle C_{ss}^{ji}\rangle\langle i_i^{s(AC)}\rangle$$

$$\begin{pmatrix} i_{a(AC)}^{j} \\ i_{b(AC)}^{j} \\ i_{c(AC)}^{j} \end{pmatrix} = \begin{pmatrix} C_{aA}^{ji} & C_{aB}^{ji} & C_{aC}^{ji} \\ C_{bA}^{ji} & C_{bB}^{ji} & C_{bC}^{ji} \\ C_{cA}^{ji} & C_{cB}^{ji} & C_{cC}^{ji} \end{pmatrix} \begin{pmatrix} i_i^{A(AC)} \\ i_i^{B(AC)} \\ i_i^{C(AC)} \end{pmatrix}$$

$$\begin{pmatrix} C_{aA}^{ji} & C_{aB}^{ji} & C_{aC}^{ji} \\ C_{bA}^{ji} & C_{bB}^{ji} & C_{bC}^{ji} \\ C_{cA}^{ji} & C_{cB}^{ji} & C_{cC}^{ji} \end{pmatrix} = \begin{pmatrix} C_{ij}^{Aa} & C_{ij}^{Ba} & C_{ij}^{Ca} \\ C_{ij}^{Ab} & C_{ij}^{Bb} & C_{ij}^{Cb} \\ C_{ij}^{Ac} & C_{ij}^{Bc} & C_{ij}^{Cc} \end{pmatrix}^T$$

References

Fijalkowski B 1987a *Modele matematyczne wybranych lotniczych i motoryzacyjnych mechano-elektro-termicznych dyskretnych nadsystemów dynamicznych (Mathematical Models of Selected Aviation and Automotive Continuous Dynamical Hyperystems)*, *Monografia* 53 (Krakow: Politechnika Krakowska imienia Tadeusza Kościuszki) 276 (In Polish)

Fijalkowski B 1987b Choice of hybrid propulsion systems - wheeled city and urban vehicle; - tracked all- terrain vehicle. Part I *Electr. Veh. Dev.* **6** 113–117 142

Fijalkowski B 2016 *Mechatronics: Dynamical Systems Approach and Theory of Holors* (Bristol, UK: IOP Publishing)

Szklarski L, Pelczewski W, Kolendowski J, Puchalka T and Komarzewska M 1963 *Dynamika ukladow nieholonomicznych (Dynamics of Electromechanical Systems)* (Warszawie: Komitet Elektrotechniki Polskiej Akademii Nauk, Panstwowe Wydawnictwo Naukowe) 259 (In Polish)

Tutaj J 2012 *Ujecie systemowe dynamiki wielofunkcyjnego prądnico-rozrusznika silnika spalinowego* (Dynamic systems approach of the polyfunctional generator-starter for a combustion

engine of the automotive vehicle). *Monografia 409, Seria Mechanika* (Krakow: Politechnika Krakowska im. Tadeusza Kosciuszki) (In Polish)
White D C and Woodson H H 1959 *Electromechanics; Energy Conversion* (New York: Wiley)

Chapter 8

Conclusion and future trends

8.1 Concluding remarks

The title of this textbook *Mechatronics: Integrated Electro-Mechanical Drive* is an abbreviation requiring some explanation, namely:

- 'Mechatronics' is the integration of physics (mechanics, fluidics, electrics and electronics, etc) and computer technologies into the research and development of physical heterogeneous continuous complex and/or simple dynamical hypersystems (Fijalkowski 2010, 2011, 2016). Thus, mechatronics consists of the synergetics of different physical disciplines like mechanics, fluidics, electrics and electronics, magnetics and thermics as well as informatics, etc. Synergetics is a very creative and therefore dynamical process, far more than the usual co-operation of the different physical disciplines and even more than a close integration of machine hardware and software. Mechatronics has offered totally new solutions and an up-to-now unknown flexibility in transportation systems, industrial production processes, aerospace, automotive and traction components, etc. Its economic success was based on functional integration, i.e., by the multiple use of **mechano-mechanical** (MM), **fluido-mechanical** (FM), and/or **electro-mechanical** (EM) actuators, the decentralisation of intelligence into physical heterogeneous continuous dynamical hypersystems, e.g. like aerospace, automotive and/or traction single-shaft propulsion mechatronical control systems, and the inherent options for sensorless self-monitoring and system protection.
- The 'dynamical systems approach' is a holistic description of the observable physical reality. Whole in the dynamical systems approach was not composed of the sum of elementary units. The multiplicity of elementory units is a result of diversification of the whole.
- The idea of a 'holor' (from Greek $\delta\lambda o\sigma$, as in the English 'holistic' pertaining to an entity or a whole) seems to have been originated by Wallis (1673) to describe a mathematical entity that was made up of one or more 'merates'

(from Greek $\mu\acute{\varepsilon}\rho o\sigma$: part of an entity) or 'coordinates' (independent quantities) of the holor, and has included complex numbers, scalars, vectors, matrices, tensors, quaternions, and other hypernumbers. Grassmann (1862, 1844, 1894) works strongly influenced the subsequent developments of holors, including the works of Gibbs (1881, 1884) and Heaviside (1893, 1925, 1950). Holors, thus defined, have been known for centuries but each has been developed more or less independently, accompanied by separate nomenclature and theory: Moon and Spencer (1986). The 'theory of holors' demonstrates how these complicated subjects can be made simple by using a single notation that applies to all holors, both tensor and non-tensor. Moon and Spencer (1965, 1986) consider all possible types of holors and developed holor algebra and holor calculus in the most general sense.
- The theory of holors should establish a method by which students and teachers can learn vector and tensor analysis via a uniform treatment (Moon and Spencer 1963, 1965, 1986; Fijalkowski 1967).

The physical heterogeneous continuous dynamical hypersystem may be treated as a whole or it may be divided into physical homogeneous continuous dynamical systems as regards their forms of energy (e.g. as kinetic energy: radiant, thermal, motion, sound and electrical homogeneous continuous dynamical systems; as potential energy: chemical, nuclear, stored mechanical, fluidical, and electrical and also gravitational homogeneous continuous systems), which may be divided into physical homogeneous continuous dynamical hyposystems, which as follows may be divisible into functional and structural physical homogeneous continuous dynamical components.

The authors' intention in writing this textbook was to demonstrate 'generalised physical-commutation-matrixer holor analyses' in the easiest practicable dynamical systems approach (systems thinking). The authors have supposed that this may be realised best through the application of physical commutation matrixers and its functional and structural physical homogeneous continuous dynamical components.

Nearly all mechanical and electrical engineering students have been educated concisely to hardly cover physical commutation matrixers in their university physics courses. In view of that, the physical model represents an exceptional structure with which to form a junction with the interruption from static to dynamical analysis.

Particularly important was probably that the generalised physical-commutation-matrixer conception itself, contributes systematically to analogy. Owing to the application of the physical-commutation-matrixer method, one (meaning any person including the writer or reader) may confirm that it was practical to consider the area under discussion in mechatronics (i.e. as kinetic energy: radiant, thermal, motion, sound and electrical, and as potential energy: chemical, nuclear, stored mechanical, fluidical and electrical and also gravitational). For instance, the analogy between charging an electrical capacitor and saturating a fluidical container of water was done by equating the physical and mathematical models of these physical processes. Naturally, one might have identified this analogy instinctively. In spite of this, one has been become aware that these physical processes were also comparable to the variation in velocity of a mass when a force is relevant. This was not practically so evident. As a result, the physical-commutation-matrixer conception proves to be exceptionally convenient in the absence of physical situations in various physical heterogeneous continuous dynamical hyposystems to a generalised physical heterogeneous continuous dynamical hyposystem from which a universal type of solution may be completed.

A new definition of the 'physical commutation matrixer', based on matrix-interconnection physical heterogeneous continuous dynamical hypersystem theory concepts, was proposed. The definition may be used to evaluate the physical commutation matrixers, which perform multivalent logical functions, and continuously operate physical commutation matrixers. The serial and parallel connections of two physical commutation matrixers were defined, and simplified formulas were given which were valid for high component physical commutation matrixers.

A comparison was made between the results arrived at by using the new and conventional definitions. It was found that results differed little in the case of equal probability output letters. For instance, a physical commutation matrixer was a general term referring to a physical heterogeneous continuous dynamical hypersystem or part of a physical heterogeneous continuous dynamical hypersystem of matrixery conductive parts and their matrixery inter-connections through which an energy-transfer holor was intended to flow.

A physical commutation matrixer; i.e. a configuration of physically (i.e. electrically, magnetically, optically, or radiationally) connected dynamical components or analogue and/or digital devices was made up of active and passive physical, functional and structural dynamical components or an assemblage of physical, functional and structural dynamical components and their matrixery interconnected row and column conductive collectors. Thus, a physical commutation matrixer was a physical heterogeneous continuous dynamical hypersystem (e.g. machine or device, etc) that produced or was powered by physical energy. The physical commutation matrixer described by a matrixer physical model (matrixery signal-flow diagram or map) has shown the active and passive physical, functional and structural dynamical components and their matrixery interconnected row and column conductive collectors (conductors).

Physical-commutation-matrixer theory included the study of all aspects of matrixers, including analysis, design, and application. In physical-commutation-matrixer theory, the fundamental quantities were the 'energy-potential-difference holors' (i.e. generalised force holors) between various points, the 'energy-transfer holors' (i.e. generalised velocity holors) flowing in a number of row and column conductive collectors, and the parameters, which describe the passive physical, functional and structural dynamical components. Other important physical-commutation-matrixer quantities such as power, energy, and time constants have been computed from the fundamental physical variables. This textbook provides a comprehensive and careful development of physical-commutation-matrixer base and using holor theory formulations of analytical methods. No attempt was made to provide the required mathematical tools: matrix theory, functions of a complex variable, Laplace transforms, numerical methods, or programming. However, some references in these areas have been included in the bibliography.

Holor analysis has been shown beyond doubt to be important in physics and engineering. Presently, every scientist and engineer must be systematically well acquainted with the symbolism and methods of manipulating holors. Because both the nomenclature and routine manipulation are rather uncomplicated and may be learned in a short time, the question emerges as to whether an independent one-term subject on holors is adequate to a science program of study. If the subject was taught on the experiential level, it may be integrated into one of the necessary physics subjects, such as electrics and electronics, magnetics, mechanics, fluidics, photonics, etc. The treatment, presented in this interdisciplinary textbook, differs from the old school method by creating 'invariance' in the theory and in providing 'general definitions' that are sensible in all coordinate systems. In this approach, holor calculus and matrix analysis is endowed with a concrete logical foundation. Irrespective of forward-thinking aspects of the textbook, one does not sense that the treatment is too difficult for the ordinary reader.

Macroelectronics is the study of ECM commutators for the conversion and control of electrical energy. The technology is a critical part of energy infrastructure, and supports almost all important electrical applications. For macroelectronics design, one considers only those matrixers and devices that, in principle, introduce no loss and can achieve near-perfect reliability. The two key characteristics of high efficiency and high reliability are implemented with ECMs, supplemented with energy storage. ECMs in turn can be organised as electrical-valve matrices. This facilitates their analysis and design.

In a macroelectronic cyber-physical continuous dynamical system, the three primary challenges are the hardware problem of implementing an ECM, the software problem of deciding how to operate that ECM, and the interface problem of removing unwanted distortion and providing the user with the desired clean electrical-energy source. The hardware is implemented with special types of power super- and/or semiconductors. These include several types of SiC and AsGa diodes, *n–p–n* and *p–n–p* transistors, especially MOSFETs, IGBTs, RB IGBT, OT IGBTs, HV IGBTs and IGCTs, as well as, several types of thyristors, especially SCRs, PPTs, PPCSTs, and GTOs. The software problem can be represented in terms of commutation (switching) functions. The frequency, duty ratio, and phase of the commutation functions are available for operational purposes. The interface problem is addressed by means of lossless filters. Most often, these are lossless LC passive filters to smooth out ripple or reduce harmonics. More recently, active filters have been applied to make dynamical corrections in power conversion waveforms.

For the time being, it is difficult to be optimistic on the future possibility of practical applications of ECM commutators. The authors believe that scientists should not promise users too much. As an example, optimistically one should note the invention of a transistor by J Bardeen, W G Brattain and W B Shockley in 1948. At that time, people assumed that the application of transistors would replace electron lamps, especially triodes.

Nobody could foresee that soon after this invention the unprecedented miniaturisation of electronics and the creation of EICs with a larger scale of integration would follow.

In the authors' opinion, such a thing could occur with ECM commutators, and for the time being it is impossible to foresee some of the future practical applications.

The results of the invention of ECM commutators and R&D works carried out in this field were that complete solid-state physics began life anew.

Not long ago people believed, that in the field of power electronics everything has already been invented, in fact this is only a half-true. Recently there has been enormous demand for R&D works in the field of solid-state physics.

Scientific research concerning ECM commutators will be transformed into a great race of scientific teams from great world consortia and the result, which will revolutionise the 'universe of macroelectronics' may be expected at any moment.

Very similar races of scientific teams have already taken place in the field of solid-state physics, when a series of discoveries of high-temperature electric-current conductivity in materials of a new kind took place.

These particular current experiments terminated in the invention of ECM commutators with bipolar ECMs, realised on 'continuous' bipolar ovonics; hybrid transistors with heterojunctions; SUPER-HETs with a superconductive base; QUITs; and very high-speed SITs. These ECM commutators will be used in newly designed energy-saving integrated MMD electrical machines with application of high-temperature superconductors, for example, made from Y–Cu–Ba–O ceramics.

Superconductivity is an exciting and rapidly developing technology, especially since the recent discovery of high T_C materials. The basic technology and performance of IECs based on Nb Josephson junctions were implemented, followed by an assessment of the prospects for IECs incorporating high T_C materials. These include uni- and bipolar semiconductor triodes (three terminal electrical valves) and hybrid semiconductor/superconductor electrical valves, the challenge being to compete successfully with current semiconductor technology, which is itself developing rapidly. ECM commutators are nowadays indispensable in modern MMD, MEM, MFD and MPD electrical machine engineering, and for energy converting processes in industrial EMD and traction propulsion IEMD cyber-physical heterogeneous dynamical hypersystems. Compared with previous poly-valvular, ECM commutators (static converters) are more practice-oriented, especially from the point of view of operating convenience, space requirement of ultra-modern semiconductor/superconductor ECM commutator techniques.

A major step forward has been taken by the authors within electrical machinery technology with the development of macroelectronic commutators named 'ECM macrocommutators'. The ECM commutator is a new product family based on the ECM and the EIC computer controller. The latest advances made within the field of macroelectronic components, such as powerful ECM commutators and in the field of 64 bit, 128 bit and 256 bit microprocessors, have been exploited during the R&D work.

Components in the form of hardware and software have been developed with the aid of these new macro-, meso-, micro- and nanoelectronic components and adapted to new modern control system structures.

Recently, the ECM on its own may be used for all the new concept integral ECM AC–AC, AC–DC–AC, AC–DC, DC–DC, DC–AC–DC and DC–AC commutator dynamotor's industrial EMD and traction propulsion IEMD cyber-physical heterogeneous dynamical hypersystems previously covered by several earlier modular cyber-physical heterogeneous dynamical hypersystems. The method used to build industrial EMD and traction propulsion IEMD cyber-physical heterogeneous dynamical hypersystems has varied over the years, depending on the types of propulsion IEMD cyber-physical heterogeneous dynamical hypersystems component available. A changeover from extremely distributed industrial EMD and traction propulsion cyber-physical heterogeneous continuous or discrete dynamical hypersystems, where each object had its own ECU, to modern industrial EMD and traction propulsion cyber-physical heterogeneous continuous or discrete dynamical hypersystems took place. The angular-velocity and torque control functions were grouped together in ECM commutator super synchronous, synchronous or asynchronous (induction) motors to provide central access to all information.

The world must solve major challenges to transform and control electrical energy in an efficient way. Examples of this are in aerospace, aviation, automotive, and railway transportation, renewable energies and industrial processing applications. These problems can be solved using ECM AC–AC, AC–DC–AC, AC–DC, DC–DC, DC–AC–DC and DC–AC commutators based on modern power super- and/or semiconductor amorphous electrical valves.

The ideal ECM AC–AC, AC–DC–AC, AC–DC, DC–DC, DC–AC–DC and DC–AC commutator in many of these applications may have the following characteristics:

- compact design with a good power-to-mass ratio;
- regeneration capability;
- operation with unity power factor;
- sinusoidal input and output currents.

ECM AC–AC, AC–DC–AC, AC–DC, DC–DC, DC–AC–DC and DC–AC commutators, can fulfil all these characteristics and this is the reason for the tremendous interest in this ECM topology.

In the last decade, many advances in the development of this ECM topology have been presented, including industrial applications up to mega-watt (MW) level.

The use of AC–AC, AC–DC–AC, AC–DC, DC–DC, DC–AC–DC and DC–AC commutators in real applications and the challenges that these applications present is very topical and important.

This textbook presents on the macro-, meso-, micro- and nanoelectronics community the most recent advances with topics such as, but not limited to, the following:

- ECM AC–AC, AC–DC–AC, AC–DC, DC–DC, DC–AC–DC and DC–AC commutators for aerospace, automotive and railway transportation, renewable energy and industrial application;
- ECM AC–AC, AC–DC–AC, AC–DC, DC–DC, DC–AC–DC and DC–AC commutator derived ECM topologies;

- new super- and/or semiconductor amorphous electrical valves for use in ECM AC–AC, AC–DC–AC, AC–DC, DC–DC, DC–AC–DC and DC–AC commutators;
- power quality issues, ECM AC–AC, AC–DC–AC, AC–DC, DC–DC, DC–AC–DC and DC–AC commutator reliability, and stability;
- implementation of **artificial intelligent** (AI) current commutation strategies in application examples;
- new control/modulation methods for ECM AC–AC, AC–DC–AC, AC–DC, DC–DC, DC–AC–DC and DC–AC commutator applications, including **direct torque control** (DTC), **passively based control** (PBC), **predictive control** (PC) with the **magnetic field oriented control** (MFOC), **sinusoidal modulation** (SINM), **vector modulation** (VECM), **holor modulation** (HOLM);
- comprehensive comparative evaluation of the ECM AC–AC, AC–DC–AC, AC–DC, DC–DC, DC–AC–DC and DC–AC commutator concepts with DC-link or AC-link energy storage components, respectively.

However, the practical applications of the 3 × 3 ECM AC–AC, commutator, as of now, are very limited.

The main reasons are:

- non-availability of the bilateral fully-controlled monolithic electrical valves capable of high-frequency operation;
- complex control law implementation;
- an intrinsic limitation of the output-input voltage ratio;
- commutation (Case 2000) and protection of the electrical valves (switches).

To date, the electrical valves are assembled from existing 'continuous' devices resulting in increased cost and complexity and only experimental ECMs of capacity well below 1 MVA have been built. However, with the advances in device technology, it is hoped that the problems will be solved eventually and the 3 × 3 ECM AC–AC super- or static-commutator will not only replace the naturally-commutated cycloconverters in all the applications but will also take over from the PWM rectifiers/inverters as well.

In Mahlein *et al* (1999), it has been shown that with **vector PWM** (VECPWM) or **holor PWM** (HOLPWM) control using over-modulation, the voltage transfer ratio may be increased to 1.05 at the expense of more harmonics and large filter capacitors.

Improvements in devices and advances in control concepts have led to steady improvements in ECM commutators, i.e., macroelectronic homogeneous continuous dynamical systems. This is driving tremendous expansion of their application. For instance, PCs, would be unwieldy and inefficient without macroelectronic DC energy supplies. Portable communication devices and computers would be impractical. High-performance HVDC transmission links, M-E/E-M generator/motor (voltage/torque) controls, lighting systems, and a wide range of industrial controls depend on macroelectronics.

In the near future, one can expect strong growth in automotive applications, in DC energy supplies for communication systems, in portable applications, and in high-end ECM commutators for advanced microprocessors. During the next generation, we will reach a time when almost all electrical energy is processed through macroelectronics somewhere in the path from generation to end use.

The 2010 Nobel Prize in Physics was awarded to the two researchers who performed the first experiments on graphene, a two-dimensional sheet of carbon atoms. The award, given to University of Manchester physicists Andre Geim and Konstantin Novoselov, recognises work that began less than a decade ago on a material that has since been used to make record-breaking transistors and stretchy electrodes.

Graphene is a one-atom-thick sheet of carbon atoms arranged in a honeycomb-like pattern. Graphene is considered to be the world's thinnest, strongest and most conductive material—of both electricity and heat. All these properties are exciting researchers and businesses around the world—as graphene has the potential to revolutionise entire industries—in the fields of electricity, conductivity, energy generation, batteries, sensors and more.

For integrated circuits, graphene has a high carrier mobility, as well as low noise, allowing it to be used as the channel in a **field-effect transistor** (FET). Single sheets of graphene are hard to produce and even harder to make on an appropriate substrate.

Graphene exhibits a pronounced response to perpendicular external electric fields, potentially forming FET. The first top-gated FET (*ON–OFF* ratio of <2) was demonstrated in 2007.

Graphene nanoribbons may prove generally capable of replacing silicon as a semiconductor.

The negative differential resistance experimentally observed in graphene field-effect transistors of conventional design allows for construction of viable non-Boolean computational architectures with graphene.

The negative differential resistance—observed under certain biasing schemes—is an intrinsic property of graphene resulting from its symmetric band structure. The results present a conceptual change in graphene research and indicate an alternative route for graphene's applications in information processing.

8.2 Future work

Further investigations into AC or DC motor IEMD cyber-physical heterogeneous dynamical hypersystems, which was presented in this textbook, may build on the promising **artificial intelligence** (AI) AC or DC motor IEMD cyber-physical heterogeneous dynamical hypersystems. They are integrations of computation, networking, and physical processes. Computers and networks monitor and control the physical processes, with feedback loops where physical processes affect computations and vice versa. The societal and economic potential of such AC or DC motor IEMD dynamical hypersystems is vastly greater than what has been realised, and major investments are being made worldwide to develop the technology. The technology builds on the principal discipline of embedded AC or DC motor IEMD dynamical hypersystems, computers and software embedded in devices whose principle mission is not computation, such as aircraft, automotive vehicles, medical devices, and scientific instruments. Cyber-physical heterogeneous dynamical hypersystem integrates the dynamics of the physical processes with those of the software and networking, providing abstractions and modelling, design, and analysis techniques for the integrated whole (Podelski and Bogomolov 2012, Lee and Seshia 2013, Sanfelice 2014).

The combination of networking and information technology with engineered AC or DC motor IEMD physical heterogeneous dynamical hypersystems and associated services enabled an innovative generation of 'smart dynamical hypersystems'. These AC or DC motor IEMD cyber-physical heterogeneous dynamical hypersystems combine distributed sensing, monitoring, actuation, and control networks. These have interoperable dynamical systems, hyposystems and component integration, very advanced analytics, and user interfaces featuring customised degrees of independence which can facilitate adaptive, predictive, and mutual optimisation of the cyber-physical heterogeneous dynamical hypersystem performance over the entire lifecycle of a machine or device (e.g. design, build, operate/use, maintain, and service). The existing and emerging applications create exciting opportunities, especially, for aerospace and automotive industry. Currently, **unmanned aircraft systems** (UAS) are exploited for search and rescue, surveillance, mapping, crop spraying, environmental monitoring. Moreover, **unmanned ground vehicles** (UGV) have very diverse applications such as security and mine clearance (Anon 2016).

Cyber-physical heterogeneous continuous dynamical hypersystems integrate analogue and/or digital devices, interfaces, networks, computer systems, and the like with the natural and/man-made physical world. The natural interrelated and heterogeneous combination of behaviours in these dynamical systems makes their analysis and design a challenging task. Safety and reliability specifications imposed in cyber-physical applications, which are normally translated into stringent robustness standards, aggravate the matter. Unfortunately, state-of-the-art tools for system analysis and design cannot cope with the intrinsic complexity in cyber-physical heterogeneous dynamical hypersystems. Tools suitable for analysis and design of cyber-physical heterogeneous dynamical hypersystems must allow a combination of physical or continuous dynamics and the cyber or computational components, as well as the ability to handle a variety of modes of perturbations, such as exogenous disturbances, time delays, and dynamical hypersystem failures (Sanfelice 2014). For instance, the AC or DC motor IEMD cyber-physical heterogeneous dynamical hypersystems demonstrate the applicability of cyber-mode control algorithms to the AC or DC motor IEMD physical heterogeneous dynamical hypersystems without significant design efforts and underline their advantage of high robustness. The remaining chattering effects, which decrease the cyber-mode position control accuracy, may be reduced by means of faster hardware or sophisticated observers and disturbance estimators. As a result, a challenging research tendency is, for instance, the investigation of an adaptive disturbance estimator for the flexible shaft system, which does not require information about the structure of a disturbance or the frequency of periodical perturbations.

The performance of the aforementioned cyber-mode position control of an AC–AC or AC–DC–AC commutator synchronous or asynchronous (induction) motor may be improved by implementing better observers as well for magnetic flux as for the angular velocity. Besides, an estimator that provides information about the unknown load may lead to a more accurate control performance (Fijalkowski 2016).

AC or DC motor IEMD cyber-physical heterogeneous dynamical hypersystems can be described as smart dynamical hypersystems that encompass computational (i.e. hardware and software) and physical AC or DC motor IEMD homogeneous dynamical systems, hyposystems and components, seamlessly integrated and closely interacting to sense the changing state of the real world (Tianbo *et al* 2015, Anon 2013). Unlike more conventional embedded AC or DC motor IEMD dynamical hypersystems, a developed AC or DC motor IEMD cyber-physical heterogeneous dynamical hypersystem is usually designed as a network (whole) of interacting dynamical systems, hyposystem and components with physical input and output instead of as standalone devices (Anon 2008, Shafi 2012). AC or DC motor IEMD cyber-physical heterogeneous dynamical hypersystems are becoming popular in many application areas: power networks, aerospace, automotive, manufacture, healthcare, critical infrastructure, etc. However, with its popularity and extensive application in critical and important applications, a cyber-physical heterogeneous dynamical hypersystem becomes increasingly more susceptible to the security vulnerabilities and targets for cyber-physical attacks (Wang *et al* 2010). Hackers can launch malicious attacks on power networks and transportation systems (Mills 2009) and be able to hack medical devices implanted in the human body which have wireless communications (Leavitt 2013).

More and more security vulnerabilities are being found in all kinds of cyber-physical heterogeneous dynamical hypersystems (Fletcher and Liu 2011). Security of AC or DC motor IEMD cyber-physical heterogeneous dynamical hypersystems has arisen as a concern of utmost importance in research and system design of AC or DC motor IEMD cyber-physical heterogeneous dynamical hypersystems.

Thus, a cyber-physical heterogeneous dynamical hypersystem integrates computing, communication and storage capabilities with the monitoring and/or control of entities in the physical world dependably, safely, securely, efficiently and in real-time (Mahapatra 2016).

- AC or DC motor IEMD cyber-physical heterogeneous dynamical hypersystems is an exciting prospect for the next decades!
- it involves multi-disciplinary research-and-development (R&D) works;
    - High confidence software;
- AC or DC motor IEMD cyber-physical heterogeneous dynamical hypersystems have the potential to change the way people interact with their surroundings;
- applications in the future AC or DC motor IEMD cyber-physical heterogeneous dynamical hypersystems are limited only by human imagination;
- affordability and ease of application will drive adoption.

AC or DC motor IEMD cyber-physical heterogeneous dynamical hypersystems are expected to play a major role in the R&D of future AC or DC motor IEMD physical heterogeneous dynamical hypersystems with new capabilities that far exceed contemporary levels of autonomy, functionality, usability, reliability, and cyber security. Advances in their R&D can be accelerated by close collaborations between academic disciplines in computation, communication, control, and other engineering and computer science disciplines, coupled with grand challenge applications.

Selected recommendations for R&D in AC or DC motor IEMD cyber-physical heterogeneous dynamical hypersystems:

- Standardised abstractions and architectures that permit modular design of AC or DC motor IEMD cyber-physical heterogeneous dynamical hypersystems are urgently needed.
- AC or DC motor IEMD cyber-physical heterogeneous dynamical hypersystems' applications involve components that interact through a complex, coupled physical environment. Reliability and security pose particular challenges in this context—new frameworks, algorithms, and tools are required.
- Future AC or DC motor IEMD cyber-physical heterogeneous dynamical hypersystems will require hardware and software components that are highly dependable, reconfigure-able, and in many applications, certifiable and trustworthiness must also extend to the system level.

In future work, the authors will focus on the AC or DC motor IEMD cyber-physical heterogeneous dynamical hypersystems of the many application areas: power networks, aerospace, automotive, manufacture, healthcare, critical infrastructure, as well as EMD etc, and measure their performance to compare with the theoretical holor analysis.

References

Anon 2008 University of California at Berkeley, Electrical Engineering and Computer Sciences, Technical Report No. UCB/EECS-2008, http://eecs.berkeley,edu/Pubs/TechRpts/2008/EECS-2008-8.htm

Anon 2013 Foundations for Innovation in Cyber Physical Systems Workshop Summary Report (January 2013) http://nist.gov/el/upload/CPS-WorkshopReport-1-30-13-Final.pdf

Anon 2016 Cranfield University Centre for Cyber-physical Systems http://Centre for Cyber-physical Systems.htm

Case M J 2000 The commutation process in the matrix converter *Conf. Proc. EPE-PEMC 2000, Kosice, Slovakia* **2** 109–12

Fijalkowski B 1967 Zastosowanie rachunku holorowego w teorii pradow przemiennych (Application of a holor calculus in the theory of alternating currents). *Przeglad elektrotechniczny* (Electrotechnics Review), Rok XLIII, Zeszyt 2, Warszawa, 1967 (In Polish)

Fijalkowski B T 2010 *Automotive Mechatronics: Operational and Practical Issues. — Volume I. International Series on Intelligent Control, and Automation Science and Engineering* vol 47 (Netherlands: Dordrecht Springer)

Fijalkowski B T 2011 *Automotive Mechatronics: Operational and Practical Issues. — Volume II. International Series on Intelligent Control, and Automotive Science and Engineering* vol 52 (Dordrecht: Springer)
Fijalkowski B 2016 *Mechatronics: Dynamical Systems Approach and Theory of Holors* (Bristol, UK: IOP Publishing)
Fletcher K K and Liu X F 2011 Security requirements analysis, specification, prioritization, and policy development in cyber-physical systems, *5th Int. Conf. on Secure Software Integration & Reliability Improvements Companion* (SSIFI-C) 106–13
Gibbs J W 1881 *Elements of Vector Analysis* (New Haven: privately printed) pp 17–50
Gibbs J W 1884 *Elements of Vector Analysis* (New Haven: privately printed) pp 50–90
Gibbs J W 1906 *The Scientific Papers of J. Willard Gibbs* vol II (London: Longmans, Green and Co.) p 17
Grassmann H 1844 *Die lineale Ausdehnungslehre, ein neuer Zweig der Mathematik* (Leipzig: Ottp Wigand)
Grassmann H 1862 *Die Ausdehnungslehre* (Berlin: Enslin) (In German)
Grassmann H 1894 *Grassmann's gesammelte math. u. phys. Werke 3 vols* (Leipzig: B G Teubner)
Heaviside O 1893 *Electromagnetic Theory* 3 vols (London: Electrician Printing and Pub; New York: Dover, 1950). Heaviside's treatment is given in vol I, ch III, pp 132–305: 'The elements of vectorial algebra and analysis'
Heaviside O 1925 *Electromagnetic Theory 3 vols* (London: The Electrician Printing and Publishing Company Ltd)
Heaviside O 1950 Electromagnetic Theory 3 vols, contains a critical and historical introduction ed E Weber (New York: Dover) see vol 11 pp 192–305 on vector analysis
Leavitt N 2013 Researchers fight to keep implanted medical devices safe from hackers *Computer* **43** 11–4
Lee E A and Seshia S A 2013 *Introduction to Embedded Systems—A Cyber-Physical Systems Approach* Lulu.com 1st edn
Lu T, Zhao J, Zhao L, Li Y and Zhang X 2015 Towards a framework for assuring cyber physical system security *Int. J. Security Applic.* **9** 25–40
Mahapatra R N 2016 Cyber-physical systems: issues and challenges (Adapted from NSF Workshops) *Presentation* Texas A&M University, Computer Science, WECON 2011March 4, 2016 pp 1–50
Mahlein J, Simon O and Braun M 1999 A matrix converter with space vector control enabling overmodulation, *Conf. Proc. EPE 99, Lussanne, Swiss* pp 1–11
Mills E 2009 Hackers broke into FAA air traffic control system *The Wall Street Journal* A6
Moon P H and Spencer D E 1963 A new mathematical representation of alternating currents *Tensor* **14** 110
Moon P H and Spencer D E 1965 *Vectors* (Princeton, NJ: van Nostrand)
Moon P H and Spencer D E 1986 *Theory of Holors. A Generalization of Tensors* (Cambridge: Cambridge University Press) (Digitally printed, 2005)
Podelski A and Bogomolov S 2012 Lectures notes http://informatik,uni-freiburg.de/teaching/532012/cps-hm
Sanfelice R G 2014 *Introduction to Cyber-physical Systems*(CMPE 142) Fall 2014 (Santa Cruz, CA: University of California)

Shafi Q 2012 Cyber-physical systems security. A brief survey *12th Int. Conf. on Computational Science and Its Applications (ICCSA), 2012* pp 146–50

Wallis J 1685 Letter to Collins, 6 May 1673 *Treatise of algebra* (London) p 264

Wang E K, Ye Y, Xu X, Yiu S M, Hui I C K and Chou K P 2010 Security issues and challenges for cyber physical systems, *2010 IEEE/ACM Conf. on Green Computers and Communications & 2010 IEEE/ACM Conference on Cyber Physical and Social Computing* pp 733–8

IOP Publishing

The Integrated Electro-Mechanical Drive
A mechatronic approach
B T Fijalkowski and J Tutaj

Appendix A

MCM and/or ECM AC–AC or AC–DC–AC or AC–DC or DC–DC or DC–AC–DC or DC–AC commutators

A.1 Introduction

In this appendix we look at examples of the **mechanical commutation matrixer** (MCM) and/or **electrical commutation matrixer** (ECM) AC–AC or AC–DC–AC or AC–DC or DC–DC or DC–AC–DC or DC–AC commutators, which are used with an IEMD, providing either DC or AC outputs, and operating from either a DC supply (storage battery or rectified), or from the conventional AC mains. The treatment is not intended to be exhaustive, but should serve to highlight the most important aspects which are common to all types of MCM and/or ECM AC–AC or AC–DC–AC or AC–DC or DC–DC or DC–AC–DC or DC–AC commutators. Although there are many different types of MCM and/or ECM AC–AC or AC–DC–AC or AC–DC or DC–DC or DC–AC–DC or DC–AC commutators, all excluding very low-power ones are based on some form of electronic commutation. The necessity to adopt a commutation strategy is emphasised in the first example, where the consequences are explored in some depth.

The reader will see that commutation is essential in order to achieve high-efficiency power conversion, but that the resulting waveforms are inevitably less than ideal from the point of view of the E-M motor. The examples have been chosen to illustrate typical practice, so for each MCM and/or ECM AC–AC or AC–DC–AC or AC–DC or DC–DC or DC–AC–DC or DC–AC commutator the most commonly used electrical valves (electronic switches), (e.g. thyristors or transistors) are shown. In many cases, several different electrical valves may be suitable (see later), so users should not identify a particular MCM or ECM as being the exclusive

 © IOP Publishing Ltd 2019

preserve of a particular electrical valve. Before discussing particular MCMs or ECMs, it will be useful to take an overall look at a typical IEMD, so that the role of the MCM or ECM AC–AC or AC–DC–AC or AC–DC or DC–DC or DC–AC–DC or DC–AC commutator can be seen in its proper context.

A complete IEMD cyber-physical heterogeneous dynamical hypersystem is shown in a structural and functional diagram form in chapter 1, figure 1.1. The task of the MCM or ECM AC–AC or AC–DC–AC or AC–DC or DC–DC or DC–AC–DC or DC–AC commutator is to draw electrical energy from the mains (at E-M motor at whatever voltage and frequency necessary to achieve the desired mechanical output). Except in the simplest AC–DC commutator (such as a simple diode AC–DC rectifier), there are usually two distinct parts to the MCM or ECM AC–AC or AC–DC–AC or AC–DC or DC–DC or DC–AC–DC or DC–AC commutator. The first is the power stage, through which the electrical energy flows to the E-M motor windings, and the second is the control section, which regulates the power flow. Control signals, in the form of low-power analogue or digital voltages, tell the ECM AC–AC or AC–DC–AC or AC–DC or DC–DC or DC–AC–DC or DC–AC commutator what it is supposed to be doing, while other low-power feedback signals are used to measure what is actually happening.

By comparing the demand and feedback signals, and adjusting the output accordingly, the target output is maintained. The simple arrangement shown in chapter 1, figure 1.1 has only one input representing the desired angular velocity, and one feedback signal indicating actual angular velocity, but the IEMD will have extra feedback signals as the reader will see later. Almost the IEMD employs closed-loop (feedback) control, so readers who are unfamiliar with the basic principles might find it helpful to read appendix B at this stage. A characteristic of MCM or ECM AC–AC or AC–DC–AC or AC–DC or DC–DC or DC–AC–DC or DC–AC commutators which is shared with most IEMD cyber-physical heterogeneous dynamical hypersystems is that they have very little capacity for storing energy. This means that any sudden change in the power supplied by the MCM and/or ECM AC–AC or AC–DC–AC or AC–DC or DC–DC or DC–AC–DC or DC–AC commutator to the E-M motor windings must be reflected in a sudden increase in the power drawn from the supply. In most cases this is not a serious problem, but it does have two drawbacks. Firstly, a sudden increase in the current drawn from the supply will cause a momentary drop in the supply voltage, because of the effect of the supply impedance. These voltage ‘spikes’ will appear as unwelcome distortion to other users on the same supply. And secondly, there may be an enforced delay before the supply can furnish extra power. With a single-phase AC mains supply, for example, there can be no sudden increase in the power supply from the mains at the instant where the mains voltage is zero, because instantaneous power is necessarily zero at this point in the cycle because the voltage is itself zero. It would be better if a significant amount of energy could be stored within the MCM and/or ECM AC–AC or AC–DC–AC or AC–DC or DC–DC or DC–AC–DC or DC–AC commutator itself: short-term energy demands could then be met instantly, thereby reducing rapid fluctuations in the power drawn from the AC mains. But unfortunately this is just not economic: most converters do have a small store of energy in their smoothing inductors and capacitors, but the amount is not sufficient to buffer the supply sufficiently to shield it from anything more than very short-term fluctuations.

In existing theory of a generalised MCM AC–AC or DC–AC commutator synchronous or asynchronous (induction) motors (formerly termed DC and/or AC motors), both rotary and linear, the role, which in these MMD electrical machines is playing by the sliding-contact MCM AC–AC or DC–AC commutator, is completely depreciated.

Up to now, the theory of a generalised MCM and/or ECM AC–AC or AC–DC–AC or AC–DC or DC–DC or DC–AC–DC or DC–AC commutator of MMD electrical machines was also not worked out and its physical mathematical models in the authors' dynamical systems approach on the MMD electrical machines and matrix notation was not formulated.

A complex of phenomena taking place with the change-over of a sense of current direction polarity in individual commutating sections of the split-ring or flat MCM AC–AC, AC–DC and/or DC–AC commutator, i.e. phase winding coils of a polyphase armature winding short circuited through the carbon brush, i.e. the matrixer-input primary collectors or a rotary sliding-contact MCM AC–AC, AC–DC and/or DC–AC–AC commutator has merely been considered.

As already mentioned above, up to now, the theory of MCMs of AC–AC, AC–DC and/or DC–AC commutator motors, both rotary and linear, has not been worked out. The consideration proposing of a mathematical model formulation and fulfilment of this theoretical blank was first carried out by the authors (Fijalkowski 1973, 1982, 1985a, 1985b, 1985c, 1987a, 1987b, 1988a, 1988b, 1988c, 1988d, 1988e, 1988f, 1988g, 2016, Tutaj 1996, 2012) and will be presented below.

The generalised MCM and/or ECMs commutators of MMD electrical machines (figure A1), both rotary and linear, sliding-contact MCM AC–AC, AC–DC and/or DC–AC commutator, and/or solid-state semi- or superconductor electrical-valve ECM AC–AC or AC–DC–AC or AC–DC or DC–DC or DC–AC–DC or DC–AC commutator is always acting as a frequency and phase number changer converting without losses the electrical energy of one form (e.g. with the voltage holor—$\boldsymbol{U}_i^P$, frequency—f^P, and phase number—m^P) into electrical energy of another form (e.g. with the voltage holor—$\boldsymbol{U}_i^s$, frequency— f^s, and phase number—m^S). At the same time, it takes into account the properties of uni- or bipolar current conduction of electrical valves. For instance, sliding-contacts (copper-bar or segment—carbon-brush sliding-contacts) establishing components of a sliding-contact MCM AC–AC, AC–DC and/or DC–AC commutator, or solid-state semi- and/or superconductor electrical valves (electronic switches), both monocrystal and amorphous, establishing components of an multi-input/multi-output (MIMO) ECM.

In the simplest way, the generalised MCM and/or ECMs commutators of MMD electrical machines, both rotary sliding-contact MCM AC–AC or DC–AC commutator and/or static ECM AC–AC, AC–DC and/or DC–AC commutator can be presented as a generalised ECM AC–AC or AC–DC–AC or AC–DC or DC–DC or DC–AC–DC or DC–AC commutator (figure A1) consisting of crossing one another, but non-galvanic connected by means of electrical valves, current

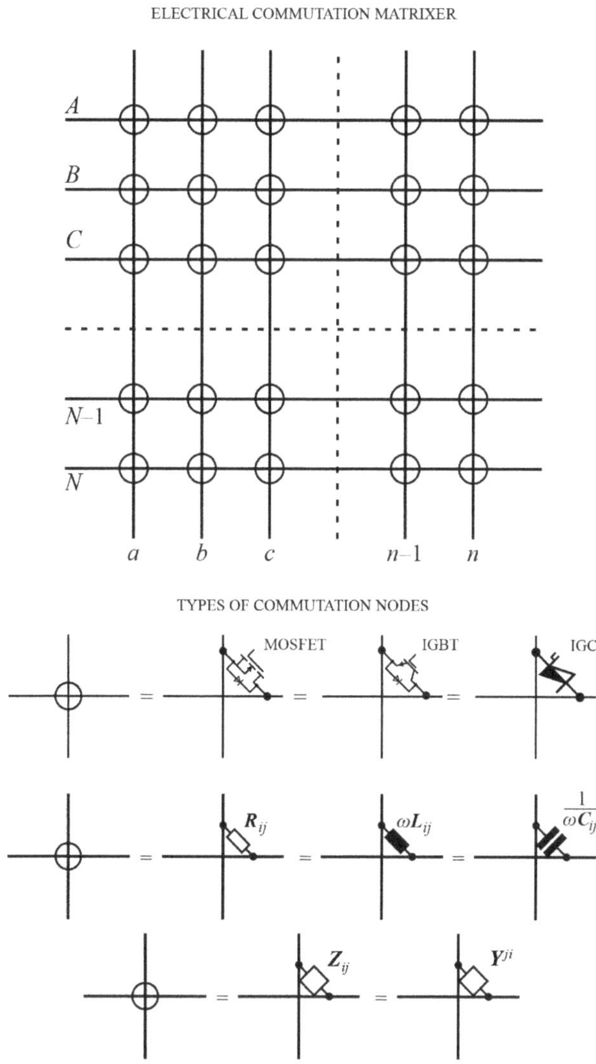

Figure A1. Generalised electrical-commutation-matrixer (ECM) AC–AC, AC–DC, DC–AC and/or DC–DC commutator. In this generalised MCM AC–AC, AC–DC, AC–DC–AC, DC–DC, DC–AC and/or DC–AC–DC, the horizontal collectors (matrix rows) *A, B, C, …, N* are called 'matrixer-input' or 'primary collectors', and the vertical collectors (matrix columns) *a, b, c, …, n*—'matrixer-output' or 'secondary collectors'.

conducting matrixer-input and matrixer-output conductive collectors (Fijalkowski 1987a, 1987b, 2016, Tutaj 1996, 2012).

Discriminating of matrixer-input or primary collectors is relative, since through the same collectors in a distinct moment of time the electrical energy can be both absorbed and recovered. This generalised MCM and/or ECM AC–AC or AC–DC–AC or DC–AC commutator may be connected with two other electroenergetic systems, namely the armature windings of an MMD electrical machine and the AC or DC supply line, only by means of self-collectors.

Considering an MCM or an ECM AC–AC or AC–DC–AC or DC–AC commutator of MMD electrical machines as the M-E/E-M cyber-physical heterogeneous continuous or discrete dynamical hypersystem it is necessary to take into account its cooperations with the armature windings of an MMD electrical machine by means of either voltage-holor sources or current holor sources leading to the matrixer-input and matrixer-output conductive collectors.

The generalised AC–AC, AC–DC, AC–DC–AC, DC–DC, DC–AC and/or DC–AC–DC commutator of MMD electrical machines is the MCM or ECM with the matrixer-input or primary collectors *A, B, C, ... , N* connected to the MMD electrical machine's polyphase armature winding terminals, which is connected poly-star or polygon (e.g. wye or delta), and matrixer-output or secondary conductive collectors *a, b, c, ..., n* to the AC or DC supply line. Individual collectors of this ECM are connected one to another by means of uni- and/or bipolar electrical valves.

Impedance of such a real electrical valve is non-linear, but impedance of such an ideal electrical valve, in turn-on state to be equal to zero, and in turn-off state—to infinity.

The most readily apparent difference between a generalised MCM and/or ECM and other types of MCMs and ECMs is the electrical-valve action. In contrast to digital ECMs, the electrical valves do not indicate a logic level. Control is affected by determining the times at which electrical valves should turn-on operate. Whether there is just one electrical valve or a large group of them, there is a complexity limit: if an E–E converter has P primary collectors (inputs) and s secondary conductive collectors (outputs), even the densest possible collection of electrical valves would have a single electrical valve between each input conductive collector (column line) and each output conductive collector (row line). The $P \times s$ electrical valves in the MCM and/or ECM AC–AC or AC–DC–AC or DC–AC commutator can be arranged according to their connections. The pattern suggests a MCM and/or an ECM, as shown in figure A2. MIMO ECM AC–AC or AC–DC–AC or DC–AC commutators fall into two classes:

- ***Direct ECM AC–AC, AC–DC and/or DC–AC commutators***. In these ECMs, energy-storage components are connected to the ECM only at the primary (input) and secondary (output) of its collectors (ports). The electrical energy-storage components effectively become part of the energy source or load. For instance, an ECM AC–DC commutator (rectifier) with an external low-pass filter is an example of a direct ECM AC–DC commutator. Recently in the literature, only direct 3 × 3 ECM AC–AC, commutators are termed 'matrix converters'.
- ***Indirect ECM AC–DC–AC and/or DC–AC–DC commutators,* also termed *embedded electrical commutators***. These ECM AC–DC–AC and/or DC–AC–DC commutators, for instance, like the polarity-reverser example, have energy-storage components connected within the ECM structure. There are usually a very small number of energy-storage components. Indirect ECM AC–DC–AC and/or DC–AC–DC commutators are most commonly analysed as a cascade connection of direct ECM AC–AC and/or AC–DC/DC–AC commutators with the energy-storage in between.

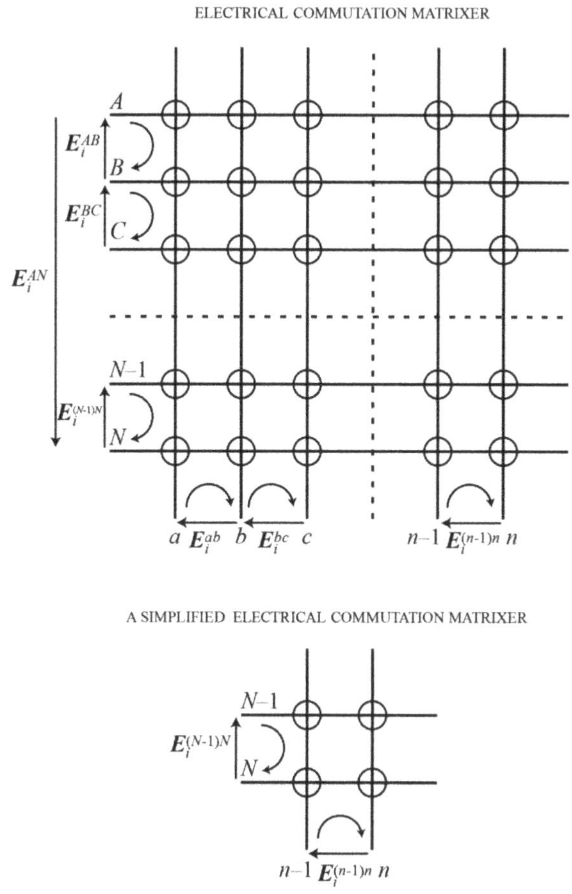

Figure A2. A multi-port generalised MIMO ECM with voltage-source holors (physical model I).

From such the dynamical systems approach on MMD electrical machines with a generalised MIMO ECM AC–AC, AC–DC, AC–DC–AC, DC–DC, DC–AC and/or DC–AC–DC commutator (Fijalkowski 1973, 1982, 1985a, 1985b, 1985c, 1987a, 1987b, 1988a, 1988b, 1988c, 1988d, 1988e, 1988f, 1988g), in referring to AC commutatorless MMD electrical machines with which this generalised MIMO ECM AC–AC, AC–DC, AC–DC–AC, DC–DC, DC–AC and/or DC–AC–DC commutator cooperates and establishes ME/EM simple dynamical hypersystem. Two different mathematical models may be formulated, and therefore, two different matrix notation manners of the generalised MIMO ECM commutator's holor equations can be obtained (Fijalkowski 1987a, 1987b, 2016).

Taking into account ECM excitations (forcing) in the form of voltage-source holor matrix, and as ECM responses—external mesh-current holor matrix, the mathematical model of a generalised MIMO ECM AC–AC, AC–DC, AC–DC–AC, DC–DC, DC–AC and/or DC–AC–DC commutator of MMD electrical machines has been formulated in the dynamical systems approach and matrix notation as follows (figure A2):

$$\begin{aligned} \boldsymbol{E}_i &= \left\| \boldsymbol{E}_i^{p^T}, \boldsymbol{E}_i^{sT} \right\|^T \\ &= \left\| \boldsymbol{E}_i^{AN^T} \boldsymbol{E}_i^{BA^T} \boldsymbol{E}_i^{CB^T} \cdots \boldsymbol{E}_i^{N(N-1)^T} \boldsymbol{E}_i^{an^T} \boldsymbol{E}_i^{ba^T} \boldsymbol{E}_i^{cb^T} \cdots \boldsymbol{E}_i^{n(n-1)^T} \right\|^T \end{aligned} \tag{A.1}$$

as responses of the multi-port MIMO generalised MIMO ECM—the column hypermatrices or the hypervectors of outer-mesh-current holors,

$$\begin{aligned} \boldsymbol{I}^j &= \left\| \boldsymbol{I}_p^{j^T} \boldsymbol{I}_s^{j^T} \right\|^T \\ &= \left\| \boldsymbol{I}_{AN}^{j\ T} \boldsymbol{I}_{BA}^{j\ T} \boldsymbol{I}_{CB}^{j\ T} \cdots \boldsymbol{I}_{N(N-1)}^{j\ \ \ T} \boldsymbol{I}_{an}^{j\ T} \boldsymbol{I}_{ba}^{j\ T} \boldsymbol{I}_{cb}^{j\ T} \cdots \boldsymbol{I}_{n(n-1)}^{j\ \ \ T} \right\|^T \end{aligned} \tag{A.2}$$

a set of the differential equations of dynamics, i.e. the mathematical model of the multi-port MIMO generalised ECM as the electrical heterogeneous continuous dynamical hypersystems may be formulated in dynamical systems approach and holor notation as follows:

$$\boldsymbol{E}_i = \boldsymbol{Z}_{ij} \boldsymbol{I}^j \tag{A.3}$$

where: $i \times j$ hypermatrix $\boldsymbol{Z}_{ij}$ has a dimension of the impedance holor.

To denote state variables associated with the primary (input) conductive collectors by an index *P*, and—associated with the secondary (output) conductive collectors by an index *s*, a set of the differential equations of dynamics, i.e. the mathematical model of the multi-port generalised MIMO ECM may be expressed as follows:

$$\left\| \begin{matrix} \boldsymbol{E}_I^P \\ \boldsymbol{E}_I^s \end{matrix} \right\| = \left\| \begin{matrix} \boldsymbol{Z}_{ij}^{PP} & \boldsymbol{Z}_{ij}^{Ps} \\ \boldsymbol{Z}_{ij}^{sP} & \boldsymbol{Z}_{ij}^{ss} \end{matrix} \right\| \left\| \begin{matrix} \boldsymbol{I}_P^j \\ \boldsymbol{I}_s^j \end{matrix} \right\| \tag{A.4a}$$

or in detail

$$\left\| \begin{matrix} \boldsymbol{E}_i^{AB} \\ \boldsymbol{E}_i^{BC} \\ \boldsymbol{E}_i^{C(N-1)} \\ \vdots \\ \boldsymbol{E}_i^{NA} \\ \boldsymbol{E}_i^{ab} \\ \boldsymbol{E}_i^{bc} \\ \boldsymbol{E}_i^{c(n-1)} \\ \vdots \\ \boldsymbol{E}_i^{na} \end{matrix} \right\| = \left\| \begin{matrix} \boldsymbol{Z}_{ij}^{AA} & \boldsymbol{Z}_{ij}^{AB} & \boldsymbol{Z}_{ij}^{AC} & & \boldsymbol{Z}_{ij}^{AN} & \boldsymbol{Z}_{ij}^{Aa} & \boldsymbol{Z}_{ij}^{Ab} & \boldsymbol{Z}_{ij}^{Ac} & & \boldsymbol{Z}_{ij}^{An} \\ \boldsymbol{Z}_{ij}^{BA} & \boldsymbol{Z}_{ij}^{BB} & \boldsymbol{Z}_{ij}^{BC} & \cdots & \boldsymbol{Z}_{ij}^{BN} & \boldsymbol{Z}_{ij}^{Ba} & \boldsymbol{Z}_{ij}^{Bb} & \boldsymbol{Z}_{ij}^{Bc} & \cdots & \boldsymbol{Z}_{ij}^{Bn} \\ \boldsymbol{Z}_{ij}^{CA} & \boldsymbol{Z}_{ij}^{CA} & \boldsymbol{Z}_{ij}^{CC} & & \boldsymbol{Z}_{ij}^{CN} & \boldsymbol{Z}_{ij}^{Ca} & \boldsymbol{Z}_{ij}^{Cb} & \boldsymbol{Z}_{ij}^{Cc} & & \boldsymbol{Z}_{ij}^{Cn} \\ & \vdots & & \ddots & \vdots & & \vdots & & \ddots & \vdots \\ \boldsymbol{Z}_{ij}^{NA} & \boldsymbol{Z}_{ij}^{NB} & \boldsymbol{Z}_{ij}^{NC} & \cdots & \boldsymbol{Z}_{ij}^{NN} & \boldsymbol{Z}_{ij}^{Na} & \boldsymbol{Z}_{ij}^{Nb} & \boldsymbol{Z}_{ij}^{Nc} & \cdots & \boldsymbol{Z}_{ij}^{Nn} \\ \boldsymbol{Z}_{ij}^{aA} & \boldsymbol{Z}_{ij}^{aB} & \boldsymbol{Z}_{ij}^{aC} & & \boldsymbol{Z}_{ij}^{aN} & \boldsymbol{Z}_{ij}^{aa} & \boldsymbol{Z}_{ij}^{ab} & \boldsymbol{Z}_{ij}^{ac} & & \boldsymbol{Z}_{ij}^{an} \\ \boldsymbol{Z}_{ij}^{bA} & \boldsymbol{Z}_{ij}^{bB} & \boldsymbol{Z}_{ij}^{bC} & \cdots & \boldsymbol{Z}_{ij}^{bN} & \boldsymbol{Z}_{ij}^{ba} & \boldsymbol{Z}_{ij}^{bb} & \boldsymbol{Z}_{ij}^{bc} & \cdots & \boldsymbol{Z}_{ij}^{bn} \\ \boldsymbol{Z}_{ij}^{cA} & \boldsymbol{Z}_{ij}^{cB} & \boldsymbol{Z}_{ij}^{cC} & & \boldsymbol{Z}_{ij}^{cN} & \boldsymbol{Z}_{ij}^{ca} & \boldsymbol{Z}_{ij}^{cb} & \boldsymbol{Z}_{ij}^{cc} & & \boldsymbol{Z}_{ij}^{cn} \\ & \vdots & & \ddots & \vdots & & \vdots & & \ddots & \vdots \\ \boldsymbol{Z}_{ij}^{nA} & \boldsymbol{Z}_{ij}^{nB} & \boldsymbol{Z}_{ij}^{nC} & \cdots & \boldsymbol{Z}_{ij}^{cN} & \boldsymbol{Z}_{ij}^{na} & \boldsymbol{Z}_{ij}^{nb} & \boldsymbol{Z}_{ij}^{nc} & \cdots & \boldsymbol{Z}_{ij}^{nn} \end{matrix} \right\| \left\| \begin{matrix} \boldsymbol{I}_{AB}^j \\ \boldsymbol{I}_{BA}^j \\ \boldsymbol{I}_{CB}^j \\ \vdots \\ \boldsymbol{I}_{N(N-1)}^j \\ \boldsymbol{I}_{an}^j \\ \boldsymbol{I}_{ba}^j \\ \boldsymbol{I}_{cb}^j \\ \vdots \\ \boldsymbol{I}_{n(n-1)}^j \end{matrix} \right\| \tag{A.4b}$$

Coefficients

$$\boldsymbol{Z}_{kl} = \frac{\partial \boldsymbol{E}_{k(k-1)}}{\partial \boldsymbol{I}^{l(l-1)}} \bigg|_{\boldsymbol{I}^{j(j-1)}} \qquad (j \neq l) \tag{A.5}$$

having a dimension of impedance holor.

To adapt a mathematical model of the multi-port generalised MIMO ECM to theory of the electrical heterogeneous continuous dynamical hypersystems' demands, one refers to a simplified and very general mathematical model of the multi-port generalised MIMO ECM, derived from a physical model that is shown in figure A1.

In order to fulfil the demands of physical energy conversion of one kind with the voltage-source holor—$\boldsymbol{E}_i^P$, current holor—$\boldsymbol{I}_P^j$, frequency—f_P, and number of phases—m_P of a primary electrical heterogeneous continuous dynamical hypersystem of the multi-port generalised MIMO ECM into another kind of electrical energy with the voltage-source holor—$\boldsymbol{E}_i^s$, current holor—$\boldsymbol{I}_s^j$, frequency—f_s, and number of phases—m_s of a secondary electrical continuous dynamical system of the multi-port generalised physical matrixer, as well as—free passage of electrical energy between these dynamical hypersystems, they ought to be connected by means of the multi-port generalised MIMO ECM's uni- and/or bipolar passive and active electrical components and devices.

A physical model of the multi-port generalised MIMO ECM, shown in figure A3, fulfils the aforementioned demands. Uni- and bipolar passive and active electrical

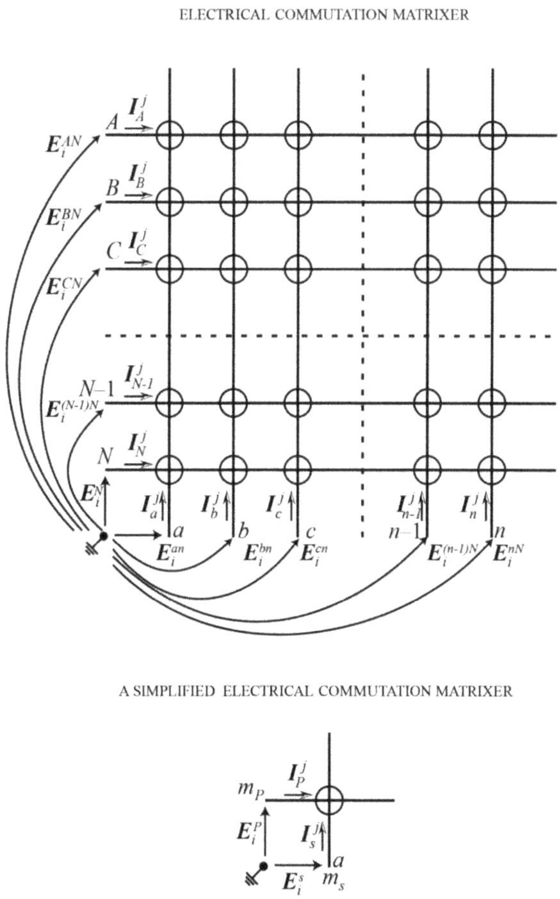

Figure A3. A multi-port generalised MIMO ECM with current-source holors (physical model I).

components and devices connect between each other both electrical heterogeneous continuous dynamical hypersystems of the multi-port generalised MIMO ECM and they make possible free passage of electrical energy between them.

In the general case, such as this MIMO ECM can connect many of electrical heterogeneous continuous dynamical hypersystems. In this textbook under consideration is the case, where only two electrical heterogeneous continuous dynamical hypersystems exist—primary and secondary. This is enough to understand a multi-port MIMO ECM concept.

Both electrical heterogeneous continuous dynamical hypersystems of the multi-port generalised MIMO ECM may be considered as the autonomous one, i.e. such as, which disconnected from the multi-port generalised MIMO ECM, have on their terminals determined potentials (Fijalkowski 1987a, 1987b, 2016).

They have voltage-source holors, current-source holors, electrical components (e.g. inductances, capacitances and resistances) as well as 'discrete' or 'continuous' uni- and/or bipolar physical valves. Assuming that excitations of a multi-port generalised MIMO ECM are given in a form of the column hypermatrices or the hypervectors of current-source holors flowing in the primary (input) and secondary (output) conductive collectors (figure A3):

$$\begin{aligned} \boldsymbol{I}^j &= \left\| \boldsymbol{I}_p^{j^T} \boldsymbol{I}_s^{j^T} \right\|^T \\ &= \left\| \boldsymbol{I}_A^{j^T} \boldsymbol{I}_B^{j^T} \boldsymbol{I}_C^{j^T} \cdots \boldsymbol{I}_N^{j^T} \boldsymbol{I}_a^{j^T} \boldsymbol{I}_b^{j^T} \boldsymbol{I}_c^{j^T} \cdots \boldsymbol{I}_n^{j^T} \right\|^T \end{aligned} \tag{A.6}$$

as responses of the multi-port MIMO ECM—the column hypermatrices or the hypervectors of node voltage holors

$$\begin{aligned} \boldsymbol{V}_i &= \left\| \boldsymbol{V}_i^{p^T} \boldsymbol{V}_i^{s^T} \right\|^T \\ &= \left\| \boldsymbol{V}_i^{A^T} \boldsymbol{V}_i^{B^T} \boldsymbol{V}_i^{T} \cdots \boldsymbol{V}_i^{N^T} \boldsymbol{V}_i^{a^T} \boldsymbol{V}_i^{b^T} \boldsymbol{V}_i^{c^T} \cdots \boldsymbol{V}_i^{n^T} \right\|^T \end{aligned} \tag{A.7}$$

a set of the differential equations of dynamics, i.e. the mathematical model of the multi-port generalised MIMO ECM as the electrical heterogeneous continuous dynamical hypersystems may be formulated in the dynamical systems approach and matrix notation as follows:

$$\boldsymbol{I}^j = \boldsymbol{Y}^{ji} \boldsymbol{V}_I \tag{A.8}$$

where: $n \times m$ hypermatrix $\boldsymbol{Y}^{ji}$ has a dimension of the admittance holor.

Node voltage holors, consisting of electrical components of the hypermatrix $\boldsymbol{V}_i$ are measured with regard to the reference node, which is located on the outside of the multi-port generalised MIMO ECM. For example, in a star connection of the multi-phase generalised electrical heterogeneous continuous dynamical hypersystem, as the reference nodes may be a null point of the multi-phase star connection.

To denote state variables associated with the primary (input) collectors by an index P, and associated with the secondary (output) collectors by an index s, the set of differential equations of dynamics, i.e. the mathematical model of the multi-port generalised MIMO ECM may be expressed as follows:

$$\left\| \begin{matrix} I_P^j \\ I_s^j \end{matrix} \right\| = \left\| \begin{matrix} Y_{PP}^{ji} & Y_{Ps}^{ji} \\ Y_{sP}^{ji} & Y_{ss}^{ji} \end{matrix} \right\| \left\| \begin{matrix} V_I^P \\ V_I^s \end{matrix} \right\| \tag{A.9a}$$

or in detail

$$\left\| \begin{matrix} I_A^j \\ I_B^j \\ I_C^j \\ \vdots \\ I_N^j \\ I_a^j \\ I_b^j \\ I_c^j \\ \vdots \\ I_n^j \end{matrix} \right\| = \left\| \begin{matrix} Y_{AA}^{ji} & Y_{AB}^{ji} & Y_{AC}^{ji} & & Y_{AN}^{ji} & Y_{Aa}^{ji} & Y_{Ab}^{ji} & Y_{Ac}^{ji} & & Y_{An}^{ji} \\ Y_{BA}^{ji} & Y_{BB}^{ji} & Y_{BC}^{ji} & \cdots & Y_{BN}^{ji} & Y_{Ba}^{ji} & Y_{Bb}^{ji} & Y_{Bc}^{ji} & \cdots & Y_{Bn}^{ji} \\ Y_{CA}^{ji} & Y_{CA}^{ji} & Y_{CC}^{ji} & & Y_{CN}^{ji} & Y_{Ca}^{ji} & Y_{Cb}^{ji} & Y_{Cc}^{ji} & & Y_{Cn}^{ji} \\ & \vdots & & \ddots & \vdots & & \vdots & & \ddots & \vdots \\ Y_{NA}^{ji} & Y_{NB}^{ji} & Y_{NC}^{ji} & \cdots & Y_{NN}^{ji} & Y_{Na}^{ji} & Y_{Nb}^{ji} & Y_{Nc}^{ji} & \cdots & Y_{Nn}^{ji} \\ Y_{aA}^{ji} & Y_{aB}^{ji} & Y_{aC}^{ji} & & Y_{aN}^{ji} & Y_{aa}^{ji} & Y_{ab}^{ji} & Y_{ac}^{ji} & & Y_{an}^{ji} \\ Y_{bA}^{ji} & Y_{bB}^{ji} & Y_{bC}^{ji} & \cdots & Y_{bN}^{ji} & Y_{ba}^{ji} & Y_{bb}^{ji} & Y_{bc}^{ji} & \cdots & Y_{bn}^{ji} \\ Y_{cA}^{ji} & Y_{cB}^{ji} & Y_{cC}^{ji} & & Y_{cN}^{ji} & Y_{ca}^{ji} & Y_{cb}^{ji} & Y_{cc}^{ji} & & Y_{cn}^{ji} \\ & \vdots & & \ddots & \vdots & & \vdots & & \ddots & \vdots \\ Y_{nA}^{ji} & Y_{aB}^{ji} & Y_{nC}^{ji} & \cdots & Y_{cN}^{ji} & Y_{na}^{ji} & Y_{nb}^{ji} & Y_{nc}^{ji} & \cdots & Y_{nn}^{ji} \end{matrix} \right\| \left\| \begin{matrix} V_i^A \\ V_i^B \\ V_i^C \\ \vdots \\ V_i^N \\ V_i^a \\ V_i^b \\ V_i^c \\ \vdots \\ V_i^n \end{matrix} \right\| \tag{A.9b}$$

Coefficients

$$Y^{kl} = \frac{\partial I^k}{\partial U_l}\bigg|U_I = 0 \qquad (i \neq l) \tag{A.9c}$$

having a dimension of admittance holor.

Summing up, a hybrid mathematical model of the multi-port generalised MIMO ECM may be written as follows:

$$\left\| \begin{matrix} E_i^{AN} \\ E_i^{BA} \\ E_i^{CB} \\ \vdots \\ E_i^{N(N-1)} \\ E_i^{an} \\ E_i^{ba} \\ E_i^{cb} \\ \vdots \\ E_i^{n(n-1)} \\ J_A^j \\ J_B^j \\ J_C^j \\ \vdots \\ J_N^j \\ J_a^j \\ J_b^j \\ J_c^j \\ \vdots \\ J_n^j \end{matrix} \right\| = \left\| \begin{matrix} \begin{matrix} Z_{ij}^{AA} & Z_{ij}^{AB} & Z_{ij}^{AC} & & Z_{ij}^{AN} & Z_{ij}^{Aa} & Z_{ij}^{Ab} & Z_{ij}^{Ac} & & Z_{ij}^{An} \\ Z_{ij}^{BA} & Z_{ij}^{BB} & Z_{ij}^{BC} & \cdots & Z_{ij}^{BN} & Z_{ij}^{Ba} & Z_{ij}^{Bb} & Z_{ij}^{Bc} & \cdots & Z_{ij}^{Bn} \\ Z_{ij}^{CA} & Z_{ij}^{CA} & Z_{ij}^{CC} & & Z_{ij}^{CN} & Z_{ij}^{Ca} & Z_{ij}^{Cb} & Z_{ij}^{Cc} & & Z_{ij}^{Cn} \\ & \vdots & & \ddots & \vdots & & \vdots & & \ddots & \vdots \\ Z_{ij}^{NA} & Z_{ij}^{NB} & Z_{ij}^{NC} & \cdots & Z_{ij}^{NN} & Z_{ij}^{Na} & Z_{ij}^{Nb} & Z_{ij}^{Nc} & \cdots & Z_{ij}^{Nn} \\ Z_{ij}^{aA} & Z_{ij}^{aB} & Z_{ij}^{aC} & & Z_{ij}^{aN} & Z_{ij}^{aa} & Z_{ij}^{ab} & Z_{ij}^{ac} & & Z_{ij}^{an} \\ Z_{ij}^{bA} & Z_{ij}^{bB} & Z_{ij}^{bC} & \cdots & Z_{ij}^{bN} & Z_{ij}^{ba} & Z_{ij}^{bb} & Z_{ij}^{bc} & \cdots & Z_{ij}^{bn} \\ Z_{ij}^{cA} & Z_{ij}^{cB} & Z_{ij}^{cC} & & Z_{ij}^{cN} & Z_{ij}^{ca} & Z_{ij}^{cb} & Z_{ij}^{cc} & & Z_{ij}^{cn} \\ & \vdots & & \ddots & \vdots & & \vdots & & \ddots & \vdots \\ Z_{ij}^{nA} & Z_{ij}^{nB} & Z_{ij}^{nC} & \cdots & Z_{ij}^{cN} & Z_{ij}^{na} & Z_{ij}^{nb} & Z_{ij}^{nc} & \cdots & Z_{ij}^{nn} \end{matrix} & \begin{matrix} 0 & 0 & 0 & & 0 & 0 & 0 & 0 & & 0 \\ 0 & 0 & 0 & \cdots & 0 & 0 & 0 & 0 & \cdots & 0 \\ 0 & 0 & 0 & & 0 & 0 & 0 & 0 & & 0 \\ & \vdots & & \ddots & \vdots & & \vdots & & \ddots & \vdots \\ 0 & 0 & 0 & \cdots & 0 & 0 & 0 & 0 & \cdots & 0 \\ 0 & 0 & 0 & & 0 & 0 & 0 & 0 & & 0 \\ 0 & 0 & 0 & \cdots & 0 & 0 & 0 & 0 & \cdots & 0 \\ 0 & 0 & 0 & & 0 & 0 & 0 & 0 & & 0 \\ & \vdots & & \ddots & \vdots & & \vdots & & \ddots & \vdots \\ 0 & 0 & 0 & \cdots & 0 & 0 & 0 & 0 & \cdots & 0 \end{matrix} \\ \begin{matrix} 0 & 0 & 0 & & 0 & 0 & 0 & 0 & & 0 \\ 0 & 0 & 0 & \cdots & 0 & 0 & 0 & 0 & \cdots & 0 \\ 0 & 0 & 0 & & 0 & 0 & 0 & 0 & & 0 \\ & \vdots & & \ddots & \vdots & & \vdots & & \ddots & \vdots \\ 0 & 0 & 0 & \cdots & 0 & 0 & 0 & 0 & \cdots & 0 \\ 0 & 0 & 0 & & 0 & 0 & 0 & 0 & & 0 \\ 0 & 0 & 0 & \cdots & 0 & 0 & 0 & 0 & \cdots & 0 \\ 0 & 0 & 0 & & 0 & 0 & 0 & 0 & & 0 \\ & \vdots & & \ddots & \vdots & & \vdots & & \ddots & \vdots \\ 0 & 0 & 0 & \cdots & 0 & 0 & 0 & 0 & \cdots & 0 \end{matrix} & \begin{matrix} Y_{AA}^{ji} & Y_{AB}^{ji} & Y_{AC}^{ji} & & Y_{AN}^{ji} & Y_{Aa}^{ji} & Y_{Ab}^{ji} & Y_{Ac}^{ji} & & Y_{An}^{ji} \\ Y_{BA}^{ji} & Y_{BB}^{ji} & Y_{BC}^{ji} & \cdots & Y_{BN}^{ji} & Y_{Ba}^{ji} & Y_{Bb}^{ji} & Y_{Bc}^{ji} & \cdots & Y_{Bn}^{ji} \\ Y_{CA}^{ji} & Y_{CA}^{ji} & Y_{CC}^{ji} & & Y_{CN}^{ji} & Y_{Ca}^{ji} & Y_{Cb}^{ji} & Y_{Cc}^{ji} & & Y_{Cn}^{ji} \\ & \vdots & & \ddots & \vdots & & \vdots & & \ddots & \vdots \\ Y_{NA}^{ji} & Y_{NB}^{ji} & Y_{NC}^{ji} & \cdots & Y_{NN}^{ji} & Y_{Na}^{ji} & Y_{Nb}^{ji} & Y_{Nc}^{ji} & \cdots & Y_{Nn}^{ji} \\ Y_{aA}^{ji} & Y_{aB}^{ji} & Y_{aC}^{ji} & & Y_{aN}^{ji} & Y_{aa}^{ji} & Y_{ab}^{ji} & Y_{ac}^{ji} & & Y_{an}^{ji} \\ Y_{bA}^{ji} & Y_{bB}^{ji} & Y_{bC}^{ji} & \cdots & Y_{bN}^{ji} & Y_{ba}^{ji} & Y_{bb}^{ji} & Y_{bc}^{ji} & \cdots & Y_{bn}^{ji} \\ Y_{cA}^{ji} & Y_{cB}^{ji} & Y_{cC}^{ji} & & Y_{cN}^{ji} & Y_{ca}^{ji} & Y_{cb}^{ji} & Y_{cc}^{ji} & & Y_{cn}^{ji} \\ & \vdots & & \ddots & \vdots & & \vdots & & \ddots & \vdots \\ Y_{nA}^{ji} & Y_{aB}^{ji} & Y_{nC}^{ji} & \cdots & Y_{cN}^{ji} & Y_{na}^{ji} & Y_{nb}^{ji} & Y_{nc}^{ji} & \cdots & Y_{nn}^{ji} \end{matrix} \end{matrix} \right\| \left\| \begin{matrix} I_{AB}^j \\ I_{BA}^j \\ I_{CB}^j \\ \vdots \\ I_{N(N-1)}^j \\ I_{an}^j \\ I_{ba}^j \\ I_{cb}^j \\ \vdots \\ I_{n(n-1)}^j \\ V_i^A \\ V_i^B \\ V_i^C \\ \vdots \\ V_i^N \\ V_i^a \\ V_i^b \\ V_i^c \\ \vdots \\ V_i^n \end{matrix} \right\| \tag{A.10}$$

$$Z_{kl} = \frac{\partial E_{k(k-1)}}{\partial I^{l(l-1)}}\bigg|I^{j(j-1)} \qquad (j \neq l) \text{ and } Y^{kl} = \frac{\partial J^k}{\partial V_l}\bigg|V_I = 0 \qquad (i \neq l) \tag{A.11}$$

having dimensions of impedance holor or admittance holor, respectively.

Similarly to mathematical models of physical heterogeneous continuous dynamical hypersystems of generalised MIMO ECMs shown in figures A2 and A3, consisting of a set of differential equations of dynamics (A.4) and (A.9), a possible formulation exists of the mathematical model of physical continuous dynamical systems of generalised MIMO ECM (figure A3) acting as a voltage-source ECM commutator (E–E converter) with an arbitrary number N of primary collectors (inputs), equals to number m of physical commutation matrixer phases and an arbitrary number n of secondary collectors (outputs).

Assuming that voltage-source holors applied across the primary collectors (inputs) $A, B, C, \ldots, N$ of a generalised MIMO ECM commutator (matrix converter), are given in a form of the voltage-source holors' column matrix or vector (input holors)

$$\boldsymbol{E}_i^P = \left\| \boldsymbol{E}_i^{A^T}\ \boldsymbol{E}_i^{B^T}\ \boldsymbol{E}_i^{C^T} \cdots \boldsymbol{E}_i^{N-1^T}\ \boldsymbol{E}_i^{N^T} \right\|^T \tag{A.12}$$

as the primary electrical state variables (inputs), and voltage-source holors, applied across the secondary collectors (outputs) $a, b, c, \ldots, n$ of a generalised MIMO ECM commutator (matrix converter), in the form of the voltage-source holors' matrix or vector (output holors)

$$\boldsymbol{E}_s^j = \left\| \boldsymbol{E}_a^{j^T}\ \boldsymbol{E}_b^{j^T}\ \boldsymbol{E}_c^{j^T} \cdots \boldsymbol{E}_{n-1}^{j^T}\ \boldsymbol{E}_n^{j^T} \right\|^T \tag{A.13}$$

as secondary electrical state variables (output holors).

A mathematical model of the electrical heterogeneous continuous dynamical hypersystem of a generalised MIMO ECM commutator (E–E converter) acting as a voltage-source ECM commutator, making dependent the outputs from the inputs, in the dynamical systems approach and a matrix notation may be written in the form:

$$\boldsymbol{E}_s^j = \boldsymbol{C}_{sP}^{ji} \boldsymbol{E}_i^P \tag{A.14}$$

or in detail

$$\left\| \begin{matrix} \boldsymbol{E}_a^j \\ \boldsymbol{E}_b^n \\ \boldsymbol{E}_c^i \\ \vdots \\ \boldsymbol{E}_{n-1}^j \\ \boldsymbol{E}_n^j \end{matrix} \right\| = \left\| \begin{matrix} \boldsymbol{C}_{Aa}^{ji} & \boldsymbol{C}_{Ba}^{ji} & \boldsymbol{C}_{Ca}^{ji} & & \boldsymbol{C}_{(N-1)a}^{ji} & \boldsymbol{C}_{Na}^{ji} \\ \boldsymbol{C}_{Ab}^{ji} & \boldsymbol{C}_{Bb}^{ji} & \boldsymbol{C}_{Cb}^{ji} & \cdots & \boldsymbol{C}_{(N-1)b}^{ji} & \boldsymbol{C}_{Nb}^{ji} \\ \boldsymbol{C}_{Ac}^{ji} & \boldsymbol{C}_{Bc}^{ji} & \boldsymbol{C}_{Cc}^{ji} & & \boldsymbol{C}_{(N-1)c}^{ji} & \boldsymbol{C}_{Nc}^{ji} \\ & \vdots & & \ddots & \vdots & \\ \boldsymbol{C}_{A(n-1)}^{ji} & \boldsymbol{C}_{B(n-1)}^{ji} & \boldsymbol{C}_{C(n-1)}^{ji} & \cdots & \boldsymbol{C}_{(N-1)(n-1)}^{ji} & \boldsymbol{C}_{N(n-1)}^{ji} \\ \boldsymbol{C}_{An}^{ji} & \boldsymbol{C}_{Bn}^{ji} & \boldsymbol{C}_{Cn}^{ji} & & \boldsymbol{C}_{(N-1)n}^{ji} & \boldsymbol{C}_{Nn}^{ji} \end{matrix} \right\| \left\| \begin{matrix} \boldsymbol{E}_i^A \\ \boldsymbol{E}_i^B \\ \boldsymbol{E}_i^C \\ \vdots \\ \boldsymbol{E}_i^{N-1} \\ \boldsymbol{E}_i^N \end{matrix} \right\| \tag{A.15}$$

however, a mathematical model of the electrical continuous dynamical system of a generalised MIMO ECM commutator (E–E converter) acting as a voltage-source ECM commutator, making dependent the inputs from the outputs may be written in the form:

$$\left\| \boldsymbol{E}_i^P \right\| = \left\| \boldsymbol{C}_{ij}^{Ps} \right\| \left\| \boldsymbol{E}_s^j \right\| \tag{A.16}$$

or in detail

$$\begin{Vmatrix} \boldsymbol{E}_i^A \\ \boldsymbol{E}_i^B \\ \boldsymbol{E}_i^C \\ \vdots \\ \boldsymbol{E}_i^{N-1} \\ \boldsymbol{E}_i^N \end{Vmatrix} = \begin{Vmatrix} \boldsymbol{C}_{ij}^{Aa} & \boldsymbol{C}_{ij}^{Ab} & \boldsymbol{C}_{ij}^{Ac} & & \boldsymbol{C}_{ij}^{A(n-1)} & \boldsymbol{C}_{ij}^{An} \\ \boldsymbol{C}_{ij}^{Ba} & \boldsymbol{C}_{ij}^{Bb} & \boldsymbol{C}_{ij}^{Bc} & \cdots & \boldsymbol{C}_{ij}^{B(n-1)} & \boldsymbol{C}_{ij}^{Bn} \\ \boldsymbol{C}_{ij}^{Ca} & \boldsymbol{C}_{ij}^{Cb} & \boldsymbol{C}_{ij}^{Cc} & & \boldsymbol{C}_{ij}^{C(n-1)} & \boldsymbol{C}_{ij}^{Cn} \\ & \vdots & & \ddots & \vdots & \\ \boldsymbol{C}_{ij}^{(N-1)a} & \boldsymbol{C}_{ij}^{(N-1)b} & \boldsymbol{C}_{ij}^{(N-1)c} & \cdots & \boldsymbol{C}_{ij}^{(N-1)(n-1)} & \boldsymbol{C}_{ij}^{(N-1)n} \\ \boldsymbol{C}_{ij}^{Na} & \boldsymbol{C}_{ij}^{Nb} & \boldsymbol{C}_{ij}^{Nc} & & \boldsymbol{C}_{ij}^{N(n-1)} & \boldsymbol{C}_{ij}^{Nn} \end{Vmatrix} \begin{Vmatrix} \boldsymbol{E}_a^j \\ \boldsymbol{E}_b^j \\ \boldsymbol{E}_c^j \\ \vdots \\ \boldsymbol{E}_{n-1}^j \\ \boldsymbol{E}_n^j \end{Vmatrix} \tag{A.17}$$

where $\boldsymbol{C}_{kL}^{ji}$ and $\boldsymbol{C}_{ij}^{Kl}$ are the voltage-source coefficients of the commutation nodes 'kL' and 'Kl', respectively, which may be expressed as follows:

$$\boldsymbol{C}_{kL}^{JI} = \frac{\partial \boldsymbol{E}_k^j}{\partial \boldsymbol{E}_i^L}\bigg|\boldsymbol{E}_i^I = 0 = \begin{cases} +1 \overset{\Delta}{\rightarrow} \boldsymbol{E}_k^j = +\boldsymbol{E}_i^L \\ \quad 0 \overset{\Delta}{\rightarrow} \boldsymbol{E}_k^j = 0 \\ -1 \overset{\Delta}{\rightarrow} \boldsymbol{E}_k^j = -\boldsymbol{E}_i^L \end{cases}$$

and

$$\boldsymbol{C}_{ij}^{Kl} = \frac{\partial \boldsymbol{E}_i^K}{\partial \boldsymbol{E}_l^j}\bigg|\boldsymbol{E}_i^j = 0 = \begin{cases} -1 \overset{\Delta}{\rightarrow} \boldsymbol{E}_i^K = -\boldsymbol{E}_l^j \\ \quad 0 \overset{\Delta}{\rightarrow} \boldsymbol{E}_i^K = 0 \\ +1 \overset{\Delta}{\rightarrow} \boldsymbol{E}_i^K = +\boldsymbol{E}_l^j \end{cases}$$

In the case of the generalised MIMO ECM with bipolar (symmetrical) electrical valves, voltage-source coefficients of the commutation nodes $\boldsymbol{C}_{kL}^{ji} = \pm 1$ i $\boldsymbol{C}_{ij}^{Kl} = \mp 1$, represent an ideal electrical-valve impedance equals zero, which is incorporated between the primary collector L and the secondary collector k, or—between the secondary collector l and the primary collector K, what corresponds to a turn-on state of the electrical valves with a sense of the current flow direction (polarisation) dependent on the sign of these coefficients; however voltage-source coefficients of the commutation nodes $\boldsymbol{C}_{kL}^{ji} = 0$ and $\boldsymbol{C}_{ij}^{Kl} = 0$—an ideal electrical-valve impedance equals infinity, which corresponds to a turn-off state of the electrical valves.

Analogically, a mathematical model of the physical heterogeneous continuous dynamical system of a generalised MIMO ECM AC–AC and/or AC–DC/DC–AC commutator (E–E converter) shown in figure A3, as a current-source ECM commutator with an arbitrary number N of primary collectors (inputs), equals a number m of physical phases and an arbitrary number n of secondary collectors (outputs).

Assuming that an energy-transfer holors' sources, flowing in primary collectors (inputs) $A, B, C, \ldots, N$ of an MIMO ECM commutator (matrix converter) are given in a form of the current-source holors' column matrix or vector (input holors)

$$\boldsymbol{I}_i^P = \left\| \boldsymbol{I}_i^{A^T}\ \boldsymbol{I}_i^{B^T}\ \boldsymbol{I}_i^{C^T} \cdots \boldsymbol{I}_i^{N-1^T}\ \boldsymbol{I}_i^{N^T} \right\|^T \tag{A.18}$$

as a primary electrical state variables (inputs), and a secondary current-source holors of a generalised MIMO ECM commutator, flowing in a secondary collectors (outputs) *a, b, c, ...,n* of an MIMO ECM, in a form of the secondary current-source holors (output holors)

$$\boldsymbol{I}_s^j = \left\| \boldsymbol{I}_a^{jT}\ \boldsymbol{I}_b^{jT}\ \boldsymbol{I}_c^{jT} \cdots \boldsymbol{I}_{n-1}^{jT}\ \boldsymbol{I}_n^{jT} \right\|^T \tag{A.19}$$

as secondary physical state variables (inputs).

A mathematical model of the physical heterogeneous continuous dynamical hypersystem of a generalised MIMO ECM commutator (E–E converter) acting as a current-source ECM commutator, making dependent the outputs from the inputs, in the dynamical systems approach and a matrix notation may be written in the form:

$$\boldsymbol{I}_s^j = \boldsymbol{C}_{sP}^{ji}\boldsymbol{I}_i^P \tag{A.20}$$

or in detail

$$\left\| \begin{matrix} \boldsymbol{I}_a^j \\ \boldsymbol{I}_b^j \\ \boldsymbol{I}_c^j \\ \vdots \\ \boldsymbol{I}_{n-1}^j \\ \boldsymbol{I}_n^j \end{matrix} \right\| = \left\| \begin{matrix} \boldsymbol{C}_{aA}^{ji} & \boldsymbol{C}_{aB}^{ji} & \boldsymbol{C}_{aC}^{ji} & & \boldsymbol{C}_{a(N-1)}^{ji} & \boldsymbol{C}_{aN}^{ji} \\ \boldsymbol{C}_{bA}^{ji} & \boldsymbol{C}_{bB}^{ji} & \boldsymbol{C}_{bC}^{ji} & \cdots & \boldsymbol{C}_{b(N-1)}^{ji} & \boldsymbol{C}_{bN}^{ji} \\ \boldsymbol{C}_{cA}^{ji} & \boldsymbol{C}_{cB}^{ji} & \boldsymbol{C}_{cC}^{ji} & & \boldsymbol{C}_{c(N-1)}^{ji} & \boldsymbol{C}_{cN}^{ji} \\ & \vdots & & \ddots & \vdots & \\ \boldsymbol{C}_{(n-1)A}^{ji} & \boldsymbol{C}_{(n-1)B}^{ji} & \boldsymbol{C}_{(n-1)C}^{ji} & \cdots & \boldsymbol{C}_{(n-1)(N-1)}^{ji} & \boldsymbol{C}_{(n-1)N}^{ji} \\ \boldsymbol{C}_{nA}^{ji} & \boldsymbol{C}_{nB}^{ji} & \boldsymbol{C}_{nC}^{ji} & & \boldsymbol{C}_{n(N-1)}^{ji} & \boldsymbol{C}_{nN}^{ji} \end{matrix} \right\| \left\| \begin{matrix} \boldsymbol{I}_i^A \\ \boldsymbol{I}_i^B \\ \boldsymbol{I}_i^C \\ \vdots \\ \boldsymbol{I}_i^{N-1} \\ \boldsymbol{I}_i^N \end{matrix} \right\| \tag{A.21}$$

however, a mathematical model of an electrical continuous dynamical system of the generalised MIMO ECM commutator (E–E converter) shown in figure A4, acting as a current-source ECM commutator, making dependent the inputs from the outputs, in the dynamical systems approach and matrix notation may be expressed in the form:

$$\boldsymbol{I}_i^P = \boldsymbol{C}_{ij}^{Ps}\boldsymbol{I}_s^j \tag{A.22}$$

in detail

$$\left\| \begin{matrix} \boldsymbol{I}_i^A \\ \boldsymbol{I}_i^B \\ \boldsymbol{I}_i^C \\ \vdots \\ \boldsymbol{I}_i^{N-1} \\ \boldsymbol{I}_i^N \end{matrix} \right\| = \left\| \begin{matrix} \boldsymbol{C}_{ij}^{Aa} & \boldsymbol{C}_{ij}^{Ab} & \boldsymbol{C}_{ij}^{Ac} & & \boldsymbol{C}_{ij}^{A(n-1)} & \boldsymbol{C}_{ij}^{An} \\ \boldsymbol{C}_{ij}^{Ba} & \boldsymbol{C}_{ij}^{Bb} & \boldsymbol{C}_{ij}^{Bc} & \cdots & \boldsymbol{C}_{ij}^{B(n-1)} & \boldsymbol{C}_{ij}^{Bn} \\ \boldsymbol{C}_{ij}^{Ca} & \boldsymbol{C}_{ij}^{Cb} & \boldsymbol{C}_{ij}^{Cc} & & \boldsymbol{C}_{ij}^{C(n-1)} & \boldsymbol{C}_{ij}^{Cn} \\ & \vdots & & \ddots & \vdots & \\ \boldsymbol{C}_{ij}^{(N-1)a} & \boldsymbol{C}_{ij}^{(N-1)b} & \boldsymbol{C}_{ij}^{(N-1)c} & \cdots & \boldsymbol{C}_{ij}^{(N-1)(n-1)} & \boldsymbol{C}_{ij}^{(N-1)n} \\ \boldsymbol{C}_{ij}^{Na} & \boldsymbol{C}_{ij}^{Nb} & \boldsymbol{C}_{ij}^{Nc} & & \boldsymbol{C}_{ij}^{N(n-1)} & \boldsymbol{C}_{ij}^{Nn} \end{matrix} \right\| \left\| \begin{matrix} \boldsymbol{I}_a^j \\ \boldsymbol{I}_b^j \\ \boldsymbol{I}_c^j \\ \vdots \\ \boldsymbol{I}_{n-1}^j \\ \boldsymbol{I}_n^j \end{matrix} \right\| \tag{A.23}$$

where $\boldsymbol{C}_{kL}^{ji}$ and $\boldsymbol{C}_{ij}^{Kl}$ are the current-source coefficients of the commutation nodes '*kL*' and '*Kl*', respectively.

In summary, it may be affirmed that an action of a generalised MIMO ECM commutator as a current-source ECM commutator (shown in figure A4), may be

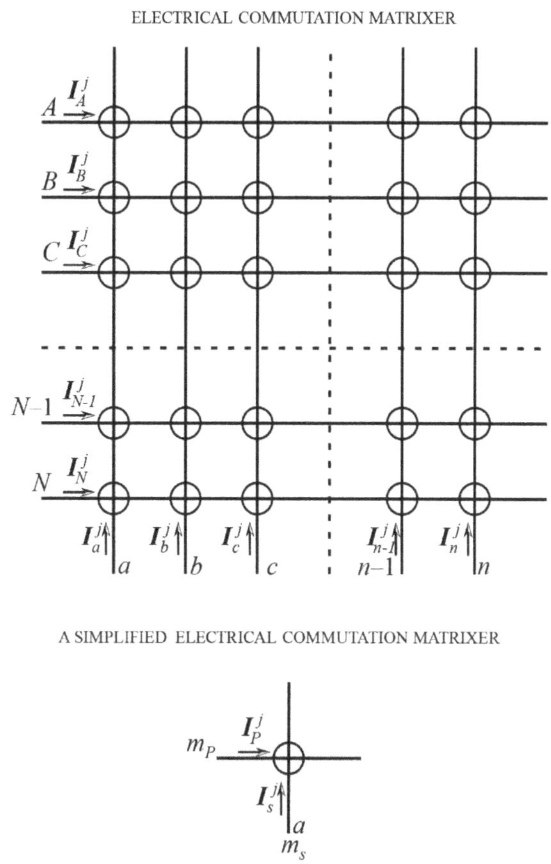

Figure A4. Physical model of a generalised MIMO ECM AC–AC and/or AC–DC/DC–AC commutator of AC and/or DC MMD electrical machine.

presented by means of its mathematical model, consisting of equations of unholonomic constraints of a current-source holor in a matrix notation (A.24) and (A.25).

For equations formulated of unholonomic constraints of current-source holors in a matrix notation, isochronal variations of the generalised coordinates are not already independent of each other and that is why we apply the dynamical systems approach in a methodology of a formulation of the mathematical models of MIMO ECM commutators (E–E converters) in a matrix notation of the Euler–Lagrange second-order differential equation of dynamics. In the case of an ideal bipolar electrical valve, the voltage-holor coefficient of a commutating junction moreover, regards the electrical-valve resistance connected between the primary collector P and the secondary collector s.

In this case the electrical-valve resistance is equal to zero, which corresponds to the turn-on state of an electrical valve with the sense of a current direction depending upon the sign of this voltage-source coefficient. On the other hand, the voltage-holor coefficient of a commutating junction regards the electrical-valve resistance to be equal to infinity, which corresponds to the turn-off state (blocking) of an electrical valve. For instance, as future spacecraft grow in size, multiple electrical energy sources may be required to supply the electrical distribution network. To supply the various electrical loads, these energy sources must be combined to derive an ECM DC–AC–DC commutator with several different load voltages. Such an ECM is important from a reliability standpoint since the electrical loads can still be supplied even if some of the electrical energy sources fail (Fijalkowski 1987a, 1987b, 2016).

One method of implementing such an ECM DC–AC–DC commutator, where several DC–AC inverters are supplied from different electrical energy sources to drive a common AC bus are also indicated by figure A4.

Because of their high output impedance, these DC–AC inverters will share the electrical load in a predictable manner, and they provide an ECM DC–AC–DC commutator that can be designed to operate into a short-circuit, and thus provide a fault tolerant system. These DC–AC inverters are operated in synchronism by a common microcontroller unit (MCU) and regulated by monitoring one of the AC–DC rectifier outputs. The output DC voltage is regulated in the usual manner by variation of the operating frequency. The use of a common AC bus appears to be very advantageous for those applications where multiple ECM outputs must be driven from multiple DC inputs. This type of ECM DC–AC–DC commutator provides the high efficiency of a resonant circuit, and at higher power levels it can still be operated in the 20 kHz range to minimise the mass of the magnet. For example, in a personal computer (PC) energy supply, there are commonly five separate DC loads, and the ECM is 2 × 10. These DC–AC inverters are operated in synchronism by a common microcontroller unit (MCU) and regulated by monitoring one of the AC–DC rectifier outputs. The output DC voltage is regulated in the usual manner by variation of the operating frequency. The use of a common AC bus appears to be very advantageous for those applications where multiple ECM outputs must be driven from multiple DC inputs. This type of ECM DC–AC–DC commutator provides the high efficiency of a resonant circuit, and at higher power levels it can still be operated in the 20 kHz range to minimise the mass of the magnetic.

A small number of applied ECM commutators have more than 36 electrical valves, and most designs use less than 12. An ECM makes available a means to organise ECM AC–AC, AC–DC–AC, AC–DC/DC–AC, DC–DC and/or DC–AC–AC commutators for a given application.

It also helps to focus the effort into three major spheres. Each of these spheres must be addressed effectively in order to create a useful macro- and microelectronic cyber-physical heterogeneous continuous dynamical system.

- ***The hardware sphere***—Create an ECM; this involves the selection of appropriate super- and/or semiconductor electrical valves and the auxiliary components that drive and protect them.
- ***The software sphere***—Activate the ECM to achieve the desired AC–AC, AC–DC–AC, AC–DC/DC–AC, DC–DC and/or DC–AC–DC conversion; all operational decisions are implemented by adjusting electrical-valve timing.
- ***The interface sphere***—Insert energy-storage components to provide the low-pass filters or intermediate energy-storage necessary to meet the application

requirements. Unlike most passive filter applications, lossless active filters with simple structures are required.

At present, the world must solve important challenges to convert electrical energy and transform information for mechatronic control systems in an efficient way. Examples of this are in transportation, renewable energies and industrial processing applications. These problems can be solved using energy converters based on modern semi- and/or superconductive ECM AC–AC, AC–DC–AC, AC–DC/DC–AC, DC–DC and/or DC–AC–DC commutators application in **magneto-mechano-dynamical** (MMD) electrical machines.

The ideal ECM commutator in many of these applications may have the following characteristics:

- sinusoidal input and output currents;
- operation with unity power factor;
- regeneration capability;
- compact design with a good power to mass ratio.

Semi- and/or superconductive ECM AC–AC, AC–DC–AC, AC–DC/DC–AC, DC–DC and/or DC–AC–DC commutators (E–E converters) can fulfil all these characteristics and this is the reason for the tremendous interest in this topology. In the last decade, many advances in the development of this topology have been presented, including industrial applications up to megawatt level. The use of the semi- and/or superconductive ECM AC–AC, AC–DC–AC, AC–DC/DC–AC, DC–DC and/or DC–AC–DC commutator (E–E converter) in real applications and the challenges that these applications present is very topical and important.

In view of the fact that sources of conventional energy are gradually being depleted, public opinion is steadily more interested in the discussion of novel types of future-oriented energy-saving industrial and traction IEMD cyber-physical heterogeneous continuous or discrete dynamical hypersystems. These energy-saving IEMD cyber-physical heterogeneous continuous or discrete dynamical hypersystems have already become an alternative to be taken seriously. The development of today's industrial and traction IEMD cyber-physical heterogeneous continuous or discrete dynamical hypersystems is closely tied up with developments in mechatronics technology.

The MMD electrical machines (dynamotors) currently in use have remained unchanged since their conception before the turn of the 20th century; modern integrated electrical machines (dynamotors) concepts have only gained a marginal acceptance.

Today, there are good possibilities available to select optimum industrial and traction IEMD cyber-physical heterogeneous continuous or discrete dynamical hypersystems with due consideration being paid to performance, lifetime costs, environment, influence on the storage battery, etc.

This very **advanced propulsion** (VAP) technology is based on two important disciplines—**electrical-integrated-matrixer** (EIM) commutator and **electrical-integrated-circuit** (EIC) processor techniques. Some of the results achieved were higher performance and other operational benefits. **Microcontroller unit** (MCU) gives the necessary conditions for further improvements such as diagnostics.

Continued **research and development** (R&D) work on the ECM commutators will lead to a greater competitiveness of AC–AC, AC–DC and/or DC–AC commutator dynamotors (generators/motors) for industrial and traction propulsion complex dynamical hypersystems.

The last years have witnessed a rapid progress not only in 'microelectronics' (i.e. an area of electronics dealing with designing, realising and applications of various low-power EICs made in a solid (continuous medium), or on its surface) but also in 'macroelectronics' (i.e. an area of electronics dealing with designing, realising, and application of various high-power ECMs made also in a solid (continuous medium), or on its surface.

Above all it refers to the developments that have taken place in the field of MCUs (i.e. a low-power EIC that has a **central processing unit** (CPU), memory and **input/output** (I/O) capability, made in the shape of individual chips containing **very large scale of integration** (VLSC) various EICs), as well as ECM commutators (i.e. electronic commutators, containing **standard scale integration** (SSI) and **medium scale integration** (MSI) ECMs). The latter implemented by the author in 1973 have been widely popularised, particularly in the field of industrial and traction IEMD cyber-physical heterogeneous dynamical hypersystems (Fijalkowski 1973, 1985a, 1985b, 1985c, 1987a, 1987b, 1988a, 1988b, 1988c, 1988d, 1988e, 1988f, 1988g, 2010, 2011, 2016).

Over 45 years ago the authors started R&D works on mono- and polycrystalline as well as amorphous semiconductor macroelectronic commutators with uni- and/or bipolar electrical valves, named for the first time by the authors 'ECM commutators', which have aroused the interest of the greatest world consortia. These consortia are aware of the fact that the authors are some of the few scientists who initialised the avalanche of concepts, the consequences of which are difficult to foresee. It is obvious that these concepts will change to a certain degree the face of not only contemporary integrated power electronics, now termed by the authors 'macroelectronics', but also—modern industrial and traction IEMD cyber-physical heterogeneous continuous or discrete dynamical hypersystems, and consequently—new designed energy-saving integrated electrical machines. The scale of these changes for the time being remains unknown. Behind these concepts hides the harbinger of the new universe—the 'universe of macroelectronics'.

Is this new universe going to be better than the old one—the 'universe of power electronics'? This is unknown. Certainly, other new possibilities are offered, of which it is difficult, not to take advantage. Power electronics, industrial and traction IEMD cyber-physical heterogeneous continuous or discrete dynamical hypersystems again are queuing for the results reached in the laboratories of solid-state physics.

Scientific revolution in the field of power electronics has been taking place very rapidly—due to to subsequent stereotypes of ideas valid in this field of knowledge being overcome.

All the progress is stimulated by thinking free from dogmas opposed to the indiscriminate approach to existing concepts. It is also known that the perfect practical solutions are born out of ingenious theory. "There is nothing more practical than a good theory"—said the German physicist Ludwig Boltzmann.

The concept of a generalised intelligent multi-input/multi-output (MIMO) ECM AC–AC, AC–DC, AC–DC–AC, DC–AC and/or DC–AC–DC commutator with 'discrete' and/or 'continuous' uni- and/or bipolar electrical valves was born due to the mathematical hypermatrix notation of the dynamical systems approach for the formation methodises of mathematical models of ECM AC–AC, AC–DC and/or DC–AC commutator dynamotors (Fijalkowski 1973, 1985a, 1985b, 1985c, 1987a, 1987b, 2016, Tutaj 1996, 2012).

MIMO ECM AC–AC, AC–DC and/or DC–AC commutators are the key to overcoming the contactless (brushless), i.e. sparkless commutation (changing the way of electrical current flow), and so the static conversion of one kind of electrical energy into another one, with the application of the phenomena inducing uncontrolled or controlled electrical current conductivity (carrying of electrical charges by positive or negative carriers into a definite medium under the action of electrical field).

MIMO ECM AC–AC, AC–DC and/or DC–AC commutators have announced the era of new designed energy-saving AC–AC, AC–DC and/or DC–AC commutator dynamotors, enabling continuous (stepless) and contactless control of their current and voltage (during generating), or their torque and speed (during motoring).

These ECM AC–AC and/or AC–DC/DC–AC commutators with uni- and/or bipolar electrical valves constitute a successive breakthrough in the development of ECM AC–AC, AC–DC and/or DC–AC commutator dynamotors, and multiply their possibilities enormously. Fantastic perspectives exist, which are likely to become reality as soon as the current century. However, this does not contain the scientific sense of concepts conceived for the first time by the authors forty years ago. It is hoped that the attention of the greatest world consortia will be attracted. The most essential value of this new concept was the achievement of a cardinal breakthrough in the thinking of the static conversion of one kind of electrical energy into another one, with the aid of continuous uni- and/or bipolar electrical valves—a change of paradigm, that was a kind of thinking standard in the field of power electronics. In order to understand it, it is necessary to look into the past. The physical science of electricity and electrical machinery is one of the youngest branches of physical knowledge.

Electrical machine building is actually over 100 years old. However, in this relatively short period a tremendous amount of R&D work has been carried out, that has revolutionised the technical and economical features of modern industry, traction and our social life.

The construction of different types of **magneto-mechano-dynamical** (MMD), and **magneto-fluido-dynamical** (MFD), **magneto-plasmo-dynamical** (MPD) and **electro-mechano-dynamical** (EMD), **electro-fluido-dynamical** (MFD), **electro-plasmo-dynamical** (EPD) electrical machines are the subject of a special course of study, which is beyond the scope of this textbook. Here the authors shall consider only this fundamental principle underlying the MMD conversion of mechanical energy into electrical energy and vice versa.

The dynamical systems approach to MMD electrical machines, classify them as cyber-physical heterogeneous continuous or discrete dynamical hypersystems, i.e. **mechano-electrical** and/or **electro-mechanical** (M-E/E-M) cyber-physical heterogeneous continuous or discrete dynamical hypersystems (Fijalkowski 1985a, 1985b, 1985c, 2016).

An MMD conversion involves the interchange of energy between a mechanical continuous dynamical system and an electrical continuous dynamical system or vice versa through the medium of coupling magnetic-field energy—and utilises the effect of electromagnetic induction and the mechanical action of the coupling magnetic-field energy for its accomplishment. The process is essentially reversible except for a small amount of thermal energy of a thermal continuous dynamical system, which is lost as heat.

Thus, the MMD electrical machines are the main types of rotary and linear converters of mechanical energy into electrical energy, and/or electrical energy into mechanical energy, as well as rotary converters of one form of electrical energy into another form, differing in voltage, current and sometimes in frequency.

When the MMD conversion takes place from mechanical into electrical energy, the MMD electrical machine is termed an 'ME generator' (or an 'ME dynamo').

When electrical energy is converted to mechanical energy, the MMD electrical machine is termed an 'EM motor'.

When bilateral energy MMD conversion is necessary, the MMD electrical machine is termed an 'ME/EM dynamotor'.

The constant improvements in the design of MMD electrical machines have made possible many new practical applications, and have served strong impulses for further progress, and the most diverse uses of electrical energy.

This explains why the MMD electrical machines were given great attention by scientists and engineers, and why the MMD electrical machines quickly attained technical perfection of design.

A.2 Status and trends

The properties of unipolar electrical-current conductivity, i.e. non-linearity of current–voltage characteristics of polycrystalline semiconductors were discovered by Adams and Day in the 1880s. In 1883 Fritsch described a construction of selenium electrical valves, which were applied on a large scale no earlier than at the end of the 1920s. In 1926, Grondahl discovered cuprous-oxide electrical valves, which became popular in the 1930s. At the same time titanium dioxide, electrical valves with an admissible working temperature up to 473 K were also applied.

In 1882, Jamin and Manevrieu published a work on the properties of unipolar electrical-arc conductivity in a mercury vapour, i.e. mercury-arc or ionic electrical valves, which were broadly applied at the beginning of the 1930s. Fleming, who named it the 'kenotron', invented the first electron diode in 1904. In parallel polyvalvular (multi-anode) mercury-arc electrical valves, named 'mutators', the manufacturing technique of 'gasotrons' and a little later—'thyratrons' expanded. In the 1950s the 'ignitrons' and 'exitrons' were also applied on a large scale. The development of manufacturing technology of monocrystalline semiconductors, and then germanium and silicon unipolar electrical valves took place in the 1940s (uncontrolled 'discrete' electrical valves—germanium and silicon diodes), and in the 1950s (controlled 'discrete' electrical valves—germanium and silicon triodes: transistors, insulated-gate bipolar transistors (IGBT), metal-oxide–semiconductor field effect transistors (MOSFET), thyristors, trigistors, magnetic field controlled thyristors (MCT) etc, as well as germanium and silicon tetrodes: spacistors, tetristors etc). Apart from silicon unipolar electrical valves, the silicon bipolar electrical valves with single or two control gates, among which there are, for example, bipolar field transistors and photo-transistors, as well as symistors (symmetrical thyristors) and photosymistors and bipolar trigistors (gate-turn-off (GTO) thyristors), **reverse-blocking insulated-gate bipolar transistors** (RB IGBT) have also been applied. Progress in a recognition of electrical properties of glassy amorphous materials is leading to series of new promising applications of amorphous semiconductors in macro- and microelectronics, and thus in the field of 'discrete' or 'continuous' uni- and bipolar (uni- and bilateral) electrical valves.

In 1958 S R Ovshinsky discovered in amorphous semiconductors the phenomenon of switching over, i.e. a rapid changeover of electrical-current conductivity attaining, by the applied electrical field, its critical value. In the case of chalkogenides glasses nearly the same phenomenon has appeared in the electrical field with the intensity of 10^5 V cm^{-1}, the changeover time yields a few nanoseconds (Holmberg and Shaw 1974), and electrical-current conductivity can be changed approximately 10^6 times. The results obtained were not published before 1970 (Fijalkowski 1988a, 1988b, 1988c, 1988d, 1988f, 1988g), while references to them can be found in the monographs of Fijalkowski (1987a, 1987b). However, some technological difficulties, as well as military applications, have contributed to the limitation of their initial development.

It is already over 50 years since S R Ovshinsky demonstrated the semiconductor device 'ovonic' (contraction of OVshinsky electrONIC), that is an electrical valve made of the amorphous semiconductor for the first time. For 50 years, although the ovonic existed and operated, nobody precisely knew the principle of its operation. There are several well-known cases, in the history of electronics, in which various semiconductor devices had been commonly used before their operation principles were exactly known, and their theory was created. That was the case for the protoplast of galena detector, as well as of the best polycrystalline semiconductor electrical valve in the past—the selenium rectifier.

Not until the 1970s did the two physicists P W Anderson and J van Vleck of the USA and English physicist Sir N F Mott work out new fundamentals of the theory and technology of amorphous semiconductors, which showed that unipolar electrical-current conductivity could exist not only in mono- and polycrystalline semiconductors, but also in amorphous semiconductors. The theory gained common approval, and its creators were honoured with the Nobel Prize in physics in 1977. If everybody had assented to an earlier formulated theory (Seitz 1940, Kittel 1956,

Mott and Davis 1971), foreseeing that the unipolar electrical-current conductivity could exist only in unipolar mono- and polycrystalline semiconductors, nobody would have an inkling of such new technologies humanity could reach for. Fortunately, there was a group of scientists who doubted the correctness of this theory (Mott and Davis 1971). Due to their obstinacy, it is known today that the earlier formulated theory was valid only for mono- and polycrystalline semiconductors. All aforementioned 'discrete' uni- and bipolar electrical valves were or are used up-to-now in MIMO ECM AC–AC, AC–DC–AC and AC–DC/DC–AC commutators with non-integrated ECMs (Fijalkowski 1973, 1982, 1985a, 1985b, 1985c, 1987a, 1987b).

In the field of macroelectronics they are named by the authors 'polyvalvular MIMO ECM AC–AC and/or AC–DC/DC–AC commutators'. It is obvious, that in the field of power electronics these polyvalvular MIMO ECM commutators are still called 'static converters'. The same also took place in the field of MCM AC–AC and AC–DC/DC–AC commutator dynamotors, which in the field of macroelectronics are named by the authors 'polysectorial MIMO MCM AC–AC and/or AC–DC/DC–AC commutators' (Fijalkowski 1973, 1985a, 1985b, 1985c, 1987a, 1987b).

Less expensive, lighter, and smaller than electrical machines' polysectorial MIMO MCM AC–AC and/or AC–DC/DC–AC commutators, polyvalvular MIMO ECM AC–AC and/or AC–DC/DC–AC commutators lie at the very heart of converting and controlling electrical energy, this in turn lies at the heart of making that energy useful. From domestic appliances to space-faring and automotive vehicles, the applications of macroelectronics are virtually limitless. Until now, however, the same could not be said for access to up-to-date reference books devoted to power electronics.

MIMO ECM AC–AC, AC–DC, AC–DC–AC, DC–DC, DC–AC and/or DC–AC–AC commutators are ECMs of controlled semiconductor electrical valves that directly connect each matrixer-input phase to each matrixer-output phase, especially without any intermediate DC link. The MIMO ECM AC–AC, AC–DC, AC–DC–AC, DC–DC, DC–AC and/or DC–AC–AC commutator is usually fed at the matrixer-input side by a voltage source and it is connected to inductive and/or capacitive loads at the matrixer-output side. The main advantages of the ECM commutator are the absence of bulky reactive components that are subject to ageing and reduce the MIMO ECM AC–AC, AC–DC, AC–DC–AC, DC–DC, DC–AC and/or DC–AC–AC commutator reliability. One significant advantage of this arrangement is that the ECM commutator does not carry current to the rotor—which eliminates the brushes and their wear-related drawbacks.

Furthermore, MIMO ECM AC–AC, AC–DC, AC–DC–AC, DC–DC, DC–AC and/or DC–AC–AC commutators provide uni- and/or bilateral (bipolar) power flow, nearly sinusoidal matrixer-input and matrixer-output waveforms, and a controllable matrixer-input power factor, regenerative capability as well as the original function to generate magnitude-frequency controllable matrixer-output voltages. Therefore, ECM AC–AC, AC–DC, AC–DC–AC, DC–DC, DC–AC and/or DC–AC–AC commutators have received considerable attention as an advantageous preference to an ECM AC–AC commutator acting in AC–AC cycloconverter topology; ECM AC–DC commutator acting in AC–DC rectifier topology; ECM AC–DC–AC commutator acting in AC–DC–AC frequency changer topology; ECM DC–DC commutator acting in DC–DC direct chopper (interrupter) topology; ECM DC–AC commutator acting in DC–AC inverter topology, and ECM DC–AC–DC commutator acting in DC–AC–DC indirect chopper (interrupter) topology.

The authors have already shown in 1973 in the work entitled: *Trends in Electric Propulsion of Locomotives* (Fijalkowski 1973), that each static converter or MIMO MCM AC–AC, AC–DC, AC–DC–AC, DC–DC, DC–AC and/or DC–AC–AC commutator of the MMD electrical machines can not only be theoretically considered, but also practically made as polyvalvular MIMO ECM AC–AC, AC–DC, AC–DC–AC, DC–DC, DC–AC and/or DC–AC–AC commutators with built-in 'discrete' and/or 'continuous' unipolar electrical valves called thyristors (see figure A5), but they did not expect then, that the results of this would have any practical consequences in the approaching years.

Besides, the authors showed once more in 1982 in the work entitled: *Trigistor Frequency Changer-Controlled Toothed Gearless Propulsion Systems for Electric and Hybrid Vehicles* (Fijalkowski 1982), that each MIMO MCM AC–AC, and/or AC–

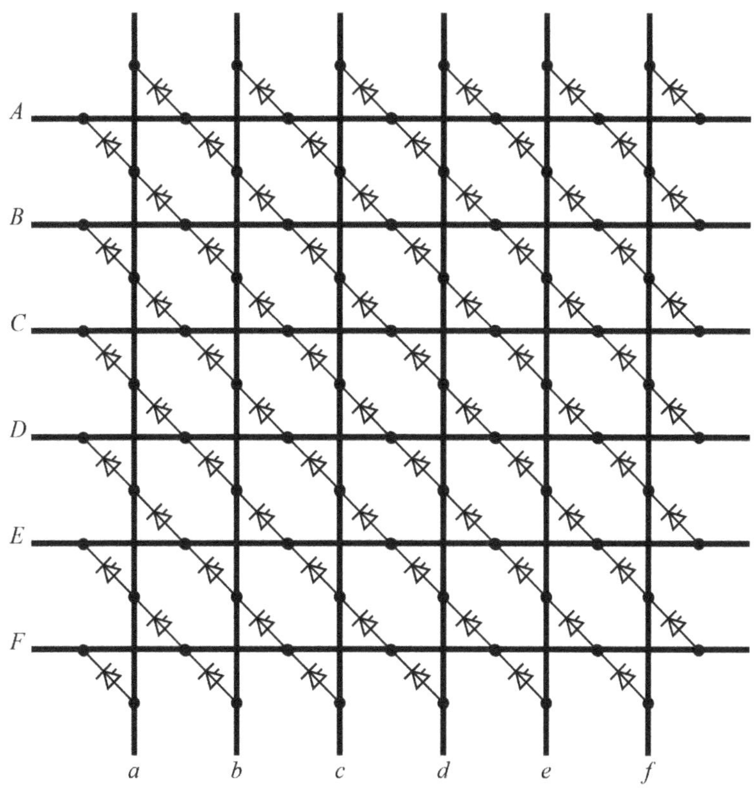

Figure A5. MIMO ECM AC–AC or AC–DC or DC–DC or DC–AC commutator with built-in 'continuous' unipolar electrical valves termed thyristors.

DC/DC–AC commutator of the electrical machines can not only be theoretically considered, but also practically made as MIMO ECM AC–AC, AC–DC, AC–DC–AC, DC–DC, DC–AC and/or DC–AC–AC commutators with built-in 'discrete' and/or 'continuous' bipolar, triggered and quenched electrical valves termed 'trigistors'.

The development of the three-phase by three-phase (3 × 3) ECM AC–AC commutators (called the 'direct frequency converters with nine electrical valves') started earliest by Zagorski (1969); and the development of each and every one $m \times n$ ECM AC–AC, AC–DC–AC, AC–DC/DC–AC, DC–DC and DC–AC–DC commutators (with $m \times n$ bipolar, triggered and quenched electrical valves) started afterwards by Fijalkowski (1973, 1982, 1985a, 1985b, 1985c, 1987a, 1987b); as well as the development of the 3 × 3 ECM AC–AC commutators (termed the *'matrix converters with nine bipolar electrical valves'*) continued soon after by Gyugyi and Pelly (1976), Cheron *et al* (1985), and Alesina and Venturini (1989), when the above mentioned authors proposed the basic principles of their operation.

The 3 × 3 ECM AC–AC commutators offer the advantage over up-to-now applied ECM AC–AC commutators (called the 'thyristor cycloconverters with eighteen or thirty-six unipolar electrical valves') of being able to produce unity power factor (PF) line current. However, compared to PWM voltage-fed AC–DC–AC ECM commutators, the parts count is significantly higher. Consequently, the R&D works in this field are continued in two trends (see references and bibliography).

On the one hand, there was the need for reliable 'discrete' and 'continuous' uni- and/or bipolar electrical valves (electronic switches). On the other hand, the initial modulation strategy was unsuitability in favour of ultra-modern ECM solutions, allowing higher energy (voltage) transfer ratio and better power (current) quality,

The performance of MIMO ECM AC–AC, AC–DC–AC, AC–DC/DC–AC, DC–DC and DC–AC–DC commutators is strictly dependent upon the pulse-width modulation (PWM) strategy employed to control uni- and/or bipolar, triggered and quenched electrical valves. Since the introduction of MIMO ECM commutators, various and numerous modulation methods have been developed to date.

In Alesina and Venturini (1989) and Tenti *et al* (1992), off-line global duty-ratio functions are mathematically derived, which can be modulated by a carrier-frequency signal. This modulation method necessitates a formidable amount of complicated computations in the stage of implementation.

Vector PWM (VECPWM) for an ECM commutator explores a more systematic approach to understand the operation of the ECM (Huber and Borojevic 1995, Casadei *et al* 2002a, 2002b, 2002c). However, the VECPWM is far from spontaneous and requires lookup tables with the *a priori* prepared and stored commutation patterns. Carrier-frequency based PWM is the latest modulation strategy for ECM AC–AC commutators (Yoon and Sul 2006, Li *et al* 2008, 2009).

The carrier-frequency based PWM employs the carrier-frequency and reference signals and can be implemented without complicated computations and lookup tables.

The modulation method described in Yoon and Sul (2006), however, necessitates proper offset voltages and discontinuous carrier-frequency signals which imply relatively indirect understanding of cannot be applied to ECM topologies with a neutral connection between matrixer-input and matrixer-output neutrals. In Li *et al* (2008), a carrier-frequency based PWM strategy, termed **direct duty-ratio PWM** (DDPWM), is presented for the 3-phase to 3-phase 3 × 3 ECM AC–AC commutator, which does not requires the reference offset and employs a continuous triangular carrier-frequency waveform. The desired matrixer-output phase voltages can be synthesised by utilising the matrixer-input phase voltages based on per-output-phase average concept over one commutation period, i.e. per-carrier cycle. A maximum gain of 0.866 can be simply obtained by injecting third harmonic components to matrixer-output voltage references in the 3-phase electrical energy system.

In Li *et al* (2009a), the DDPWM method is extended in order to use it to the various ECM commutator topologies, e.g. such as 3-phase to 1-phase 3 × 1 or 3 × 2 ECM AC–AC and/or AC–DC/DC–AC commutators; 3-phase to 2-phase 3 × 2 or 3 × 3 ECM AC–AC commutators; 3-phase to 3-phase 3 × 3 or 3 × 4 ECM AC–AC commutators; 3-phase to 5-phase 3 × 5 or 3 × 6 ECM AC–AC commutators; and 5-phase to 5-phase 5 × 5 or 5 × 6 ECM AC–AC commutators, etc.

In Li *et al* (2009b), a systematic method to generate the reference signals in various ECM commutators is presented. Prior to the DDPWM is used to synthesise the matrixer-output voltages, the per phase matrixer-output voltage references should be given at each commutation time. The global minimum of the matrixer-input voltage range is used to determine the matrixer-output voltage magnitude.

The feasibility and validity of the proposed DDPWM method applied in various ECM commutators was verified with simulation and experiment results. When over 40 years ago, the authors had the idea of the MCM and ECM commutators, realised on 'discrete' and 'continuous' uni- and/or bipolar electrical valves, only a few people believed in the sense of the R&D programme started by them. It is clear that amorphous semiconductor electrical valves will be the basis of the next great advance in macro-, meso-, micro- and nanoelectronics.

The authors intend to show why the field of power electronics is presently in a state of crisis, and therefore historically ripe for a basic new approach. A characteristic component of a large group of MMD electrical machines is the integral MCM or ECM AC–AC and/or AC–DC/DC–AC commutator that is in the dynamical systems approach on MMD electrical machines, i.e. IEMD cyber-physical heterogeneous continuous dynamical hypersystem is classified as a **generalised commutation matrixer** (GCM) (Fijalkowski 1973, 1985a, 1985b, 1985c, 2016).

The concept of design fully-static power electronic electrical-valve ECM AC–DC/DC–AC commutators by the authors named 'ECM commutators' for the conversion of polyphase alternating voltages, constant both in amplitude and frequency or direct voltage constant in magnitude to polyphase alternating voltages, modulated both in amplitude and frequency in order to be able to control the torque and speed of ECM AC–AC and AC–DC/DC–AC commutator dynamotors is by no means new, if one studies old literature on the static converters, one can observe that such static power electronic electrical-valve ECM AC–DC/DC–AC commutators were described as early as the beginning of the 1890s.

The chronological background of MIMO ECM AC–DC commutators goes back to the end of the 19th century, when in 1895 (Date of Application) the notable Polish electrician Charles (Karol Franciszek) Pollak (the honorary doctor of the Warsaw University of Technology), was issued an English Patent (24 398 AD 1895) and a German Patent (DRP 96564) for an 'aluminium' electrical valve bridge rectifier (Pollak 1895, 1896).

Descriptions of the patents also included a scheme of the Pollak bridge rectifier (figure A6) as a single-phase, full-wave MIMO ECM AC–DC commutator in a single-phase double-way connection, known at present as the Graetz bridge rectifier. These descriptions were published in the *Elektronische Zeitung*, No 25, 1897, with notation from the editor that at that time Leo Graetz was working on rectifiers of similar principle of operation. However, the solution of Leo Graetz was published in 1897, a year and half after the patent for Karol Pollak had been issued (Strzelecki and Benysek 2008).

The single-phase, full-wave MIMO ECM AC–DC commutators in realistic applications are major. A 2 × 2 ECM AC–DC commutator, e.g. covers all possible cases with a single-port input energy source and a two-terminal load.

The ECM AC–DC commutator is commonly drawn with a single homogeneous either anode or cathode commutation group, or a two heterogeneous anode and cathode commutation groups, i.e. the well-known in literature H-bridge converter (AC–DC rectifier) shown in figure A7.

In 1901, the Belorussian academician Vladimir Fedorovich Mitkevich (1872–951) published two-phase and three-phase, half-wave ECM AC–DC commutators in a two- and/or three-phase single-way connection, known at present as the Mitkevich rectifiers (Neiman 1972). Soon after, in 1924, the Russian academician Andrei Nikolaevich Larionov (1889–963) published a three-phase, full-wave ECM AC–DC commutator in a three-phase double-way connection, known at present as the Larionov bridge rectifier (Anon 1950). Other successive inventions that are of

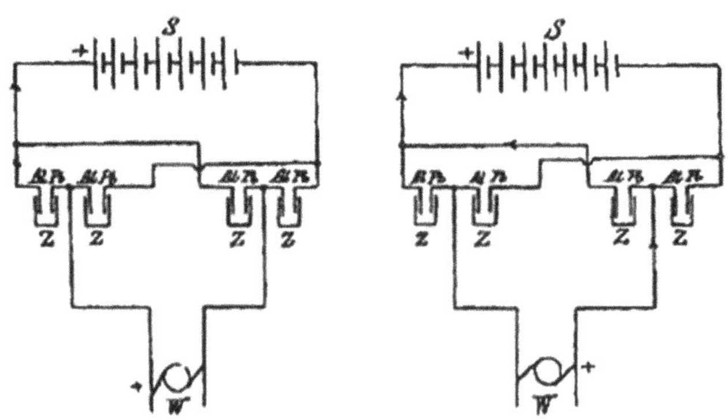

Figure A6. The Pollak ECM AC–DC commutator acting as an AC–DC rectifier with four aluminium electrical valves (English Patent (24 398 AD 1895; German Patent (DRP 96564)).

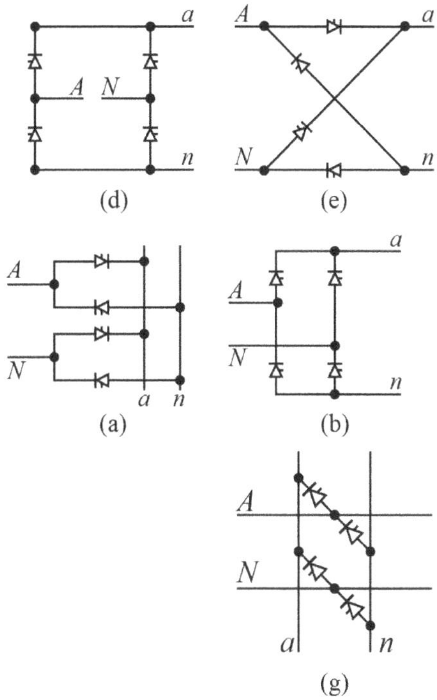

Figure A7. Seven well-known schematic configurations of the Pollak 2 × 2 ECM AC–DC commutator with a single homogeneous either anode or cathode commutation group, or a two heterogeneous anode and cathode commutation groups.

importance to ECM AC–AC commutators were developed in 1930s (Kloninger 1932, Willis 1933, Alexanderson and Mittag 1934, Kern 1938, Schilling 1940).

Improved ECM AC–AC, AC–DC and DC–AC commutators were frequently proposed during the 1940s and 1950s (Schilling 1940, Rémy 1954). However, such ECM AC–AC, AC–DC and DC–AC commutators did not gain wide acceptance and most of the concept probably did not leave the laboratory. The fact that this was so must naturally be attributed to the characteristics of the electrical valves available to the scientists and engineers in those days—electronic tubes, mercury-arc electrical valves of different kinds. It was not until the 1960s that scientists were given the possibility, both technically and economically, to solve this problem by introduction of the **semiconductor-controlled rectifier** (SCR) in the power electronics world termed the ‘thyristor’ later on.

In France, the first ECM commutator for an MMD electrical machine, probably the first in the world was conceived by Kloninger (1932), while in England the first complete ECM commutator based on ‘thyratron’ electrical valves was developed by Alexanderson and Mittag (1934).

The authors started R&D work on ECM AC–DC/DC–AC commutator dynamotors (see figure A8), for high-power E-M propulsion systems in the early 1960s (Fijalkowski 1962, 1964).

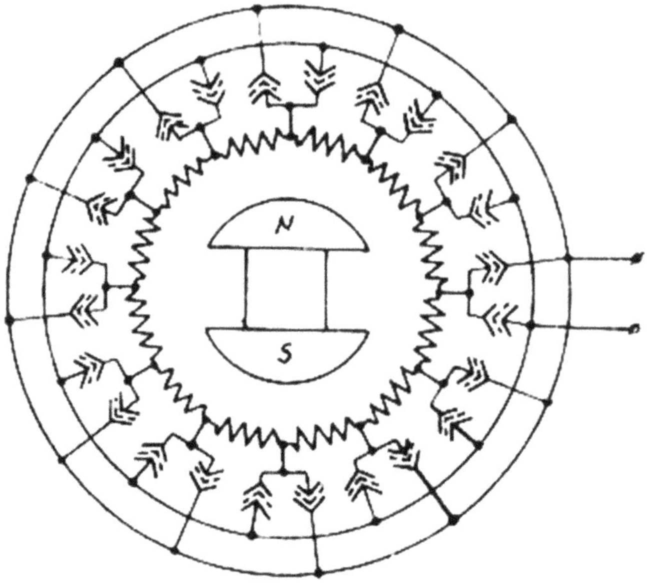

Figure A8. 2 × 12/12 × 2 electrical-commutation-matrixer (ECM) AC–DC/DC–AC commutator dynamotor.

The last years of the 1960s witnessed rapid progress in ECM high-power electronics, called 'macroelectronics' by the authors, above all when it comes to the developments that have taken place in the field of the bipolar, triggered and quenched electrical valves such as trigistor, gate-turn-off (GTO) thyristor and/or bipolar transistor ECMs.

New applications for macroelectronics have thus evolved and established applications have undergone further advancements. This textbook briefly presents some of the new developments and progress that have occurred within the macroelectronics field over the previous century.

By definition 'macroelectronics' is the specific application of electronics using electrical-commutation-matrixer (ECM) commutators, by the authors named multi-input/ multi-output (MIMO) 'ECM commutators' to the electrical power field, mainly new concept ECM AC–AC and AC–DC/DC–AC commutator dynamotors for industrial and traction E-M propulsion dynamical systems. It has followed the development that has taken place within macro- and microelectronics as a whole.

The introduction of the ECM AC–AC, AC–DC–AC, AC–DC/DC–AC, DC–DC and/or DC–AC–DC commutator combined with macro- and microelectronic based technology—has been the most significant factor in the tremendous improvement of the power handling capacity, voltage and current data, etc, of the new concept power handling components, i.e. ECM commutators.

At the same time, it has been possible to exploit fully the rapid integration of components by miniaturising the static high-power MIMO ECM AC–AC, AC–DC–AC, AC–DC/DC–AC, DC–DC and/or DC–AC–DC commutators, in the MMD electrical machines as well as low-power (small signal) electronic computer processor, i.e. microprocessor in the higher- and lower-level controllers included in macro- and microelectronics equipment or ECM AC–AC, AC–DC–AC, AC–DC/DC–AC, DC–DC and/or DC–AC–DC commutators, as they are usually termed in this way, the ECM AC–AC, AC–DC–AC, AC–DC/DC–AC, DC–DC and/or DC–AC–DC commutator has evolved from a rather un-qualified integral static high-power ECM AC–AC, AC–DC–AC, AC–DC /DC–AC, DC–DC and/or DC–AC–DC commutator to EM propulsion process equipment incorporating qualified electrical integrated circuits (EIC) for control, monitoring, operation and protection.

Up to now, the use of analogue EIC techniques has predominated, but the trend towards digital EIC techniques is both justified for cost reasons and essential in this case. For example, very advanced ECM AC–AC, AC–DC–AC, AC–DC/DC–AC, DC–DC and/or DC–AC–DC commutators of the **sinusoidal pulse-width-modulated** (SINPWM) type.

In the 1970s, as ECM AC–AC, AC–DC–AC, AC–DC/DC–AC, DC–DC and/or DC–AC–DC commutators were developed and put into practical use, the demands for fast-switching thyristor ECMs with high performance, such a higher blocking voltage, and larger current capability as well as short turn-off time (Fijalkowski 1973).

Current trends in aerospace avionics and automotive vetronics are towards intelligent electronically-commutated (reciprocal and rotational) EM actuators, namely electrical machines and other digital devices with much higher performance and greater compactness, not only at the interconnection stage.

The increasing complexity of intelligent electronically-commutated actuators and digital devices is leading to large chip sizes and the greater degree of integration means that the power dissipated by the working chip is also rising. Increasingly, the tendency is to package several chips together in the form of multichip module, using techniques such as wire bonding, **tape automated bonding** (TAB) and flip chip bonding to reduce the inter-connections length and hence the overall size.

Continuing demands for high-performance intelligent, MIMO ECM AC–AC, AC–DC–AC, AC–DC/DC–AC, DC–DC and/or DC–AC–DC commutators of far greater complexity than before ought to be required.

Artificial **neural networks and fuzzy logic** (NF) are increasingly being incorporated in intelligent, MIMO ECMs to provide robust and effective control in a wide range of aerospace and automotive applications. The method is based on a many-valued-logic, which enables general principles and expert knowledge to be used to provide control rules and procedures.

The intrinsic non-linearity and variability of operating conditions make NF control an ideal method for this area.

The authors are actively pursuing the integration of matrixery and circuitry on the same transparent substrate as the intelligent, MIMO ECM AC–AC, AC–DC–AC, AC–DC/DC–AC, DC–DC and/or DC–AC–DC commutators and NF micro-controllers, respectively.

The most obvious advantage of integrated matrixery and circuitry on the same transparent substrate is the reduction in the ‘work-horse’ components that are mono- and polycrystalline as well as amorphous Si or GaAs super- and/or semiconductor, fast-switching ‘discrete’ and/or ‘continuous’ uni- and/or bipolar electrical valves, e.g. diodes, triodes, tetrodes etc, namely **charge-injection transistors** (CHINT), **complementary metal-oxide–semiconductor** (CMOS) transistors, **hetero-junction bipolar transistors** (HBT), **high-electron injection transistors** (HEIT), **metal–semiconductor field effect transistors** (MESFET), **metal–insulator–semiconductor FETs** (MISFET), **metal-oxide–semiconductor FETs** (MOSFET), **organic FETs** (OFET), **junction-gate field effect transistor** (JFET or JUGFET), bipolar **quantum interference transistors** (QUIT), **static induction transistors** (SIT), bipolar **super-conductor-base hot electron transistors** (SUPER-HET), **insulated-gate bipolar transistors** (IGBT), **gate-turn-off** (GTO) thyristors, **light-triggered-and-quenched** (LTQ) thyristors, **MOS controlled thyristors** (MCT), **reverse-blocking insulated-gate bipolar transistors** (RB IGBT), ovonics, etc, as well as in external connections. The 'discrete' electrical valves suffer from having a performance that is strongly dependent upon temperature.

The authors have therefore carried out a detailed programme of research into intelligent MIMO ECM commutators using 'continuous' electrical valves, in which electrical-current conduction is much less temperature-dependent than in 'discrete' electrical valves.

The external connections are also one of the dominating causes of the failure of MIMO ECM commutators. They will be lighter and more compact because of the proportion of their size and mass.

It is difficult to estimate the cost and reliability of these ECM commutators, as they are not yet in manufacture.

Although R&D as well as initial manufacture costs are high, the leading electronics manufacturers will be investigating significantly this key technology and the intelligent MIMO ECM AC–AC, AC–DC, AC–DC–AC, DC–DC DC–AC, and/or DC–AC–DC commutators capable of true inaudible (> 20 kHz) operation, and NF controllers will soon extend their application from the current aerospace avionics and automotive vetronics intelligent electronically-commutated E-M actuators and other digital devices markets in other areas, and they have every chance of completely revolutionising the ECM universe.

Recently, attention has been focused on the bipolar **reverse-conducting** (RC) GTO thyristor, MCT, MOSFET, and IGBT ECM, termed the 'RC trigistor ECM' as fast-switching semiconductor controlled electrical-valve ECM, which have the feasibility of overcoming these difficulties. The electrical-valve ECMs proved to be successful for practical ECM AC–AC, AC–DC, AC–DC–AC, DC–DC DC–AC, and/or DC–AC–DC commutator applications, and important application data were obtained. These bipolar trigistor and bipolar transistor ECMs have been used in ECM AC–AC, AC–DC, AC–DC–AC, DC–DC DC–AC, and/or DC–AC–DC commutators for industrial ME drives and traction propulsion systems and have the advantage of making them more compact, less expensive, and of lighter mass than the unipolar **reverse-blocking** (RB) thyristor with freewheeling diode ECMs.

The Nobel Prize in Physics 2010 was awarded jointly to Andre Geim and Konstantin Novoselov 'for groundbreaking experiments regarding the two-dimensional material graphene'. The most promising and versatile material nowadays, graphene has attracted attention as a promising two-dimensional honeycomb lattice composed of sp^2-bonded carbon-atoms material for very advanced ECM AC–AC, AC–DC, AC–DC–AC, DC–DC DC–AC, and/or DC–AC–DC commutators including high-speed **single electron transistors** (SET) which play a central role in LSI technology.

The SET is based on the idea of controlling the number of electrons in a quantum dot by its electrical potential. The energy consumption of SET is thought to be very small, because the unit of information is a single electron.

Graphene is a one-atom-thick layer of carbon atoms arranged in a hexagonal lattice. It is the building-block of graphite (which is used, among others things, in pencil tips), but graphene is a remarkable substance on its own—with a multitude of astonishing properties which repeatedly earn it the title 'wonder material'.

Graphene is the thinnest material known to man at one atom thick, and also incredibly strong—about 200 times stronger than steel. On top of that, graphene is an excellent conductor of heat and electricity and has interesting light absorption abilities. It is truly a material that could change the world, with unlimited potential for integration in almost any industry. Graphene has a potential advantage that quantum dots can be easily formed just by cutting out small regions by etching, because graphene originally has only a single atomic layer thickness. However, controllable manufacture with a precision of several nanometers still remains as a challenge. For example, the SET commutation (switching) performance at room temperature is not well reproducible yet due to the lack of precise manufacture technology. Therefore, SET is expected to be an electrical valve (electronic switch) which will be used widely in the near future, and many research efforts have been focussed on it.

Researchers at MIT and Harvard University found that graphene can be tuned to behave at two electrical extremes: as an insulator, in which electrons are completely blocked from flowing; and as a superconductor, in which electrical current can stream through without resistance. Researchers in the past, including this team, have been able to synthesise graphene superconductors by placing the material in contact with other superconducting metals—an arrangement that allows graphene to inherit some superconducting behaviours. In this new work, the team found a way to make graphene superconduct on its own, demonstrating that superconductivity can be an intrinsic quality in the purely carbon-based material.

Today, the authors are designing ECM commutators, which compared with the first thyristor ECM AC–AC, AC–DC, AC–DC–AC, DC–DC, DC–AC, and/or DC–AC–DC commutators will be designed for a 40 times higher ECM AC–AC, AC–DC, AC–DC–AC, DC–DC, DC–AC, and/or DC–AC–DC commutator power rating.

The integrated ECM commutators now being conceived and developed for current projects of ECM AC–AC, AC–DC–AC and DC–AC commutator dynamotors belong to the third generation. The first two generations were air-cooled, air-circulated thyristor ECM AC–AC, AC–DC, AC–DC–AC, DC–DC, DC–AC, and DC–AC–DC commutator, while oil has been selected as coolant for the third generation.

The changeover to oil cooling was a natural consequence of the fact that ECM AC–AC, AC–DC, AC–DC–AC, DC–DC, DC–AC, and/or DC–AC–DC commutator having a considerably better performance were developed. Oil-cooling systems have substantially lower inherent losses, which gave significant gains with the rapid escalation of lower evaluation costs.

The performance and efficiency of MIMO ECM commutators as well as MMD electrical machines (M-E/E-M dynamotors and E–E transformers) is limited due to thermal constraints in the cooling ECM channels and MMD electrical-machine windings. These constraints can be diminished somewhat by circulating liquid through the liquid cooling ECM channels or air through the air gaps in the MMD electrical machines. However, such approaches necessitate the addition of pumping mechanisms and are limited due to the thermal conductivity of air or liquid.

Recently, a novel approach has been made to cool the MIMO ECM commutators as well as MMD electrical machines by using **ferro fluids** (FF), as the cooling media. The aim is to take advantage of the thermal and magnetic characteristics of FFs, respectively, such that not only will the FF fluid circulate through the MIMO ECM commutators as well as MMD electrical machines, but existing thermal and magnetic fields in the MIMO ECM commutators as well as MMD electrical machines will provide the energy required for pumping the FF.

The cooling mechanism will therefore necessitate no additional moving mechanical parts, M-F pumps, M-P compressors or sensors. The behaviour of the FF will provide the method of pumping and cooling of the MIMO ECM commutators as well as MMD electrical machines (thus increasing the MIMO ECM commutators as well as MMD electrical machines' potential energy density and efficiency).

It is generally the need for a high short-circuit capability and low losses rather than the performance during normal operation that determines, e.g. the choice of trigistor cathode area for the ECM AC–AC, AC–DC, AC–DC–AC, DC–DC, DC–AC, and DC–AC–DC commutator's electrical valves. A high short-circuit current capability allows one to optimise the armature-winding reactance concerning reactive power consumption, MMD electrical machine costs and its losses.

Today, there are no plans for MIMO ECM AC–AC, AC–DC, AC–DC–AC, DC–DC, DC–AC, and DC–AC–DC commutators where the data such that the parallel connection of electrical valves will be necessary.

The bipolar 'trigistors' with a cathode area exceeding 100 cm^2 now available give a single chain of trigistors connected in series. This results in the minimum of complexity, high reliability, the lowest cost and the lowest losses.

The triggering pulses for MIMO ECM AC–AC, AC–DC, AC–DC–AC, DC–DC, DC–AC, and DC–AC–DC commutator's electrical valves, where a large number of electrical valves are connected in series to obtain a sufficient voltage rating must be transmitted by non-galvanic means.

For instance, for this purpose, one uses light guides, which transmit light pulses for the electrical-valves' **electronic control unit** (ECU) at earth potential, to the individual electrical valves. Because the electrical valves requires an electrical triggering pulse, conversion from light to electrical current pulses is done in an ECU by each electrical valve (electronic switch).

A.3 Hybrid ECM commutators

The hybrid ECM AC–AC, AC–DC, AC–DC–AC, DC–DC, DC–AC, and/or DC–AC–DC commutator, realised in 'continuous' uni- and/or bipolar electrical valves, established a mediate component amidst the polyvalvular, ECM AC–AC, AC–DC, AC–DC–AC, DC–DC, DC–AC, and/or DC–AC–DC commutator, realised also on 'continuous' uni- and/or bipolar electrical valves. Reactive components of ECM, and so primary (matrixer-input) and secondary (matrixer-output) collectors, i.e. the connection between the 'continuous' uni- and/or bipolar electrical valves are made on a ceramic or glassy base with the vacuum evaporation, printing or injection casting technique. Active components of ECMs, and so 'continuous' uni- and/or bipolar (bilateral) electrical valves: diodes, triodes (transistors, thyristors, trigistors etc) or tetrodes (spacistors, tetristors etc) without package are soldered or pressed into the ECM manufactured in this manner.

The hybrid ECM AC–AC, AC–DC, AC–DC–AC, DC–DC, DC–AC, and/or DC–AC–DC commutator, especially made in the thick-film technique, has a number of essential advantages:

- not too long time of designing and preparation of production;
- moderate cost of production, even in small series;
- possibility of integration of 'continuous' uni- and/or bipolar electrical valves and protection circuits;
- high over-current capability;
- high over-voltage capability;
- high output power.

By way of example the rotary ECM AC–DC/DC–AC commutators, designed to be built into the rotor of the ECM AC–DC/DC–AC commutator dynamotor, which work in the most severe conditions are nowadays, most frequently made using the hybrid technique.

A.4 Monolithic ECM commutators

The monolithic ECM AC–AC, AC–DC, AC–DC–AC, DC–DC, DC–AC, and/or DC–AC–DC commutator, realised on 'continuous' uni- and/or bipolar electrical valves is an intelligent MIMO ECM commutator, in which all active elements ('continuous' uni- and/or bipolar electrical valves) are made in a single wafer of the crystalline or amorphous material in the shape of single chips. Thus, the technique has assured high packing density of active components.

According to the number of individual single active components integrated on the single wafer of crystalline or amorphous material, ECMs with a SSI, comprising several to 36 'continuous' uni- and/or bipolar electrical valves, or ECMs with a MSI from 24 to hundreds 'continuous' uni- and/or bipolar electrical valves can be distinguished.

The monolithic, intelligent, MIMO ECM AC–AC, AC–DC, AC–DC–AC, DC–DC, DC–AC, and/or DC–AC–DC commutator, realised on monocrystalline semiconductor 'continuous' uni- and/or bipolar electrical valves, used in the static converter technique and ECM AC–AC, AC–DC, AC–DC–AC, DC–DC, DC–AC, and/or DC–AC–DC commutator dynamotors show far series of limitations, namely:

- difficulties with the integration of 'continuous' uni- and/or bipolar electrical valves;
- limited resistance against over-voltages;
- low cost only for series production on a large scale (100 000 of ECMs yearly).

Recently in hybrid and monolithic ECM AC–AC, AC–DC, AC–DC–AC, DC–DC, DC–AC, and/ or DC–AC–DC commutators, two basic types of ECMs can be used: unipolar and/or bipolar.

The authors announced the expansion of the pioneering family of compact, intelligent, MIMO ECM AC–AC, AC–DC, AC–DC–AC, DC–DC, DC–AC, and/ or DC–AC–DC commutators with integral press pack ECMs.

These novel macroelectronic devices have, recently, a 'continuous' electrical valve's voltage rating of 8 kV and a current rating of 2.5 kA. The compact matrix-built, package design gives the new macroelectronic device unrivalled 'megawatt' power density with the advantages of high reliability and efficiency.

The unique matrix-built, hermetic ceramic package and internal construction provides double side cooling for the electrical valves, resulting in superior thermal power cycling and power ratings, when compared to alternate module technologies.

The compact matrix-built, fully-hermetic package design is mechanically compatible with cooling systems, including total immersion systems such as oil, currently employed for 3 or 4 kA GTO thyristors, this feature allowing for either retrofitting or design continuation of existing systems.

The high-current rating allows the conversion of old bipolar static converter EMDs to state-of-the-art ECM technology, without the need to replace the basic mechanical structure, Adopting a refurbishment approach to component obsolescence is both lower cost and environmentally friendly; as it reduces consumption of both materials and energy, when compared to replacement of the equipment.

The high-power density of the compact, intelligent, MIMO AC–AC, AC–DC, AC–DC–AC, DC–DC, DC–AC, and/or DC–AC–DC commutator with an integral press pack ECM, and its unique matrix-built, mechanical construction, also make them ideal for new designs not only in MCM AC–AC and AC–DC/DC–AC commutator dynamotors but also in green power applications such as normally used matrix converters for wind and/or wave power generation in the high 'megawatt' range.

The novel ECM AC–AC, AC–DC, AC–DC–AC, DC–DC, DC–AC, and/or DC–AC–DC commutator dynamotors with an intelligent self-protecting compact, MIMO ECMs offer size and mass reduction and improved reliability, among others benefits.

The ECM AC–AC, AC–DC, AC–DC–AC, DC–DC, DC–AC, and/or DC–AC–DC commutator dynamotors are highly integrated and provide key functions that include over-current, over-temperature, short-circuit protections and supply **ultra-violet** (UV) lock functions.

The novel ECM AC–AC, AC–DC, AC–DC–AC, DC–DC, DC–AC, and/or DC–AC–DC commutator dynamotors are high-performance MMD electrical machines with built-in current-sense electrical valves in the ECM stage, which is housed in a thin and compact matrix-built package, thanks to their insulated base-plate with high thermal conductivity. They are the first ECMs capable of true inaudible (> 20 kHz) operation.

The high reliability and exceptional thermal ruggedness, are also well suited to the demands of high-power ECM AC–AC, AC–DC, AC–DC–AC, DC–DC, DC–AC, and/or DC–AC–DC commutator dynamotors for industrial and traction propulsion mechatronical control systems, e.g. both for high rail and main line applications such as high-speed **electrical multiple units** (EMU). For instance, full-diffusion technology with loose mono- and/or poly-silicon wafer may be used for manufacturing **MOS controlled thyristor** (MCT) type intelligent, MIMO ECM commutators. Up-to-now, neutron-doped silicon of the so-calle*d 'float-zone' typ*e is mainly used. Neutron doping means that the silicon rod is irradiated with neutrons in a nuclear reactor prior is being sliced into wafers. A number of silicon atoms are then converted into phosphorus atoms and the material becomes weakly *n*-doped.

The major advantage, however, is that the silicon rod after heat treatment will have a very homogeneous resistivity. This is valid for achieving a high-voltage capability of the ECMs. To obtain an acceptable yield in the ECM manufacturing, i.e. the number of MCT type ECMs fulfilling the specification in relation to the number of MCT type ECMs manufactured, extreme cleanness requirements are made on the environment.

Diffusion technology means that the silicon wafer is coated with gallium, aluminium or phosphorus. These substances are then allowed to diffuse into the silicon wafer in diffusion furnaces under high temperature. In certain ECM manufacturing stages only parts of the silicon wafer surface are coated for subsequent diffusion. For this purpose, the photo-lithography is utilised.

After the completion of the diffusion and the coating of an aluminium contact, the silicon wafer is sliced into ECM chips with the aid of a high-power laser.

Depending on the silicon wafer area of the chip, the intelligent MIMO ECMs, e.g. for AC–AC commutator dynamotors (generators/motors) incorporated 3 × 3 or 3 × 4 ECMs, and AC–DC/DC–AC commutator dynamotors incorporated 3 × 2/2 × 3 ECMs. By controlling the lifetime of the charge carriers, it is possible to determine the position of the MCT type ECM on the recovery charge—minimum on-state voltage drop curve. For this purpose, the gold irradiation is applied, which gives considerably closer tolerances. With electron irradiation, the ECM components are bombarded with high-energy electrons (> 10 MeV) from electron guns.

On the basis of the large ECM commutator (formerly known as a static converter) manufacturing volume of primarily diode or thyristor type MIMO ECMs for AC–AC, AC–DC, AC–DC–AC, DC–DC, DC–AC, and/or DC–AC–

DC commutators that occurred in the 20th century, it has been possible to introduce more reliable dimensioning criteria. In combination with closer MIMO ECM manufacturing tolerances, this has substantially raised above all the maximum reverse voltage/off-state voltage. Lowering the on-state voltage drop, as a consequence reducing the on-state charge and consequently high-power losses in protective circuits around the intelligent, MIMO ECM AC–AC, AC–DC, AC–DC–AC, DC–DC, DC–AC, and/or DC–AC–DC commutators.

Recently, the authors have also worked out a concept of the ECM commutator with the bipolar ECM, realised on super- and semiconductor hybrid 'continuous' bipolar electrical valves, based on the use of heterojunctions, i.e. a contact of amorphous and mono- and/or polycrystalline materials. The properties of these heterojunctions seem to be for various reasons competitive on relation to structures made from mono- and/or polycrystalline semiconductors.

A concept of the bipolar ECM, realised on 'continuous' bipolar power transistors of *n–p–n* type, in which the emitters will be made of chalcogenides glass and the bases—of the monocrystalline silicon; or 'continuous' bipolar hetero-junction and dielectrically insulated-gate in field effect transistors is being worked out.

As well, a concept of bipolar ECM, realised on 'continuous' bipolar power **superconductor-base hot electron transistors** (SUPER-HET) on the base of GaAs/Nab/Ins; 'continuous' bipolar power (QUIT); or 'continuous' bipolar power and very high-speed **static induction transistors** (SIT) and/or graphene **single electron transistors** (SET) with low-power-delay product and very high transconductance.

The bipolar ECMs of that type can be used, owing to specific conditions of the electrical-current super- and/or semiconductivity, as basic trivalent-logic functions of the ECM AC–AC, AC–DC, AC–DC–AC, DC–DC, DC–AC, and/or DC–AC–DC commutator, which seem to be the simplest and most effective static converter for conversion of one kind of electrical energy into another one. This trivalent-logic is based on the Lukasiewicz–Tarski algebra.

Semi- and/or superconductive ECM AC–AC, AC–DC, AC–DC–AC, DC–DC, DC–AC, and/or DC–AC–DC commutators, nowadays, are particularly indispensable in modern EM actuator engineering, and for energy converting purposes, mainly in **fly-by-wire** (FBW) aircraft and **ride-by-wire** (RBW) or **x-by-wire** (XBW) automotive vehicles. Compared with previous polysectorial (polysegmental) MCM commutators and polyvalvular (sliding-contacts) ECM AC–AC, AC–DC, AC–DC–AC, DC–DC, DC–AC, and/or DC–AC–DC commutator (static converters) they are more practice-oriented, especially from the point of view of operating convenience, space requirements of ultra-modern super- and/or semiconductor ECM techniques.

Numerous semi- and/or superconductive ECM AC–AC, AC–DC, AC–DC–AC, DC–DC, DC–AC, and/or DC–AC–DC commutators have been developed for modern MMD electrical machines and power applications. When selecting these commutators various criteria have to be borne in mind, i.e. the cost of the semi- and/or superconductive ECM AC–AC, AC–DC, AC–DC–AC, DC–DC, DC–AC, and/or DC–AC–DC commutator, the driver requirements, switching and network, mounting and wiring needs, together with the ECM commutators electrical characteristics and performance. Repeatedly, the semi- and/or superconductive ECM AC–AC, AC–DC, AC–DC–AC, DC–DC, DC–AC, and/or DC–AC–DC commutator's cost, when compared to that of the non-MCM wiring requirements, has been and will continue to reduce significantly.

The development aims of semi- and/or superconductive ECM AC–AC, AC–DC, AC–DC–AC, DC–DC, DC–AC, and/or DC–AC–DC commutators, are to manufacture ECM commutators that have uncomplicated drivers, a reduction in circuitry and a simplified method of the ECM's wiring and mounting, with realistic processes.

Newly developed insulated ECMs meet the requirements for modern MMD electrical machines and power applications, namely:

Insulated ECMs reduce mounting costs

- ECM—wiring through printed circuit;
- ECM wiring using copper bus-bars;
- a common heat-sink for ECM's all components.

Insulated ECMs limit RFI phenomena

- extremely low parasitic capacitance;
- the heat-sink is grounded to the ECM's shield and chassis.

Insulated ECMs increase safety

- internal insulation 2.5 kV (50 Hz, 1 min);
- safer construction, no mica insulation as is needed for MCMs;
- very low thermal resistance between junction and case of electrical valves.

The semi- and/or superconductive ECM AC–AC, AC–DC, AC–DC–AC, DC–DC, DC–AC, and/ or DC–AC–DC commutators for modern MMD electrical machines and power applications, are specifically designed to reduce the number of components, especially electrical valves.

In the case of very high-power semi- and/or superconductive ECM AC–AC, AC–DC, AC–DC–AC, DC–DC, DC–AC, and/or DC–AC–DC commutators for modern MMD electrical machines and power applications, the use of 'continuous' electrical valves, permits a reduction in parasitical inductance and an improvement in cooling characteristics.

The semi- and/or superconductive ECM AC–AC, AC–DC, AC–DC–AC, DC–DC, DC–AC, and/or DC–AC–DC commutators for modern MMD electrical machines and power applications operate at high-current values and, due to their high-current gain, can be controlled by low-power gate or base drivers. Their switching safe operating areas, namely **reverse-biased safe-operating area** (RBSOA) allows repeatable operation without use of snubbers.

For high-performance ECM AC–AC, AC–DC, AC–DC–AC, DC–DC, DC–AC, and/or DC–AC–DC commutator dynamotors, ECM-mode power supplies, storage battery chargers and welding equipment, the main requirements are:

- noiseless operation;
- large band-pass;
- a reduction in the size and mass of the passive components.

The result is the choice of switching frequencies $f \geqslant 15$ kHz or even more for amorphous electrical valves, e.g. MOSFETs and ovonics.

The ECM AC–AC, AC–DC, AC–DC–AC, DC–DC, DC–AC, and/or DC–AC–DC commutator has a number of advantages over conventional power converter technologies when used on an AC bus system. As higher temperature power semi- or superconductive electrical valves, such as silicon carbide ones, become available, the technology will become a good candidate for power conversion in applications where volume and mass are at a premium.

It should be noted that power density of ECMs (including auxiliaries, switchgear, cooling system) has improved even more dramatically over the past 40 years. This can be explained by the fact that electronic commutators (macrocommutators) are a more recent technology development having more opportunities to improve in several areas.

Key to improving power density was the development of improved (lossless) turn-off electrical valves (electronic switches), improved heat-sink technology, improved control minimising losses, compact microelectronic controller (microcontroller) design and better design tools which allow the E–E converter designer to 'push for the limits' without compromising the life of the ECM. However, in the authors' opinion, deploying voltage source topologies (instead of current source topologies) and improving design and technology of passive components, in particular of capacitors, had an equally strong impact on power density of ECMs over the past five decades.

Over the past 50 years volumetric power density of industrial air-cooled ECMs improved from 50 up to 500 kVA m^{-3}. The most important limiting factor in designing high-power density ECM modules is thermal, i.e. maximum operating temperature and thermal cycle life of electrical valves (switches) and their packages. A significant difference can be noticed between disc type ECM modules and plastic ECM modules with bond wires. Hence, ECM module density depends greatly on the specific losses of the ECM modules (less losses requires less heatsinks), coolant temperature and operating conditions.

A.5 MCM DC–AC/AC–DC commutators

In 1831, the great experimentalist Michael Faraday discovered that changes in a magnetic field could induce an electromotive force and current in a nearby circuit—a phenomenon known as electromagnetic induction. MCM DC–AC/AC–DC commutators are used in DC electrical machines: M-E generators and many E-M motors as well as universal E-M motors.

In a DC–AC commutator synchronous motor (formerly known as a DC motor) the MCM DC–AC commutator applies electric current to the windings. By reversing the sense of current direction in the rotating windings each half turn, a steady rotating force (torque) is produced.

In an AC–DC commutator synchronous generator (formerly known as a DC generator) the MCM AC–DC commutator picks off the current generated in the windings, reversing the sense of current direction with each half turn, serving as a mechanical AC–DC rectifier to convert the alternating current from the windings to unilateral (unipolar) direct current in the external load circuit. The first MCM AC–DC commutator-type DC electrical machine, the DC dynamo, was built by French instrument maker Hippolyte Pixii in 1832, based on a suggestion by André-Marie Ampère. The 'Gramme machine' or 'Gramme dynamo', an AC–DC commutator synchronous generator, was designed by Belgian inventor Zénobe Gramme in 1871. It was the first **magneto-mechano-dynamical** (MMD) electrical machine to use industrially generated DC electrical energy. Gramme's innovation was to use many armature windings, wound on a doughnut shaped armature, and commutated (switched) with a many segmented mechanical commutator, to rectify (smooth) the output AC waveforms, generating nearly constant DC electrical energy. A hand-cranked physical model from the 1870s designed for use in laboratories used laminated permanent magnets invented by Jamin (Fontaine 1878).

The phenomenon of unipolar electrical-current conductivity was already applied in 1883, when English physicist W Ritchie constructed an MMD electrical machine, i.e. an AC–DC/DC–AC commutator dynamotor with a rotary **mechanical-commutation-matrixer** (MCM) commutator, i.e. an MCM AC–DC/DC–AC commutator, which can be taken for the prototype of a contemporary split-ring or flat MCM AC–DC/DC–AC commutator.

Up-to-now, characteristic components common to a large group of DC electrical machines are the MCM AC–AC and AC–DC/DC–AC commutators. The MCM AC–AC and AC–DC/DC–AC commutators (figure A9) are conventionally made by assembling hard-drawn copper sectors (bars or segments) interleaved with 0.7–2.0 mm sheet mica or micanite, these separators being 'undercut' by about 1 mm (Fijalkowski 1973, 2016).

The stationary or movable brushes of a suitable carbon/graphite content, are mounted in boxes with spring loading to hold them against the MCM commutator surface with a medium to strong pressure depending on the application. The circumferential brush width is typically 2–3 sector width (10–20 mm) and about 30 mm axially. One brush-arm per pole is employed except for certain four-pole wave-wound MCM AC–DC/DC–AC commutator dynamotors (generator/motors) which have two brush-arms in adjacent positions to facilitate maintenance.

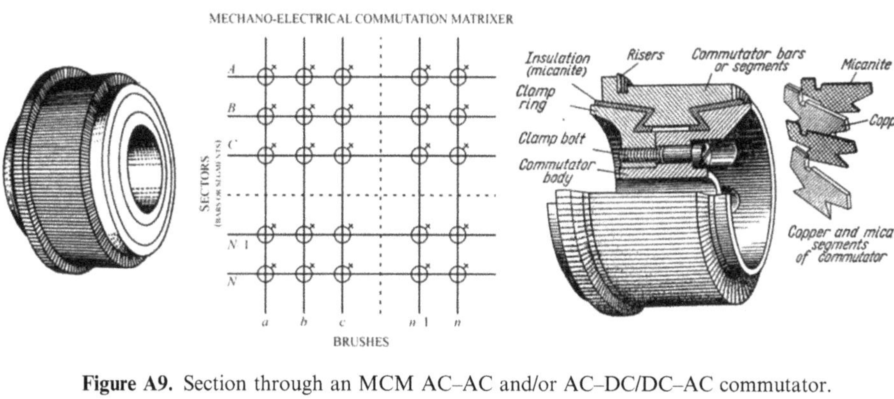

Figure A9. Section through an MCM AC–AC and/or AC–DC/DC–AC commutator.

The MCM AC–DC/DC–AC commutator is a necessary part of a DC commutator electrical machine; besides the use of an MCM AC–AC commutator made possible AC commutator motors that have certain valuable features. Finally, MCM AC–DC/DC–AC commutators are used in single-armature converters, the function of which is to convert AC–DC or vice versa DC–AC. Thus, the first MCM with 'mechanical contact' electrical valves have became an MCM connection of sliding-contacts, setting up, for example, collector phosphorus-bronze springs or carbon/graphite brushes, sliding respectively on the surface of the collector split-rings or hard-drawn copper sectors (bars or segments).

Rotary sliding-contact, mechanical AC–DC commutators, by the authors termed 'MIMO MCM DC–AC commutators' (figures A10(a)–A10(e)), which are widely used in MCM AC–DC commutator generators, according to the dynamical systems approach physical heterogeneous continuous dynamical hypersystems, that is, MMD electrical machines are a series of copper sectors (bars or segments), i.e. insulated from one another and forming matrixer-input primary conductive collectors of an MCM AC–DC commutators. The copper sectors (bars or segments), i.e. matrixer-input primary conductive collectors are connected to the commutating sections, i.e. phase winding terminals of an armature winding mounted on the rotor of electrical machine.

Stationary carbon brushes are placed on the MCM DC–AC commutator cylindrical surface with copper sectors.

Carbon brushes forming matrixer-output secondary conductive collectors of an MCM commutator dynamical system are connected by means of sliding-contacts, that is, electrical valves (copper-sector/carbon-brush contacts) to the copper sectors, i.e. matrixer-input primary conductive collectors of an MCM DC–AC commutator.

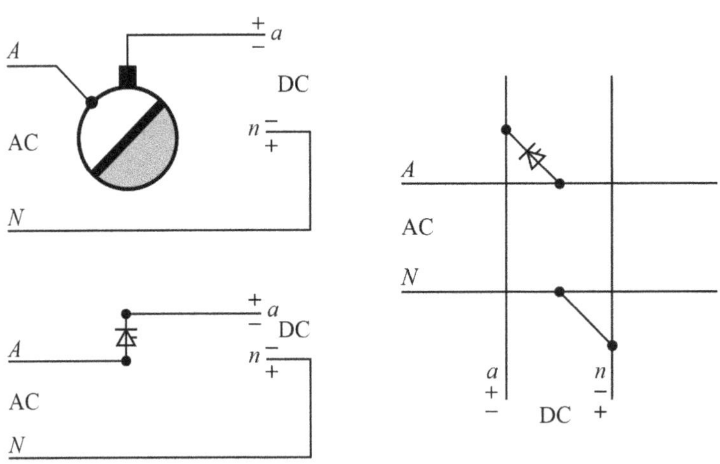

a)

Figure A10a. The single-phase, half-wave MCM AC–DC commutator, with a single homogeneous cathode commutation group and a neutral, acting as a one-way MCM AC–DC/DC–AC rectifier/inverter.

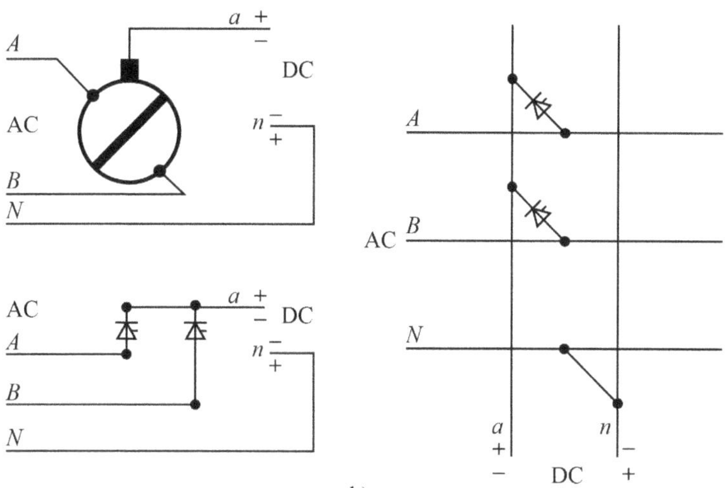

b)

Figure A10b. The two-phase, half-wave MCM AC–DC commutator, with a single homogeneous cathode commutation group and a neutral, acting as a one-way MCM AC–DC/DC–AC rectifier/inverter.

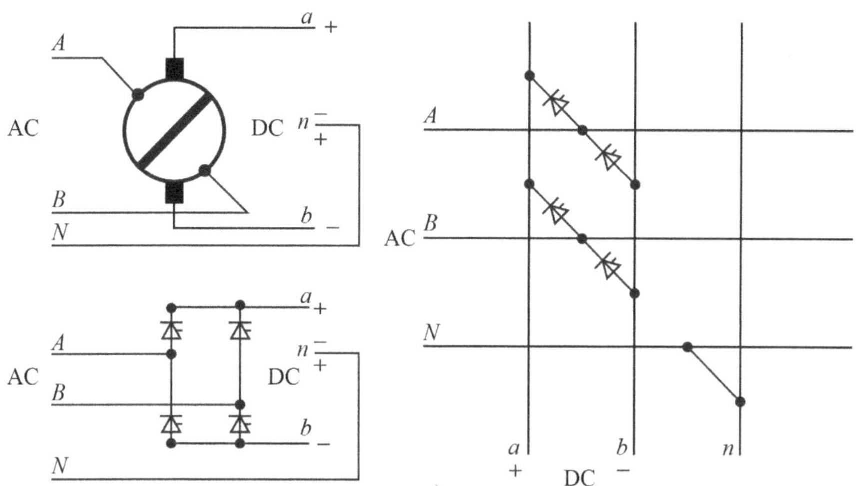

c)

Figure A10c. The two-phase, full-wave MCM AC–DC commutator, with double heterogeneous cathode and anode commutation groups and a neutral, acting as a double-way MCM AC–DC/DC–AC rectifier/inverter.

By means of these sliding-contacts, acting as electrical valves it is possible to connect the phase windings of the rotary armature winding in the electrical machine to an external electrical-commutation-matrixer dynamical system.

The operating principle of the MIMO ECM AC–DC/DC–AC commutator will be considered in more detail later on. The MIMO MCM AC–DC/DC–AC commutator was in the past and is still now a necessary component of yesterday

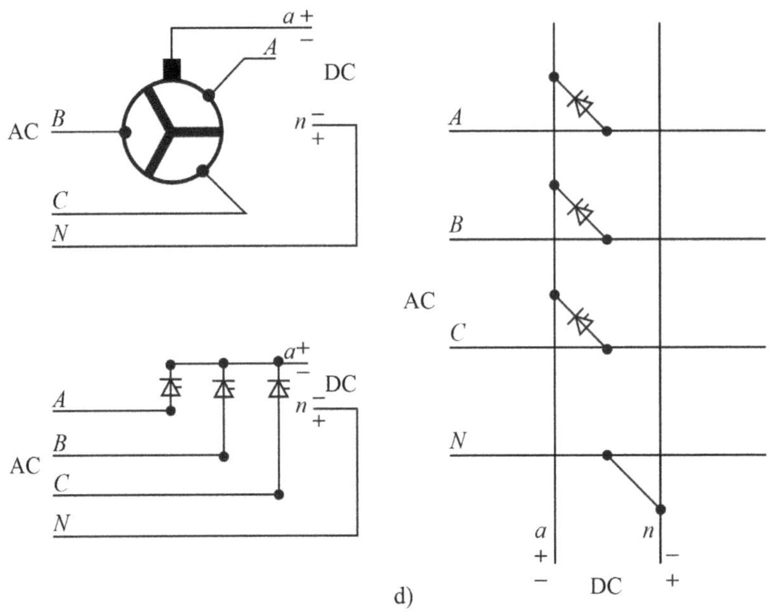

Figure A10d. The three-phase, half-wave MCM AC–DC commutator, with a single homogeneous cathode commutation group and a neutral, acting as a one-way MCM AC–DC/DC–AC rectifier/inverter.

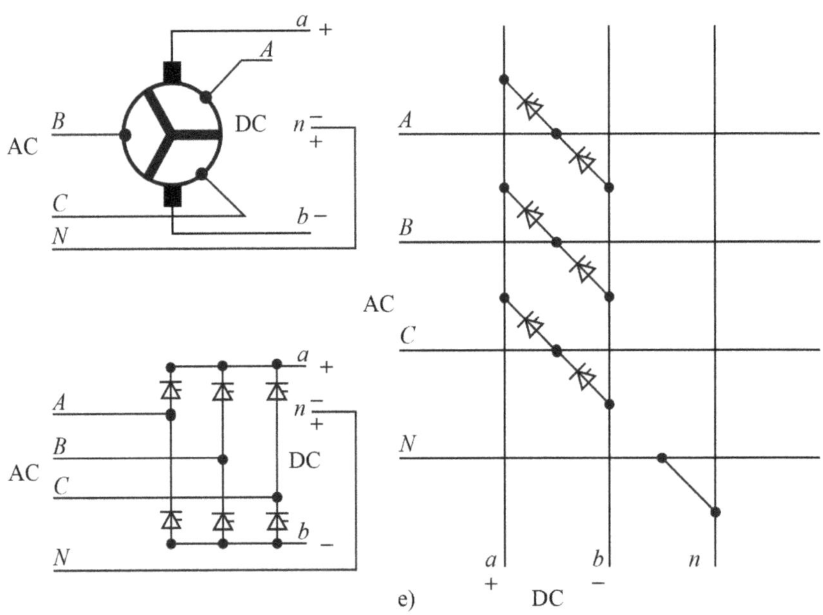

Figure A10e. The three-phase, full-wave MCM AC–DC commutator, with double heterogeneous cathode and anode commutation groups and a neutral, acting as a double-way MCM AC–DC/DC–AC rectifier/inverter.

and today's AC–DC/DC–AC commutator dynamotors. Besides the use of an MIMO MCM AC–AC and/or AC–DC/DC–AC commutator made possible yesterday's and today's polyphase MCM AC–AC and/or AC–DC/DC–AC commutator electrical machines.

Finally, MIMO MCM AC–DC commutators are used in yesterday's and today's single-armature MCM AC–DC commutator rotary converters, the function of which is to convert three-phase alternating voltages to direct voltage, or vice versa.

On the other hand, MCM AC–AC and AC–DC/DC–AC commutator dynamotors are very expensive to manufacture compared with AC and DC commutatorless machines, demand special care, and are not so rugged.

MCM AC–AC and AC–DC/DC–AC commutator dynamotors are not as maintenance-free as AC and DC commutatorless dynamotors, due to brush and MCM commutator wear.

The roundness of the MIMO MCM AC–AC and AC–DC/DC–AC commutator is the most important design parameter to reduce the brush noise of the MCM AC–DC/DC–AC commutator dynamotor. The roundness of the MIMO MCM AC–AC or AC–DC/DC–AC commutator needs to be maintained in order to maintain the brush noise under a certain level throughout the durability life span of the MCM AC–AC and AC–DC/DC–AC commutator dynamotor.

AC and DC commutatorless electrical machines do not permit economical and smooth controls of voltage and current (for MMD generating) or torque, and speed control (for MMD motoring) of this kind is necessary for some forms of industrial and traction propulsion complex dynamical hypersystems.

The use of integral MCM commutator in MCM AC–AC and AC–DC/DC–AC commutator dynamotors makes it possible to produce variable-voltage and variable-current or variable-torque and variable-speed dynamotors with a flat or drooping voltage-current and torque-speed characteristics and with power factor in some of them brought to unity.

All these useful characteristics of MCM AC–AC and AC–DC/DC–AC commutator dynamotors are achieved at the expense of complexity and extra costs of the MMD electrical machine, with reduced reliability, and worse operating conditions.

The main difficulty in building of MCM AC–AC and AC–DC/DC–AC commutator dynamotors is to obtain satisfactory ME commutation of alternating and direct currents, especially under starting conditions.

In order to make a MCM AC–DC/DC–AC commutator dynamotor suitable for AC supply it is necessary to adapt its magnetic circuit to an alternating magnetic field, to reduce the reactance of the windings to a minimum, and finally, to improve the conditions of contacted commutation in MCM AC–AC/DC–AC commutator dynamotors, the first two requirements may be considered as suitable fulfilled true at the expense of considerable complexity and increased cost of the MMD electrical machine.

As for securing good commutation conditions, nothing that has been done in this sense of a direction satisfies scientists, who consider the problem as not properly resolved and are looking for a better solution.

An MCM DC–AC/AC–DC commutator is an MCM of bars or segments (MCM rows) so connected to armature winding coils of a DC motor or DC generator that rotation of the armature will in conjunction with fixed brushes (MCM columns) result in unilateral (unipolar) current output in the case of a DC generator and in the reversal of the current into the winding coils in the case of a DC motor.

In the other words, the MCM DC–AC/AC–DC commutator is a moving part of a rotary ECM in certain types of DC electrical machines (DC motors and DC generators) that periodically reverses the sense of current direction between the rotor and the external circuit. It consists of a cylinder composed of multiple metal electrical-contact collectors termed 'bars or segments' (MCM rows) on the rotating armature of the electrical machine.

Two or more electrical-contact collectors termed 'brushes' (MCM columns) made of a soft conductive material like carbon press against the MCM DC–AC/AC–DC commutator, making sliding electrical-contact with successive bars or segments of the MCM DC–AC/AC–DC commutator as it rotates. The windings (coils of wire) on the armature are connected to the MCM DC–AC/AC–DC commutator bars or segments.

MCM DC–AC/AC–DC commutators are relatively inefficient, and also require periodic maintenance such as brush replacement. Therefore, MCM DC–AC/AC–DC commutated electrical machines are declining in use, being replaced by ECM AC–AC or AC–DC–AC commutated electrical machines, and in recent years by DC–AC commutator synchronous or asynchronous (induction) motors which use ECM DC–AC commutators.

Principle of operations—An MCM DC–AC/AC–DC commutator consists of a set of electrical-contact bars or segments (MCM rows) fixed to the rotating shaft of an electrical machine, and connected to the armature windings. As the shaft rotates, the MCM DC–AC/AC–DC commutator reverses the flow of current in a winding.

For a single-phase armature winding (figure A11), when the shaft has made one half complete turn, the winding is now connected so that current flows through it in the opposite of the initial sense of direction.

In a DC–AC commutator synchronous motor (DC motor), the armature current causes the fixed magnetic field to exert a rotational force, or a torque, on the winding to make it turn. In an AC–DC commutator synchronous generator (DC generator), the mechanical torque applied to the shaft maintains the motion of the armature winding through the stationary magnetic field, inducing a current in the winding.

In both the EM motor and ME generator case, the MCM DC–AC/AC–DC commutator periodically reverses the sense of direction of current flow through the winding so that current flow in the circuit external to the electrical machine continues in only one sense of direction.

Basic MCM DC–AC/AC–DC commutators—have at least three electrical-contact bars or segments (MCM rows), to prevent a 'dead' spot where two brushes (MCM columns) simultaneously bridge only two MCM DC–AC/AC–DC commutator bars or segments.

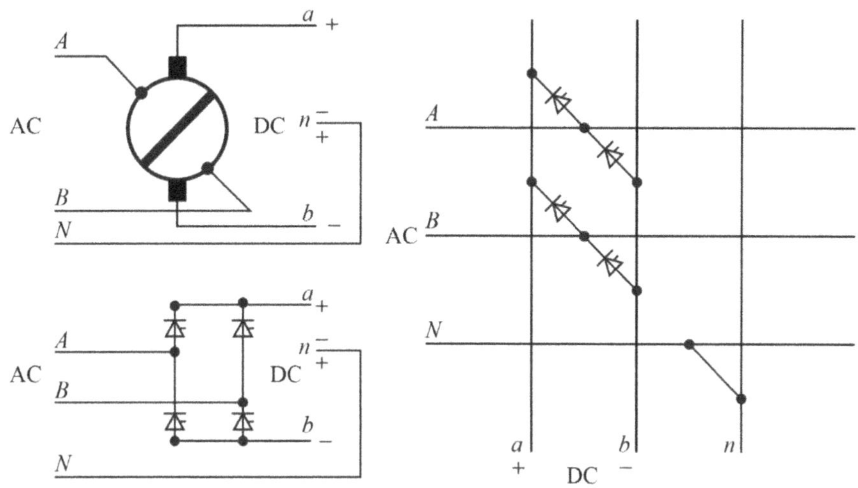

Figure A11. The single-phase, full-wave MCM AC–DC commutator, with double heterogeneous cathode and anode commutation groups and a neutral, acting as a double-way MCM AC–DC/DC–AC rectifier/inverter.

Brushes are made wider than the insulated gap, to ensure that brushes are always in contact with an armature winding coil. For MCM DC–AC/AC–DC commutators with at least three bars or segments, although the rotor can potentially stop in a position where two MCM DC–AC/AC–DC commutator bars or segments touch one brush, this only de-energises one of the rotor arms while the others will still function correctly.

With the remaining rotor arms, a DC–AC commutator synchronous motor (DC motor) can produce sufficient torque to begin spinning the rotor, and an AC–DC commutator synchronous generator (DC generator) can provide useful power to an external circuit.

Ring/sector (bar or segment) construction—A MCM DC–AC/AC–DC commutator consists of a set of copper bars or segments (MCM rows), fixed around the part of the circumference of the rotating electrical machine, or the rotor, and a set of spring-loaded carbon brushes (MCM columns) fixed to the stationary frame of the electrical machine.

MCM DC–AC/AC–DC commutator bars or segments are connected to the winding coils of the armature, with the number of winding coils (and MCM DC–AC/AC–DC commutator bars or segments) depending on the angular velocity and voltage of the electrical machine.

Large DC–AC commutator synchronous motors (DC motors) may have hundreds of bars or segments. Each conducting bar or segment of the MCM DC–AC/AC–DC commutator is insulated from adjacent bars or segments. Mica was used on early electrical machines and is still used on large ones. Many other insulating materials are used to insulate smaller electrical machines; plastics allow quick manufacture of an insulator, for example. The bars or segments are held onto the shaft using a dovetail shape on the edges or underside of each bar or segment.

Insulating wedges around the perimeter of each bar or segment are pressed so that the MCM DC–AC/AC–DC commutator maintains its mechanical stability throughout its normal operating range. In small appliance and tool DC motors the bars or segments are typically crimped permanently in place and cannot be removed. When the DC motor fails it is discarded and replaced.

On large industrial electrical machines (e.g. from several kilowatts to thousands of kilowatts in rating) it is economical to replace individual damaged bars or segments, and so the end-wedge can be unscrewed and individual bars or segments removed and replaced. Replacing the copper and mica bars or segments is commonly referred to as 'refilling'. Refillable dovetailed MCM DC–AC/AC–DC commutators are the most common construction of larger industrial type MCM DC–AC/AC–DC commutators, but refillable MCM DC–AC/AC–DC commutators may also be constructed using external bands made of fibre-glass (glass banded construction) or forged steel rings (external steel shrink ring type construction and internal steel shrink ring type construction). Disposable, moulded type MCM DC–AC/AC–DC commutators commonly found in smaller DC motors are becoming increasingly more common in larger EM motors.

In addition to the commonly used heat, torque, and tonnage methods of seasoning MCM DC–AC/AC–DC commutators, some high-performance MCM DC–AC/AC–DC commutator applications require a more expensive, specific 'spin seasoning' process or over-speed spin-testing to guarantee stability of the individual bars or segments and prevent premature wear of the carbon brushes. Such requirements are common with aerospace, automotive, traction, mining, metallurgy, nuclear, military and high-speed applications where premature failure can lead to serious negative consequences.

Friction between the bars or segments and the brushes eventually causes wear to both surfaces. Carbon brushes, being made of a softer material, wear faster and may be designed to be replaced easily without dismantling the electrical machine.

Older copper brushes caused more wear to the MCM DC–AC/AC–DC commutator, causing deep grooving and notching of the surface over time. The MCM DC–AC/AC–DC commutator on small DC motors (e.g. less than a kilowatt rating) is not designed to be repaired through the life of the device. On large industrial equipment, the MCM DC–AC/AC–DC commutator may be resurfaced with abrasives, or the rotor may be removed from the frame, mounted in a large metal lathe, and the MCM DC–AC/AC–DC commutator resurfaced by cutting it down to a smaller diameter. The largest of equipment can include a lathe turning attachment directly over the MCM DC–AC/AC–DC commutator.

Brush construction—Early electrical machines used brushes made from strands of copper wire to contact the surface of the MCM DC–AC/AC–DC commutator. However, these hard metal brushes tended to scratch and groove the smooth MCM DC–AC/AC–DC commutator bars or segments, eventually requiring resurfacing of the MCM DC–AC/AC–DC commutator. As the copper brushes wore away, the dust and pieces of the brush could wedge between MCM DC–AC/AC–DC commutator bars or segments, shorting them and reducing the efficiency of the device. Fine copper wire mesh or gauze provided better surface contact with less bar or segment wear, but gauze brushes were more expensive than strip or wire copper brushes.

Emerging and future rotating MMD electrical machines with MCM DC–AC/AC–DC commutators almost exclusively use carbon brushes, which may have copper powder mixed in to improve conductivity.

Metallic copper brushes can be found in toy or very small E-M motors, and some E-M motors which only operate very intermittently, such as automotive starter DC motors.

DC motors and DC generators suffer from a phenomenon known as 'armature reaction', one of the effects of which is to change the position at which the current reversal through the windings should ideally take place as the loading varies.

Early rotating MMD electrical machines had the brushes mounted on a ring that was provided with a handle. During operation, it was necessary to adjust the position of the brush ring to adjust the commutation to minimise the sparking at the brushes. This process was known as 'rocking the brushes'.

Various developments took place to automate the process of adjusting the commutation and minimising the sparking at the brushes. One of these was the development of 'high resistance brushes', or brushes made from a mixture of copper powder and carbon. Although described as high resistance brushes, the resistance of such a brush was of the order of milliohms, the exact value dependent on the size and function of the MMD electrical machine. Also, the high resistance brush was not constructed like a brush but in the form of a carbon block with a curved face to match the shape of the MCM DC–AC/AC–DC commutator.

The high resistance or carbon brush is made large enough that it is significantly wider than the insulating bar or segment that it spans (and on large MMD electrical machines may often span two insulating bars or segments). The result of this is that as the MCM DC–AC/AC–DC commutator bar or segment passes from under the carbon brush, the current passing to it ramps down more smoothly than had been the case with pure copper brushes where the contact broke suddenly.

Similarly the bar or segment coming into contact with the carbon brush has a similar ramping up of the current. Thus, although the current passing through the carbon brush was more or less constant, the instantaneous current passing to the two MCM DC–AC/AC–DC commutator bars or segments was proportional to the relative area in contact with the carbon brush.

The introduction of the carbon brush had convenient side effects. Carbon brushes tend to wear more evenly than copper brushes, and the soft carbon causes far less damage to the MCM DC–AC/AC–DC commutator bars or segments. There is less sparking with carbon as compared to copper, and as the carbon wears away, the higher resistance of carbon results in fewer problems from the dust collecting on the MCM DC–AC/AC–DC commutator bars or segments.

The ratio of copper to carbon can be changed for a particular purpose. Brushes with higher copper content perform better with very low voltages and high current, while brushes with a higher carbon content are better for high voltage and low current. High copper content brushes normally carry 60–80 A cm^{-2} of contact surface, while higher carbon content only carries 15–30 A cm^{-2}. The higher resistance of carbon also results in a greater voltage drop of 0.8–1.0 V per electrical-contact, or 1.6–2.0 V across the MCM DC–AC/AC–DC commutator.

Brush holders—A spring is typically used with the carbon brush, to maintain constant electrical-contact with the MCM DC–AC/AC–DC commutator. As the carbon brush and MCM DC–AC/AC–DC commutator wear down, the spring steadily pushes the carbon brush down-wards towards the MCM DC–AC/AC–DC commutator. Eventually the carbon brush wears small and thin enough that steady electrical-contact is no longer possible or it is no longer securely held in the carbon-brush holder, and so the carbon brush must be replaced. It is common for a flexible power cable to be directly attached to the carbon brush, because current flowing through the support spring would cause heating, which may lead to a loss of metal temper and a loss of the spring tension. When a DC–AC/AC–DC commutator synchronous motor (DC motor) uses more power than a single carbon brush is capable of conducting, an assembly of several carbon-brush holders is mounted in parallel across the surface of the very large MCM DC–AC/AC–DC commutator. This parallel holder distributes current evenly across all the carbon brushes, and permits a careful operator to remove a bad carbon brush and replace it with a new one, even as the electrical machine continues to spin fully powered and under load. High-power, high-current MCM AC–DC commutators are now uncommon, due to the less complex design of AC–DC commutator synchronous generators (AC generators) that permits a low-current, high-voltage spinning magnetic-field winding coil to energise high-current fixed-position stator winding coils. This permits the use of very small singular carbon brushes in the AC generator (alternator) design. In this instance, the rotating contacts are continuous rings, termed split-rings, and no commutation (switching) happens.

Emerging and future MCM DC–AC/AC–DC commutators using carbon brushes usually have a maintenance-free design that requires no adjustment throughout the life of the MCM DC–AC/AC–DC commutator, using a fixed-position carbon-brush holder slot and a combined brush-spring-cable assembly that fits into the slot. The worn carbon brush is pulled out and a new carbon brush inserted.

Limitations and alternatives—The resistance of the MCM DC–AC/AC–DC commutator contacts causes inefficiency in low-voltage, high-current MMD electrical machines like this, requiring a huge elaborate MCM DC–AC/AC–DC commutator dynamo from the late 1800s for electroplating generated 7 V at 310 A.

Although DC motors and DC dynamos once dominated industry, the disadvantages of the MCM DC–AC/AC–DC commutator have caused a decline in the use of commutated electrical machines in the last century.

These disadvantages are:

- Due to friction, the brushes and copper MCM DC–AC/AC–DC commutator bars or segments wear out. In small consumer products such as power tools and appliances the brushes may last as long as the product, but larger electrical machines require regular replacement of brushes and occasional resurfacing of the MCM DC–AC/AC–DC commutator. So commutated

electrical machines are not used in equipment at remote locations that must operate for long periods without maintenance.
- The resistance of the sliding contact between brush and MCM DC–AC/AC–DC commutator causes a voltage drop called the 'brush drop'. This may be several volts, so it can cause large power losses in low-voltage, high-current electrical machines. The friction of the brush against the MCM DC–AC/AC–DC commutator also absorbs some of the energy of the electrical machine. AC–AC, AC–DC–AC or DC–AC/AC–DC commutator synchronous or asynchronous (induction) motors, with the ECM AC–AC, AC–DC–AC or DC–AC/AC–DC commutators instead of use MCM DC–AC/AC–DC commutators, are much more efficient.
- There is a limit to the maximum current density and voltage which can be commutated (switched) with a MCM DC–AC/AC–DC commutator. Very large DC electrical machines, say, more than several mega-watts rating, cannot be built with MCM DC–AC/AC–DC commutators. The largest DC motors and DC generators are really all AC electrical machines.
- The commutation (switching) action of the MCM DC–AC/AC–DC commutator can cause sparking at the contacts, generating electromagnetic interference.

With the wide availability of alternating current, DC motors have been replaced by more efficient AC–AC or AC–DC–AC synchronous or asynchronous (induction) motors.

In recent years, with the widespread availability of power semiconductors, in many remaining applications commutated DC motors have been replaced with ECM DC–AC synchronous or asynchronous (induction) motors. These have an ECM DC–AC commutator; instead the sense of current direction is commutated (switched) electronically. A sensor keeps track of the rotor position and electrical valves (electronic switches), e.g. such as **reverse-blocking insulated-gate bipolar transistors** (RB IGBT) reverse the current. Operating life of these electrical machines is much longer, limited mainly by bearing wear.

Commutation in DC electrical machines—The voltage generated in the armature, placed in a rotating magnetic field, of a DC generator is alternating in nature. The commutation in DC electrical machines or more specifically commutation in AC–DC commutator synchronous generator is the process in which generated alternating current in the armature winding of a DC electrical machine is converted (rectified) into direct current and the stationary carbon brushes after going through the ring or disc MCM DC–AC/AC–DC commutator's copper bars or segments MCM rows or MCM columns). Again in DC–AC commutator synchronous motor, the input DC is to be converted (inverted) in AC form in armature and that is also done through commutation (figure A14). This transformation of current from the rotating armature of a DC electrical machine to the stationary carbon brushes (MCM columns) needs to maintain continuously moving contact between the ring or disc MCM DC–AC/AC–DC commutator's copper bars or segments and the carbon brushes.

Again in DC–AC commutator synchronous motors, the input DC is to be converted (inverted) in AC form in armature and that is also done through commutation (figure A12). This transformation of current from the rotating armature of a DC electrical machine to the stationary carbon brushes (MCM columns) needs to maintain continuously moving contact between the ring or disc MCM DC–AC/AC–DC commutator's copper bars or segments and the carbon brushes.

When the armature (rotor) starts to rotate, then the armature-winding coils situated under one pole (let it be N magnetic pole) rotates between a positive carbon brush and its consecutive negative carbon brush and the current flows. Then the armature-winding coil is short circuited with the help of a carbon brush for a very short fraction of time (1/500 s). It is termed 'commutation period'. After this short-circuit time the armature-winding coils rotates under S magnetic pole and rotates between a negative carbon brush and its succeeding positive carbon brush. Then the sense of current direction is reversed which is in the direction away from the MCM DC–AC/AC–DC commutator's copper bars or segments. This phenomena of the reversal of current is termed the 'commutation processes'. One gets direct current from the carbon-brush terminal.

The commutation is termed 'ideal' if the commutation process or the reversal of current is completed by the end of the short-circuit time or the commutation period. If the reversal of current is completed during the short-circuit time then sparking occurs at the carbon-brush contacts and the MCM DC–AC commutator surface is damaged due to over-heating and the DC electrical machine is termed 'poorly commutated'.

The mechanical split-ring 2 × 12 MCM DC–AC commutator with a bipolar MCM, realised on 24 electrical valves. i.e. sector/brush sliding-contacts in 2 × 12 ECM connection is shown in figure A13. However, the approach may be applicable

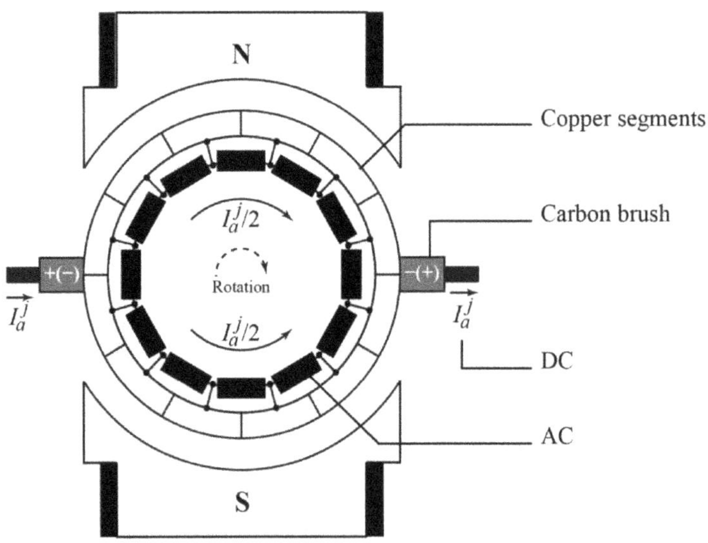

Figure A12. AC–DC/DC–AC commutator synchronous electrical machine (DC machine): the twelve-phase, full-wave MCM AC–DC/DC–AC commutator, with a single homogeneous cathode commutation group and a single homogeneous anode commutation group, acting as a two-way MCM AC–DC/DC–AC rectifier/inverter.

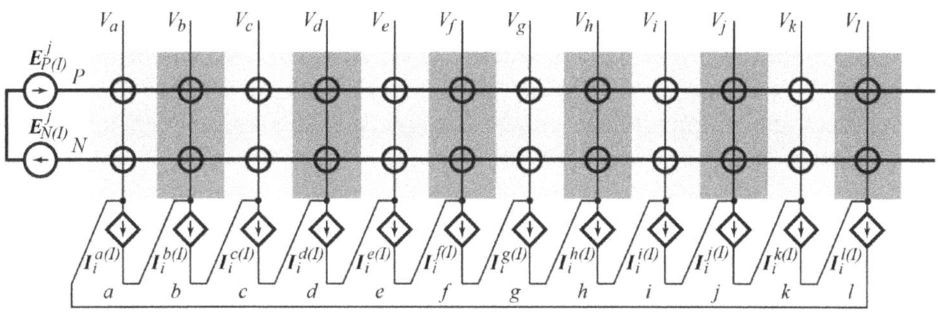

Figure A13. AC–DC/DC–AC commutator synchronous electrical machine (DC machine): the mechanical split-ring 2 × 12 MCM DC–AC commutator with a bipolar MCM, realised on 24 electrical valves. i.e. sector/brush sliding-contacts in 2 × 12 ECM connection.

for the mechanical split-ring 2 × 12 MCM (24 electrical valves) topology well, as may be demonstrated through appropriate topological mapping.

Figure A13 illustrates the 2 × 12 MCM DC–AC commutator acting as a DC–AC inverter, using twelve single-pole-double-valvular (SPDV) commutation groups V_a, V_b, V_c, V_d, V_e, V_f, V_g, V_h, V_i, V_j, V_k and V_l with 24 electrical valves i.e. sector/brush sliding-contacts. If the commutation function of the electrical valves i.e. sector/brush sliding-contacts, as illustrated in figure A13 is defined to be $\boldsymbol{C}_{ij}^{sP(I)}$, the input–output transposition properties of the MCM can be expressed as

$$\boldsymbol{E}_i^{s(I)} = \boldsymbol{C}_{ij}^{sP(I)} \boldsymbol{E}_{P(I)}^{j} \tag{A.24}$$

or in detail

$$\left\| \begin{matrix} \boldsymbol{E}_i^{a(I)} \\ \boldsymbol{E}_i^{b(I)} \\ \boldsymbol{E}_i^{c(I)} \\ \boldsymbol{E}_i^{d(I)} \\ \boldsymbol{E}_i^{e(I)} \\ \boldsymbol{E}_i^{f(I)} \\ \boldsymbol{E}_i^{g(I)} \\ \boldsymbol{E}_i^{h(I)} \\ \boldsymbol{E}_i^{i(I)} \\ \boldsymbol{E}_i^{j(I)} \\ \boldsymbol{E}_i^{k(I)} \\ \boldsymbol{E}_i^{l(I)} \end{matrix} \right\| = \left\| \begin{matrix} \boldsymbol{C}_{ij}^{Pa} & \boldsymbol{C}_{ij}^{Na} \\ \boldsymbol{C}_{ij}^{Pb} & \boldsymbol{C}_{ij}^{Nb} \\ \boldsymbol{C}_{ij}^{Pc} & \boldsymbol{C}_{ij}^{Nc} \\ \boldsymbol{C}_{ij}^{Pd} & \boldsymbol{C}_{ij}^{Nd} \\ \boldsymbol{C}_{ij}^{Pe} & \boldsymbol{C}_{ij}^{Ne} \\ \boldsymbol{C}_{ij}^{Pf} & \boldsymbol{C}_{ij}^{Nf} \\ \boldsymbol{C}_{ij}^{Pg} & \boldsymbol{C}_{ij}^{Ng} \\ \boldsymbol{C}_{ij}^{Ph} & \boldsymbol{C}_{ij}^{Nh} \\ \boldsymbol{C}_{ij}^{Pi} & \boldsymbol{C}_{ij}^{Ni} \\ \boldsymbol{C}_{ij}^{Pj} & \boldsymbol{C}_{ij}^{Mj} \\ \boldsymbol{C}_{ij}^{Pk} & \boldsymbol{C}_{ij}^{Nk} \\ \boldsymbol{C}_{ij}^{Pl} & \boldsymbol{C}_{ij}^{Nl} \end{matrix} \right\| \left\| \begin{matrix} \boldsymbol{E}_{P(I)}^{j} \\ \boldsymbol{E}_{N(I)}^{j} \end{matrix} \right\| \tag{A.25}$$

however, a mathematical model of an electrical continuous dynamical system of the mechanical split-ring 2 × 12 MCM DC–AC commutator shown in figure A13, acting as a current-source MCM DC–AC commutator, making dependent the input holors from the output holors, in the dynamical systems approach and matrix notation may be expressed in the form:

$$\boldsymbol{I}^{j}_{P(I)} = \boldsymbol{C}^{ji}_{Ps(I)}\boldsymbol{I}^{s(I)}_{i}$$

or in detail

$$\left\|\begin{matrix} \boldsymbol{I}^{j}_{P(I)} \\ \boldsymbol{I}^{j}_{N(I)} \end{matrix}\right\| = \left\|\begin{matrix} \boldsymbol{C}^{ji}_{Pa}\,\boldsymbol{C}^{ji}_{Pb}\,\boldsymbol{C}^{ji}_{Pc}\,\boldsymbol{C}^{ji}_{Pd}\,\boldsymbol{C}^{ji}_{Pe}\,\boldsymbol{C}^{ji}_{Pf}\,\boldsymbol{C}^{ji}_{Pg}\,\boldsymbol{C}^{ji}_{Ph}\,\boldsymbol{C}^{ji}_{Pi}\,\boldsymbol{C}^{ji}_{Pj}\,\boldsymbol{C}^{ji}_{Pk}\,\boldsymbol{C}^{ji}_{Pl} \\ \boldsymbol{C}^{ji}_{Na}\,\boldsymbol{C}^{ji}_{Nb}\,\boldsymbol{C}^{ji}_{Nc}\,\boldsymbol{C}^{ji}_{Nd}\,\boldsymbol{C}^{ji}_{Ne}\,\boldsymbol{C}^{ji}_{Nf}\,\boldsymbol{C}^{ji}_{Ng}\,\boldsymbol{C}^{ji}_{Nh}\,\boldsymbol{C}^{ji}_{Ni}\,\boldsymbol{C}^{ji}_{Nj}\,\boldsymbol{C}^{ji}_{Nk}\,\boldsymbol{C}^{ji}_{Nl} \end{matrix}\right\| \left\|\begin{matrix} \boldsymbol{I}^{a(I)}_{i} \\ \boldsymbol{I}^{b(I)}_{i} \\ \boldsymbol{I}^{c(I)}_{i} \\ \boldsymbol{I}^{d(I)}_{i} \\ \boldsymbol{I}^{e(I)}_{i} \\ \boldsymbol{I}^{f(I)}_{i} \\ \boldsymbol{I}^{g(I)}_{i} \\ \boldsymbol{I}^{h(I)}_{i} \\ \boldsymbol{I}^{i(I)}_{i} \\ \boldsymbol{I}^{j(I)}_{i} \\ \boldsymbol{I}^{k(I)}_{i} \\ \boldsymbol{I}^{l(I)}_{i} \end{matrix}\right\|$$

where:

$$\boldsymbol{E}^{s(I)}_{i} = \left\|\boldsymbol{E}^{a(I)}_{i}\,\boldsymbol{E}^{b(I)}_{i}\,\boldsymbol{E}^{c(I)}_{i}\,\boldsymbol{E}^{d(I)}_{i}\,\boldsymbol{E}^{e(I)}_{i}\,\boldsymbol{E}^{f(I)}_{i}\,\boldsymbol{E}^{g(I)}_{i}\,\boldsymbol{E}^{h(I)}_{i}\,\boldsymbol{E}^{i(I)}_{i}\,\boldsymbol{E}^{j(I)}_{i}\,\boldsymbol{E}^{k(I)}_{i}\,\boldsymbol{E}^{l(I)}_{i}\right\|^{T}$$

$$\boldsymbol{E}^{j}_{P(I)} = \left\|\boldsymbol{E}^{j}_{p(I)}\,\boldsymbol{E}^{j}_{n(I)}\right\|^{T}$$

$$\left\|\boldsymbol{I}^{j}_{P(I)}\right\| = \left\|\boldsymbol{I}^{j}_{p(I)}\,\boldsymbol{I}^{j}_{n(I)}\right\|^{T}$$

$$\|\boldsymbol{I}^{s(I)}_{i}\| = \left\|\boldsymbol{I}^{a(I)}_{i}\,\boldsymbol{I}^{b(I)}_{i}\,\boldsymbol{I}^{c(I)}_{i}\,\boldsymbol{I}^{d(I)}_{i}\,\boldsymbol{I}^{e(I)}_{i}\,\boldsymbol{I}^{f(I)}_{i}\boldsymbol{I}^{g(I)}_{i}\,\boldsymbol{I}^{h(I)}_{i}\,\boldsymbol{I}^{i(I)}_{i}\,\boldsymbol{I}^{j(I)}_{i}\,\boldsymbol{I}^{k(I)}_{i}\,\boldsymbol{I}^{l(I)}_{i}\right\|^{T}$$

$\boldsymbol{C}^{sP(I)}_{ij}$ is the commutation-matrix holor of the voltage-source 2 × 12 MMC DC–AC commutator connected with the stiff voltage holors, $\boldsymbol{C}^{ji}_{Ps(I)}$ is the commutation-matrix holor of the current-source 2 × 12 MMC DC–AC commutator connected with the stiff current holors. They are given by

$$
C_{ij}^{sP(I)} = \left\| \begin{array}{cc}
C_{ij}^{Pa} & C_{ij}^{Na} \\
C_{ij}^{Pb} & C_{ij}^{Nb} \\
C_{ij}^{Pc} & C_{ij}^{Nc} \\
C_{ij}^{Pd} & C_{ij}^{Nd} \\
C_{ij}^{Pe} & C_{ij}^{Ne} \\
C_{ij}^{Pf} & C_{ij}^{Nf} \\
C_{ij}^{Pg} & C_{ij}^{Ng} \\
C_{ij}^{Ph} & C_{ij}^{Nh} \\
C_{ij}^{Pi} & C_{ij}^{Ni} \\
C_{ij}^{Pj} & C_{ij}^{Mj} \\
C_{ij}^{Pk} & C_{ij}^{Nk} \\
C_{ij}^{Pl} & C_{ij}^{Nl}
\end{array} \right\|
$$

$$
= \left\| \begin{array}{cccccccccccc}
C_{ij}^{Pa} & C_{ij}^{Pb} & C_{ij}^{Pc} & C_{ij}^{Pd} & C_{ij}^{Pe} & C_{ij}^{Pf} & C_{ij}^{Pg} & C_{ij}^{Ph} & C_{ij}^{Pi} & C_{ij}^{Pj} & C_{ij}^{Pk} & C_{ij}^{Pl} \\
C_{ij}^{Na} & C_{ij}^{Nb} & C_{ij}^{Nc} & C_{ij}^{Nd} & C_{ij}^{Ne} & C_{ij}^{Nf} & C_{ij}^{Ng} & C_{ij}^{Nh} & C_{ij}^{Ni} & C_{ij}^{Nj} & C_{ij}^{Nk} & C_{ij}^{Nl}
\end{array} \right\|^{T}
$$

and

$$
C_{Ps(I)}^{ji} = \left\| \begin{array}{cccccccccccc}
C_{ij}^{Pa} & C_{ij}^{Pb} & C_{ij}^{Pc} & C_{ij}^{Pd} & C_{ij}^{Pe} & C_{ij}^{Pf} & C_{ij}^{Pg} & C_{ij}^{Ph} & C_{ij}^{Pi} & C_{ij}^{Pj} & C_{ij}^{Pk} & C_{ij}^{Pl} \\
C_{ij}^{Na} & C_{ij}^{Nb} & C_{ij}^{Nc} & C_{ij}^{Nd} & C_{ij}^{Ne} & C_{ij}^{Nf} & C_{ij}^{Ng} & C_{ij}^{Nh} & C_{ij}^{Ni} & C_{ij}^{Nj} & C_{ij}^{Nk} & C_{ij}^{Nl}
\end{array} \right\| .
$$

The mechanical split-ring 12 × 2 MCM DC–AC commutator acting as an AC–DC rectifier with a bipolar MCM, realised on 24 electrical valves. i.e. sector/brush sliding-contacts in 12 × 2 MCM connection is shown in figure A14. However, the approach may be applicable for the mechanical split-ring 12 × 2 ECM (24 electrical valves) topology well, as may be demonstrated through appropriate topological mapping.

Figure A14 illustrates the 12 × 2 MCM using two single-pole-twelve-valvular (SPTV) commutation groups V_p, and V_n with 24 electrical valves i.e. sector/brush sliding-contacts.

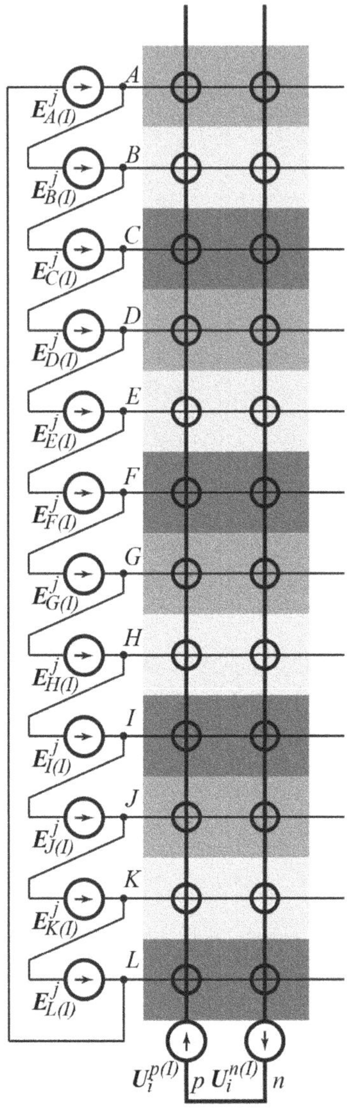

Figure A14. AC–DC/DC–AC commutator synchronous electrical machine (DC machine): the mechanical split-ring 12 × 2 MCM DC–AC commutator acting as an AC–DC rectifier with a bipolar MCM, realised on 24 electrical valves, i.e. sector/brush sliding-contacts in 12 × 2 MCM connection.

If the commutation function of the electrical valves i.e. sector/brush sliding-contacts, as illustrated in figure A14 is defined to be $\boldsymbol{C}_{ij}^{Ps(E)}$, the input–output transposition properties of the MCM can be expressed as

$$\boldsymbol{E}_i^{s(E)} = \boldsymbol{C}_{ij}^{sP(E)} \boldsymbol{E}^j_{P(E)} \tag{A.26}$$

or in detail

$$\left\| \begin{matrix} E_i^{p(E)} \\ E_i^{n(E)} \end{matrix} \right\| = \left\| \begin{matrix} C_{ij}^{Ap}\, C_{ij}^{Bp}\, C_{ij}^{Cp}\, C_{ij}^{Dp}\, C_{ij}^{Ep}\, C_{ij}^{Fp}\, C_{ij}^{Gp}\, C_{ij}^{Hp}\, C_{ij}^{Ip}\, C_{ij}^{Jp}\, C_{ij}^{Kp}\, C_{ij}^{Lp} \\ C_{ij}^{An}\, C_{ij}^{Bn}\, C_{ij}^{Cn}\, C_{ij}^{Dn}\, C_{ij}^{En}\, C_{ij}^{Fn}\, C_{ij}^{Gn}\, C_{ij}^{Hn}\, C_{ij}^{In}\, C_{ij}^{Jn}\, C_{ij}^{Kn}\, C_{ij}^{Ln} \end{matrix} \right\| \left\| \begin{matrix} E^{j}_{A(E)} \\ E^{j}_{B(E)} \\ E^{j}_{C(E)} \\ E^{j}_{D(E)} \\ E^{j}_{R(E)} \\ E^{j}_{F(E)} \\ E^{j}_{G(E)} \\ E^{j}_{H(E)} \\ E^{j}_{I(E)} \\ E^{j}_{J(E)} \\ E^{j}_{K(E)} \\ E^{j}_{L(E)} \end{matrix} \right\| \qquad \text{(A.27)}$$

however, a mathematical model of an electrical continuous dynamical system of the 12 × 2 ECM AC–DC commutator acting as an AC–DC rectifier shown in figure A14, acting as a voltage-source ECM AC–DC commutator, making dependent the input holors from the output holors, in the dynamical systems approach and matrix notation may be expressed in the form:

$$I^{j}_{p(E)} = C^{ji}_{Ps(E)} I_i^{s(E)}$$

or in detail

$$\left\| \begin{matrix} I^{j}_{A(E)} \\ I^{j}_{B(E)} \\ I^{j}_{C(E)} \\ I^{j}_{D(E)} \\ I^{j}_{E(E)} \\ I^{j}_{F(E)} \\ I^{j}_{G(E)} \\ I^{j}_{H(E)} \\ I^{j}_{I(E)} \\ I^{j}_{J(E)} \\ I^{j}_{K(E)} \\ I^{j}_{L(E)} \end{matrix} \right\| = \left\| \begin{matrix} C^{ji}_{Ap}\, C^{ji}_{An} \\ C^{ji}_{Bp}\, C^{ji}_{Bn} \\ C^{ji}_{Cp}\, C^{ji}_{Cn} \\ C^{ji}_{Dp}\, C^{ji}_{Dn} \\ C^{ji}_{Ep}\, C^{ji}_{Ep} \\ C^{ji}_{Fp}\, C^{ji}_{Fn} \\ C^{ji}_{Gp}\, C^{ji}_{Gn} \\ C^{ji}_{Hp}\, C^{ji}_{Hn} \\ C^{ji}_{iIp}\, C^{ji}_{In} \\ C^{ji}_{Jp}\, C^{ji}_{Jn} \\ C^{ji}_{Kp}\, C^{ji}_{Kn} \\ C^{ji}_{Lp}\, C^{ji}_{Ln} \end{matrix} \right\| \left\| \begin{matrix} I_i^{p(E)} \\ I_i^{n(E)} \end{matrix} \right\|$$

where:

$$\|\boldsymbol{E}_i^{s(E)}\| = \left\|\boldsymbol{E}_i^{p(E)}\ \boldsymbol{E}_i^{n(E)}\right\|^T$$

$$\|\boldsymbol{E}_i^{P(E)}\| = \left\|\boldsymbol{E}_i^{A(E)}\ \boldsymbol{E}_i^{B(E)}\ \boldsymbol{E}_i^{C(E)}\ \boldsymbol{E}_i^{D(E)}\ \boldsymbol{E}_i^{E(E)}\ \boldsymbol{E}_i^{F(E)}\boldsymbol{E}_i^{G(E)}\ \boldsymbol{E}_i^{H(E)}\ \boldsymbol{E}_i^{I(E)}\ \boldsymbol{E}_i^{J(E)}\ \boldsymbol{E}_i^{K(E)}\ \boldsymbol{E}_i^{L(E)}\right\|^T$$

$$\left\|\boldsymbol{I}_{s(E)}^j\right\| = \left\|\boldsymbol{I}_{p(E)}^j\ \boldsymbol{I}_{n(E)}^j\right\|^T$$

$$\|\boldsymbol{I}_i^{P(E)}\| = \left\|\boldsymbol{I}_i^{A(E)}\ \boldsymbol{I}_i^{B(E)}\ \boldsymbol{I}_i^{C(E)}\ \boldsymbol{I}_i^{D(E)}\ \boldsymbol{I}_i^{E(E)}\ \boldsymbol{I}_i^{F(E)}\boldsymbol{I}_i^{G(E)}\ \boldsymbol{I}_i^{H(E)}\ \boldsymbol{I}_i^{I(E)}\ \boldsymbol{I}_i^{J(E)}\ \boldsymbol{I}_i^{K(E)}\ \boldsymbol{I}_i^{L(E)}\right\|^T$$

$\boldsymbol{C}_{ij}^{sP(E)}$ is the commutation-matrix holor of the voltage-source 12 × 2 ECM AC–DC commutator connected with the stiff voltage holors. $\boldsymbol{C}_{Ps(I)}^{ji}$ is the commutation-matrix holor of the current-source 12 × 2 ECM AC–DC commutator connected with the stiff current holors. They are given by

$$\boldsymbol{C}_{ij}^{sP(E)} = \left\|\begin{matrix} \boldsymbol{C}_{ij}^{Ap}\ \boldsymbol{C}_{ij}^{Bp}\ \boldsymbol{C}_{ij}^{Cp}\ \boldsymbol{C}_{ij}^{Dp}\ \boldsymbol{C}_{ij}^{Ep}\ \boldsymbol{C}_{ij}^{Fp}\ \boldsymbol{C}_{ij}^{Gp}\ \boldsymbol{C}_{ij}^{Hp}\ \boldsymbol{C}_{ij}^{Ip}\ \boldsymbol{C}_{ij}^{Jp}\ \boldsymbol{C}_{ij}^{Kp}\ \boldsymbol{C}_{ij}^{Lp} \\ \boldsymbol{C}_{ij}^{An}\ \boldsymbol{C}_{ij}^{Bn}\ \boldsymbol{C}_{ij}^{Cn}\ \boldsymbol{C}_{ij}^{Dn}\ \boldsymbol{C}_{ij}^{En}\ \boldsymbol{C}_{ij}^{Fn}\ \boldsymbol{C}_{ij}^{Gn}\ \boldsymbol{C}_{ij}^{Hn}\ \boldsymbol{C}_{ij}^{In}\ \boldsymbol{C}_{ij}^{Jn}\ \boldsymbol{C}_{ij}^{Kn}\ \boldsymbol{C}_{ij}^{Ln} \end{matrix}\right\|$$

and

$$\boldsymbol{C}_{Ps(I)}^{ji} = \left\|\begin{matrix} \boldsymbol{C}_{Ap}^{ji}\ \boldsymbol{C}_{An}^{ji} \\ \boldsymbol{C}_{Bp}^{ji}\ \boldsymbol{C}_{Bn}^{ji} \\ \boldsymbol{C}_{Cp}^{ji}\ \boldsymbol{C}_{Cn}^{ji} \\ \boldsymbol{C}_{Dp}^{ji}\ \boldsymbol{C}_{Dn}^{ji} \\ \boldsymbol{C}_{Ep}^{ji}\ \boldsymbol{C}_{Ep}^{ji} \\ \boldsymbol{C}_{Fp}^{ji}\ \boldsymbol{C}_{Fn}^{ji} \\ \boldsymbol{C}_{Gp}^{ji}\ \boldsymbol{C}_{Gn}^{ji} \\ \boldsymbol{C}_{Hp}^{ji}\ \boldsymbol{C}_{Hn}^{ji} \\ \boldsymbol{C}_{iIp}^{ji}\ \boldsymbol{C}_{In}^{ji} \\ \boldsymbol{C}_{Jp}^{ji}\ \boldsymbol{C}_{Jn}^{ji} \\ \boldsymbol{C}_{Kp}^{ji}\ \boldsymbol{C}_{Kn}^{ji} \\ \boldsymbol{C}_{Lp}^{ji}\ \boldsymbol{C}_{Ln}^{ji} \end{matrix}\right\|$$

A.6 ECM AC–AC commutators

The monolithic 3 × 3 ECM AC–AC commutator acting as an AC–AC cycloconverter with a bipolar ECM, realised on only nine 'continuous' **reverse-blocking**

insulated-gate bipolar transistors (RB IGBT), or **metal-oxide semiconductor field effect transistors** (MOSFET), or **integrated-gate commutated thyristor** (IGCT), or bipolar symmetrical trigistors, that is, **gate-turn-off** (GTO) symmetrical thyristors (symistors) in 3 × 3 ECM connection is shown in figure A15.

However, the approach may be applicable for the monolithic 3 × 3 ECM (nine electrical valves) topology well, as may be demonstrated through appropriate topological mapping.

Figure A15 illustrates the 3 × 3 ECM using three monolithic **single-pole-triple-valvular** (SPTV) commutation groups V_a, V_b, V_c with nine (nine RB IGBTs).

If the commutation function of the RB IGBTs, as illustrated in figure A15, is defined to be $\boldsymbol{C}_{ij}^{sP}$, the input–output transposition properties of the ECM can be expressed as

$$\boldsymbol{E}_i^s = \boldsymbol{C}_{ij}^{sP} \boldsymbol{E}_P^j \tag{A.28}$$

or in detail

$$\begin{Vmatrix} \boldsymbol{E}_i^a \\ \boldsymbol{E}_i^b \\ \boldsymbol{E}_i^c \end{Vmatrix} = \begin{Vmatrix} \boldsymbol{C}_{ij}^{Aa} & \boldsymbol{C}_{ij}^{Ba} & \boldsymbol{C}_{ij}^{Ca} \\ \boldsymbol{C}_{ij}^{Ab} & \boldsymbol{C}_{ij}^{Bb} & \boldsymbol{C}_{ij}^{Cb} \\ \boldsymbol{C}_{ij}^{Ac} & \boldsymbol{C}_{ij}^{Bc} & \boldsymbol{C}_{ij}^{Cc} \end{Vmatrix} \begin{Vmatrix} \boldsymbol{E}_A^j \\ \boldsymbol{E}_B^j \\ \boldsymbol{E}_C^j \end{Vmatrix} \tag{A.29}$$

however, a mathematical model of an electrical continuous dynamical system of the 3 × 3 ECM AC–AC commutator shown in figure A15, acting as a current-source ECM AC–AC commutator, making dependent the input holors from the output holors, in the dynamical approach and matrix notation may be systems expressed in the form:

$$\boldsymbol{I}_P^j = \boldsymbol{C}_{Ps}^{ji} \boldsymbol{I}_i^s \tag{A.30}$$

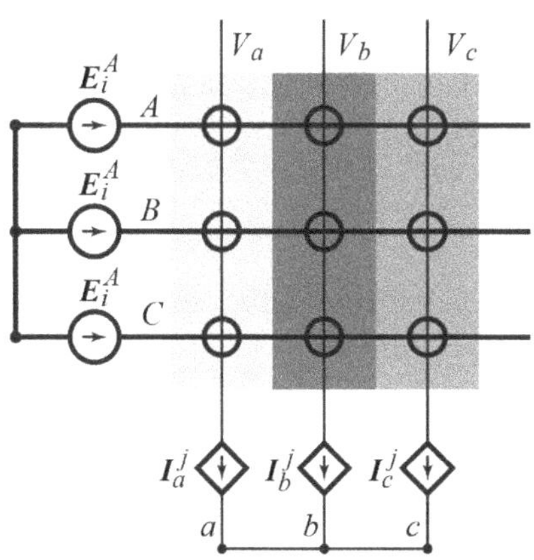

Figure A15. The monolithic 3 × 3 ECM AC–AC commutator, i.e. with nine RB IGBTs.

in detail

$$\left\| \begin{matrix} I_A^j \\ I_B^j \\ I_C^j \end{matrix} \right\| = \left\| \begin{matrix} C_{Aa}^{ji} & C_{Ab}^{ji} & C_{Ac}^{ji} \\ C_{Ba}^{ji} & C_{Bb}^{ji} & C_{Bc}^{ji} \\ C_{Ca}^{ji} & C_{Cb}^{ji} & C_{Cc}^{ji} \end{matrix} \right\| \left\| \begin{matrix} I_i^a \\ I_i^b \\ I_i^c \end{matrix} \right\| \tag{A.31}$$

where:

$$\|E_i^s\| = \left\| E_i^a \; E_i^b \; E_c^c \right\|^T;$$

$$\|E_P^j\| = \left\| E_A^j \; E_B^j \; E_C^j \right\|^T;$$

$$\|I_P^j\| = \left\| I_A^J \; I_B^J \; I_C^J \right\|^T$$

$$\|I_i^s\| = \left\| I_i^a \; I_i^b \; I_i^c \right\|^T$$

C_{ij}^{sP} is the voltage-throws shown in figure A15 source holors commutation matrix with elements being commutation functions of various throws

$$C_{ij}^{sP} = \left\| \begin{matrix} C_{ij}^{Aa} & C_{ij}^{Ba} & C_{ij}^{Ca} \\ C_{ij}^{Ab} & C_{ij}^{Bb} & C_{ij}^{Cb} \\ C_{ij}^{Ac} & C_{ij}^{Bc} & C_{ij}^{Cc} \end{matrix} \right\|$$

C_{Ps}^{ji} is the current-source holors commutation matrix with elements being commutation of various throws shown in figure A15.

$$C_{Ps}^{ji} = \left\| \begin{matrix} C_{Aa}^{ji} & C_{Ab}^{ji} & C_{Ac}^{ji} \\ C_{Ba}^{ji} & C_{Bb}^{ji} & C_{Bc}^{ji} \\ C_{Ca}^{ji} & C_{Cb}^{ji} & C_{Cc}^{ji} \end{matrix} \right\| = \left\| \begin{matrix} C_{ij}^{Aa} & C_{ij}^{Ba} & C_{ij}^{Ca} \\ C_{ij}^{Ab} & C_{ij}^{Bb} & C_{ij}^{Cb} \\ C_{ij}^{Ac} & C_{ij}^{Bc} & C_{ij}^{Cc} \end{matrix} \right\|^T$$

C_{kL}^{ji} are the voltage-source coefficients of the commutation nodes 'kL', which may be expressed as follows:

$$C_{kL}^{ji} = \frac{\partial E_k^j}{\partial E_i^L} \bigg| E_i^I = 0 = \begin{cases} +1 \overset{\Delta}{\rightarrow} E_k^j = +E_i^L \\ 0 \overset{\Delta}{\rightarrow} E_k^j = 0 \\ -1 \overset{\Delta}{\rightarrow} E_k^j = -E_i^L \end{cases}$$

C_{ij}^{Kl} are the current-source coefficients of the commutation nodes 'Kl', which may be expressed as follows:

$$C_{ij}^{Kl} = \frac{\partial E_i^K}{\partial E_l^j} \bigg| E_i^j = 0 = \begin{cases} -1 \overset{\Delta}{\rightarrow} E_i^K = -E_l^j \\ 0 \overset{\Delta}{\rightarrow} E_i^K = 0 \\ +1 \overset{\Delta}{\rightarrow} E_i^K = +E_l^j \end{cases}$$

This monolithic 3 × 3 ECM AC–AC commutator makes the bilateral flow of secondary electrical energy possible as well during generating (absorbing) as during motoring (driving) of the AC–AC commutator dynamotor with the variable angular velocity. Six operating modes are possible in this MMD electrical machine.

Potential attractive features of the monolithic 3 × 3 ECM, such as the absence of DC link energy storage with reactive components, high-power density, high quality of input and output waveforms are leading to continued research efforts aimed at translating their promise into reality (Fijalkowski 1973, 1982, 1985a, 1985b, 1985c 2016, Tutaj 2012, Wheeler *et al* 2004).

Although the monolithic 3 × 3 ECMs were first introduced as a class of frequency and phase number changer realised using controlled electrical valves (electronic switches) by Fijalkowski (1973, 1982, 1988a, 1988b, 1988c, 1988d, 1988e, 1988f, 1988g) and later by Gyugyi (1976), significant progress began with the introduction of high-frequency synthesis proposed by Alesina and Venturini (1989).

Although the initial approach was limited to a voltage transfer ratio of one half, improved modulation strategies with voltage transfer ratio of $\sqrt{3/2}$ was published by Alesina and Venturini (1989). Further along, indirect modulation methods based on fictitious DC link concept have been proposed (Rodriguez 1983, Alesina and Venturini 1989, Oyama *et al* 1989, Neft and Schauder 1988a, 1988b, Ziogas *et al* 1986). Except the programmed commutation (switching) pattern in (Ziogas *et al* 1986), all these modulation strategies for the monolithic 3 × 3 ECM, based on high-frequency synthesis have not been able to break the upper bound given by $\sqrt{3/2}$ pu.

Since AC–AC commutator motor output power is proportional to the square of applied voltage, this limitation leads to reduced output power, limited to 3/4 of the AC–AC commutator motor rating, when fed from a monolithic 3 × 3 ECM AC–AC commutator operating from the line nominally rated at AC–AC commutator motor voltage. Otherwise, a custom designed AC–AC commutator motor designed for the appropriate output voltage level would be required. This remains one among the many roadblocks limiting the viability of the monolithic 3 × 3 ECM AC–AC commutator.

A low commutation (switching) frequency six-step modulation strategy for frequency conversion using 3 × 3 ECM AC–AC commutators capable of providing more than unity voltage transfer factor was presented by Wang and Venkataramanan (2006).

A.7 ECM AC–DC commutators

The monolithic 3 × 2 ECM AC–DC commutator acting as an AC–DC rectifier with a bipolar ECM, realised on only six 'continuous' RB IGBTs, or MOSFETs, or bipolar symmetrical trigistors, that is, GTO symmetrical thyristors (symistors) in 3 × 2 ECM connection is shown in figure A16.

However, the approach may be applicable for the monolithic 3 × 2 ECM (six electrical valves) topology well, as may be demonstrated through appropriate topological mapping.

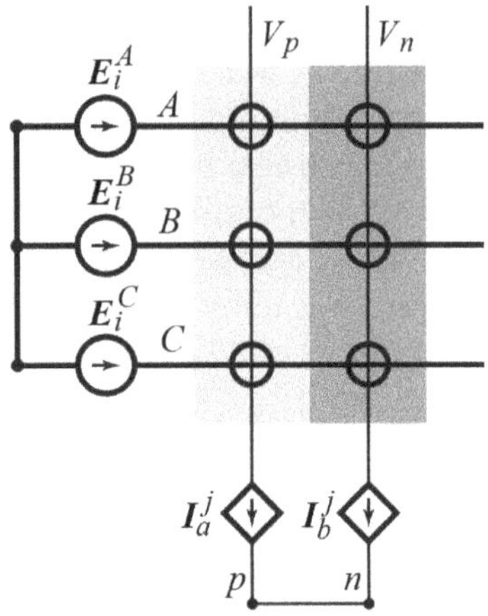

Figure A16. The monolithic 3 × 2 ECM AC–DC commutator.

Figure A16 illustrates the 3 × 2 ECM using three monolithic single-pole-double-valvular (SPDV) commutation groups V_p V_n, i.e. with six RB IGBTs.

If the commutation function of the RB IGBTs, as illustrated in figure A16, is defined to be $C_{ij}^{sP(E)}$, the input–output transposition properties of the ECM can be expressed as

$$E_i^{s(E)} = C_{ij}^{sP(E)} E_{P(E)}^j \tag{A.32}$$

or in detail

$$\begin{Vmatrix} E_i^{p(E)} \\ E_i^{n(E)} \end{Vmatrix} = \begin{Vmatrix} C_{ij}^{Ap} & C_{ij}^{Bp} & C_{ij}^{Cp} \\ C_{ij}^{An} & C_{ij}^{Bn} & C_{ij}^{Cn} \end{Vmatrix} \begin{Vmatrix} E_{A(E)}^j \\ E_{B(E)}^j \\ E_{C(E)}^j \end{Vmatrix} \tag{A.33}$$

However, a mathematical model of an electrical continuous dynamical system of the 3 × 2 ECM AC–AC commutator shown in figure A16, acting as a voltage-source ECM AC–DC commutator, making dependent the input holors from the output holors, in the dynamical systems approach and matrix notation may be expressed in the form:

$$I_{P(E)}^j = C_{Ps(E)}^{ji} I_i^{s(E)}$$

or in detail

$$\begin{Vmatrix} I_{A(E)}^j \\ I_{B(E)}^j \\ I_{C(E)}^j \end{Vmatrix} = \begin{Vmatrix} C_{Ap}^{ji} & C_{An}^{ji} \\ C_{Bp}^{ji} & C_{Bn}^{ji} \\ C_{Cp}^{ji} & C_{Cn}^{ji} \end{Vmatrix} \begin{Vmatrix} I_i^{p(E)} \\ I_i^{n(E)} \end{Vmatrix}$$

where:

$$\|\boldsymbol{E}_i^{s(E)}\| = \left\|\boldsymbol{E}_i^{p(E)}\ \boldsymbol{E}_i^{n(E)}\right\|^T$$

$$\left\|\boldsymbol{I}_{P(E)}^{j}\right\| = \left\|\boldsymbol{I}_{A(E)}^{j}\ \boldsymbol{I}_{B(E)}^{j}\ \boldsymbol{I}_{C(E)}^{j}\right\|^T$$

$$\|\boldsymbol{I}_i^{s(E)}\| = \left\|\boldsymbol{I}_i^{p(E)}\ \boldsymbol{I}_i^{n(E)}\right\|^T$$

$\boldsymbol{C}_{ij}^{sP(E)}$ is the commutation-matrix holor of the voltage-source 3 × 2 ECM AC–DC commutator connected with the stiff voltage holors. $\boldsymbol{C}_{Ps(I)}^{ji}$ is the commutation-matrix holor of the current-source 3 × 2 ECM AC–DC commutator connected with the stiff current holors. They are given by

$$\boldsymbol{C}_{ij}^{sP(E)} = \begin{Vmatrix} \boldsymbol{C}_{ij}^{Ap}\ \boldsymbol{C}_{ij}^{Bp}\ \boldsymbol{C}_{ij}^{Cp} \\ \boldsymbol{C}_{ij}^{An}\ \boldsymbol{C}_{ij}^{Bn}\ \boldsymbol{C}_{ij}^{Cn} \end{Vmatrix}$$

and

$$\boldsymbol{C}_{Ps(I)}^{ji} = \begin{Vmatrix} \boldsymbol{C}_{Ap}^{ji}\ \boldsymbol{C}_{An}^{ji} \\ \boldsymbol{C}_{Bp}^{ji}\ \boldsymbol{C}_{Bn}^{ji} \\ \boldsymbol{C}_{Cp}^{ji}\ \boldsymbol{C}_{Cn}^{ji} \end{Vmatrix} = \begin{Vmatrix} \boldsymbol{C}_{ij}^{Ap}\ \boldsymbol{C}_{ij}^{Bp}\ \boldsymbol{C}_{ij}^{Cp} \\ \boldsymbol{C}_{ij}^{An}\ \boldsymbol{C}_{ij}^{Bn}\ \boldsymbol{C}_{ij}^{Cn} \end{Vmatrix}^T$$

A.8 ECM AC–DC–AC commutators

The modulation strategy may be easier developed on the basis of operation of the dual monolithic 3 × 2 and 2 × 3 ECM topology with a fictitious DC link, by using commutation-matrix holors multiplication of these two cascaded connected ECMs earlier introduced by the authors of this textbook (Fijalkowski 1967, 1982, 1985a, 1985b, 1985c, 1987a, 1987b, Tutaj 2012).

Figure A17 illustrates the reduced electrical valve (switch) two monolithic SPTV commutation groups V_p and V_n with six IGBTs and three monolithic SPDV commutation groups V_a, V_b, V_c with six IGBTs. The two realisations are labelled as conventional ECM and indirect ECM, respectively.

The input–output transposition properties of the 3 × 2 and 2 × 3 ECM AC–DC–AC commutator can be expressed as

$$\boldsymbol{E}_i^s = \boldsymbol{C}_{ij}^{Ps(E)T}\boldsymbol{C}_{ij}^{Ps(I)T}\boldsymbol{E}_P^j \tag{A.34}$$

or in detail

$$\begin{Vmatrix} \boldsymbol{E}_i^a \\ \boldsymbol{E}_i^b \\ \boldsymbol{E}_i^c \end{Vmatrix} = \begin{Vmatrix} \boldsymbol{C}_{ij}^{Pa}\ \boldsymbol{C}_{ij}^{Na} \\ \boldsymbol{C}_{ij}^{Pb}\ \boldsymbol{C}_{ij}^{Nb} \\ \boldsymbol{C}_{ij}^{Pc}\ \boldsymbol{C}_{ij}^{Nc} \end{Vmatrix} \begin{Vmatrix} \boldsymbol{C}_{ij}^{Ap}\ \boldsymbol{C}_{ij}^{Bp}\ \boldsymbol{C}_{ij}^{Cp} \\ \boldsymbol{C}_{ij}^{An}\ \boldsymbol{C}_{ij}^{Bn}\ \boldsymbol{C}_{ij}^{Cn} \end{Vmatrix} \begin{Vmatrix} \boldsymbol{E}_A^j \\ \boldsymbol{E}_B^j \\ \boldsymbol{E}_C^j \end{Vmatrix} \tag{A.35}$$

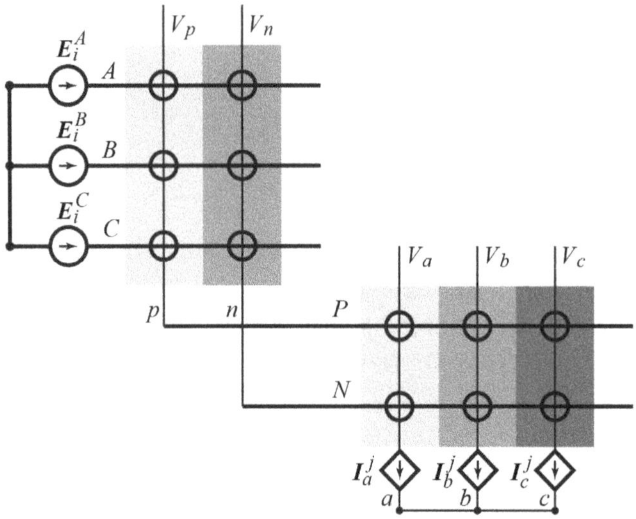

Figure A17. Representation of the monolithic 3 × 2 and 2 × 3 ECM AC–DC–AC commutator.

where $\boldsymbol{C}_{ij}^{Ps(E)}\boldsymbol{C}_{ij}^{Ps(I)}$ is the commutation matrix of the current-source 2 × 3 ECM DC–AC commutator connected with the stiff voltage holors. $\boldsymbol{C}_{ij}^{Ps(E)}$ is the commutation matrix of the voltage-source 3 × 2 ECM AC–DC commutator connected with the stiff current holors. They are given by

$$\boldsymbol{C}_{ij}^{Ps(I)T}\boldsymbol{C}_{ij}^{Ps(E)T} = \begin{Vmatrix} \boldsymbol{C}_{ij}^{Pa} & \boldsymbol{C}_{ij}^{Na} \\ \boldsymbol{C}_{ij}^{Pb} & \boldsymbol{C}_{ij}^{Nb} \\ \boldsymbol{C}_{ij}^{Pc} & \boldsymbol{C}_{ij}^{Nc} \end{Vmatrix} \begin{Vmatrix} \boldsymbol{C}_{ij}^{Ap} & \boldsymbol{C}_{ij}^{Bp} & \boldsymbol{C}_{ij}^{Cp} \\ \boldsymbol{C}_{ij}^{An} & \boldsymbol{C}_{ij}^{Bn} & \boldsymbol{C}_{ij}^{Cn} \end{Vmatrix}$$

$$\boldsymbol{C}_{ij}^{sP(E)} = \begin{Vmatrix} \boldsymbol{C}_{ij}^{Ap} & \boldsymbol{C}_{ij}^{Bp} & \boldsymbol{C}_{ij}^{Cp} \\ \boldsymbol{C}_{ij}^{An} & \boldsymbol{C}_{ij}^{Bn} & \boldsymbol{C}_{ij}^{Cn} \end{Vmatrix}$$

and

$$\boldsymbol{C}_{ij}^{sP(I)} = \begin{Vmatrix} \boldsymbol{C}_{ij}^{Pa} & \boldsymbol{C}_{ij}^{Na} \\ \boldsymbol{C}_{ij}^{Pb} & \boldsymbol{C}_{ij}^{Nb} \\ \boldsymbol{C}_{ij}^{Pc} & \boldsymbol{C}_{ij}^{Nc} \end{Vmatrix}$$

The input–output transposition properties of the 3 × 2 ECM AC–DC commutator can be expressed as

$$\boldsymbol{E}_i^{s(E)} = \boldsymbol{C}_{ij}^{sP(E)}\boldsymbol{E}_{P(E)}^{j} \tag{A.36}$$

or in detail

$$\left\| \begin{matrix} \boldsymbol{E}_i^{p(E)} \\ \boldsymbol{E}_i^{n(E)} \end{matrix} \right\| = \left\| \begin{matrix} \boldsymbol{C}_{ij}^{Ap} & \boldsymbol{C}_{ij}^{Bp} & \boldsymbol{C}_{ij}^{Cp} \\ \boldsymbol{C}_{ij}^{An} & \boldsymbol{C}_{ij}^{Bn} & \boldsymbol{C}_{ij}^{Cn} \end{matrix} \right\| \left\| \begin{matrix} \boldsymbol{E}_{A(E)}^{j} \\ \boldsymbol{E}_{B(E)}^{j} \\ \boldsymbol{E}_{C(E)}^{j} \end{matrix} \right\| \tag{A.37}$$

and the input–output transposition properties of the 2 × 3 ECM DC–AC commutator can be expressed as

$$\boldsymbol{E}_i^{s(I)} = \boldsymbol{C}_{ij}^{sP(I)} \boldsymbol{E}_{P(I)}^{j} \tag{A.38}$$

or in detail

$$\left\| \begin{matrix} \boldsymbol{E}_i^{a(I)} \\ \boldsymbol{E}_i^{b(I)} \\ \boldsymbol{E}_i^{c(I)} \end{matrix} \right\| = \left\| \begin{matrix} \boldsymbol{C}_{ij}^{Pa} & \boldsymbol{C}_{ij}^{Na} \\ \boldsymbol{C}_{ij}^{Pb} & \boldsymbol{C}_{ij}^{Nb} \\ \boldsymbol{C}_{ij}^{Pc} & \boldsymbol{C}_{ij}^{Nc} \end{matrix} \right\| \left\| \begin{matrix} \boldsymbol{E}_{P(I)}^{j} \\ \boldsymbol{E}_{N(I)}^{j} \end{matrix} \right\| \tag{A.39}$$

However, a mathematical model of an electrical continuous dynamical system of the 3 × 2 and 2 × 3 ECM AC–DC–AC commutator shown in figure A17, acting as a current-source ECM AC–DC–AC commutator, making dependent the input holors from the output holors, in the dynamical systems approach and matrix notation may be expressed in the form:

$$\boldsymbol{I}_P^j = \boldsymbol{C}_{Ps(I)}^{ji} \boldsymbol{C}_{Ps(E)}^{ji} \boldsymbol{I}_i^s \tag{A.40}$$

or in detail

$$\left\| \begin{matrix} \boldsymbol{I}_i^A \\ \boldsymbol{I}_i^B \\ \boldsymbol{I}_i^C \end{matrix} \right\| = \left\| \begin{matrix} \boldsymbol{C}_{Ap}^{ji} & \boldsymbol{C}_{An}^{ji} \\ \boldsymbol{C}_{Bp}^{ji} & \boldsymbol{C}_{Bn}^{ji} \\ \boldsymbol{C}_{Cp}^{ji} & \boldsymbol{C}_{Cn}^{ji} \end{matrix} \right\| \left\| \begin{matrix} \boldsymbol{C}_{Pa}^{ji} & \boldsymbol{C}_{Pb}^{ji} & \boldsymbol{C}_{Pc}^{ji} \\ \boldsymbol{C}_{Na}^{ji} & \boldsymbol{C}_{Nb}^{ji} & \boldsymbol{C}_{Nc}^{ji} \end{matrix} \right\| \left\| \begin{matrix} \boldsymbol{I}_a^j \\ \boldsymbol{I}_b^j \\ \boldsymbol{I}_c^j \end{matrix} \right\| \tag{A.41}$$

The input–output transposition properties of the 3 × 2 ECM AC–DC commutator can be expressed as

$$\boldsymbol{I}_{P(E)}^{j} = \boldsymbol{C}_{Ps(E)}^{ji} \boldsymbol{I}_i^{s(E)}$$

or in detail

$$\left\| \begin{matrix} \boldsymbol{I}_i^{A(E)} \\ \boldsymbol{I}_i^{B(E)} \\ \boldsymbol{I}_i^{C(E)} \end{matrix} \right\| = \left\| \begin{matrix} \boldsymbol{C}_{Ap}^{ji} & \boldsymbol{C}_{An}^{ji} \\ \boldsymbol{C}_{Bp}^{ji} & \boldsymbol{C}_{Bn}^{ji} \\ \boldsymbol{C}_{Cp}^{ji} & \boldsymbol{C}_{Cn}^{ji} \end{matrix} \right\| \left\| \begin{matrix} \boldsymbol{I}_{p(E)}^{j} \\ \boldsymbol{I}_{n(E)}^{j} \end{matrix} \right\|$$

and the input–output transposition properties of the 2 × 3 ECM DC–AC commutator can be expressed as

$$\boldsymbol{I}^{j}_{P(I)} = \boldsymbol{C}^{ji}_{Ps(I)}\boldsymbol{I}^{s(I)}_{i}$$

or in detail

$$\left\|\begin{matrix}\boldsymbol{I}^{P(I)}_{i}\\ \boldsymbol{I}^{N(I)}_{i}\end{matrix}\right\| = \left\|\begin{matrix}\boldsymbol{C}^{ji}_{Pa} & \boldsymbol{C}^{ji}_{Pb} & \boldsymbol{C}^{ji}_{Pc}\\ \boldsymbol{C}^{ji}_{Na} & \boldsymbol{C}^{ji}_{Nb} & \boldsymbol{C}^{ji}_{Nc}\end{matrix}\right\| \left\|\begin{matrix}\boldsymbol{I}^{j}_{a(I)}\\ \boldsymbol{I}^{j}_{b(I)}\\ \boldsymbol{I}^{j}_{c(I)}\end{matrix}\right\|$$

where:

$$\|\boldsymbol{E}^{s(E)}_{I}\| = \left\|\boldsymbol{E}^{p(E)}_{i}\ \boldsymbol{E}^{n(E)}_{i}\right\|^{T}$$

$$\left\|\boldsymbol{E}^{j}_{P(E)}\right\| = \left\|\boldsymbol{E}^{j}_{A(E)}\ \boldsymbol{E}^{j}_{B(E)}\ \boldsymbol{E}^{j}_{C(E)}\right\|^{T}$$

$$\|\boldsymbol{E}^{s(I)}_{i}\| = \left\|\boldsymbol{E}^{a(I)}_{i}\ \boldsymbol{E}^{b(I)}_{i}\ \boldsymbol{E}^{c(I)}_{i}\right\|^{T}$$

$$\left\|\boldsymbol{E}^{j}_{P(I)}\right\| = \left\|\boldsymbol{E}^{j}_{P(I)}\ \boldsymbol{E}^{j}_{N(I)}\right\|^{T}$$

$$\left\|\boldsymbol{I}^{j}_{s(I)}\right\| = \left\|\boldsymbol{I}^{j}_{a(I)}\ \boldsymbol{I}^{j}_{b(I)}\ \boldsymbol{I}^{j}_{c(I)}\right\|^{T}$$

$$\|\boldsymbol{I}^{P(I)}_{i}\| = \left\|\boldsymbol{I}^{P(I)}_{i}\ \boldsymbol{I}^{N(I)}_{i}\right\|^{T}$$

$$\|\boldsymbol{I}^{P(E)}_{i}\| = \left\|\boldsymbol{I}^{A(E)}_{i}\ \boldsymbol{I}^{B(E)}_{i}\ \boldsymbol{I}^{C(E)}_{i}\right\|^{T}$$

$$\left\|\boldsymbol{I}^{j}_{s(E)}\right\| = \left\|\boldsymbol{I}^{j}_{p(E)}\ \boldsymbol{I}^{j}_{n(E)}\right\|^{T}$$

$\boldsymbol{C}^{ji}_{sP(E)}$ is the commutation-matrix holor of the voltage-source 3 × 2 ECM AC–DC commutator connected with the stiff voltage holors, $\boldsymbol{C}^{ji}_{sP(I)}$ is the commutation-matrix holor of the current-source 2 × 3 ECM DC–AC commutator connected with the stiff current holors. They are given by

$$\boldsymbol{C}^{sP(E)}_{ij}\boldsymbol{C}^{sP(I)}_{ij} = \left\|\begin{matrix}\boldsymbol{C}^{Pa}_{ij} & \boldsymbol{C}^{Na}_{ij}\\ \boldsymbol{C}^{Pb}_{ij} & \boldsymbol{C}^{Nb}_{ij}\\ \boldsymbol{C}^{Pc}_{ij} & \boldsymbol{C}^{Nc}_{ij}\end{matrix}\right\| \left\|\begin{matrix}\boldsymbol{C}^{Ap}_{ij} & \boldsymbol{C}^{Bp}_{ij} & \boldsymbol{C}^{Cp}_{ij}\\ \boldsymbol{C}^{An}_{ij} & \boldsymbol{C}^{Bn}_{ij} & \boldsymbol{C}^{Cn}_{ij}\end{matrix}\right\|$$

$$\boldsymbol{C}^{sP(E)}_{ij} = \left\|\begin{matrix}\boldsymbol{C}^{Pa}_{ij} & \boldsymbol{C}^{Na}_{ij}\\ \boldsymbol{C}^{Pb}_{ij} & \boldsymbol{C}^{Nb}_{ij}\\ \boldsymbol{C}^{Pc}_{ij} & \boldsymbol{C}^{Nc}_{ij}\end{matrix}\right\|$$

and

$$\boldsymbol{C}^{sP(I)}_{ij} = \left\|\begin{matrix}\boldsymbol{C}^{Ap}_{ij} & \boldsymbol{C}^{Bp}_{ij} & \boldsymbol{C}^{Cp}_{ij}\\ \boldsymbol{C}^{An}_{ij} & \boldsymbol{C}^{Bn}_{ij} & \boldsymbol{C}^{Cn}_{ij}\end{matrix}\right\|$$

By matrix multiplication of the commutation-matrix holor of the voltage-source 3×2 ECM AC–DC commutator and the commutation-matrix holor of the current-source 2×3 ECM DC–AC commutator may receive a mathematical confirmation that a product of the commutation-matrixer holors of the 3×2 and 2×3 ECM commutator is equal the commutation-matrix holor of the 3×3 ECM AC–AC commutator.

$$\boldsymbol{C}_{ij}^{sP(E)}\boldsymbol{C}_{ij}^{sP(I)} = \begin{Vmatrix} \boldsymbol{C}_{ij}^{Pa} & \boldsymbol{C}_{ij}^{Na} \\ \boldsymbol{C}_{ij}^{Pb} & \boldsymbol{C}_{ij}^{Nb} \\ \boldsymbol{C}_{ij}^{Pc} & \boldsymbol{C}_{ij}^{Nc} \end{Vmatrix} \begin{Vmatrix} \boldsymbol{C}_{ij}^{Ap} & \boldsymbol{C}_{ij}^{Bp} & \boldsymbol{C}_{ij}^{Cp} \\ \boldsymbol{C}_{ij}^{An} & \boldsymbol{C}_{ij}^{Bn} & \boldsymbol{C}_{ij}^{Cn} \end{Vmatrix}$$

$$= \begin{Vmatrix} \boldsymbol{C}_{ij}^{Pa}\boldsymbol{C}_{ij}^{Ap} + \boldsymbol{C}_{ij}^{Na}\boldsymbol{C}_{ij}^{An} & \boldsymbol{C}_{ij}^{Pa}\boldsymbol{C}_{ij}^{Bp} + \boldsymbol{C}_{ij}^{Na}\boldsymbol{C}_{ij}^{Bn} & \boldsymbol{C}_{ij}^{Pa}\boldsymbol{C}_{ij}^{Cp} + \boldsymbol{C}_{ij}^{Na}\boldsymbol{C}_{ij}^{Cn} \\ \boldsymbol{C}_{ij}^{Pb}\boldsymbol{C}_{ij}^{Ap} + \boldsymbol{C}_{ij}^{Nb}\boldsymbol{C}_{ij}^{An} & \boldsymbol{C}_{ij}^{Pb}\boldsymbol{C}_{ij}^{Bp} + \boldsymbol{C}_{ij}^{Nb}\boldsymbol{C}_{ij}^{Bn} & \boldsymbol{C}_{ij}^{Pb}\boldsymbol{C}_{ij}^{Cp} + \boldsymbol{C}_{ij}^{Nb}\boldsymbol{C}_{ij}^{Cn} \\ \boldsymbol{C}_{ij}^{Pc}\boldsymbol{C}_{ij}^{Ap} + \boldsymbol{C}_{ij}^{Nc}\boldsymbol{C}_{ij}^{An} & \boldsymbol{C}_{ij}^{Pc}\boldsymbol{C}_{ij}^{Bp} + \boldsymbol{C}_{ij}^{Nc}\boldsymbol{C}_{ij}^{Bn} & \boldsymbol{C}_{ij}^{Pc}\boldsymbol{C}_{ij}^{Cp} + \boldsymbol{C}_{ij}^{Nc}\boldsymbol{C}_{ij}^{Cn} \end{Vmatrix}$$

$$= \begin{Vmatrix} \boldsymbol{C}_{ij}^{Aa} & \boldsymbol{C}_{ij}^{Ba} & \boldsymbol{C}_{ij}^{Ca} \\ \boldsymbol{C}_{ij}^{Ab} & \boldsymbol{C}_{ij}^{Bb} & \boldsymbol{C}_{ij}^{Cb} \\ \boldsymbol{C}_{ij}^{Ac} & \boldsymbol{C}_{ij}^{Bc} & \boldsymbol{C}_{ij}^{Cc} \end{Vmatrix}.$$

Similarly, by matrix multiplication of the commutation-matrix holor of the current-source 2×3 ECM AC–DC commutator and the commutation-matrix holor of the voltage-source 3×2 ECM DC–AC commutator may receive a mathematical confirmation that a product of the commutation-matrixer holors of the 3×2 and 2×3 ECM commutator is equal as well the commutation-matrix holor of the 3×3 ECM AC–AC commutator.

$$\boldsymbol{C}_{Ps(E)}^{ji}\boldsymbol{C}_{Ps(I)}^{ji} = \begin{Vmatrix} \boldsymbol{C}_{Ap}^{ji} & \boldsymbol{C}_{An}^{ji} \\ \boldsymbol{C}_{Bp}^{ji} & \boldsymbol{C}_{Bn}^{ji} \\ \boldsymbol{C}_{Cp}^{ji} & \boldsymbol{C}_{Cn}^{ji} \end{Vmatrix} \begin{Vmatrix} \boldsymbol{C}_{Pa}^{ji} & \boldsymbol{C}_{Pb}^{ji} & \boldsymbol{C}_{Pc}^{ji} \\ \boldsymbol{C}_{Na}^{ji} & \boldsymbol{C}_{Nb}^{ji} & \boldsymbol{C}_{Nc}^{ji} \end{Vmatrix}$$

$$= \begin{Vmatrix} \boldsymbol{C}_{Ap}^{ji}\boldsymbol{C}_{Pa}^{ji} + \boldsymbol{C}_{An}^{ji}\boldsymbol{C}_{Na}^{ji} & \boldsymbol{C}_{Ap}^{ji}\boldsymbol{C}_{Pb}^{ji} + \boldsymbol{C}_{An}^{ji}\boldsymbol{C}_{Nb}^{ji} & \boldsymbol{C}_{Ap}^{ji}\boldsymbol{C}_{Pc}^{ji} + \boldsymbol{C}_{An}^{ji}\boldsymbol{C}_{Nc}^{ji} \\ \boldsymbol{C}_{Bp}^{ji}\boldsymbol{C}_{Pa}^{ji} + \boldsymbol{C}_{Bn}^{ji}\boldsymbol{C}_{Na}^{ji} & \boldsymbol{C}_{Bp}^{ji}\boldsymbol{C}_{Pb}^{ji} + \boldsymbol{C}_{Bn}^{ji}\boldsymbol{C}_{Nb}^{ji} & \boldsymbol{C}_{Bp}^{ji}\boldsymbol{C}_{Pc}^{ji} + \boldsymbol{C}_{An}^{ji}\boldsymbol{C}_{Nc}^{ji} \\ \boldsymbol{C}_{Cp}^{ji}\boldsymbol{C}_{Pa}^{ji} + \boldsymbol{C}_{Cn}^{ji}\boldsymbol{C}_{Na}^{ji} & \boldsymbol{C}_{Cp}^{ji}\boldsymbol{C}_{Pb}^{ji} + \boldsymbol{C}_{Cn}^{ji}\boldsymbol{C}_{Nb}^{ji} & \boldsymbol{C}_{Cp}^{ji}\boldsymbol{C}_{Pc}^{ji} + \boldsymbol{C}_{An}^{ji}\boldsymbol{C}_{Nc}^{ji} \end{Vmatrix}$$

$$= \begin{Vmatrix} \boldsymbol{C}_{Aa}^{ji} & \boldsymbol{C}_{Ab}^{ji} & \boldsymbol{C}_{Ac}^{ji} \\ \boldsymbol{C}_{Ba}^{ji} & \boldsymbol{C}_{Bb}^{ji} & \boldsymbol{C}_{Bc}^{ji} \\ \boldsymbol{C}_{Ca}^{ji} & \boldsymbol{C}_{Cb}^{ji} & \boldsymbol{C}_{Cc}^{ji} \end{Vmatrix},$$

where:

$$\left\| \begin{matrix} C^{ji}_{Aa} & C^{ji}_{Ab} & C^{ji}_{Ac} \\ C^{ji}_{Ba} & C^{ji}_{Bb} & C^{ji}_{Bc} \\ C^{ji}_{Ca} & C^{ji}_{Cb} & C^{ji}_{Cc} \end{matrix} \right\| = \left\| \begin{matrix} C^{ji}_{Aa} & C^{ji}_{Ba} & C^{ji}_{Ca} \\ C^{ji}_{Ab} & C^{ji}_{Bb} & C^{ji}_{Cb} \\ C^{ji}_{Ac} & C^{ji}_{Bc} & C^{ji}_{Cc} \end{matrix} \right\|^T .$$

Commutation functions and voltage/current holors—Let the three-phase matrixer-input-voltage holors and three-phase matrixer-output-current holors be assumed to be balanced stiff AC holor quantities and expressed as

$$\begin{aligned} &E_i^A = E\angle\alpha && I_a^j = I\angle\beta \\ &E_i^B = E\angle\,\alpha + 2\frac{\pi}{3} && I_b^j = I\angle\beta\; +\; 2\pi/3 \\ &E_i^C = E\angle\alpha - \frac{2\pi}{3} && I_c^j = I\angle\beta - 2\pi/3 \end{aligned} \tag{A.42}$$

where:

E is voltage-source holor amplitude;
I is the current-source holor amplitude;
α_P is the phase angle of the voltage-source holor, given by $\alpha_P = \omega_P t + \alpha_{P0}$;
β_s is the phase angle of the current-source holor, given by $\beta_s = \omega_s t + \beta_{s0}$.

Let the modulation functions for the current-source 2 × 3 ECM and voltage-source 3 × 2 ECM will be defined as

$$\begin{aligned} &M_i^A = 1\angle\alpha && M_a^j = 1\angle\beta \\ &M_i^B = 1\angle\alpha\; +\; 2\frac{\pi}{3} && M_b^j = 1\angle\beta\; +\; 2\pi/3 \\ &M_i^C = 1\angle\alpha{-}2\pi/3 && M_c^j = \angle\beta{-}2\pi/3 \end{aligned} \tag{A.43}$$

The displacement angle φ_P on the current-source 2 × 3 ECM side and the displacement angle φ_s on the voltage-source 3 × 2 ECM side are defined as

$$\begin{aligned} \varphi_P &= \alpha_{P0} - \beta_{P0} \\ \varphi_s &= \alpha_{s0} - \beta_{s0} \end{aligned} \tag{A.44}$$

The commutation (switching) functions for the various throws of the current-source 3 × 2 ECM and voltage-source 2 × 3 ECM are determined by equations (A.43) and (A.44), respectively.

$$
\begin{aligned}
C_{pA}^{ji} &= \begin{cases} 1 & \text{if } (M_i^A > M_i^B) \wedge (M_i^A > M_i^C) \\ 0 & \text{otherwise} \end{cases} \\
C_{pB}^{ji} &= \begin{cases} 1 & \text{if } (M_i^B > M_i^C) \wedge (M_i^B > M_i^A) \\ 0 & \text{otherwise} \end{cases} \\
C_{pC}^{ji} &= \begin{cases} 1 & \text{if } (M_i^C > M_i^A) \wedge (M_i^C > M_i^B) \\ 0 & \text{otherwise} \end{cases} \\
C_{nA}^{ji} &= \begin{cases} 1 & \text{if } (M_i^A < M_i^B) \wedge (M_i^A < M_i^C) \\ 0 & \text{otherwise} \end{cases} \\
C_{nB}^{ji} &= \begin{cases} 1 & \text{if } (M_i^B < M_i^C) \wedge (M_i^B < M_i^A) \\ 0 & \text{otherwise} \end{cases} \\
C_{nC}^{ji} &= \begin{cases} 1 & \text{if } (M_i^C < M_i^A) \wedge (M_i^C < M_i^B) \\ 0 & \text{otherwise} \end{cases}
\end{aligned}
\tag{A.45}
$$

$$
\begin{aligned}
C_{ij}^{ap} &= \begin{cases} 1 & \text{if } M_i^A > 0 \\ 0 & \text{otherwise} \end{cases} \quad ; C_{ij}^{an} = 1 - C_{ij}^{ap} \\
C_{ij}^{bp} &= \begin{cases} 1 & \text{if } M_i^B > 0 \\ 0 & \text{otherwise} \end{cases} \quad ; C_{ij}^{bn} = 1 - C_{ij}^{bp} \\
C_{ij}^{cp} &= \begin{cases} 1 & \text{if } M_i^C > 0 \\ 0 & \text{otherwise} \end{cases} \quad ; C_{ij}^{cn} = 1 - C_{ij}^{cp}
\end{aligned}
\tag{A.46}
$$

The commutation functions for the current-source 3×2 ECM and voltage-source 2×3 ECM are depicted in figure A18. With this choice of commutation functions, the resulting output voltage holors and input-current holors for unity displacement-factor operation ($\cos \varphi = 1$) at input and output simultaneously.

Summary—Conventional modulation strategies ECM commutators based on trigonometric transformations and holor techniques have their voltage transfer capability limited at about 0.866 pu. This leads to a derating of AC–AC commutator motor capability by 25% in the realisation of an IEMD using **off-the-shelf** (OTS) line-fed commutatorless AC induction motors, or requires custom designed AC–AC commutator asynchronous (induction) motors for operation at reduced voltage. Thus, a six-step variable-frequency modulation strategy for the ECM commutators that is capable of providing more than 1 pu fundamental component of output voltage is presented.

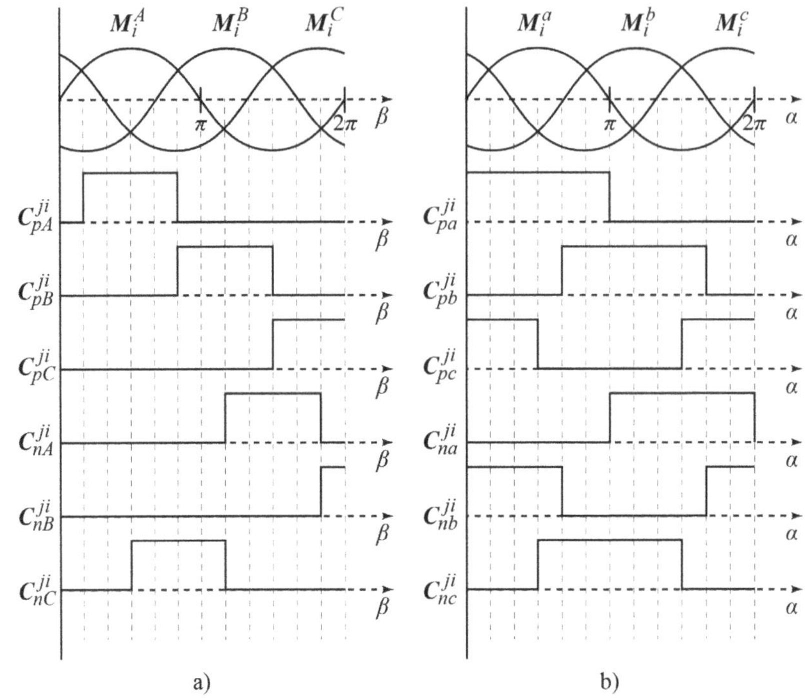

Figure A18. Waveforms of various commutation functions for current-source 3 × 2 ECM AC–DC commutator (a) and voltage-source 2 × 3 ECM DC–AC commutator (b) for unity power factor operation at both sets of the matrixer terminals or both matrixer ports.

In this subsection a new modulation/control strategy for the 3 × 2 and 2 × 3 ECM commutator featuring up to unity voltage transfer ratio is presented. Using such a low frequency modulation approach combined with proper filter design and/or multi-pulse techniques, practical designs may be realised for applications using electrical valves as IGBTs or IGCTs.

The combination of the presented modulation together with conventional **pulse-width modulation** (PWM) provides independent specification of power factor at both the matrixer-input and matrixer-output ports of the MCM commutator. Furthermore, the presented approach makes it possible to develop the limits of conventional scalar and vector PWM approaches in the transition region between linear operation and six-step operation.

Six-step operation suitable for the conventional 9-electrical-valve topology and the reduced electrical valve (switch) dual monolithic ECM topologies is presented, with capability of unity displacement-factor operation ($\cos \varphi = 1$) at the matrixer-input and matrixer-output simultaneously.

It is demonstrated that by varying the commutation (switching) angles, control of fundamental component of output voltage holor and input-current holor is possible, with concomitant generation of reactive power at the matrixer-input and matrixer-output terminals of the monolithic ECM AC–DC–AC commutator, with bipolar (bilateral) power flow.

A.9 ECM DC–DC commutators

The monolithic 2 × 2 ECM DC–DC commutator with a bipolar ECM, realised on only four (4) 'continuous' **reverse-blocking insulated-gate bipolar transistors** (RB IGBT), or **metal-oxide semiconductor field effect transistors** (MOSFET), or **integrated-gate commutated thyristor** (IGCT), or bipolar symmetrical trigistors, that is, **gate-turn-off** (GTO) symmetrical thyristors (symistors) in 2 × 2 ECM connection is shown in figure A19.

However, the approach may be applicable for the monolithic 2 × 2 ECM (4-electrical valves) topology well, as may be demonstrated through appropriate topological mapping.

Figure A19 illustrates the 2 × 2 ECM using three monolithic single-pole-double-valvular (SPDV) commutation groups V_p, V_n, i.e. with four IGBTs.

If the commutation function of the IGBTs, as illustrated in figure A19, is defined to be $\boldsymbol{C}_{ij}^{sP}$, the input–output transposition properties of the ECM can be expressed as

$$\boldsymbol{E}_i^s = \boldsymbol{C}_{ij}^{sP} \boldsymbol{E}_P^j \tag{A.47}$$

or in detail

$$\begin{Vmatrix} \boldsymbol{E}_i^p \\ \boldsymbol{E}_i^n \end{Vmatrix} = \begin{Vmatrix} \boldsymbol{C}_{ij}^{Pp} & \boldsymbol{C}_{ij}^{Np} \\ \boldsymbol{C}_{ij}^{Pn} & \boldsymbol{C}_{ij}^{Nn} \end{Vmatrix} \begin{Vmatrix} \boldsymbol{E}_P^j \\ \boldsymbol{E}_N^j \end{Vmatrix} \tag{A.48}$$

however, a mathematical model of an electrical continuous dynamical system of the 2 × 2 ECM AC–AC commutator shown in figure A19, acting as a current-source ECM DC–DC commutator, making dependent the input holors from the output holors, in the dynamical approach and matrix notation may be systems expressed in the form:

$$\boldsymbol{I}_P^j = \boldsymbol{C}_{Ps}^{ji} \boldsymbol{I}_i^s \tag{A.49}$$

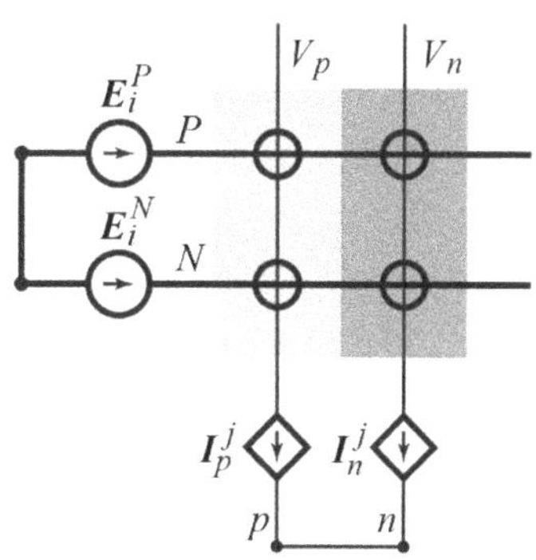

Figure A19. The monolithic 2 × 2 ECM DC–DC commutator, i.e. with four IGBTs.

in detail

$$\left\| \begin{matrix} I_P^j \\ I_N^j \end{matrix} \right\| = \left\| \begin{matrix} C_{Pp}^{ji} & C_{Pn}^{ji} \\ C_{Np}^{ji} & C_{Nn}^{ji} \end{matrix} \right\| \left\| \begin{matrix} I_i^p \\ I_i^n \end{matrix} \right\| \tag{A.50}$$

where:

$$\|E_i^s\| = \left\|E_i^p \; E_i^n\right\|^T;$$
$$\|E_P^j\| = \left\|E_P^j \; E_N^j\right\|^T;$$
$$\|I_P^j\| = \left\|I_P^J \; I_N^J\right\|^T;$$
$$\|I_i^s\| = \left\|I_I^p \; I_I^n\right\|^T.$$

C_{ij}^{sP} is the voltage-throws shown in figure A19 source holors commutation matrix with elements being commutation functions of various throws

$$C_{ij}^{sP} = \left\| \begin{matrix} C_{ij}^{Pp} \; C_{ij}^{Np} \\ C_{ij}^{Pn} \; C_{ij}^{Nn} \end{matrix} \right\|$$

C_{Ps}^{ji} is the current-source holors commutation matrix with elements being commutation of various throws shown in figure A19.

$$C_{Ps}^{ji} = \left\| \begin{matrix} C_{Pp}^{ji} \; C_{Pn}^{ji} \\ C_{Np}^{ji} \; C_{Nn}^{ji} \end{matrix} \right\| = \left\| \begin{matrix} C_{Pp}^{ji} \; C_{Np}^{ji} \\ C_{Pn}^{ji} \; C_{Nn}^{ji} \end{matrix} \right\|^T$$

C_{ij}^{Ps} are the voltage-source coefficients of the commutation nodes 'kL', which may be expressed as follows:

$$C_{ij}^{Kl} = \frac{\partial E_i^K}{\partial E_l^j}\bigg| E_i^j = 0 = \begin{cases} -1 \overset{\Delta}{\rightarrow} E_i^K = -E_l^j \\ 0 \overset{\Delta}{\rightarrow} E_i^K = 0 \\ +1 \overset{\Delta}{\rightarrow} E_i^K = +E_l^j \end{cases}$$

C_{sP}^{ji} are the current-source coefficients of the commutation nodes '*Kl*', which may be expressed as follows:

$$C_{kL}^{ji} = \frac{\partial E_k^j}{\partial E_i^L}\bigg| E_i^I = 0 = \begin{cases} +1 \overset{\Delta}{\rightarrow} E_k^j = +E_i^L \\ 0 \overset{\Delta}{\rightarrow} E_k^j = 0 \\ -1 \overset{\Delta}{\rightarrow} E_k^j = -E_i^L \end{cases}$$

This monolithic 2 × 2 ECM DC–DC commutator makes the bilateral flow of secondary electrical energy possible both during generating (absorbing) as well as during motoring (driving) of the DC–DC commutator dynamotor with the variable angular velocity. Four operating modes are possible in this MMD electrical machine.

A.10 ECM DC–AC commutators

ECM DC–AC commutators, acting as DC–AC inverters, are electrical power converters that change **direct current** (DC) to **alternating current** (AC). The converted AC can be at any required voltage and frequency with the use of appropriate transformers, commutation (switching), and control circuits.

Solid-state ECM DC–AC commutators have no moving parts and are used in a wide range of applications, from small commutation (switching) power supplies in computers, to large electric utility **high-voltage direct current** (HVDC) applications that transport bulk power. ECM DC–AC commutators are commonly used to supply AC power from DC sources such as solar panels or **chemo-electrical** (Ch-E) storage batteries.

The ECM DC–AC commutator, acting as a DC–AC inverter, performs the opposite function of an ECM AC–DC commutator, acting as an AC–DC rectifier. The ECM DC–AC commutator is a high-power macroelectronic oscillator. It is so named because early MCM AC–DC commutator converters were made to operate in reverse, and thus were 'inverted', to convert DC–AC.

ECM DC–AC commutators, acting as DC–AC inverters, depending on the kind of the source of feeding and the related ECM topology of the power electrical circuit, are classified as **voltage source** (VS) DC–AC inverters and **current source** (CS) DC–AC inverters.

Figure A20 shows a type of ECM DC–AC commutator, acting as a DC–AC inverter.

Three-phase counterparts of the single-phase half and full ECM DC–AC commutators, acting as VS DC–AC inverters, are shown in figures A21 and A22.

Single-phase VS DC–AC inverters cover low-power range applications and three-phase VS DC–AC inverters cover medium to high-power range applications. The main purpose of these ECM topologies is to provide a three-phase voltage source, where the amplitude, phase and frequency of the voltages can be controlled. The three-phase VS DC–AC inverters are extensively being used in the DC as well as AC IEMD, active filters and unified power flow controllers in power systems and **uninterrupted power supplies** (UPS) to generate controllable frequency and AC voltage magnitudes using various **pulse-width modulation** (PWM) strategies. The standard three-phase ECM DC–AC commutator, acting as a DC–AC inverter, shown in figure A23 has six electrical valves (electronic switches) the commutation (switching) of which depends on the modulation scheme.

The input DC is usually obtained from a single-phase or three-phase utility power supply through an ECM commutator, acting as a diode AC–DC rectifier and *LC* or *C* filter. There are many applications required DC–AC conversion especially in industry. For example, E-M motor control and renewable energy where the DC source will be inverted to AC output to suit the E-M motor rating. The speed of the

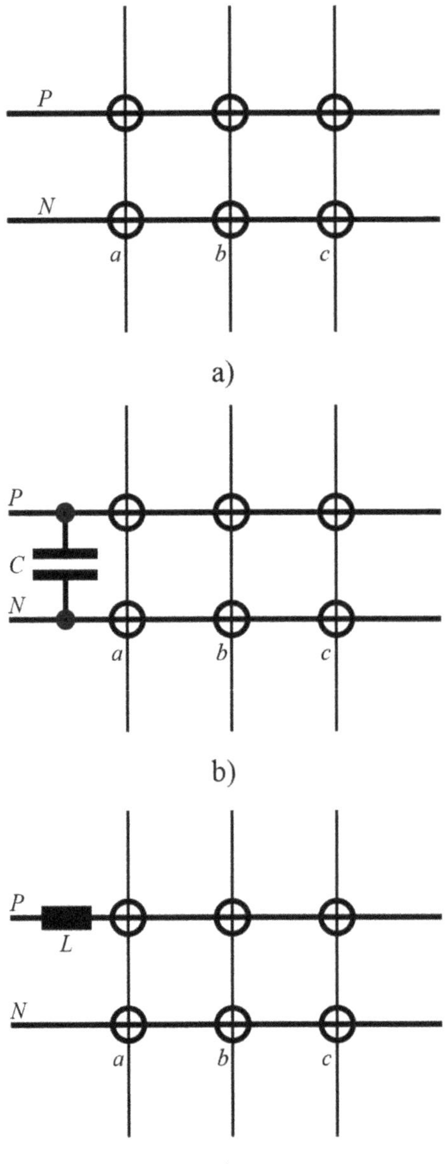

Figure A20. A type of ECM DC–AC commutator, acting as a DC–AC inverter: (a) general DC–AC inverter; (b) VS DC–AC inverter; (c) CS DC–AC inverter.

AC–DC–AC commutator synchronous or asynchronous (induction) motor can be varied by controlling the output voltage frequency and amplitude.

The load is fed from a VS DC–AC inverter with current control. The control is performed by regulating the flow of current to load. Current controllers are used to generate gate signals for the DC–AC inverter. Proper selection of the DC–AC

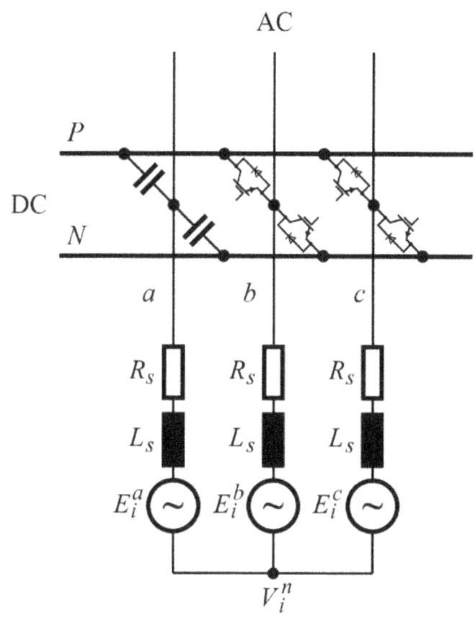

Figure A21. Single-phase half ECM DC–AC commutators, acting as the VS DC–AC inverter.

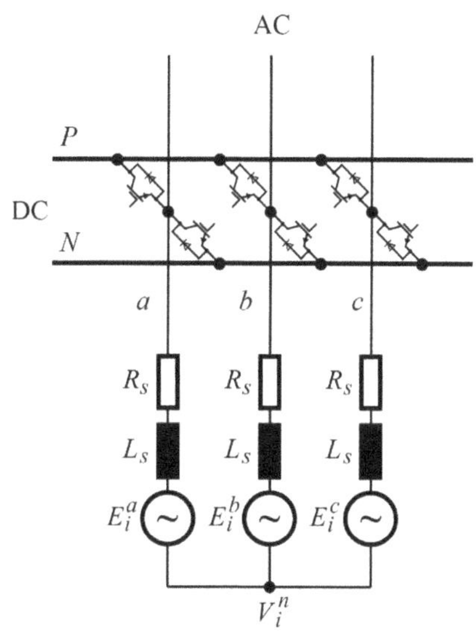

Figure A22. Single-phase full ECM DC–AC commutators, acting as the VS DC–AC inverter.

inverter's electrical valves and selection of the control technique will guarantee the efficacy of the DC motor IEMD. VS DC–AC inverters are macroelectronic devices that convert a DC voltage to AC voltage of variable frequency and magnitude. They are very commonly used in adjustable velocity DC motor IEMDs and are characterised by a well-defined commutated (switched) voltage waveform in the terminals. Figure A20(b) shows a VS DC–AC inverter. The AC voltage frequency can be variable or constant depending on the application.

Three-phase ECM DC–AC commutators, acting as DC–AC inverters, consist of six electrical valves (electronic switches) connected to a DC voltage source, as shown in figure A24.

The DC–AC inverter's electrical valves must be carefully selected based on the requirements of operation, ratings and the application. There are several electrical valves available today and these are thyristors, **bipolar junction transistors** (BJTs), **MOS field effect transistors** (MOSFETs), **insulated-gate bipolar transistors** (IGBTs), **gate-turn-off** (GTO) thyristors and **integrated-gate commutated thyristors**. MOSFETs and IGBTs are preferred by industry because of the MOS gating permits high-power gain and control advantages.

While MOSFET is considered a universal electrical valve for low-power and low-voltage applications, IGBT has wide acceptance for the IEMD and other

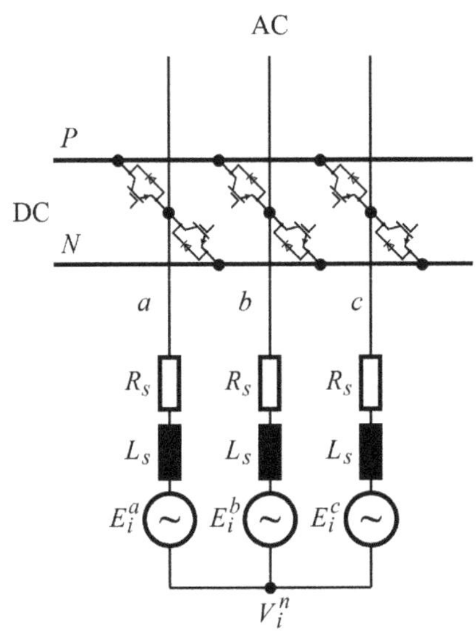

Figure A23. Standard three-phase ECM DC–AC commutator, acting as a DC–AC inverter.

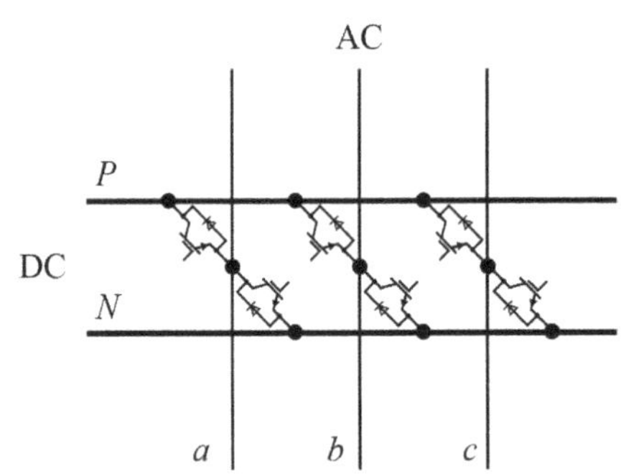

Figure A24. Three-phase ECM DC–AC commutators, acting as DC–AC inverters, consist of six electrical valves connected to a DC voltage source.

application in the low- and medium-power range. The electrical valves, when used in IEMD applications, require an inductive E-M motor current path provided by antiparallel diodes when it is turned off.

IGBTs provide high-input impedance and are used for high-voltage applications. The high-input impedance allows the electrical valve to commute (switch) with a small amount of energy and for high-voltage applications the electrical valve must have large blocking voltage ratings. The electrical valve behaviour is described by parameters like voltage drop or on resistance, turn-on time and turn-off time. The ECM DC–AC commutator, acting as a DC–AC inverter with IGBTs, is shown in figure A24.

The monolithic 2 × 3 ECM DC–AC commutator with a bipolar ECM, realised on only six (6) 'continuous' MOSFETs in 2 × 3 ECM connection is shown in figure A25.

However, the approach may be applicable for the monolithic 2 × 3 ECM (six electrical valves) topology well, as may be demonstrated through appropriate topological mapping.

Figure A26 illustrates the 2 × 3 ECM using three monolithic SPTV commutation groups V_a V_b and V_c, i.e. with six RB IGBTs.

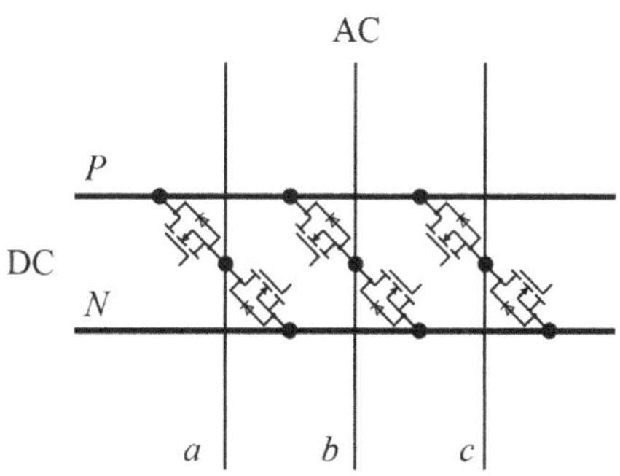

Figure A25. Monolithic 2 × 3 ECM DC–AC commutator with a bipolar ECM, realised on only six 'continuous' MOSFETs in a 2 × 3 ECM connection.

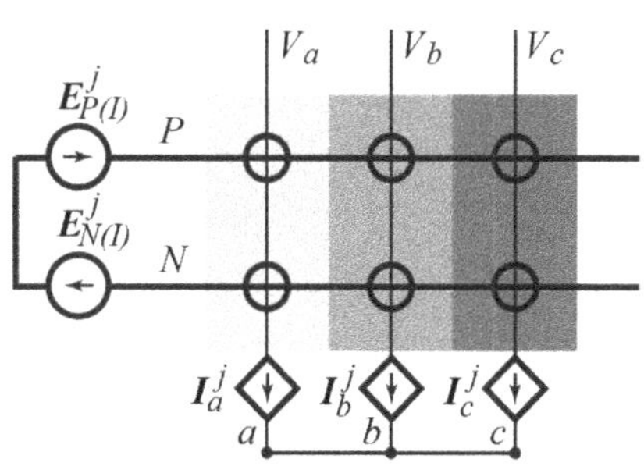

Figure A26. The monolithic 2 × 3 ECM DC–AC commutator, i.e. with six RB IGBTs.

If the commutation function of the RB IGBTs, as illustrated in figure A26, is defined to be $C_{ij}^{sP(I)}$, the input–output transposition properties of the ECM can be expressed as

$$E_i^{s(I)} = C_{ij}^{sP(I)} E_{P(I)}^j \tag{A.51}$$

or in detail

$$\begin{Vmatrix} E_i^{a(I)} \\ E_i^{b(I)} \\ E_i^{c(I)} \end{Vmatrix} = \begin{Vmatrix} C_{ij}^{Pa} & C_{ij}^{Na} \\ C_{ij}^{Pb} & C_{ij}^{Nb} \\ C_{ij}^{Pc} & C_{ij}^{Nc} \end{Vmatrix} \begin{Vmatrix} E_{P(I)}^j \\ E_{N(I)}^j \end{Vmatrix} \tag{A.52}$$

however, a mathematical model of an electrical continuous dynamical system of the 2×3 ECM AC–AC commutator shown in figure A26, acting as a current-source ECM DC–AC commutator, making dependent the input holors from the output holors, in the dynamical systems approach and matrix notation may be expressed in the form:

$$I_{P(I)}^j = C_{Ps(I)}^{ji} I_I^{s(I)}$$

or in detail

$$\begin{Vmatrix} I_{P(I)}^j \\ I_{N(I)}^j \end{Vmatrix} = \begin{Vmatrix} C_{Pa}^{ji} & C_{Pb}^{ji} & C_{Pc}^{ji} \\ C_{Na}^{ji} & C_{Nb}^{ji} & C_{Nc}^{ji} \end{Vmatrix} \begin{Vmatrix} I_i^{a(I)} \\ I_i^{b(I)} \\ I_i^{c(I)} \end{Vmatrix}$$

where:

$$\|E_i^{s(I)}\| = \begin{Vmatrix} E_i^{a(I)} & E_i^{b(I)} & E_i^{c(I)} \end{Vmatrix}^T$$

$$\begin{Vmatrix} E_{P(I)}^j \end{Vmatrix} = \begin{Vmatrix} E_{P(I)}^j & E_{N(I)}^j \end{Vmatrix}^T$$

$$\begin{Vmatrix} I_{s(I)}^j \end{Vmatrix} = \begin{Vmatrix} I_{a(I)}^j & I_{b(I)}^j & I_{c(I)}^j \end{Vmatrix}^T$$

$$\|I_i^{P(I)}\| = \begin{Vmatrix} I_i^{P(I)} & I_i^{P(I)} \end{Vmatrix}^T$$

$C_{ij}^{sP(I)}$ is the commutation-matrix holor of the voltage-source 2×3 ECM AC–DC commutator connected with the voltage holors, $C_{Ps(I)}^{ji}$ is the commutation-matrix holor of the current-source 2×3 ECM DC–AC commutator connected with the current holors. They are given by

$$C_{ij}^{sP(I)} = \begin{Vmatrix} C_{ij}^{Pa} & C_{ij}^{Na} \\ C_{ij}^{Pb} & C_{ij}^{Nb} \\ C_{ij}^{Pc} & C_{ij}^{Nc} \end{Vmatrix}$$

and

$$C^{ji}_{Ps(I)} = \begin{Vmatrix} C^{ji}_{Pa} & C^{ji}_{Pb} & C^{ji}_{Pc} \\ C^{ji}_{Na} & C^{ji}_{Nb} & C^{ji}_{Nc} \end{Vmatrix}$$

A.11 ECM DC–AC–DC commutators

The modulation strategy may be more easily developed on the basis of operation of the dual monolithic 2 × 3 and 3 × 2 ECM topology with a fictitious AC link, by using commutation-matrix holors multiplication of these two cascade connected ECMs.

Figure A27 illustrates the reduced electrical valve (switch) three monolithic SPDV commutation groups V_a, V_b, V_c with six (6) RB IGBTs and two monolithic SPTV commutation groups V_p and V_n with six RB IGBTs. The two realisations are labelled as conventional ECM and indirect ECM, respectively.

The input–output transposition properties of the 2 × 3 and 3 × 2 ECM DC–AC–DC commutator can be expressed as

$$E_i^s = C_{ij}^{sP(E)} C_{ij}^{sP(I)} E_P^j \tag{A.53}$$

or in detail

$$\begin{Vmatrix} E_i^p \\ E_i^n \end{Vmatrix} = \begin{Vmatrix} C_{ij}^{Ap} & C_{ij}^{Bp} & C_{ij}^{Cp} \\ C_{ij}^{An} & C_{ij}^{Bn} & C_{ij}^{Cn} \end{Vmatrix} \begin{Vmatrix} C_{ij}^{Pa} & C_{ij}^{Na} \\ C_{ij}^{Pb} & C_{ij}^{Nb} \\ C_{ij}^{Pc} & C_{ij}^{Nc} \end{Vmatrix} \begin{Vmatrix} E_P^j \\ E_N^j \end{Vmatrix} \tag{A.54}$$

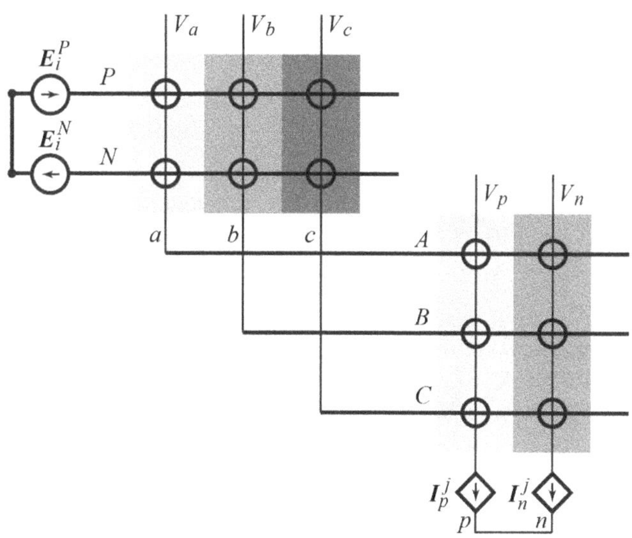

Figure A27. Representation of the monolithic 2 × 3 and 3 × 2 ECM AC–DC–AC commutator.

where $\boldsymbol{C}_{ij}^{sP(I)}$ is the commutation matrix of the voltage-source 2×3 ECM DC–AC commutator connected with the stiff voltage holors. $\boldsymbol{C}_{ij}^{sP(E)}$ is the commutation matrix of the voltage-source 3×2 ECM AC–DC commutator connected with the stiff voltage holors. They are given by

$$\boldsymbol{C}_{ij}^{sP(E)} = \left\| \begin{matrix} \boldsymbol{C}_{ij}^{Ap} & \boldsymbol{C}_{ij}^{Bp} & \boldsymbol{C}_{ij}^{Cp} \\ \boldsymbol{C}_{ij}^{An} & \boldsymbol{C}_{ij}^{Bn} & \boldsymbol{C}_{ij}^{Cn} \end{matrix} \right\|$$

and

$$\boldsymbol{C}_{ij}^{sP(I)} = \left\| \begin{matrix} \boldsymbol{C}_{ij}^{Pa} & \boldsymbol{C}_{ij}^{Na} \\ \boldsymbol{C}_{ij}^{Pb} & \boldsymbol{C}_{ij}^{Nb} \\ \boldsymbol{C}_{ij}^{Pc} & \boldsymbol{C}_{ij}^{Nc} \end{matrix} \right\|$$

The input–output transposition properties of the 3×2 ECM AC–DC commutator can be expressed as

$$\boldsymbol{E}_{i}^{s(E)} = \boldsymbol{C}_{ij}^{sP(E)} \boldsymbol{E}_{P(E)}^{j} \tag{A.55}$$

or in detail

$$\left\| \begin{matrix} \boldsymbol{E}_{i}^{p(E)} \\ \boldsymbol{E}_{i}^{n(E)} \end{matrix} \right\| = \left\| \begin{matrix} \boldsymbol{C}_{ij}^{Ap} & \boldsymbol{C}_{ij}^{Bp} & \boldsymbol{C}_{ij}^{Cp} \\ \boldsymbol{C}_{ij}^{An} & \boldsymbol{C}_{ij}^{Bn} & \boldsymbol{C}_{ij}^{Cn} \end{matrix} \right\| \left\| \begin{matrix} \boldsymbol{E}_{A(E)}^{j} \\ \boldsymbol{E}_{B(E)}^{j} \\ \boldsymbol{E}_{C(E)}^{j} \end{matrix} \right\| \tag{A.56}$$

and the input–output transposition properties of the 2×3 ECM DC–AC commutator can be expressed as

$$\boldsymbol{E}_{i}^{s(I)} = \boldsymbol{C}_{ij}^{sP(I)} \boldsymbol{E}_{P(I)}^{j} \tag{A.57}$$

or in detail

$$\left\| \begin{matrix} \boldsymbol{E}_{i}^{a(I)} \\ \boldsymbol{E}_{i}^{b(I)} \\ \boldsymbol{E}_{i}^{c(I)} \end{matrix} \right\| = \left\| \begin{matrix} \boldsymbol{C}_{ij}^{Pa} & \boldsymbol{C}_{ij}^{Na} \\ \boldsymbol{C}_{ij}^{Pb} & \boldsymbol{C}_{ij}^{Nb} \\ \boldsymbol{C}_{ij}^{Pc} & \boldsymbol{C}_{ij}^{Nc} \end{matrix} \right\| \left\| \begin{matrix} \boldsymbol{E}_{P(I)}^{j} \\ \boldsymbol{E}_{N(I)}^{j} \end{matrix} \right\| \tag{A.58}$$

however, a mathematical model of an electrical continuous dynamical system of the 2×3 and 3×2 ECM AC–DC–AC commutator shown in figure A27, acting as a current-source ECM AC–DC–AC commutator, making dependent the input holors from the output holors, in the dynamical systems approach and matrix notation may be expressed in the form:

$$\boldsymbol{I}_{P}^{j} = \boldsymbol{C}_{sP(E)}^{ji} \boldsymbol{C}_{sP(I)}^{ji} \boldsymbol{I}_{i}^{s} \tag{A.59}$$

or in detail

$$\left\| \begin{matrix} I_P^j \\ I_N^j \end{matrix} \right\| = \left\| \begin{matrix} C_{Ap}^{ji} & C_{An}^{ji} \\ C_{Bp}^{ji} & C_{Bn}^{ji} \\ C_{Cp}^{ji} & C_{Cn}^{ji} \end{matrix} \right\| \left\| \begin{matrix} C_{Pa}^{ji} & C_{Pb}^{ji} & C_{Pc}^{ji} \\ C_{Na}^{ji} & C_{Nb}^{ji} & C_{Nc}^{ji} \end{matrix} \right\| \left\| \begin{matrix} I_i^p \\ I_i^n \end{matrix} \right\| \tag{A.60}$$

The input–output transposition properties of the 3 × 2 ECM AC–DC commutator can be expressed as

$$I_{P(E)}^j = C_{Ps(E)}^{ji} I_i^{s(E)}$$

or in detail

$$\left\| \begin{matrix} I_{A(E)}^j \\ I_{B(E)}^j \\ I_{C(E)}^j \end{matrix} \right\| = \left\| \begin{matrix} C_{Ap}^{ji} & C_{An}^{ji} \\ C_{Bp}^{ji} & C_{Bn}^{ji} \\ C_{Cp}^{ji} & C_{Cn}^{ji} \end{matrix} \right\| \left\| \begin{matrix} I_i^{p(E)} \\ I_i^{n(E)} \end{matrix} \right\|$$

and the input–output transposition properties of the 2 × 3 ECM DC–AC commutator can be expressed as

$$I_{P(I)}^j = C_{Ps(I)}^{ji} I_i^{s(I)}$$

or in detail

$$\left\| \begin{matrix} I_{P(I)}^j \\ I_{N(I)}^j \end{matrix} \right\| = \left\| \begin{matrix} C_{Pa}^{ji} & C_{Pb}^{ji} & C_{Pc}^{ji} \\ C_{Na}^{ji} & C_{Nb}^{ji} & C_{Nc}^{ji} \end{matrix} \right\| \left\| \begin{matrix} I_{a(I)}^j \\ I_{b(I)}^j \\ I_{c(I)}^j \end{matrix} \right\|$$

where:

$$E_I^{s(I)} = \left\| E_i^{p(I)} \; E_i^{n(I)} \right\|^T$$

$$E_i^{s(E)} = \left\| E_i^{a(E)} \; E_i^{b(E)} \; E_i^{c(E)} \right\|^T$$

$$E_{P(I)}^j = \left\| E_{P(I)}^j \; E_{N(I)}^j \right\|^T$$

$$I_{s(I)}^j = \left\| I_{p(I)}^j \; I_{n(I)}^j \right\|^T$$

$$I_i^{P(I)} = \left\| I_i^{A(I)} \; I_i^{B(I)} \; I_i^{C(I)} \right\|^T$$

$$I_{s(E)}^j = \left\| I_{a(E)}^j \; I_{b(E)}^j \; I_{c(E)}^j \right\|^T$$

$$I_i^{P(E)} = \left\| I_i^P \; I_i^N \right\|^T$$

$C_{ij}^{sP(E)}$ is the commutation-matrix holor of the voltage-source 3 × 2 ECM AC–DC commutator connected with the stiff voltage holors, $C_{Ps(I)}^{ji}$ is the commutation-matrix holor of the current-source 2 × 3 ECM DC–AC commutator connected with the stiff current holors. They are given by

$$\boldsymbol{C}_{ij(I)}^{sP} = \begin{Vmatrix} \boldsymbol{C}_{ij}^{Pa} & \boldsymbol{C}_{ij}^{Na} \\ \boldsymbol{C}_{ij}^{Pb} & \boldsymbol{C}_{ij}^{Nb} \\ \boldsymbol{C}_{ij}^{Pc} & \boldsymbol{C}_{ij}^{Nc} \end{Vmatrix}$$

and

$$\boldsymbol{C}_{ij}^{sP(E)} = \begin{Vmatrix} \boldsymbol{C}_{ij}^{Ap} & \boldsymbol{C}_{ij}^{Bp} & \boldsymbol{C}_{ij}^{Cp} \\ \boldsymbol{C}_{ij}^{An} & \boldsymbol{C}_{ij}^{Bn} & \boldsymbol{C}_{ij}^{Cn} \end{Vmatrix}$$

By matrix multiplication of the commutation-matrix holor of the voltage-source 3 × 2 ECM AC–DC commutator and the commutation-matrix holor of the current-source 2 × 3 ECM DC–AC commutator may be received a mathematical confirmation that a product of the commutation-matrixer holors of the 2 × 3 and 3 × 2 ECM commutator is equal to the commutation-matrix holor of the 2 × 2 ECM AC–AC commutator.

$$\boldsymbol{C}_{ij}^{sP(E)}\boldsymbol{C}_{ij}^{sP(I)} = \begin{Vmatrix} \boldsymbol{C}_{ij}^{Ap} & \boldsymbol{C}_{ij}^{Bp} & \boldsymbol{C}_{ij}^{Cp} \\ \boldsymbol{C}_{ij}^{An} & \boldsymbol{C}_{ij}^{Bn} & \boldsymbol{C}_{ij}^{Cn} \end{Vmatrix} \begin{Vmatrix} \boldsymbol{C}_{ij}^{Pa} & \boldsymbol{C}_{ij}^{Na} \\ \boldsymbol{C}_{ij}^{Pb} & \boldsymbol{C}_{ij}^{Nb} \\ \boldsymbol{C}_{ij}^{Pc} & \boldsymbol{C}_{ij}^{Nc} \end{Vmatrix}$$

$$= \begin{Vmatrix} \boldsymbol{C}_{ij}^{Ap}\boldsymbol{C}_{ij}^{Pa} + \boldsymbol{C}_{ij}^{Bp}\boldsymbol{C}_{ij}^{Pb} + \boldsymbol{C}_{ij}^{Cp}\boldsymbol{C}_{ij}^{Pc} & \boldsymbol{C}_{ij}^{Ap}\boldsymbol{C}_{ij}^{Na} + \boldsymbol{C}_{ij}^{Bp}\boldsymbol{C}_{ij}^{Nb} + \boldsymbol{C}_{ij}^{Cp}\boldsymbol{C}_{ij}^{Nc} \\ \boldsymbol{C}_{ij}^{An}\boldsymbol{C}_{ij}^{Pa} + \boldsymbol{C}_{ij}^{Bn}\boldsymbol{C}_{ij}^{Pb} + \boldsymbol{C}_{ij}^{Cn}\boldsymbol{C}_{ij}^{Pc} & \boldsymbol{C}_{ij}^{An}\boldsymbol{C}_{ij}^{Na} + \boldsymbol{C}_{ij}^{Bn}\boldsymbol{C}_{ij}^{Nb} + \boldsymbol{C}_{ij}^{Cn}\boldsymbol{C}_{ij}^{Nc} \end{Vmatrix}$$

$$= \begin{Vmatrix} \boldsymbol{C}_{ij}^{Pp} & \boldsymbol{C}_{ij}^{Np} \\ \boldsymbol{C}_{ij}^{Pn} & \boldsymbol{C}_{ij}^{Nn} \end{Vmatrix}$$

Similarly, by matrix multiplication of the commutation-matrix holor of the current-source 2 × 3 ECM AC–DC commutator and the commutation-matrix holor of the voltage-source 3 × 2 ECM DC–AC commutator may be received a mathematical confirmation that a product of the commutation-matrixer holors of the 2 × 3 and 3 × 2 ECM commutator is equal as well the commutation-matrix holor of the 2 × 2 ECM AC–AC commutator.

$$
C^{ji}_{Ps(I)}C^{ji}_{Ps(E)} = \left\| \begin{matrix} C^{ji}_{Pa} & C^{ji}_{Pb} & C^{ji}_{Pc} \\ C^{ji}_{Na} & C^{ji}_{Nb} & C^{ji}_{Nc} \end{matrix} \right\| \left\| \begin{matrix} C^{ji}_{Ap} & C^{ji}_{An} \\ C^{ji}_{Bp} & C^{ji}_{Bn} \\ C^{ji}_{Cp} & C^{ji}_{Cn} \end{matrix} \right\|
$$

$$
= \left\| \begin{matrix} C^{ji}_{Pa}C^{ji}_{Ap} + C^{ji}_{Pb}C^{ji}_{Bp} + C^{ji}_{Pc}C^{ji}_{Cp} & C^{ji}_{Pa}C^{ji}_{An} + C^{ji}_{Pb}C^{ji}_{Bn} + C^{ji}_{Pc}C^{ji}_{Cn} \\ C^{ji}_{Na}C^{ji}_{Ap} + C^{ji}_{Nb}C^{ji}_{Bp} + C^{ji}_{Nc}C^{ji}_{pC} & C^{ji}_{Na}C^{ji}_{An} + C^{ji}_{Nb}C^{ji}_{Bn} + C^{ji}_{Nc}C^{ji}_{Cn} \end{matrix} \right\|
$$

$$
= \left\| \begin{matrix} C^{ji}_{Pp} & C^{ji}_{Pn} \\ C^{ji}_{Np} & C^{ji}_{Nn} \end{matrix} \right\|
$$

where:

$$
\left\| \begin{matrix} C^{ji}_{Pp} & C^{ji}_{Pn} \\ C^{ji}_{Np} & C^{ji}_{Np} \end{matrix} \right\| = \left\| \begin{matrix} C^{ji}_{Pp} & C^{ji}_{Np} \\ C^{ji}_{Pn} & C^{ji}_{Nn} \end{matrix} \right\|^{T}.
$$

References

Alesani A and Venturini M G B 1989 Analysis and design of optimum-amplitude nine-switch direct AC–AC converters *IEEE Trans. Power Electron.* **4** 101–12

Alexanderson E F W and Mittag A H 1934 The 'thyratron' motor *Electr. Eng.* **53** 1817

Anon 1950 Professor A.N. Larionov k 60-letiiu so dnia rozhdeniia i 30-letiiu nauchno-pedagogicheskoi deiatelnosti *Elektrichestvo* **9** (In Russian)

Casadei D, Serra G, Tani A and Zarri L 2002a Matrix converter modulation strategies: a new general approach on space-vector representation of the switch state *IEEE Trans. Ind. Electron.* **49** 2

Casadei D, Serra G, Tani A and Zarri L 2002b Stability analysis of electrical drives fed by matrix converters, *Proc. of IEEE-ISIE, L'Aquila, Italy, July 8–11, 2002*

Casadei D, Trentin A, Matteini M and Calvini M 2002c Matrix converter commutation strategy using both output current and input voltage sign measurement, *EPE'2003* 2–4 Sept. 2002, *Toulouse, France* Paper 1101, CD-ROM.

Cheron Y, Foch H and Perin A 1985 An improved direct frequency changer using power transistors, *First European Conf. On Power Electronics and Applications Proc.* **vol 1** pp 1.123–8

Fijalkowski B 1962 Sterowanie maszyny wyciagowej pradu stalego z przeksztaltnikami rteciowymi (Control of the DC Mine Winder Using Mercury-Aarc Converters) *Zeszyty naukowe Politechniki Szczecinskiej, Nr 36, Prace Monograficzne, Nr 9* (Szczecin: Politechnika Szczecinska) p 59 (In Polish)

Fijalkowski B 1964 Sterowany przeksztaltnikowy naped elektryczny i perspektywy jego rozwoju (Converter drive and perspectives of its development) *Zeszyty Naukowe Politechniki Szczecinskiej* Elektryka VI, Nr **53** pp 6–34 (In Polish)

Fijalkowski B 1967 Zastosowanie rachunku holorowego w teorii pradow przemiennych (Application of a holor calculus in the theory of alternating currents) *Przeglad elektrotechniczny (Electrotechnics Review)* Rok XLIII, Zeszyt 2, Warszawa, 1967 (In Polish)

Fijalkowski B 1973 Kierunki rozwoju napedu elektrycznego lokomotyw (Development Trends of the Electric Drive of Locomotives) *Materiały konferencyjne—I Krajowa Konferencja—Pojazdy Szynowe* (Krakow: Instytut Pojazdow Szynowych Politechniki Krakowskiej) pp 109–28 (In Polish)

Fijalkowski B T 1982 Trigistor frequency changer-controlled toothed gearless propulsion systems for electric and hybrid vehicles, *Proc. Europe's Int. Conf. on Electric Road Vehicle Systems Drive Electric Amsterdam, Netherlands, 25–28 October 1982* pp 554–67

Fijalkowski B 1985a New concept MACRO- and MICRO-electronics cradled dynamometer systems for testing of combustion engines, *Proc. ISATA 85: Int. Symp. on Automotive Technology and Automation, Graz, Austria, 23-27 September 1985* **vol 2** pp 587–616

Fijalkowski B T 1985b On the new concept hybrid and bimodal vehicles for the 1980s and 1990s, *Proc. Drive Electric Italy '85, Sorrento (Naples), Italy, 1–4 October 1985* pp 4.04.2–8

Fijalkowski B T 1985c On the new concept MMD electrical machines with integral macro-electronic commutators—Development for the future, *Proc. First European Conf. on Power Electronics And Applications, Brussels, Belgium, 16–18 October 1985* **vol 2** pp 3.377–84

Fijalkowski B 1987a *Modele matematyczne wybranych lotniczych i motoryzacyjnych mechano-elektro-termicznych dyskretnych nadsystemów dynamicznych (Mathematical Models of Selected Aviation and Automotive Continuous Dynamical Hyperystems), Monografia* 53 (Krakow: Politechnika Krakowska imienia Tadeusza Kościuszki) 276 p (In Polish)

Fijalkowski B 1987b Choice of hybrid propulsion systems-wheeled city and urban vehicle; -tracked all-terrain vehicle. Part I *Electr. Veh. Dev.* **6** 113–7

Fijalkowski B 1988a Choice of hybrid propulsion systems-wheeled city and urban vehicle; -tracked all-terrain vehicle. Part II *Electr. Veh. Dev.* **7** 31–4

Fijalkowski B 1988b Odnochipnyi makrokommutator—Tehnologiia v stadii razvitiia—Uzhe nie teoriia, no eshche ne promyshlennost', *Zbornik Prednasok z VIII celostatnej konferencje so zahranicnou ucastou elektricke pohony a vykonova elektronika, I. DIEL, Kosice, Kosicka Bela, Czecho-Slovakia*6.-9. 9.1988 pp 182–94 (In Russian)

Fijalkowski B 1988c Niekonwencjonalne uklady napedowe lokomotyw spalinowo-elektry-cznych z akumulacja energii—Spojrzenie w przyszlosc, *Materialy IV Konferencji Elektrotechnika, Elektronika I Automatyka w Transporcie Szynowym Semtrak '88* Czesc II—Naped i sterowanie pojazdow trakcyjnych, *Krakow—Zakopane* 28-30 wrzesnia pp 9–17 (In Polish)

Fijalkowski B 1988d Unconventional internal combustion engines for automotive vehicles, *Papers: Motor Vehicles And Motors '88 3–5 October 1988* [also in *Motorna Vozila—Motori Saopstenja* (XIV-83)—Kragujevac, November 1988] pp 265–74

Fijalkowski B 1988e Single-chip static-commutator—An indispensable module of new concept AC and /or DC dynamotors for automotive very advanced propulsion systems, *Proc. EVS9: The 9th Int. Electric Vehicle Symp., Toronto, Ontario, Canada, 13–16 November 1988* EVS88-031 pp 1–6

Fijalkowski B 1988f Automotive very advanced propulsion systems, *Proc. EVS9: the 9th Electric Vehicle Symp., Toronto, Ontario, Canada, 13-16 November 1988*, EVS88-081 pp 1–9

Fijalkowski B 1988g Amorphous and polycrystalline semiconductor single-chip super-commutator. An indispensable module of new concept AC and/or DC electrical machines, *Proc. Int. Conf. and Intensive Tutorial Course on Semiconductor Materials, University of Delhi, New Delhi, India, 8–16 December 1988*

Fijalkowski B T 2010 *Automotive Mechatronics: Operational and Practical Issues. — Volume I. International Series on Intelligent Control, and Automation Science and Engineering* vol 47 (Berlin: Springer)

Fijalkowski B T 2011 *Automotive Mechatronics: Operational and Practical Issues. — Volume II. International Series on Intelligent Control, and Automotive Science and Engineering* vol 52 (Berlin: Springer)

Fijalkowski B 2016 *Mechatronics: Dynamical Systems Approach and Theory of Holors* (Bristol, UK: IOP Publishing)

Fontaine H 1878 *Electric Lighting: A Practical Treatise* (Translated from French by P Higgs) (London: E. & F.N. Spon)

Gyugyi L and Pelly B R 1976 *Static Power Frequency Changers: Theory, Performance and Application* (New York: Wiley)

Holmberg S N and Shaw M P 1974 *Proc. Int. Conf. Amorphous Liquid Semiconductor* 5th edn (London: Taylor and Francis)

Huber L and Borojevic D 1995 Space vector modulated three-phase to three-phase matrix converter with input power factor correction *IEEE Trans. Ind. Appl.* **31** 1234–46

Kern E 1938 Der Dreiphasenstromrichtermotor *u*nd seine Steurerung bei Betrieb als Umkehrmotor ETZ 59 467 (in German)

Kittel C 1956 *Introduction to Solid State Physics* (New York: Wiley)

Kloninger M J 1932 Applications des redresseurs á grilles polarisées. Séance S.F.E. 6 Fevrier 1932, *Bulletin S.F.E.* (In French)

Li Y, Coi N S, Han B-M, Kim K-M, Lee B and Park J-H 2008 Direct duty ratio pulse width modulation method for matrix converter *Int. J. Control Autom. Syst.* **6** 660–9

Li Y and Choi N-S 2009a Carrier based pulse width modulation for matrix converter, *Conf. Record of IEEE Applied Power Electronics Conf.* 1709–15

Li Y, Choi N-S and Han B-M 2009b DDPWM based control of matrix converters *Power Electron.* **9** 535–43

Mott M F and Davis E A 1971 *Electronic Processes in Mono-crystalline Materials* (London and New York: Oxford University Press)

Neft C L and Schauder C D 1988a Theory and design of a 30-hp matrix converter, *Conf. Record of the 1988 Industry Applications Society Annual Meeting (IEEE Cat. No.88CH 2565–0), Pittsburgh, PA, USA, 1988* p 934

Neft C L and Schauder C D 1988b Theory and design of a 30-hp matrix converter, *Proc. of the IEEE Industrial Application Society Annual Meeting, 1988* pp 248–53

Neiman L R 1972 Akademik Vladimir Fedorovich Mitkevich ego trudy i progresssivnye idei. K 100-letiiu so dnia rozhdeniya *Elektrichestvo* **8** (In Russian)

Oyama J, Higuchi T, Yamada E, Koga T and Lipo T 1989 New control strategy for matrix converter, *Record of the 20th Annual IEEE Power Electronics Specialists Conf.—PESC '89 (Cat. No.89 CH2721–9), Milwaukee, WI, USA, 1989* pp 360–7

Pollak C 1895 Improvements in means for controlling and directing electric currents. English Patent No 24,398 A.D. 1985, Date of Application 19th Dec., 1895, Accepted 5th Dec., 1896

Pollak C 1896 Elektrische Flüssigkeitskondensator mit Aluminium-elektroden. Kaiser-liches Patentamt, Patentschrift No 92564, Klasse 21: Elektriche Apparate, Patentirt im Deutschen Reiche vom 14. Januar 1896 ab. Ausgaben den 19. Mai 1897

Rodriguez J 1983 A new control technique for AC-AC converters *Proc. of the 3rd IFAC Symp. on Control in Power Electronics and Electrical Drives (Lausanne)* **16** 203–8

Schilling W 1940 *Die Wechselrichter und Umrichter—Ihre Berechnung und Arbeitweise.* (Munich: von R. Oldenbourg) (In German)

Seitz P 1940 *Modern Theory of Solids* (New York: McGraw-Hill)

Strzelecki R M and Benysek G (ed) 2008 *Power Electronics in Smart Electrical Energy Networks* (Berlin: Springer)

Tenti P, Malesani L and Rossetto L 1992 Optimum control of N-input K-output matrix converters *IEEE Trans. Power Electron* **7** 707–13

Tutaj J 1996 Estymacja jakosci energii samochodowych komutatorowych pradnic mechano-elektrycznych pradu stalego z komutatorami elektronicznymi (Energy quality estimation of DC mechano-electrical generators with electronic commutators) *PhD thesis* (Promotor: Bogdan Fijalkowski, Automotive Vehicles and Combustion Engines Institute, Cracow University of Technology, Krakow) (In Polish)

Tutaj J 2012 *Ujecie systemowe dynamiki wielofunkcyjnego pradnico-rozrusznika silnika spalino-wego pojazdu samochodowego*(Dynamical Systems Approach of a Multifunctional Generator/Starter for the Combustion Engine of an Automotive Vehicle) Seria Mechanika, Monografia 409 ed B Fijalkowski Wydawnictwo Naukowe (Krakow: Politechniki Krakowskiej im. Tadeusza Kosciuszki) 2012 (In Polish)

Wang B and Venkataramanan G 2006 Six step modulation of matrix converter with increased voltage transfer ratio, *Proc. of the 37th IEEE Power Electronics Specialists Conf. June 18–22, 2006, Jeju, Korea, PS2 49* 930–6

Wheeler P W, Clare J C, Empringham L, Bland M and Kerris K G 2004 Matrix converters *IEEE Ind. Appl. Mag.* **10**(1) 59

Willis C H 1933 A study of the thyratron commutation motor *Gen. Electr. Rev.* **33** 76

Yoon Y-D and Sul S-K 2006 Carrier-based modulation technique for matrix converter *IEEE Trans. Power Electron.* **21** 1691–703

Ziogas P D, Khan S I and Rashid M H 1986 Analysis and design of forced commutated cycloconverter structures with improved transfer characteristics *IEEE Trans. Ind. Electron.* **IE-33** 271

Appendix B

Pulse width modulation (PWM)

B.1 Pulse width modulation fundamentals

Pulse width modulation (PWM) is an elaborate term for describing a type of digital signal. PWM is used in a variety of applications including sophisticated control matrixery and circuitry. Users can accomplish a range of results in both applications because PWM allows them to vary how much time the signal is *HIGH* in an analog fashion. While the signal can only be *HIGH* (usually 5 V) or *LOW* (ground) at any time, users can change the proportion of time the signal is *HIGH* compared to when it is *LOW* over a consistent time interval.

PWM is a technique to encode data such that it corresponds to the width of the pulse given a fixed frequency (Brocker *et al* 1986, Holtz 1992, Kazmierkowski *et al* 2002, Zhang *et al* 2002, Zhou and Wang 2002, Holmes *et al* 2003, Perales *et al* 2003). It is also a method to control EM motors' angular velocity and torque, power circuits, etc, using the 'width' of the pulse. PWM has numerous uses like motion control, dimming, encoding analog signal into its digital form, in power regulation, etc.

PWM period and frequency

PWM period (T) can be thought of as the time required for a new pulse to arrive—it is really the sum of *ON* time (T_{ON}) and *OFF* time (T_{OFF}) of a PWM cycle. PWM frequency is the rate at which a PWM pulse is repeated. It can be also termed as the 'repetition rate'. Frequency is the inverse of the period. PWM frequency is fixed depending on the application.

doi:10.1088/2053-2563/aae7d7ch10

© IOP Publishing Ltd 2019

T_{ON}, T_{OFF} and duty cycle

Each period of PWM signal is divided into T_{ON} and T_{OFF}. Where T_{ON} is the time required or taken for the pulse to remain *ON,* i.e. in a *HIGH* (1) state and similarly T_{OFF} is the time required or taken for the pulse to remain *OFF* i.e. in a *LOW* (0) state. Now, the duty cycle is the percentage of period required for T_{ON}. For example, if $T_{ON} = T_{OFF}$ then the period is divided into two equal parts, i.e. 50% (T_{ON}): 50% (T_{OFF}). Hence the duty cycle will be 50%. If T_{OFF} is three times T_{ON} then the period is divided into two unequal parts, i.e. 25% (T_{ON}): 75% (T_{OFF}), hence duty cycle in this case will be 25%. Duty cycle can be expressed as a simple formula given below:

Duty cycle as a ratio is:

$$D = \frac{T_{ON}}{T_{ON} + T_{OFF}}$$

Duty cycle in percentage is:

$$D = \frac{T_{ON}}{T_{ON} + T_{OFF}} 100\%$$

where D is duty cycle.

When the signal is *HIGH*, users term this 'ON time'. To describe the amount of 'ON time', users use the concept of duty cycle. Duty cycle is measured as a ratio or percentage.

The percentage duty cycle specifically describes the percentage of time a digital signal is *ON* over an interval or period of time. This period is the inverse of the frequency of the waveform.

If a digital signal spends half of the time *ON* and the other half *OFF*, users would say the digital signal has a duty cycle of 50% and resembles an ideal square wave. If the percentage is higher than 50%, the digital signal spends more time in the *HIGH* (1) state than the *LOW* (0) state and vice versa if the duty cycle is less than 50%. A graph that illustrates these three scenarios is as shown in figure B1:

100% duty cycle would be the same as setting the voltage to 5 V (*HIGH*). 0% duty cycle would be the same as grounding the signal.

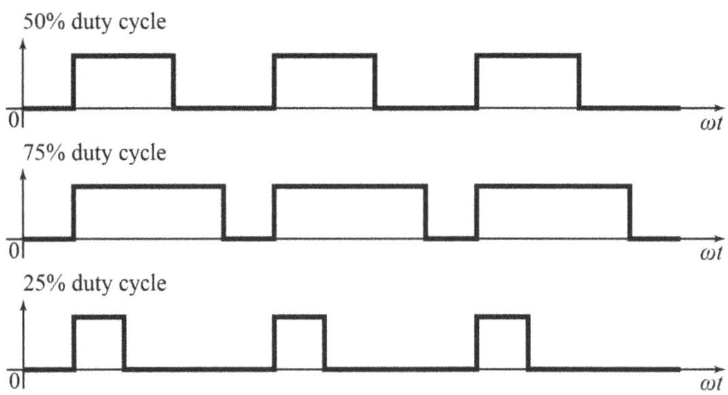

Figure B1. 50%, 75%, and 25% duty cycle examples.

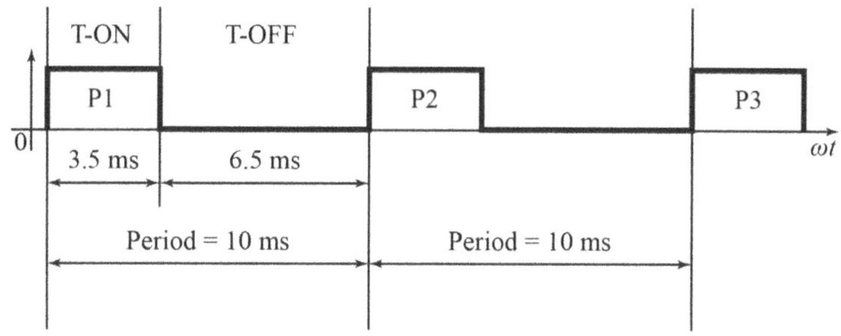

Figure B2. PWM diagram.

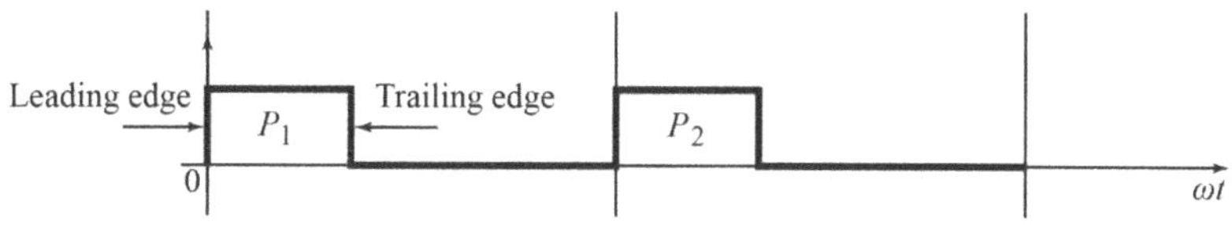

Figure B3. Types of PWM edges.

Example
PWM diagram shown on figure B2 the readers have:
Period, i.e. $T = 10$ ms.
Hence, frequency = 100 Hz.
$T_{ON} = 3.5$ ms.
$T_{OFF} = 6.5$ ms.
Hence, duty cycle = 35%.

PWM edges

A PWM signal contains two types of edges which are termed 'leading edge' and 'trailing edge'. The PWM diagram shown in figure B3 explains this:

Types of PWM

PWM signal can be classified in different ways. However, one would like to classify PWM as:

- single edge PWM;
- double edge PWM.

Single edge PWM

For single edge PWM the pulse can either be at the beginning or the end of the period and hence can be further classified into leading edge (right aligned) PWM and trailing edge (left aligned) PWM.

In trailing edge PWM the leading edge is fixed at the beginning of a period and the trailing edge is modulated, i.e. varied. A PWM diagram for trailing edge (left aligned) PWM is shown in figure B4:

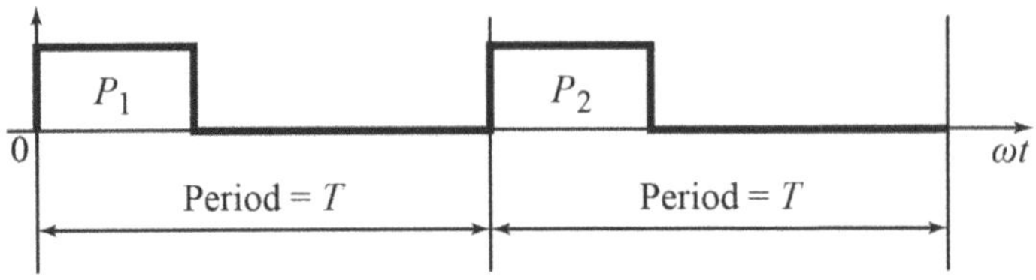

Figure B4. Trailing edge PWM, i.e. left aligned PWM.

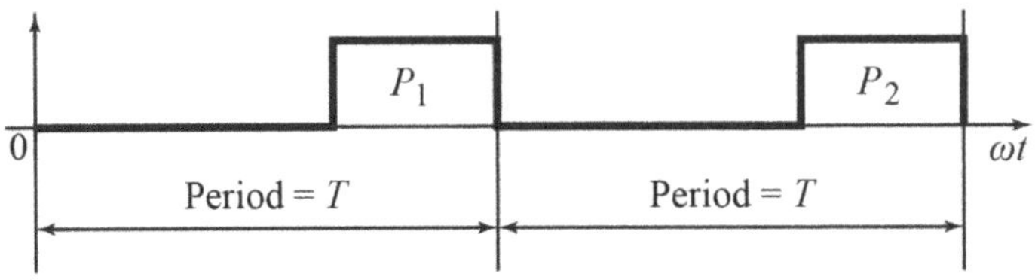

Figure B5. Leading edge PWM, i.e. right aligned PWM.

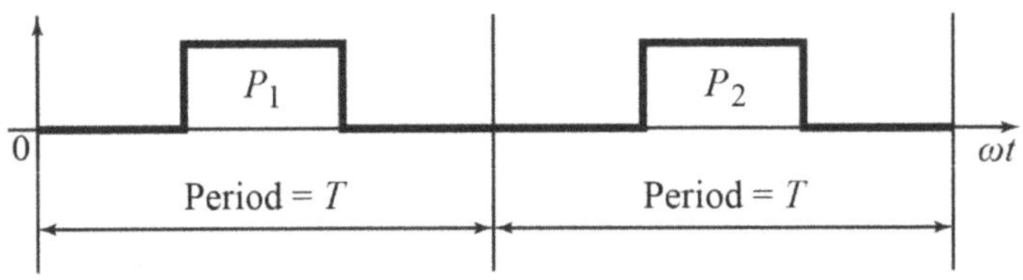

Figure B6. Double edge PWM (centre aligned PWM).

In leading edge PWM the trailing edge is fixed at the end of a period and the leading edge is modulated, i.e. varied. PWM diagram for leading edge (right aligned) PWM is shown in figure B5:

Double edge PWM

In double edge PWM the pulse can be positioned anywhere within the period. It is termed 'double edge' because both the edges are modulated or varied.

In applications like multi-phase EM motor control double edge PWM is used where the pulse is center aligned to reduce harmonics. A PWM diagram for a double edge centre aligned PWM is shown in figure B6:

PWM voltage

The medium (average) voltage of a PWM output depends on its duty cycle. The lower the duty cycle, the lower will be the voltage and vice-versa. The maximum voltage will be equal to the value of the *HIGH* (1) state and the minimum voltage will be equal to *LOW* (0) state of the pulse. For example, if *HIGH* (1) state is 5 V and

LOW (0) state is 0 V the maximum voltage will be 5 V at 100% duty cycle and minimum voltage will be 0 V at 0% duty cycle.

$$V_{med} = D\ V_H$$

In the case where the *LOW* (0) state represents a negative voltage then the above equation can be generalised as follows:

$$V_{med} = D\ V_H + (1 - D)V_L$$

where: V_H—voltage for *HIGH* (1) state and V_L—voltage for *LOW* (0) state.

This can be converted into an analog voltage by using a simple RC filter.

If users were working with an ECM DC–AC or DC–AC–DC commutator synchronous or asynchronous (induction) motor control, it probably has a set of single ECM stages. Each of these has two electrical valves (electronic switches) connected in series across a DC link that carries some voltage, V_{DC}.

At any given instant, users have three choices for controlling the single ECM stage: users can pull the output phase low to the negative side of the DC link (let us cal this the *LOW* (0) state), users can pull the output phase high to the positive side of the DC link (let us call this the *HIGH* (1) state), and users can leave the output phase open-circuited. For the moment let us ignore this last choice, and any transients involved going between the *LOW* (0) and *HIGH* (1) states. Users can output *LOW* (0) or *HIGH* (1), and commute (switch) back and forth any way users like over time.

Figure B7 shows the *LOW* (0) state and figure B8 the *HIGH* (1) state with commute (switch) between the *LOW* (0) and *HIGH* (1) state at a fixed frequency.

In each period of time *T*, the electrical valve is at the *HIGH* (1) state some fraction of the time, and at the *LOW* (0) state the rest of the time.

The average frequency *f* PWM fraction of time is *D* (the duty cycle), and this creates an average voltage on the single ECM stage output of

$$V_{out} = D\ V_{DC}.$$

Early ECM DC–AC commutator synchronous or asynchronous (induction) motors generally used an analog circuit technique for generating the PWM waveforms, termed the 'sine-triangle' PWM. The idea is that if users want to generate a duty cycle with an average value that approximates a sine wave, they create a

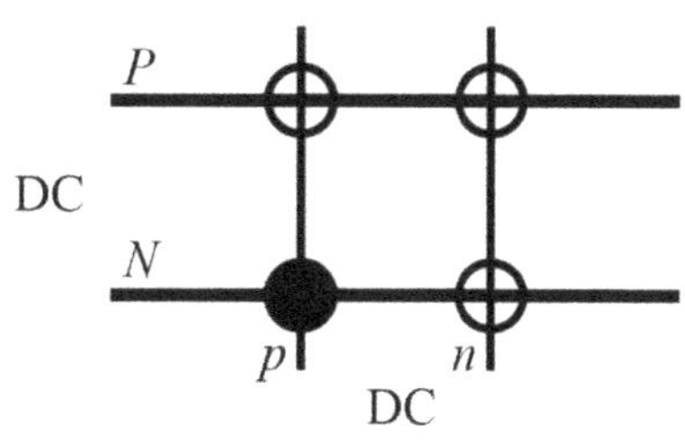

Figure B7. *LOW* (0).

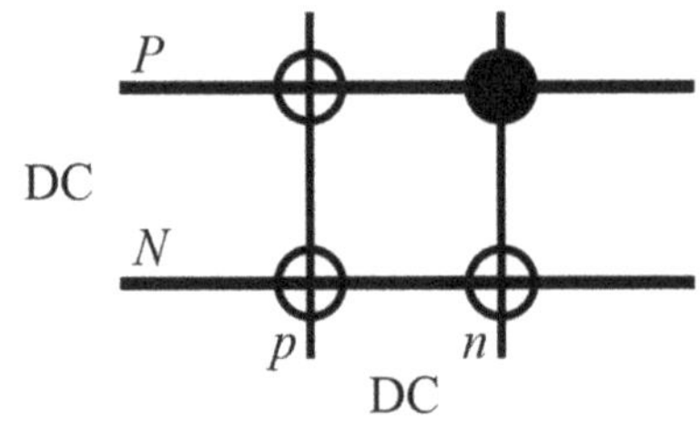

Figure B8. *HIGH* (1) state.

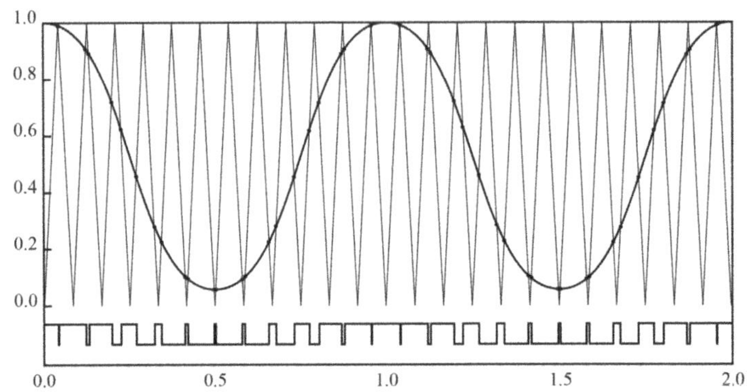

Figure B9. Generation of a duty cycle with an average value that approximates a sine wave by creation of a triangle waveform and comparing it to the desired output waveform (sine wave).

triangle waveform and compare it to the desired output waveform (sine wave), using the results of an analog comparator to determine whether or not to be in the *LOW* (0) or *HIGH* (1) commutation states (see figure B9).

Sine-triangle PWM will act for three sine waves that are $2\pi/3$ out of phase (see figure B10).

The maximum line-to-line voltage users can get in this case, without creating distorted waveforms, is $3\sqrt{2}\ V_{DC}$, or 86.6% of the DC link voltage. But users can do better than this (figure B11).

The distorted waveform shown in figure B12 was introduced by K G King (1974) when he received US Patent #3839 667 on adding harmonics of the operating frequency to increase the line-to-line voltage of an ECM DC–AC commutator, acting as a DC–AC inverter. In particular he described adding 3rd, 6th, 9th, etc harmonics, also known as 'triplens', to each of the phase voltages. This affects the line-to-ground voltage, but the line-to-line voltages are unaffected, because the triplen harmonics on each phase line up. The resulting line-to-ground voltages appear squashed (see figure B12 from the US Patent #3839 667):

This general idea involves adding what is known as a zero-sequence component, namely the same voltage added identically to each phase.

Note: the term 'zero-sequence' is from symmetrical components to analyse three-phase power systems: positive sequence represents equal-amplitude sine waves in the normal phase order, e.g. *A* leading *B* leading *C*; negative sequence represents equal-

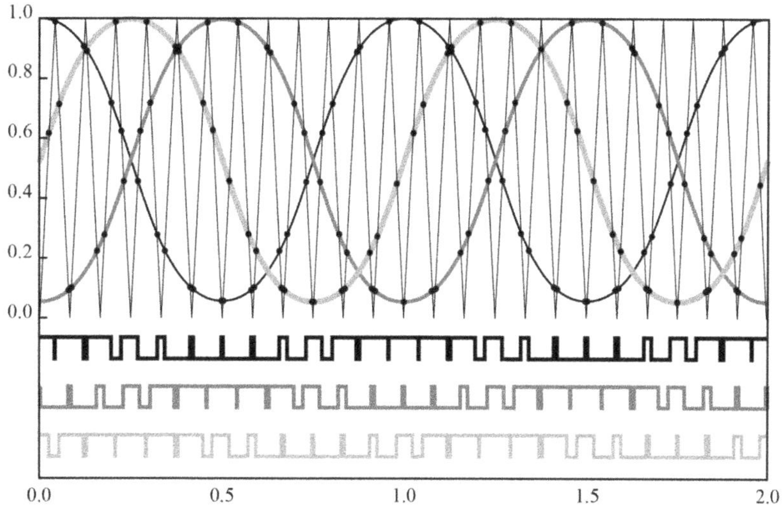

Figure B10. Sine-triangle PWM action for three sine waves that are $2\pi/3$ out of phase.

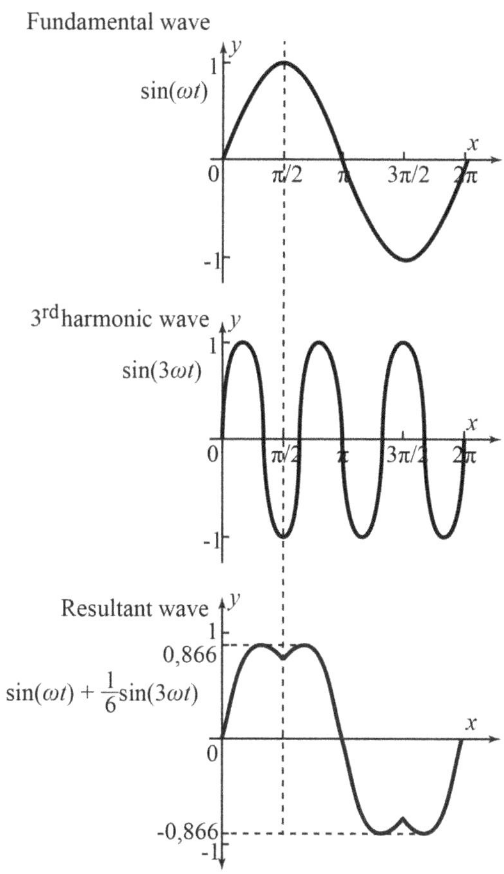

Figure B11. Direct injection of third harmonic component.

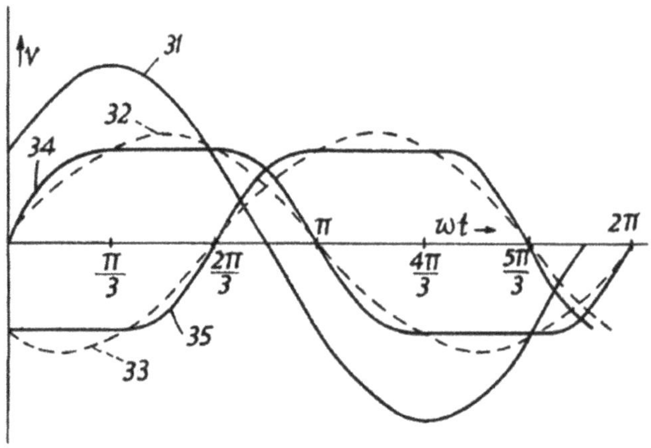

Figure B12. Adding 3rd, 6th, 9th, etc harmonics affects the line-to-ground voltage, but the line-to-line voltages are unaffected (US Patent #3839 667).

amplitude sine waves in the reverse phase order, e.g. *A* leading *C* leading *B*; and zero-sequence represents equal-amplitude sine waves that are aligned in phase. It has been known that any set of three-phase sinusoidal voltages can be expressed as the sum of a positive, negative, and zero-sequence component. Only the positive and negative sequence components affect the voltage measured between any two phases.

It is also well known that a 3rd-harmonic component, with 1/6 the amplitude of the fundamental, can be added to each phase, to produce waveforms that have line-to-line voltages reaching the full DC link voltage (15.5% higher than without any triplen harmonics).

PWM of the ECM commutator

- This is most efficient method of the ECM DC–AC commutator, acting as the DC–AC inverter output voltage control.
- The constant DC voltage is applied the input of ECM DC–AC commutator, acting as the DC–AC inverter and output voltage is controlled by electrical valves of the ECM DC–AC commutator, acting as the DC–AC inverter in this method.

There are the following advantages of PWM techniques:

- There are no extra components necessary to control output voltage of the ECM DC–AC commutator, acting as the DC–AC inverter.
- As the low-order harmonics (3rd, 5th) reduce, whereas higher order harmonics (7th, 9th and 11th) are filtered out, there is less requirement to filter.

Disadvantages of PWM techniques:

- As the electrical valves require low turn *ON* and turn *OFF* time, cost of electrical valves increase.

When an ECM DC–AC commutator, acting as a DC–AC inverter, is connected to load, the output voltage of a DC–AC inverter is controlled for the following reasons:

- in order to adjust output voltage of an ECM DC–AC commutator, acting as a DC–AC inverter;
- when there is any change in input supply source, the output of an ECM DC–AC commutator, acting as a DC–AC inverter, is also changed;
- when the frequency or angular velocity changes, the output of an ECM DC–AC commutator, acting as a DC–AC inverter is also changed; this is done particularly in the DC–AC or AC-DC–AC commutator synchronous or asynchronous (induction) motor's adjustable-velocity or torque control by the ECM DC–AC commutator, acting as the DC–AC inverter.

The output voltage of an ECM DC–AC commutator, acting as a DC–AC inverter, is controlled by the following methods:

- External control of output AC voltage (figure B13):
    - the AC voltage microcontroller is connected between the load and output of the ECM DC–AC commutator, acting as the DC–AC inverter;

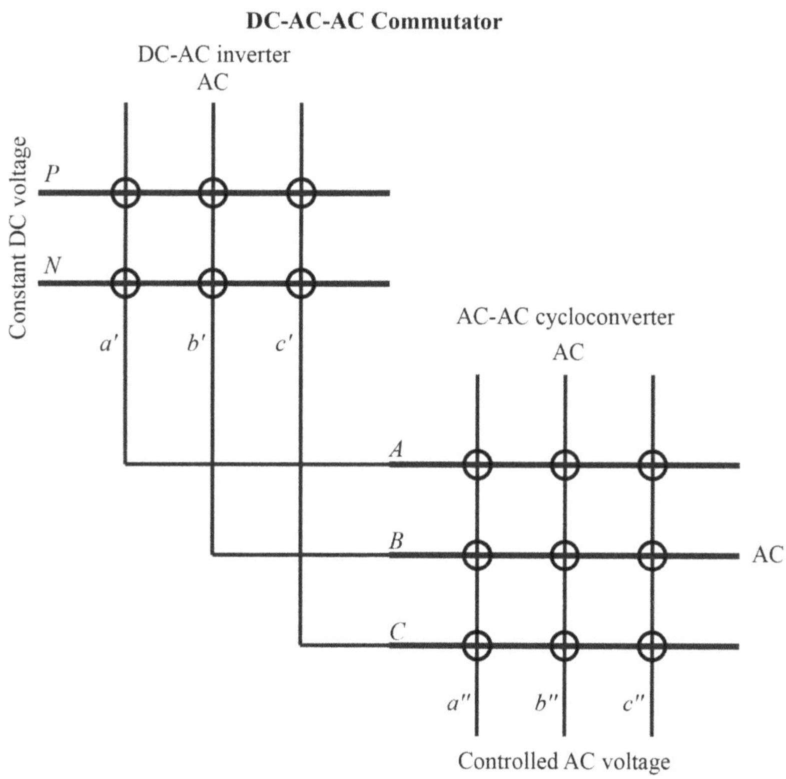

Figure B13. External control of output AC voltage.

- the load voltage is controlled by the triggering-pulse angle of AC voltage microcontroller;
- as the harmonics is produced in the output voltage, this method is used for low power application.

- External control of input DC voltage (figure B14):
  - The ECM DC–DC commutator, acting as the chopper, is connected between an ECM DC–AC commutator, acting as a DC–AC inverter and DC input source when the input voltage is DC. The output voltage of an ECM DC–AC commutator, acting as a DC–AC inverter change due to change in ECM DC–DC commutator, acting as the chopper output voltage.
  - When the input supply is AC, the AC–DC conversion is done by an ECM AC–DC commutator, acting as the controlled AC–DC rectifier. The DC voltage is adjusted by controlling the triggering angle of the ECM AC–DC commutator, acting as the controlled AC–DC rectifier. This will result in output voltage of the ECM DC–AC commutator, acting as the DC–AC inverter, also being adjustable (figures B15–B17).
  - When the input voltage is AC, the AC output voltage is controlled by AC voltage microcontroller. The AC–DC conversion is done by an

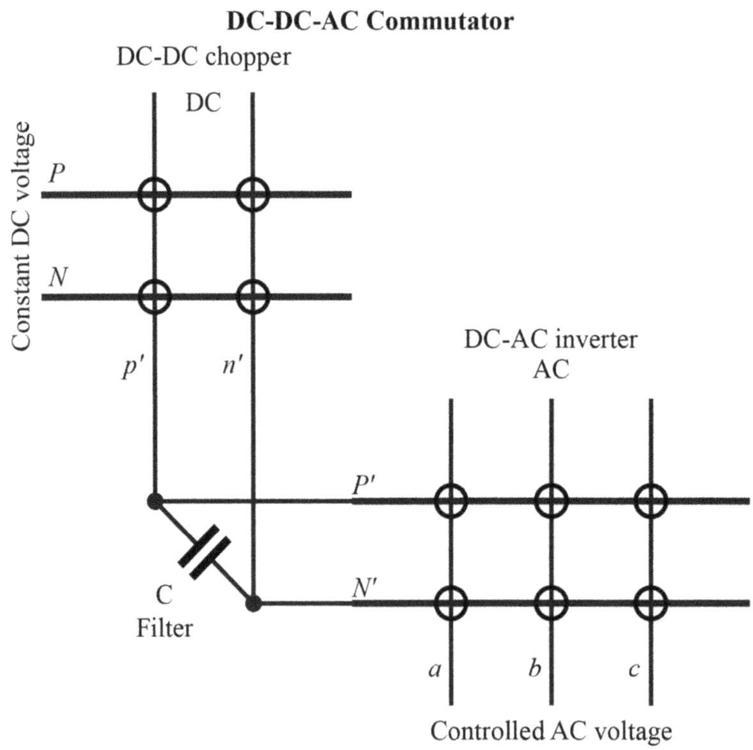

Figure B14. External control of input DC voltage.

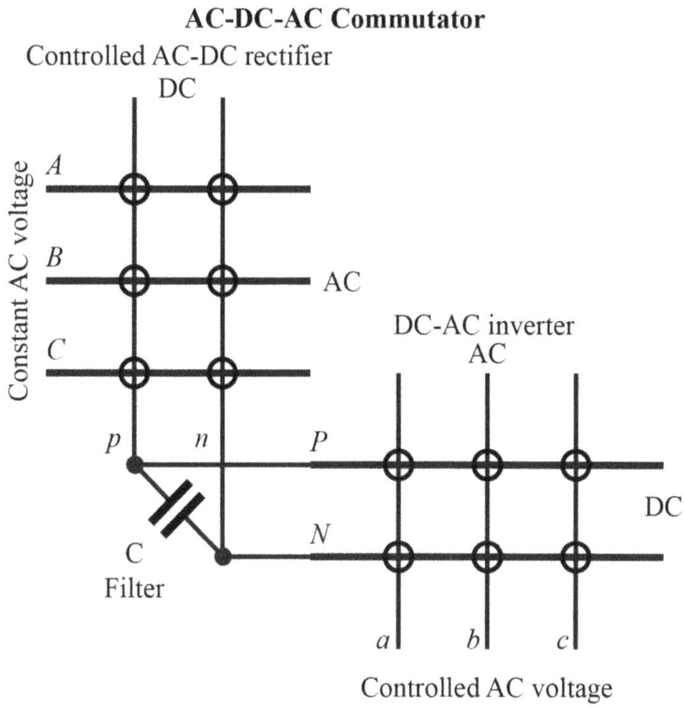

Figure B15. External control of input AC voltage.

ECM commutator, acting as an uncontrolled AC–DC rectifier therefore output voltage of the ECM commutator, acting as the DC–AC inverter is adjustable.

  - When the input voltage is AC, the ECM commutator, acting as the uncontrolled AC–DC rectifier coverts AC into DC. The ECM commutator, acting as the DC–DC chopper converts fixed DC into variable DC supply and the output of the ECM commutator, acting as the DC–DC chopper is feed to input if the ECM DC–AC commutator, acting as a DC–AC inverter.

- There are the following advantages and disadvantages of the above mentioned methods.

  Advantages

  - When the ECM DC–AC commutator, acting as the DC–AC inverter is adjusted by controlling the DC input voltage, there is no change in output waveforms and harmonics;
  - When the ECM DC–AC commutator, acting as the DC–AC inverter is adjusted by controlling the input supply source, the design of the ECM DC–AC commutator, acting as the DC–AC inverter is done for specific voltage limit and its efficiency increases due to small power loss.

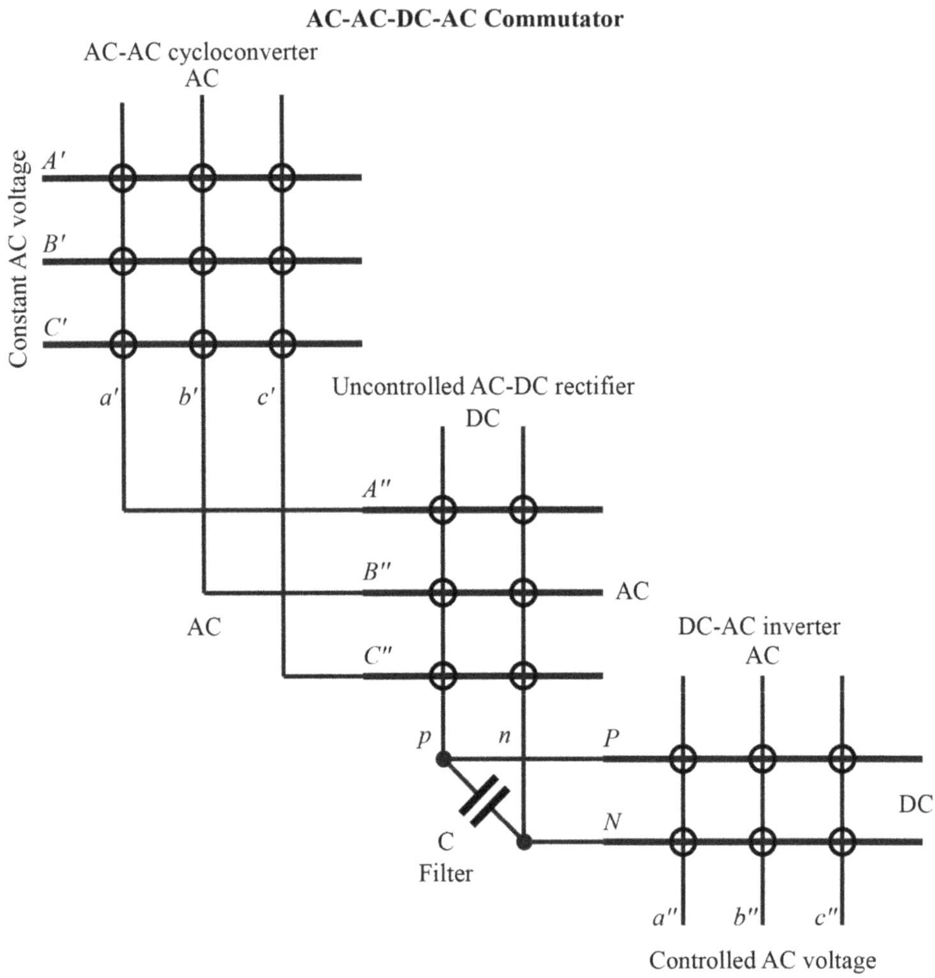

Figure B16. External control of input AC voltage.

Disadvantages

- There are filter requirements at the input side of the ECM DC–AC commutator, acting as the DC–AC inverter in order to reduce DC voltage ripple. This filter circuit increases the mass, volume and cost of the ECM DC–AC commutator, acting as the DC–AC inverter.
- The ECM DC–AC commutator, acting as the DC–AC inverter's efficiency decreases as the power stages increase more than one.
- When it is require to control output voltage for constant current, the DC input voltage control method is used because the commutating capacitor voltage decreases as the DC input voltage decreases and this will result in the turn-off time of electrical valves decreasing.

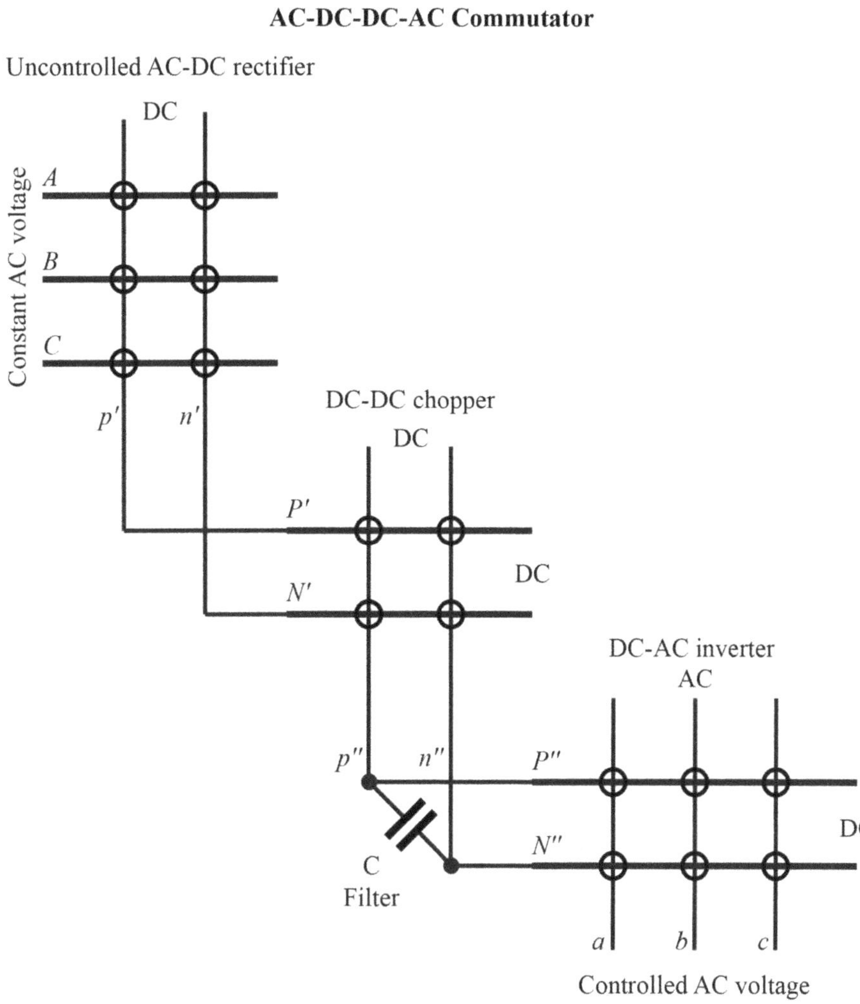

Figure B17. External control of input AC voltage.

- Internal control of the ECM DC–AC commutator, acting as the DC–AC inverter:
    - When the load is connected at the output of the ECM DC–AC commutator, acting as the DC–AC inverter its output voltage is controlled by internal control of the ECM DC–AC commutator, acting as the DC–AC inverter.

There are the following methods of internal control of the ECM DC–AC commutator, acting as the DC–AC inverter's voltage.

- Series ECM DC–AC commutator, acting as the DC–AC inverter control (figure B18)

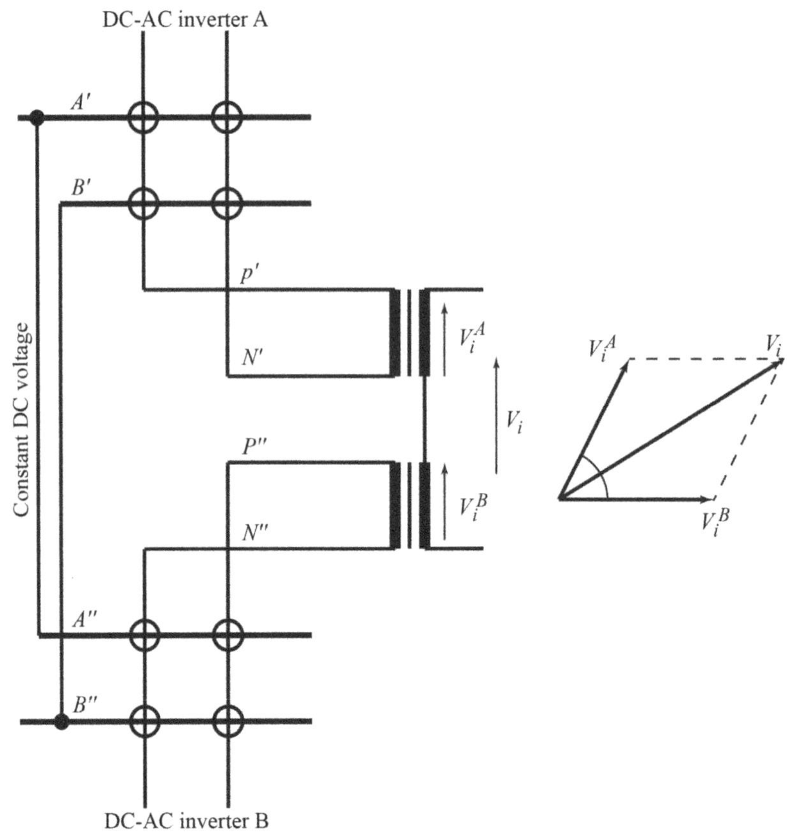

Figure B18. Series voltage control of two DC–AC inverters.

- This method is also termed as the multiple ECM DC–AC commutator, acting as the DC–AC inverter 0 control.
- There are two or more ECM DC–AC commutators, acting as the DC–AC inverters connected in series in this method.
- The output of the ECM DC–AC commutator, acting as the DC–AC inverter is connected with primary winding of the transformer, whereas the load is connected to the secondary windings of the transformer.
- Let the secondary voltages of the transformer be V_i^A and V_i^B, therefore the load voltage is

$$V_i = \sqrt{V_i^{A^2} + 2V_i^A V_i^B \sin\alpha + V_i^{B^2}}$$

where α is triggering angle of the ECM DC–AC commutator, acting as the DC–AC inverter.

If $\alpha = 0$

$$V_i = \sqrt{V_i^{A^2} + 2V_i^A V_i^B \sin\alpha + V_{sB}^2} = \sqrt{\left(V_i^A + V_i^B\right)^2}$$

or

$$V = V_i^A + V_i^B$$

If $\alpha = \pi$

$$V = \sqrt{V_i^{A^2} - 2V_i^A V_i^B \sin\alpha + V_i^{B^2}} = \sqrt{\left(V_i^A - V_i^B\right)^2}$$

or

$$V = V_i^A - V_i^B$$

The load voltage can be changed by changing the triggering angle of the ECM DC–AC commutator, acting as the DC–AC inverter.

B.2 Methods of PWM control: single pulse modulation (SPWM)

- As the electrical valve receives only one pulse during one half cycle, one electrical valve is commutated on.
- The output voltage of the ECM DC–AC commutator, acting as the DC–AC inverter can be controlled by controlling width of pulse.
- Figure B19 shows the gate signal and output voltage waveform for a single phase ECM DC–AC commutator, acting as the DC–AC inverter.
- The gate signal is generated by comparing V_r and V_c amplitude control signal.
- The width of gate pulse can be varied from 0 to π by controlling the reference signal from 0 to V_r.
- This will control the output voltage of the ECM DC–AC commutator, acting as the DC–AC inverter.
- The frequency of the output voltage depends on the frequency of the reference signal.

The amplitude modulation depth or index M is ratio of modulation reference-frequency signal $V_i^r{}_r$ and carrier-frequency signal V_i^c

$$M = V_i^r / V_i^c.$$

The analysis of the waveform shown in figure B19(a) is done by Fourier series. The output voltage becomes maximum when the width of pulse becomes π radian

$$V_L^j = 4V_{DC}^j/\pi. \tag{B.1}$$

RMS output voltage

$$V_{RMS}^j = V_{DC}^j\sqrt{d}/\pi \tag{B.2}$$

and maximum value the of nth harmonic

$$V_{Ln}^j = 4V_{RMS}^j/n\pi(\sin nd/2) \tag{B.3}$$

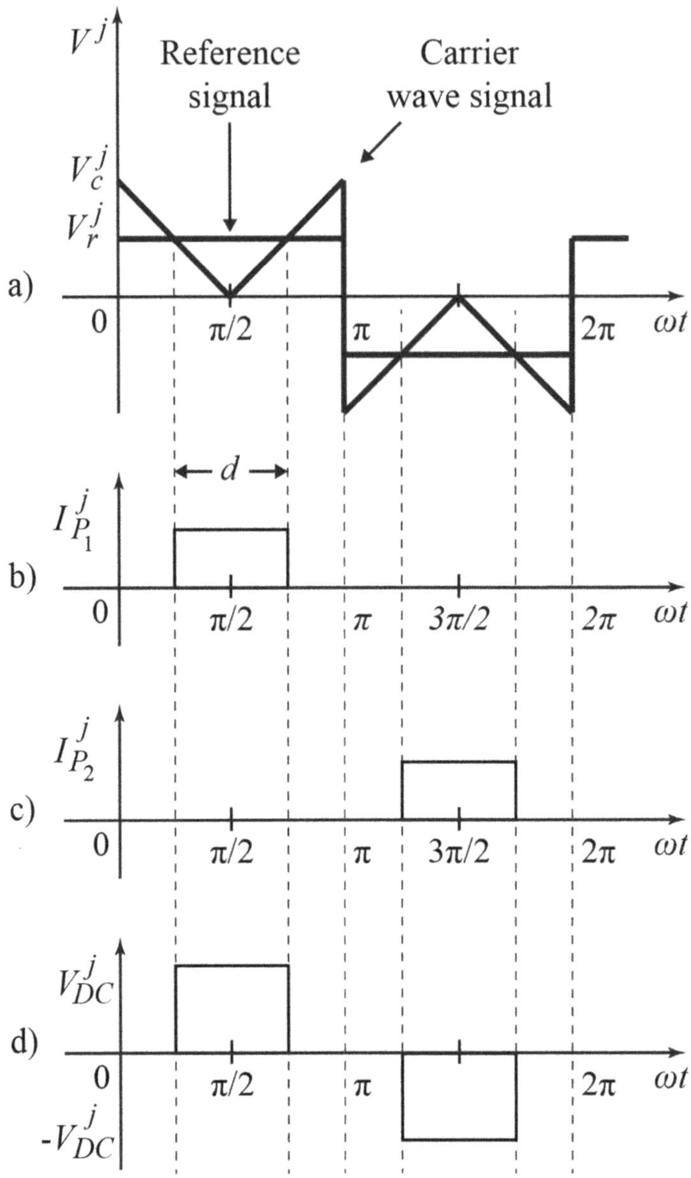

Figure B19. Single pulse width modulation.

From equations (B.1) and (B.3)

$$V_{Ln}^j / V_L^j = \sin(nd/2)/n \tag{B.4}$$

The graphical representation of pulse width on degree (x-axis) and $n = 1, 3, 5$ and 7 (y-axis) is shown in figure B20.

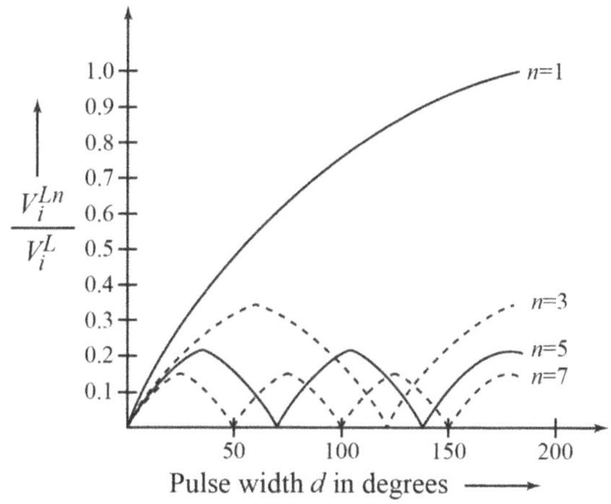

Figure B20. Single pulse width modulation.

B.3 Methods of PWM control: multiple pulse width modulation (MPWM)

- There is more than one pulse per half cycle in the multiple pulse width modulation (MPWM).
- The gate pulses of the electrical valves are used to control output voltage of the ECM DC–AC commutator, acting as the DC–AC inverter as well as to reduce harmonics.
- The magnitude and width of the pulses are equal in this method.
- The reference signal and higher frequencies carrier signal are compared in this method in order to generate more than one gate pulse.
- The number of gate pulses depends upon carrier frequencies, whereas the output voltage depends on the frequencies of the reference signal.

From figure B21:
Carrier frequency = f_c in Hz.
Reference frequency = f_r in Hz.

$$1/f_c = \pi/3 \tag{B.5}$$

or

$$T_c = \pi/3$$

Similarly

$$1/2f_r = \pi \tag{B.6}$$

or

$$T_r = 2\pi$$

There the number of pulses per half cycle

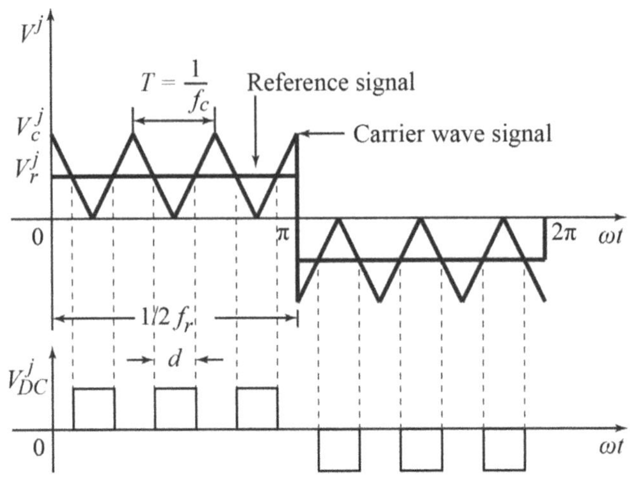

Figure B21. Multiple pulse width modulation.

N_p = length of half cycle reference signal/length of one cycle of triangular waveform = $(f_r/2)/(1/f_c)$.

$$N_p = f_c/2f_r.$$

Number of generated pulses from equations (B.5) and (B.6)

$$N_p = (3/\pi) \times \pi = 3.$$

The RMS voltage when the pulse width is equal to d

$$V_{RMS}^j = V_{DC}^j \sqrt{N_p \frac{d}{\pi}}.$$

- As the number of pulses increases in each half cycle, lower order harmonics reduce but higher order harmonics increase.
- The higher order harmonics are reduced by using filter. It should be noted that the commutation losses of the electrical valves increase as there are a greater number of pulses per half cycle.

The modulation technique is also termed symmetrical modulation control.

B.4 Methods of PWM control: sinusoidal pulse width modulation (SINPWM)

Currently, **sinusoidal pulse width modulation** (SINPWM) is playing a major role in the generation of pure sinusoidal waveforms using microcontroller-based ECM DC–AC commutators, acting as DC–AC inverters. **Digital SINPWM** (DSINPWM) is a type of 'carrier-based' **pulse width modulation** (PWM). Carrier based PWM uses pre-defined modulation signals to determine output voltages. In SINPWM, the modulation signal is sinusoidal, with the peak of the modulating signal always less than the peak of the carrier signal. SINPWM ECM DC–AC commutator (acting as DC–AC inverter) ECM column and line–line voltages are illustrated below.

Amplitude modulation depth or index

The amplitude modulation depth or index, M is defined using the ratio of peak modulation reference and carrier signals as follows

$$M = V_r/V_c,$$

where V_r, V_c, are the modulation reference and carrier signal voltages, respectively.
For SINPWM, the amplitude modulation depth or index must be less than 1.0.

Output voltages

The fundamental output voltage for one ECM column of the ECM DC–AC commutator, acting as a DC–AC inverter is given by

$$V_{col} = M\frac{V_{DC}}{2}$$

$$V_{col} = \frac{1}{2\sqrt{2}}MV_{DC}$$

and the fundamental line–line voltage is given by

$$V_{LL} = \frac{\sqrt{3}}{2\sqrt{2}}MV_{DC}$$

Drawbacks

There are two significant drawbacks with SINPWM.

I. *Available output voltage*
Assuming that the DC voltage is created using an uncontrolled ECM AC–DC commutator, acting as a diode AC–DC rectifier and capacitor DC link, the maximum available DC voltage is given by

$$V_{DCmax} = \sqrt{2}\,V_{LLS}$$

where V_{LLS} is the line–line supply voltage,
The maximum output using SINPWM ($M = 1$) is

$$V_{LLmax} = \frac{\sqrt{3}}{2\sqrt{2}}\sqrt{2}\,V_{LLS} = \frac{\sqrt{3}}{2}V_{LLS}.$$

i.e. a SINPWM IEMD cannot produce a line–line output voltage as high as the line supply. One option to mitigate this discrepancy is to use higher supply voltages (e.g. 480 V supply, 460 V EM motor; 600 V supply, 575 V EM motor).

II. *Short pulses*
If the output is to be truly SINPWM, it is important to include very small pulses when the peak modulation signal is close to the peak carrier voltage. These small pulses can contribute significantly to inverter losses, while not significantly affecting the output voltage. In addition, small pulses may be impractical due to the time required to switch one device off and another device on. As a rule, industrial IEMD 'drop' small pulses to improve efficiency.

Number of pulses per half cycle when the amplitude of triangular waveform becomes maximum and sinusoidal waveform becomes zero.

Therefore

$$N_p = f_c / 2f_r$$

where: $f_c = 3/\pi$ — carrier wave frequency;
$f_r = 1/2\pi$ — reference wave frequency.

- The reference signal is taken as a sinusoidal waveform whereas the carrier signal is taken as a triangular waveform in this method.
- The width of pulse in SINPWM is not equal because the reference signal is taken as a sinusoidal waveform.
- The amplitude of the sinusoidal waveform is also not constant.
- The width of the gate pulse of the electrical valve is determined by the intersect point of the sinusoidal waveform and triangular waveform.
- The frequency of the ECM DC–AC commutator (acting as the DC–AC inverter) output voltage depends upon the frequency of the reference signal f_r and the amplitude of the reference signal V_r controls the modulation index M.
- The number of pulses per half cycle

$$N_p = (3/\pi) \times (2\pi/2) = 3.$$

- The number of pulses per half cycle when the amplitude of triangular waveform and sinusoidal waveform becomes zero at same time (figure B22 and B23),

$$N_p = f_c / 2f_r - 1 = 2.$$

- The modulation depth or index

$$M = V_r / V_c.$$

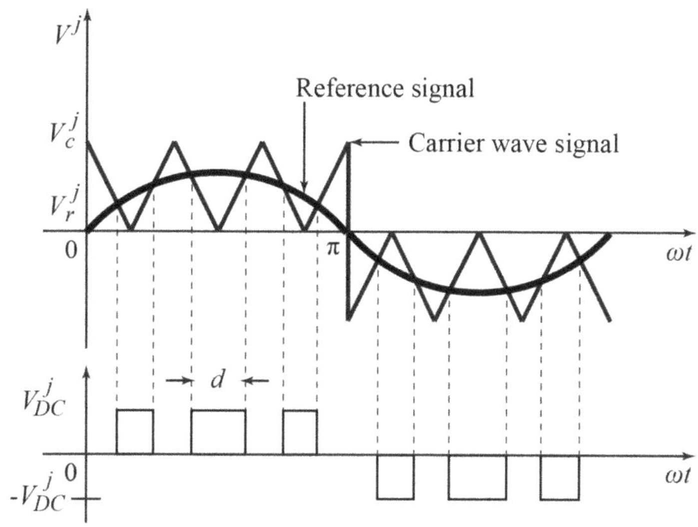

Figure B22. Sinusoidal pulse width modulation.

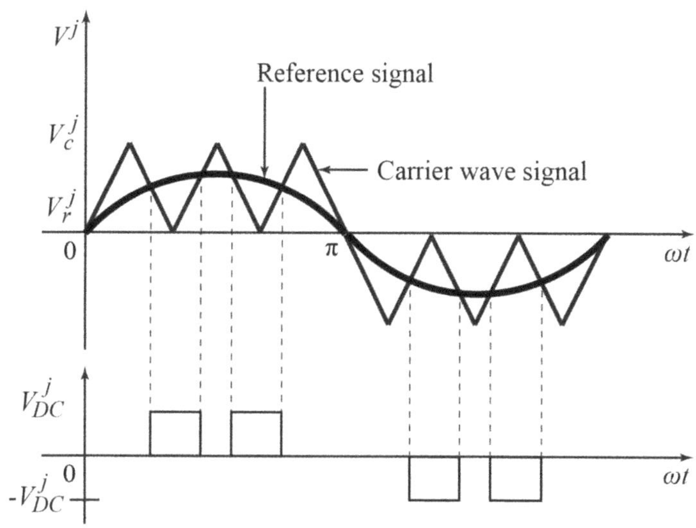

Figure B23. Sinusoidal pulse width modulation.

The analysis of harmonics is done in the SINPWM control as below.

- When the value of modulation depth or index is less than one ($M < 1$), the maximum harmonic number in the output voltage is

$$f_c/f_r \pm 1$$

or

$$2N_p \pm 1.$$

where N_p—number of pulses per half cycle.

- As the number of pulses per half cycle increases, the higher order harmonics also increases.
- Let $N_p = 4$, it will generate 7th harmonic and 9th harmonic but higher order harmonics are easily filtered out.
- As the number of pulses increases per half cycle the commutation losses also increase and it will affect the efficiency of the ECM DC–AC commutator, acting as the DC–AC inverter.
- When the value of modulation depth or index is greater than one ($M > 1$), lower order harmonics are induced in the output of the ECM DC–AC commutator, acting as the DC–AC inverter.

In SINPWM the pulse-width instead of being uniform, as in the waveform of a sinusoidal function of its angular position with respect to a reference sine wave, results in a reduction in the harmonic content. The control function consists of a sinusoidal wave obtained from an oscillator of variable amplitude $\boldsymbol{A}$ and of fundamental output DC–AC inverter frequency $\boldsymbol{f} = 1/T$ as well as a triangular wave of fixed amplitude $\boldsymbol{A_c}$ and frequency $\boldsymbol{f_c}$ with a direct component of amplitude $\boldsymbol{A_c}$ as shown in figure B24. The biased triangular waveform is reversed in polarity at the end of each half-cycle of the output voltage. Electrical valves' pulsed gating signals are generated by comparison of the sinusoidal and triangular wave-forms—gating pulses are obtained for time intervals in which the sinusoidal signal is more positive than the triangular signal.

It is easily established from the figure that the number of gate pulses (sinusoidally modulated) per half-cycle is

$$N_p = f_c / 2f = \text{integer} \tag{B.7}$$

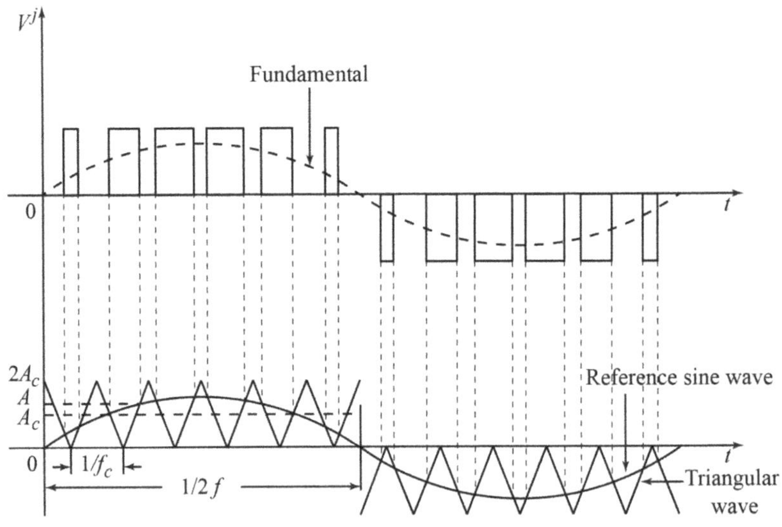

Figure B24. Output voltage with sinusoidal pulse modulation.

It follows from the above that the angles for *TURN-ON* and commutation of the electrical valves are determined by the intersections of the two signals referred to above. It can be easily seen from figure B24 that the amplitude of the fundamental voltage can be controlled by varying amplitude $\boldsymbol{A}$ of the sine wave over the range $0 < \boldsymbol{A} < A_{max}$, where $A_{max} = A_c$. If $\boldsymbol{A}$ is made larger than $\boldsymbol{A}_{max}$, the number of output pulses becomes less than $\boldsymbol{N_p}$ (B.7). In the limit the output becomes a single rectangular pulse of half-cycle width as in the case of ordinary (unmodulated) ECM DC–AC inverter.

The gating signals obtained as the above are employed to trigger the electrical valves in the ECM DC–AC inverter of figure B25. Electrical valves C_{ij}^{Pa} 1 and C_{ij}^{Pb} are gated during the interval of a positive pulse and C_{ij}^{Na} and C_{ij}^{Nb} are gated during the negative pulse. During the zero-value intervals of the pulsed-gating signals G_{ij}^{Pa} and G_{ij}^{Nb} or G_{ij}^{Na} and G_{ij}^{Pb} are gated resulting in zero output voltage.

Thus, the output voltage as shown in figure B24, is a replica of the pulsed-gating signals generated by comparison of sinusoidal and triangular waveforms fed to the commutating matrixery.

Examination of the matrix of the ECM DC–AC commutator, acting as the DC–AC inverter of figure B25 reveals that there is no way in which the load voltage can be reduced to zero. At all times, $V_0 = \pm V_{DC}/2$. The above method of SINPWM, therefore, cannot be applied as such. A modified method of modulation that can be employed in the ECM DC–AC commutator, acting as the DC–AC inverter is

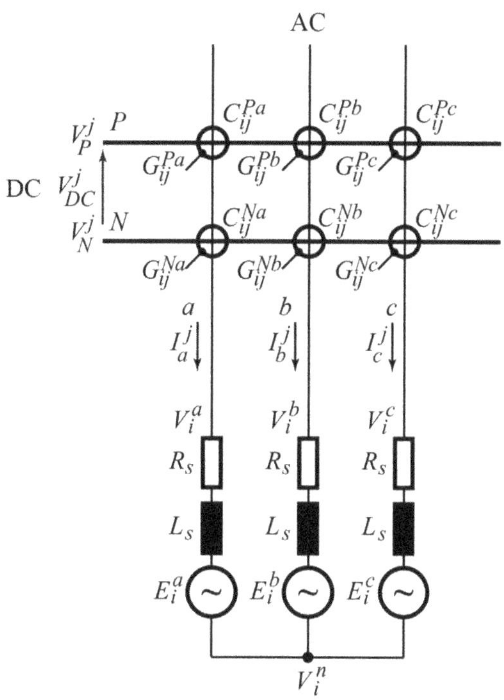

Figure B25. ECM DC–AC inverter.

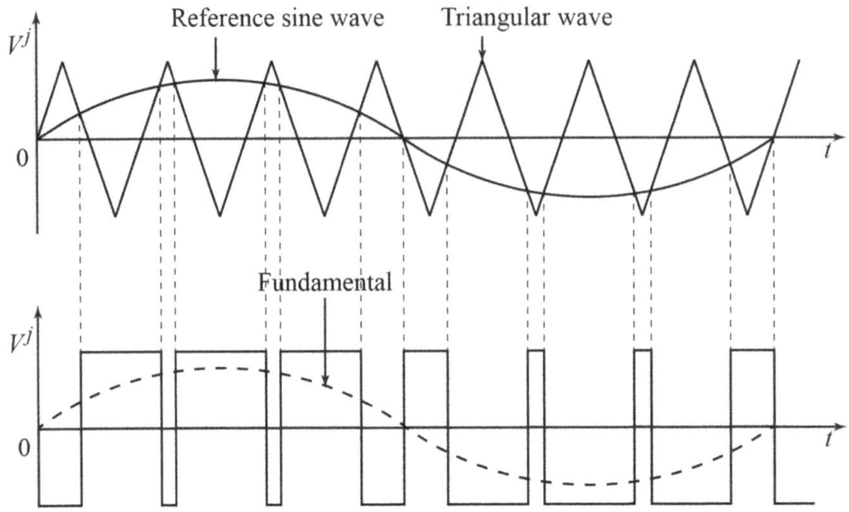

Figure B26. Modified method of modulation that can be employed in the ECM DC–AC commutator, acting as the DC–AC inverter.

illustrated in figure B26. Voltage control is achieved by varying the amplitude of the reference sine wave. This type of voltage control is rarely employed because of its high ratio of harmonic amplitudes to the amplitude of the fundamental component of the output voltage wave.

The use of a PWM ECM AC–DC–AC commutator for DC or AC motor control is illustrated in figure B25 where the DC–AC inverter is fed from a substantially constant DC voltage uncontrolled AC–DC rectifier, while the control of AC voltage at EM motor terminals is achieved through the DC–AC inverter. The chief advantage of this scheme of control lies in the fact that it requires a single controllable power conversion so that the power factor of AC supply to the uncontrolled AC–DC rectifier is high. Further, the low-frequency harmonic content of the PWM DC–AC inverter is lower than that of an ordinary (unmodulated DC–AC inverter). These advantages of the PWM drive are obtained at the cost of greater sophistication in the control logic.

In PWM IEMD the DC–AC inverter efficiency is somewhat decreased because of many commutations per half-cycle. The commutation frequency should be increased as permitted by the electrical valves so as to obtain a good balance between increase in DC–AC inverter loss and estimates of MMD electrical machine loss (due to reduced harmonic content).

Of late, the technique of selected harmonic elimination in PWM DC–AC inverters has received considerable attention. In this method, notches are created at predetermined angles of the square wave which permits voltage control with elimination of selected harmonics.

Currently, DSINPWM is playing a major role in the generation of pure sinusoidal waveforms using microcontroller-based DC–AC commutators, acting as DC–AC inverters. Below we consider the benefits and limitations of three major DSINPWMs for a DC–AC inverter and investigate their performance.

B.5 Methods of PWM control: modified sinusoidal pulse width modulation (MSINPWM)

When considering sinusoidal PWM waveform, the pulse width does not change significantly with the variation of modulation depth or index. The reason for this is due to the characteristics of the sine wave. Hence, this sinusoidal PWM technique is modified so that the carrier signal is applied during the first and last $\pi/3$ intervals per half cycle, as shown in figure B27.

The fundamental component is increased and its harmonic characteristics are improved. The main advantages of this technique are increased fundamental component, improved harmonic characteristics, reduced number of electrical valves (electronic switches) and decreased commutation (switching) losses.

Advantages of PWM

- The output voltage control with this method can be obtained without any additional components.
- With this method, lower order harmonic can be eliminated or minimized along with it is output voltage control.
- It reduces the filtering requirements.

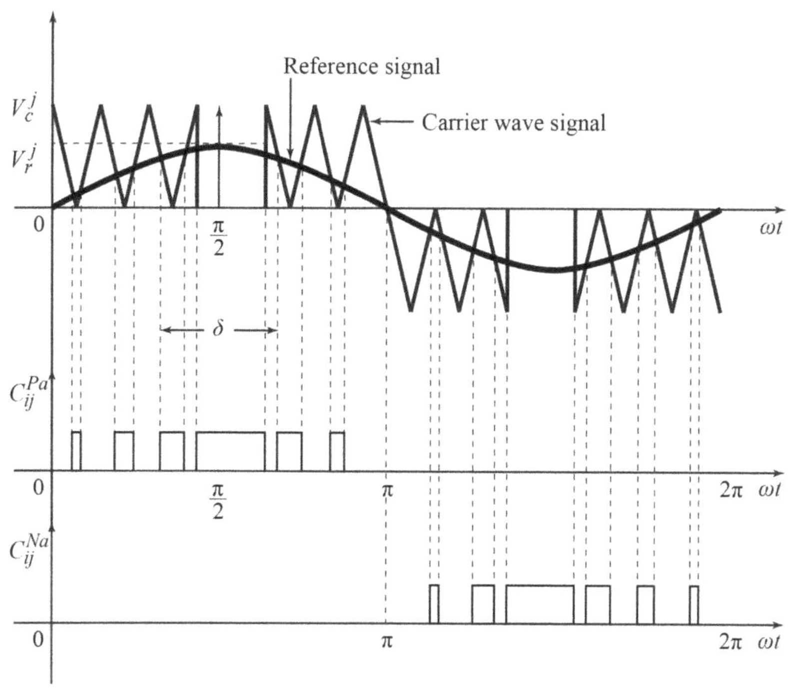

Figure B27. Generation of modified sinusoidal pulse width modulation.

B.6 Methods of PWM control: vector pulse width modulation (VECPWM)

B.6.1 Introduction

Vector pulse width modulation (VECPWM) is a means of generating a three-phase variable voltage, variable frequency PWM output voltage. The ECM DC–AC commutator, acting as the DC–AC inverter comprises six solid-state electrical valves (electronic switches), two for each phase with one electrical valve on each phase connecting to the positive ECM row and one electrical valve connecting to the negative ECM row. By a combination of commutation (switching) states of these output electrical valves, users can create a sinusoidal output current.

In effect, there are eight commutation states that define six output *ACTIVE* commutation vectors and two *NULL* commutation vectors, namely:

$V_i^0 = 000$: NULL.
$V_i^1 = 100$: Vector 1.
$V_i^2 = 110$: Vector 2.
$V_i^3 = 010$: Vector 3.
$V_i^4 = 011$: Vector 4.
$V_i^5 = 001$: Vector 5.
$V_i^6 = 101$: Vector 6.
$V_i^7 = 111$: NULL.

Maximum output voltage would be achieved by stepping through the six major commutation states in sequence (as shown in figure B28). The output pattern would be V_i^0, V_i^1, V_i^2, V_i^3, V_i^4, V_i^5, V_i^6, V_i^7 and would result in maximum voltage and also quite a high level of distortion. The output voltage at any vector angle can be reduced by PWM techniques. If users consider the vector V_i^1, they can reduce the voltage at this angle by commuting between V_i^1 and V_i^0. Half voltage would be with 50% time on V_i^0 and 50% time on V_i^1.

Hence, users can have a variable voltage output waveform that steps round the six commutation vectors by using PWM with the *ACTIVE* commutation vectors and the *NULL* commutation vectors V_i^0 and V_i^7 (see figure B29).

Intermediate commutation vectors can be generated by using PWM techniques between the adjacent *ACTIVE* commutation vectors and the *NULL* commutation vector. The angle is changed by the ratio between the *ACTIVE* commutation vectors and the voltage is reduced by increasing the *NULL* time. For example, a $\pi/2$rad vector at half voltage would be achieved by 50% time with *NULL* commutation vector, 25% time with a V_i^3.

If users consider the six non-*NULL* voltage vectors V_i^1 to V_i^6, then they describe six sectors 1 to 6. Within each sector, users can derive any voltage vector V_i^s of reduced voltage and an angle within the sector. There are a number of strategies for generating the resultant vectors, each strategy has advantages and disadvantages, affecting **total harmonic distortion** (THD), commutation losses and bearing currents in EM motors.

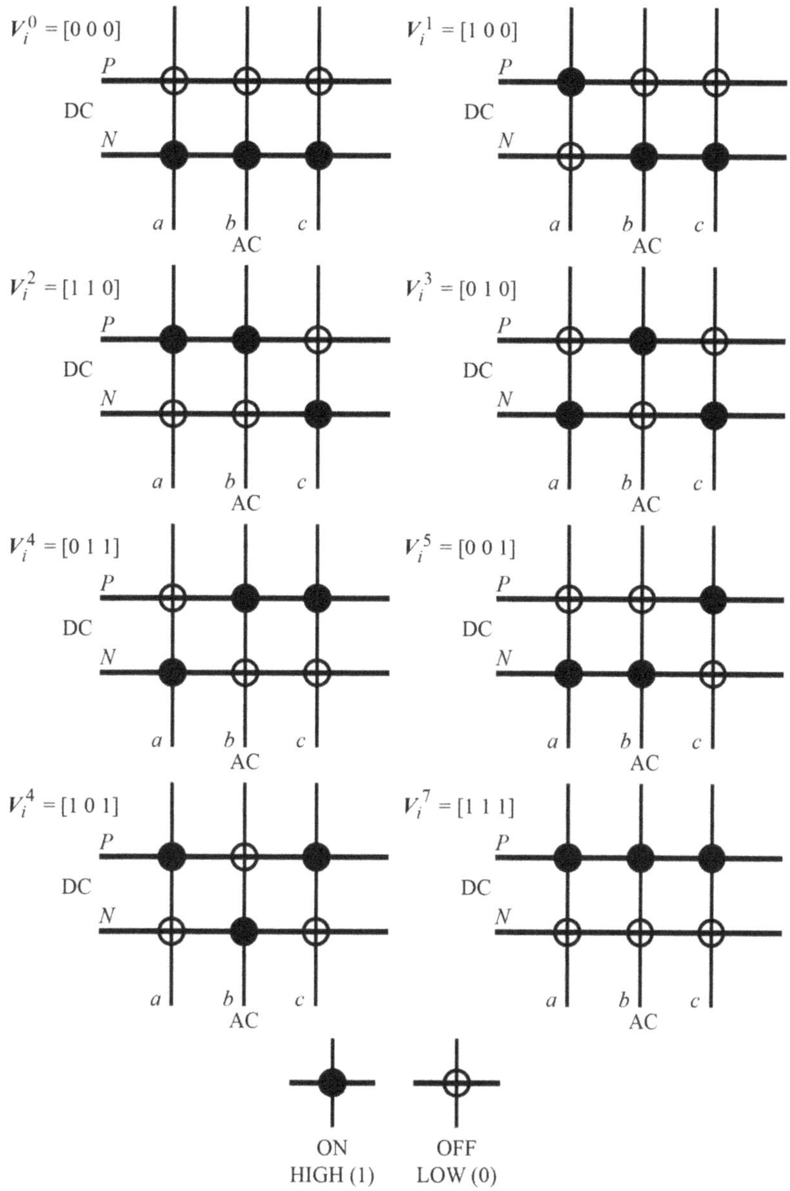

Figure B28. Six major commutation states in sequence.

Right aligned sequence. $V_i^0 - V_i^1 - V_i^2 - V_i^7 - V_i^0 - V_i^1 - V_i^2 - V_i^7$. The angle and magnitude of the vector is determined by the ratios of the periods T_0, T_1 and T_2 (figure B30).

Symmetric sequence. $V_i^0 - V_i^1 - V_i^2 - V_i^7 - V_i^2 - V_i^1 - V_i^0 - V_i^1 - V_i^2 - V_i^1 - V_i^0$ etc. The angle and magnitude of the vector is determined by the ratios of the periods T_0, T_1 and T_2 (figure B31).

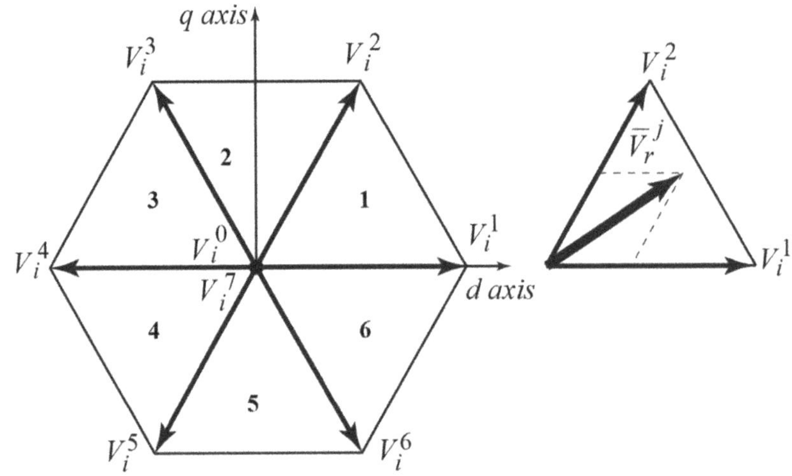

Figure B29. Six commutation vectors by using PWM with the *ACTIVE* commutation vectors and the *NULL* commutation vectors V_i^0 and V_i^7.

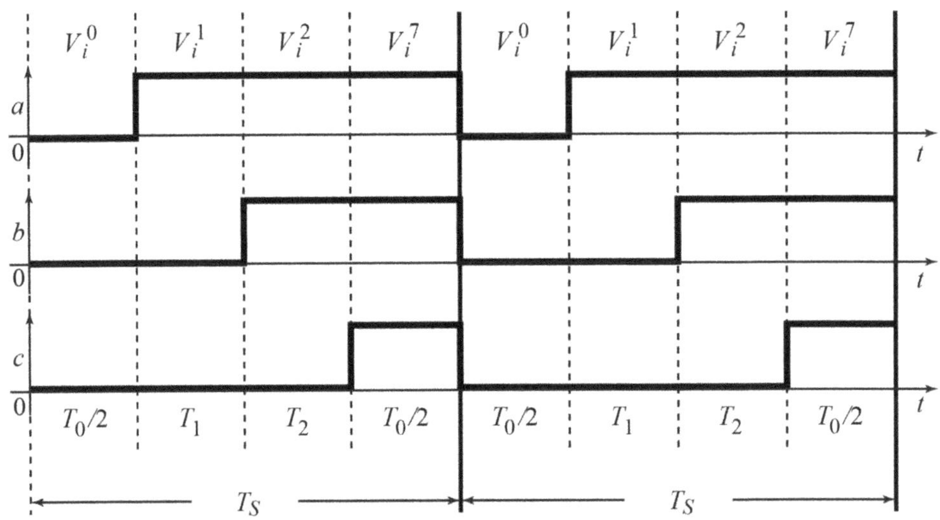

Figure B30. Right aligned sequence.

The fundamental pattern construction is repeated for all six sectors. There are other VECPWM sequences that can be used to generate the required vector patterns.

VECPWM is actually just a modulation algorithm which translates phase voltage (phase to neutral) references, coming from the microcontroller, into modulation times/duty cycles to be applied to the PWM peripheral. It is a general technique for any three-phase load, although it has been developed for EM motor control. VECPWM maximises DC-link voltage exploitation and uses the 'nearest' vectors, which translates into a minimisation of the harmonic content.

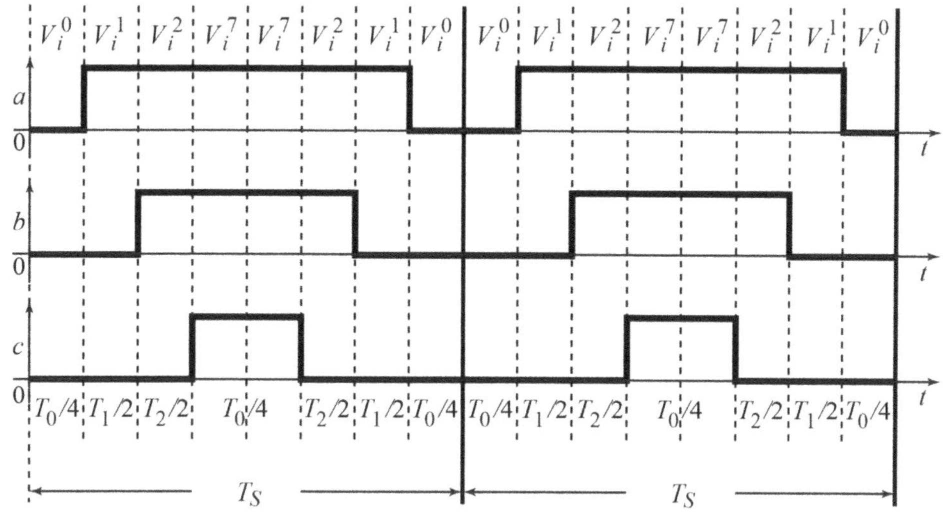

Figure B31. Symmetric sequence.

The classical application of VECPWM is AC–DC–AC, DC–AC commutator synchronous or asynchronous (induction) motors' vector control, which is based on the control of the currents' projection on two orthogonal coordinates (direct and quadrature, *d–q*), termed **magnetic-field oriented control** (MFOC).

For AC-DC–AC, DC–AC commutator synchronous or synchronous (induction) motors, the most common choice for the direct axis is to align it to the **rotor magnetic-field** (RMF) or to the **stator magnetic-field** (SMF). The basic concept is that with a known EM motor and known voltage output pulses users can accurately determine rotor slip by monitoring current and phase shift. The microcontroller can then modify the PWM 'sine' wave shape, frequency or amplitude to achieve the desired result. For example the desired speed is 200 rpm and the control senses there is 2 rpm of slip so it increases the frequency slightly to bring the speed up. Since torque can also be determined it can also be controlled. VECPWM just does a lot of sampling, calculating and wave form manipulation. The specific algorithms and deciding what the best output solution is for different situations could fill up several textbooks, for instance in Kazmierkowski *et al* 2002, in which SV means space vector, as in space vector modulation.

Vector modulation (VECM) principally allows a 3 × 2 and 2 × 3 or 2 × 3ECM commutator VECPWM IEMD to supply about 15% higher peak voltage to an EM motor than the standard sine-triangle modulation scheme by allowing the neutral point of the EM motor to move away from the nominal 1/2 of the supply ECM row (it will look like a triangle wave). The characteristic voltage output of a vector-modulated sine wave is a sine wave with a double-hump on the peaks. To be more precise, VECM output about 15% more 'non over-modulated' voltage than the traditional carrier based PWM. That is the maximum voltage before over-modulation happens. The shape of the phase voltage has a double-hump as mentioned (similar to a standard PWM plus third harmonic addition). The line-to-line voltage will not show the humps though.

VECM is an algorithm for the control of PWM. It is used for the creation of AC waveforms; most commonly to drive phase AC–DC–AC or DC–AC commutator synchronous or asynchronous (induction) motors at varying speeds from DC using multiple class-D amplifiers. There are variations of VECM that result in different quality and computational requirements. One active area of development is in the reduction of **total harmonic distortion** (THD) created by the rapid commutation inherent to these algorithms.

B.6.2 Principle of vector PWM (VECPWM)

The physical model of a classical ECM DC–AC commutator, acting as a **voltage-source** (VS) DC–AC inverter is shown in figure B32 G_{ij}^{Pa}, G_{ij}^{Pb}, G_{ij}^{Pc} to G_{ij}^{Na}, G_{ij}^{Nb}, G_{ij}^{Nc} are the six electrical valves (electronic switches) that shape the output, which are controlled by the commutation (switching) variables G_{ij}^{Pa}, G_{ij}^{Na}, G_{ij}^{Pb}, G_{ij}^{Na}, G_{ij}^{Pc} and G_{ij}^{Na}. When an upper electrical valve is commuted (switched) *ON*, i.e. when G_{ij}^{Pa}, G_{ij}^{Pb}or G_{ij}^{Pc} is 1, the corresponding lower electrical valve is commuted *OFF*, i.e. the corresponding G_{ij}^{Na}, G_{ij}^{Nb} or G_{ij}^{Nc} is 0. Therefore, the *ON* and *OFF* commutation states of the upper electrical valves, G_{ij}^{Pb} and G_{ij}^{Pc} can be used to determine the output voltage.

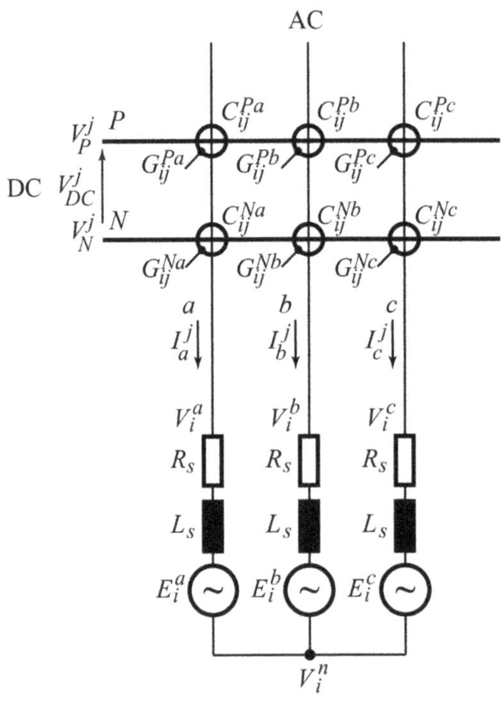

Figure B32. ECM DC–AC commutator, acting as a voltage-source DC–AC inverter.

The relationship between the commutation variable vector $\|a\ b\ c\|^T$ and the line-to-line voltage vector $\| V_i^{ab}\ V_i^{bc}\ V_i^{ca}\|^T$ is given in the following:

$$\begin{Vmatrix} V_i^{ab} \\ V_i^{bc} \\ V_i^{ca} \end{Vmatrix} = V_{DC}^j \begin{Vmatrix} 1 & -1 & 0 \\ 0 & 1 & -1 \\ -1 & 0 & 1 \end{Vmatrix} \begin{Vmatrix} a \\ b \\ c \end{Vmatrix} \tag{B.8}$$

Also, the relationship between the commutation (switching) variable vector $\|a\ b\ c\|^T$ and the phase voltage vector $\| V_i^{an} V_i^{bn}\ V_i^{cn}\|^T$ can be expressed below.

$$\begin{Vmatrix} V_i^{an} \\ V_i^{bn} \\ V_i^{cn} \end{Vmatrix} = V_{DC}^j \begin{Vmatrix} 2 & -1 & -1 \\ -1 & 2 & -1 \\ -1 & -1 & 2 \end{Vmatrix} \begin{Vmatrix} a \\ b \\ c \end{Vmatrix} \tag{B.9}$$

As illustrated in figure B33, there are eight possible combinations of *ON* and *OFF* patterns for the three upper electrical valves. The *ON* and *OFF* commutation states of the lower electrical valves are opposite to the upper one and so are easily determined once the commutation states of the upper electrical valves are determined.

According to equations (B.8) and (B.91), the eight commutation vectors, output line to neutral voltage (phase voltage), and output line-to-line voltages in terms of DC link *VDC*, are given in table B1 and figure B33 shows the eight voltage vectors (V_i^0 to V_i^7).

Vector PWM (VECPWM) refers to a special commutation (switching) sequence of the upper three electrical valves of an ECM DC–AC commutator, acting as a DC–AC inverter. It has been shown to generate less harmonic distortion in the output voltages and/or currents applied to the phases of an EM motor and to provide more efficient use of supply voltage compared with **sinusoidal pulse width modulation** (SINPWM) technique as shown in figure B34.

To implement the VECPWM, the voltage equations in the *a–b–c* reference frame can be transformed into the stationary *d–q* reference frame that consists of the horizontal *d* and vertical *q* axes, as depicted in figure B35. From this figure, the relation between these two reference frames is below

$$\left\|F_{dq0}^j\right\| = \|H_s^{ji}\|\|F_i^{abc}\| \tag{B.10}$$

where

$$\|H_s^{ji}\| = \frac{2}{3}\begin{Vmatrix} 1 & -1/2 & 1/2 \\ 0 & \sqrt{3}/2 & -\sqrt{3}/2 \\ 1/2 & 1/2 & 1/2 \end{Vmatrix};$$

$$\left\|F_{dq0}^j\right\| = \left\|F_d^j\ \ F_q^j\ \ F_0^j\right\|^T;$$

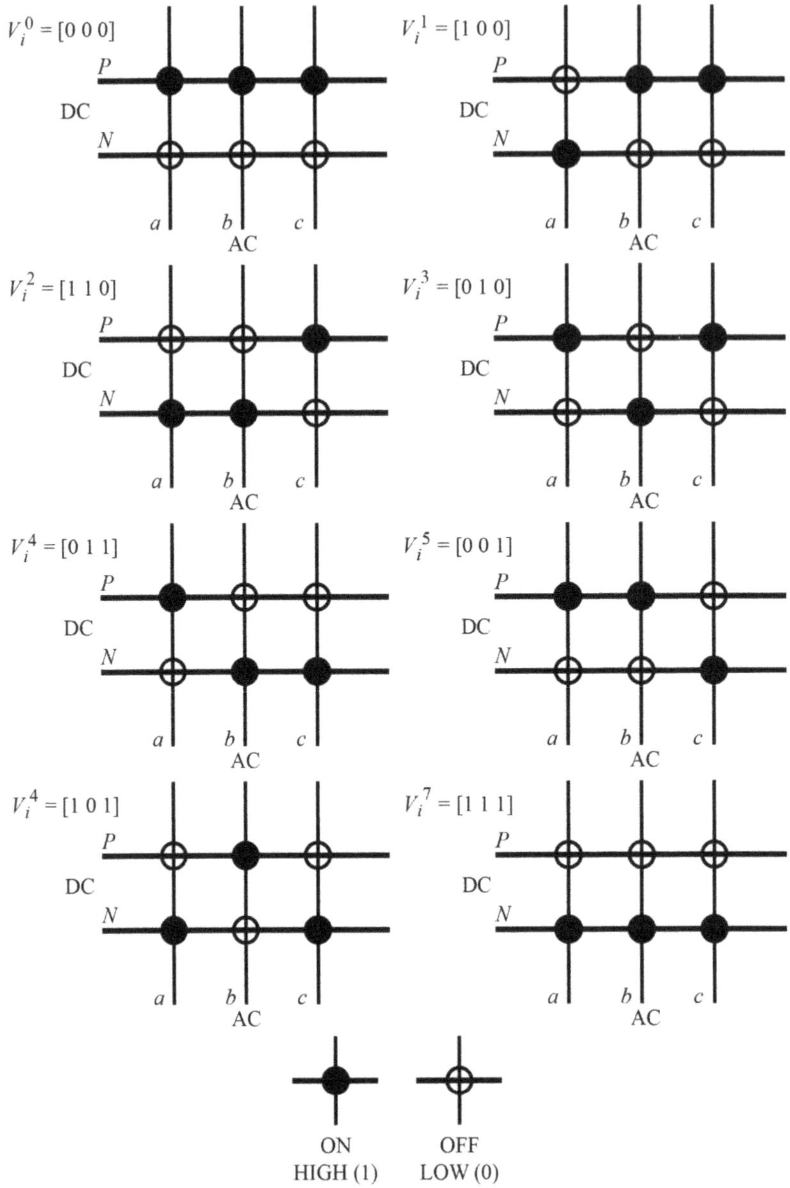

Figure B33. The eight voltage vectors (V_i^0 to V_i^7).

$$\|F_i^{abc}\| = \left\| F_i^a \quad F_i^b \quad F_i^c \right\|^T$$

and F denotes either a voltage or a current variable.

As described in figure B35, this transformation is equivalent to an orthogonal projection of $\|a\ b\ c\|^T$ onto the two-dimensional perpendicular to the vector $\|111\|^T$ (the equivalent d–q plane) in a three-dimensional coordinate system.

Table B1. Commutation vectors, phase voltages and output line to line voltages.

	Commutating vectors			Line to neutral voltage			Line to line voltage		
Voltage vectors	a	b	c	V_i^{an}	V_i^{bn}	V_i^{cn}	V_i^{ab}	V_i^{bc}	V_i^{ca}
V_i^0	0	0	0	0	0	0	0	0	0
V_i^1	1	0	0	2/3	−1/3	−1/3	1	0	−1
V_i^2	1	1	0	1/3	1/3	−2/3	0	1	−1
V_i^3	0	1	0	−1/3	2/3	−1/3	−1	1	0
V_i^4	0	1	1	−2/3	1/3	1/3	−1	0	1
V_i^5	0	0	1	−1/3	−1/3	2/3	0	−1	1
V_i^6	1	0	1	1/3	−2/3	1/3	1	−1	0
V_i^7	1	1	1	0	0	0	0	0	0

(Note that the respective voltage should be multiplied by V_i^{DC}.)

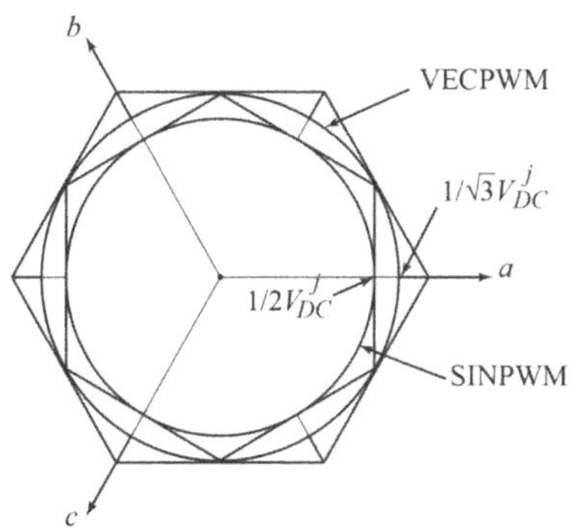

Figure B34. Locus comparison of maximum linear control voltage in SINPWM and VECPWM.

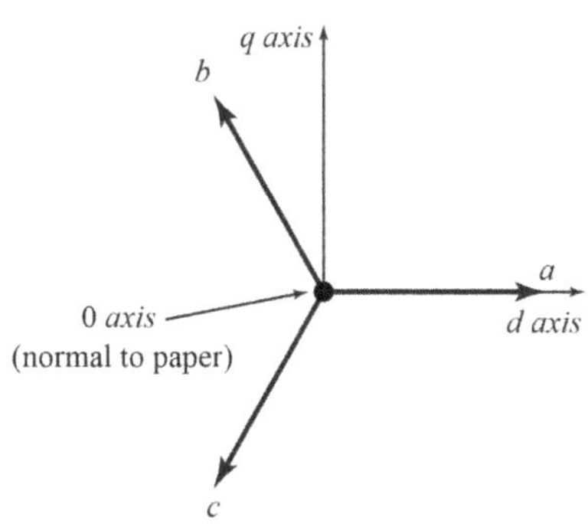

Figure B35. The relationship of *a–b–c* reference frame and stationary *d–q* reference frame.

As a result, six vectors and two *NULL* vectors are possible. Six *ACTIVE* vectors (V_i^1–V_i^6) shape the axes of a hexagonal as depicted in figure B36, and feed electrical energy to the EM motor.

The angle between any adjacent two *ACTIVE* vectors is $\pi/3$ rad. Meanwhile, two *NULL* vectors (V_i^0 and V_i^7) are at the origin and apply zero voltage to the EM motor. The eight vectors are termed the fundamental commutation vectors and are denoted by $V_i^0, V_i^1, V_i^2, V_i^3, V_i^4, V_i^5, V_i^6$, and V_i^7. The same transformation can be applied to the desired output voltage to get the desired reference voltage vector V_r^j in the d–q plane.

The objective of VECPWM technique is to approximate the reference voltage vector V_r^j using the eight commutation patterns.

One simple method of approximation is to generate the average output of the ECM AC–DC–AC or DC–AC commutator, acting as a DC–AC inverter in a small period, T to be the same as that of V_r^j in the same period.

Therefore, VECPWM can be implemented by the following steps:

- Step 1. determine V_d^j, V_q^j, V_0^j and angle α;
- Step 2. determine time duration $T1$, $T2$, $T0$;
- Step 3. determine the commutation time of each electrical valve (C_{ij}^{Pa}, C_{ij}^{Pb}, C_{ij}^{Pc} to C_{ij}^{Na}, C_{ij}^{Nb}, C_{ij}^{Nc}).

Step 1: Determine V_d^j, V_q^j, V_r^j *and angle* α

From figure B37 —V_d^j, V_q^j, V_r^j, and angle α can be determined as follows:

$$V_d^j = V_i^{an} - V_i^{bn}\cos\pi/3 - V_i^{cn}\cos\pi/3 = V_i^{an} - \frac{1}{2}V_i^{bn} - \frac{1}{2}V_i^{cn};$$

$$V_q^j = V_i^{an} + V_i^{bn}\cos\pi/6 - V_i^{cn}\cos\pi/6 = V_i^{an} + \frac{\sqrt{3}}{2}V_i^{bn} - \frac{\sqrt{3}}{2}V_i^{cn};$$

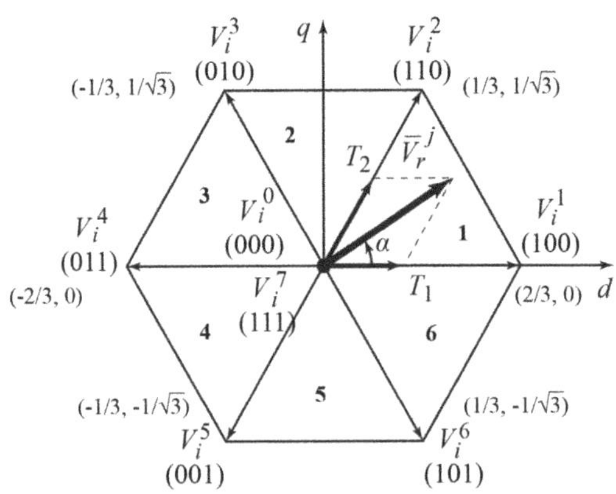

Figure B36. Fundamental commutation vectors and sectors.

$$\begin{Vmatrix} V_d^j \\ V_q^j \end{Vmatrix} = \frac{2}{3}\begin{Vmatrix} 1 & -\frac{1}{2} & -\frac{1}{2} \\ 0 & \frac{\sqrt{3}}{2} & -\frac{\sqrt{3}}{2} \end{Vmatrix}\begin{Vmatrix} V_i^{an} \\ V_i^{bn} \\ V_i^{bn} \end{Vmatrix}.$$

$$\overline{V}_r^j = \sqrt{(V_d^j)^2 + (V_q^j)^2};$$

$$\alpha = \tan^{-1}\left(\frac{V_q^j}{V_d^j}\right) = \omega t = 2\pi f t$$

where f—fundamental frequency.

Step 2: Determine time duration $T1$, $T2$, $T0$

From figure B38 the commutation time duration can be calculated as follows:

- Commutation (switching) time duration at sector 1

$$\int_0^{T_z} \overline{V}_r^j dt = \int_0^{T_1} \overline{V}_i^1 dt + \int_{T_1}^{T_1+T_2} \overline{V}_i^2 dt + \overline{V}_i^0 dt;$$

$$T_z \overline{V}_r^j = T_1 \overline{V}_i^1 + T_2 \overline{V}_i^2;$$

$$T_z = |\overline{V}_r^j| \begin{Vmatrix} \cos\alpha \\ \sin\alpha \end{Vmatrix} = T_1 \frac{2}{3} \overline{V}_{DC}^j \begin{Vmatrix} 1 \\ 0 \end{Vmatrix} + T_2 \frac{2}{3} \overline{V}_{DC}^j \begin{Vmatrix} \cos\pi/3 \\ \sin\pi/3 \end{Vmatrix};$$

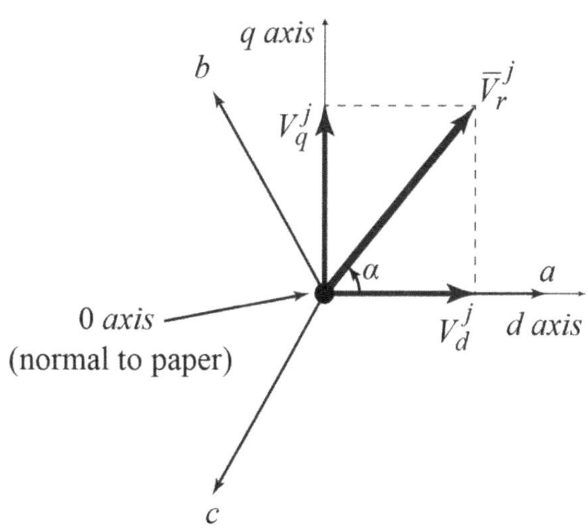

Figure B37. Voltage vector and its components in d–q axes.

where: $0 \leqslant \alpha \leqslant \pi/3$.

$$T_1 = T_z a \frac{\sin(\pi/3 - \alpha)}{\sin \pi/3};$$

$$T_2 = T_z a \frac{\sin \alpha}{\sin \pi/3};$$

$$T_0 = T_z - (T_1 + T_2)$$

where:

$$T_z = \frac{1}{f_z};$$

and

$$a = \frac{|\overline{V}_r^j|}{\frac{2}{3}\overline{V}_{DC}^j}.$$

- Commutation time duration at any sector

$$T_1 = \frac{\sqrt{3}\,T_z|\overline{V}_r^j|}{\overline{V}_{DC}^j}\sin\left(\frac{\pi}{3} - \alpha + \frac{n-1}{3}\pi\right) = \frac{\sqrt{3}\,T_z|\overline{V}_r^j|}{\overline{V}_{DC}^j}\sin\left(\frac{\pi}{3}n - \alpha\right)$$

$$= \frac{\sqrt{3}\,T_z|\overline{V}_r^j|}{\overline{V}_{DC}^j}\left(\sin\frac{\pi}{3}n\cos\alpha - \cos\frac{\pi}{3}n\sin\alpha\right);$$

$$T_2 = \frac{\sqrt{3}\,T_z|\overline{V}_r^j|}{\overline{V}_{DC}^j}\sin\left(\alpha - \frac{n-1}{3}\pi\right) = \frac{\sqrt{3}\,T_z|\overline{V}_r^j|}{\overline{V}_{DC}^j}\left(-\cos\alpha\sin\frac{n-1}{3}\pi + \sin\alpha\cos\frac{n-1}{3}\pi\right);$$

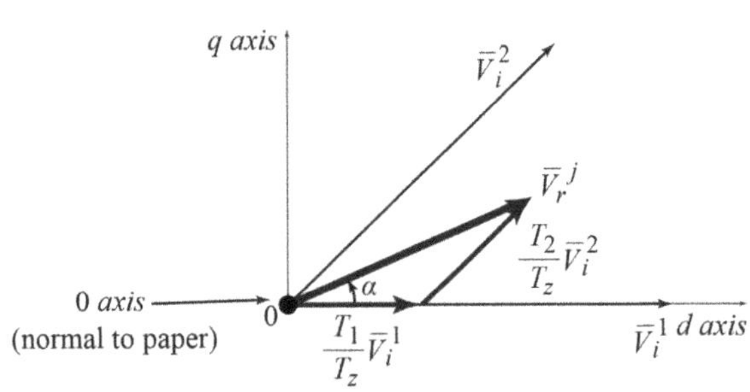

Figure B38. Reference vector as a combination of adjacent vectors at sector 1.

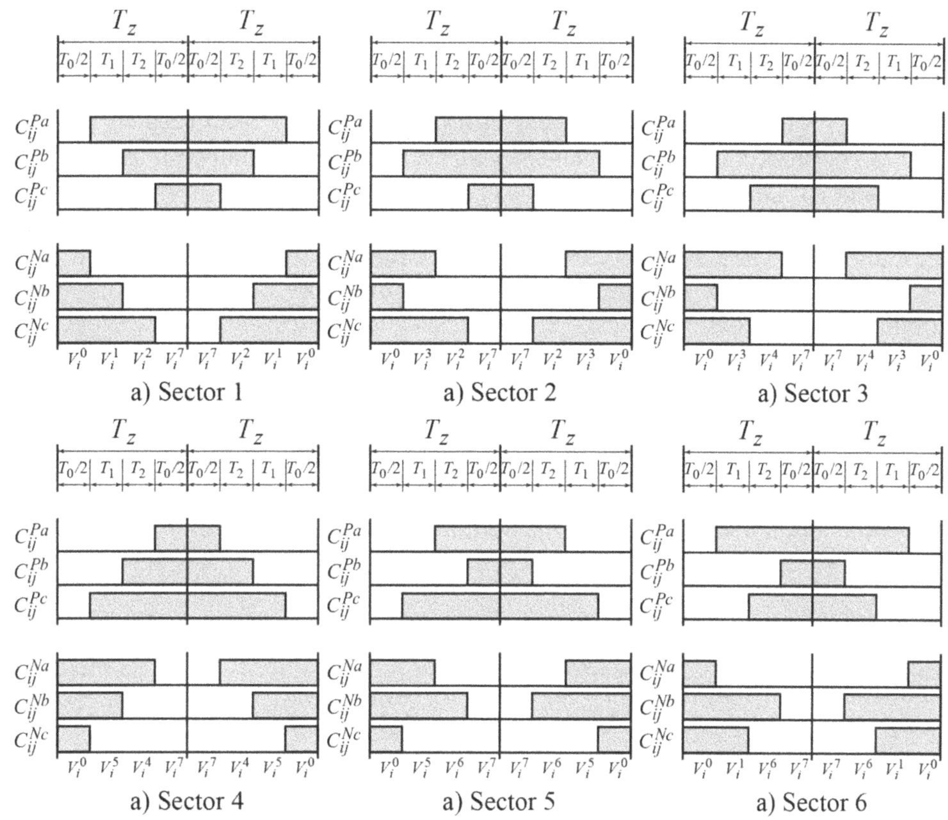

Figure B39. VECPWM commutation patterns at each sector.

$$T_0 = T_z - T_1 - T_2$$

where: n – 1 through 6, i.e. sectors 1–6;

$$0 \leqslant \alpha \leqslant \pi/3.$$

Step 3: Determine the commutation time of each electrical valve (C_{ij}^{Pa}, C_{ij}^{Pb}, C_{ij}^{Pc} to C_{ij}^{Na}, C_{ij}^{Nb}, C_{ij}^{Nc}).

Figure B39 shows VECPWM commutation patterns at each sector.

Based on figure B39, the commutation time at each sector is summarised in table B2, and it will be built in the physical model to implement VECPWM.

An ECM AC–DC–AC or DC–AC commutator, acting as a DC–AC inverter as shown in figure B40 converts a DC supply, via a series of electrical valves (electronic switches), to three output ECM columns which could be connected to an EM motor.

The electrical valves must be controlled so that at no time are both commutes (switches) in the same ECM column turned *ON* or else the DC supply would be shorted. This requirement may be met by the complementary operation of the electrical valves within an ECM column. I.e. if C_{ij}^{Pa+} is *ON* then C_{ij}^{Na-} is *OFF* and vice versa. This leads to eight possible commutation vectors for the ECM AC–DC–

AC or DC–AC commutator, acting as the DC–AC inverter, V_i^0 through V_i^7 with six *ACTIVE* commutation vectors and two *NULL* commutation vectors.

Note that looking down the columns for the *ACTIVE* commutation (switching) vectors V_i^{1-6}, the output voltages vary as a pulsed sinusoid, with each leg offset by $2\pi/3$ rad of phase angle. To implement VECM, a reference signal V_r^j is sampled with a frequency f_s ($T_s = 1/f_s$).

The reference signal may be generated from three separate phase references using the transform. The reference vector is then synthesised using a combination of the two adjacent *ACTIVE* commutation vectors and one or both of the *NULL* commutation vectors.

Various strategies of selecting the order of the vectors and which commutation vector(s) to use exist as it is shown in figure B41. Strategy selection will affect the harmonic content and the commutation losses.

All eight possible commutation vectors for an ECM DC–AC commutator, acting as a DC–AC inverter using VECM. An example V_r^j is shown in the first sector. V_{rmax}^j is the maximum amplitude of V_r^j before non-linear over-modulation is reached.

Table B2. Commutation time calculation at each sector.

Sector	Upper electrical valves	Lower electrical valves
1	$C_{ij}^{Pa}: T_1 + T_2 + T_0/2$	$C_{ij}^{na}: T_0/2$
	$C_{ij}^{Pb}: T_2 + T_0/2$	$C_{ij}^{nb}: T_1 + T_0/2$
	$C_{ij}^{Pc}: T_0/2$	$C_{ij}^{nc}: T_1 + T_2 + T_0/2$
2	$C_{ij}^{Pa}: T_1 + T_0/2$	$C_{ij}^{na}: T_2 + T_0/2$
	$C_{ij}^{Pb}: T_1 + T_2 + T_0/2$	$C_{ij}^{nb}: T_0/2$
	$C_{ij}^{Pc}: T_0/2$	$C_{ij}^{nc}: T_1 + T_2 + T_0/2$
3	$C_{ij}^{Pa}: T_0/2$	$C_{ij}^{na}: T_1 + T_2 + T_0/2$
	$C_{ij}^{Pb}: T_1 + T_2 + T_0/2$	$C_{ij}^{nb}: T_0/2$
	$C_{ij}^{Pc}: T_2 + T_0/2$	$C_{ij}^{nc}: T_1 + T_0/2$
4	$C_{ij}^{Pa}: T_0/2$	$C_{ij}^{na}: T_1 + T_2 + T_0/2$
	$C_{ij}^{Pb}: T_1 + T_0/2$	$C_{ij}^{nb}: T_2 + T_0/2$
	$C_{ij}^{Pc}: T_1 + T_2 + T_0/2$	$C_{ij}^{nc}: T_0/2$
5	$C_{ij}^{Pa}: T_2 + T_0/2$	$C_{ij}^{na}: T_1 + T_0/2$
	$C_{ij}^{Pb}: T_0/2$	$C_{ij}^{nb}: T_1 + T_2 + T_0/2$
	$C_{ij}^{Pc}: T_1 + T_2 + T_0/2$	$C_{ij}^{nc}: T_0/2$
6	$C_{ij}^{Pa}: T_1 + T_2 + T_0/2$	$C_{ij}^{na}: T_0/2$
	$C_{ij}^{Pb}: T_0/2$	$C_{ij}^{nb}: T_1 + T_2 + T_0/2$
	$C_{ij}^{Pc}: T_1 + T_0/2$	$C_{ij}^{nc}: T_2 + T_0/2$

Table B3. Eight possible commutation vectors for the ECM DC–AC commutator, acting as the DC–AC inverter.

Vector	C_{ij}^{Pa}	C_{ij}^{Pb}	C_{ij}^{Pc}	C_{ij}^{Na}	C_{ij}^{Nb}	C_{ij}^{Nc}	V_i^{ab}	V_i^{bc}	V_i^{ca}	
$V_0 = \{000\}$	*OFF*	*OFF*	*OFF*	*ON*	*ON*	*ON*	0	0	0	Null vector
$V_1 = \{100\}$	*ON*	*OFF*	*OFF*	*OFF*	*ON*	*ON*	$+V_{DC}$	0	$-V_{DC}$	Active vector
$V_2 = \{110\}$	*ON*	*ON*	*OFF*	*OFF*	*OFF*	*ON*	0	$+V_{DC}$	$-V_{DC}$	Active vector
$V_3 = \{010\}$	*OFF*	*ON*	*OFF*	*ON*	*OFF*	*ON*	$-V_{DC}$	$+V_{DC}$	0	Active vector
$V_4 = \{011\}$	*OFF*	*ON*	*ON*	*ON*	*OFF*	*OFF*	$-V_{DC}$	0	$+V_{DC}$	Active vector
$V_5 = \{001\}$	*OFF*	*OFF*	*ON*	*ON*	*ON*	*OFF*	0	$-V_{DC}$	$+V_{DC}$	Active vector
$V_6 = \{101\}$	*ON*	*OFF*	*ON*	*OFF*	*ON*	*OFF*	$+V_{DC}$	$-V_{DC}$	0	Active vector
$V_7 = \{111\}$	*ON*	*ON*	*ON*	*OFF*	*OFF*	*OFF*	0	0	0	Null vector

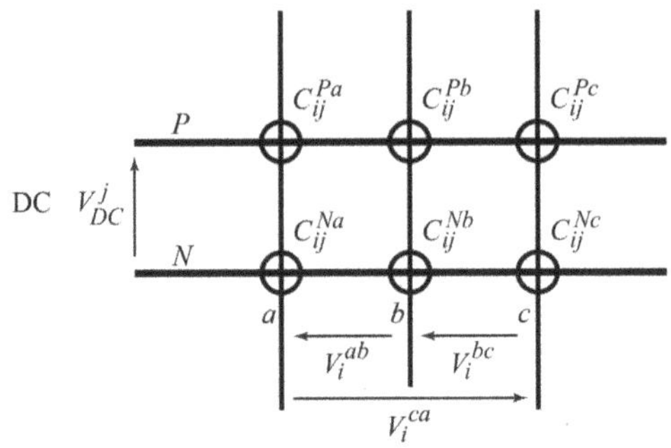

Figure B40. Topology of an ECM DC–AC commutator, acting as a DC–AC inverter.

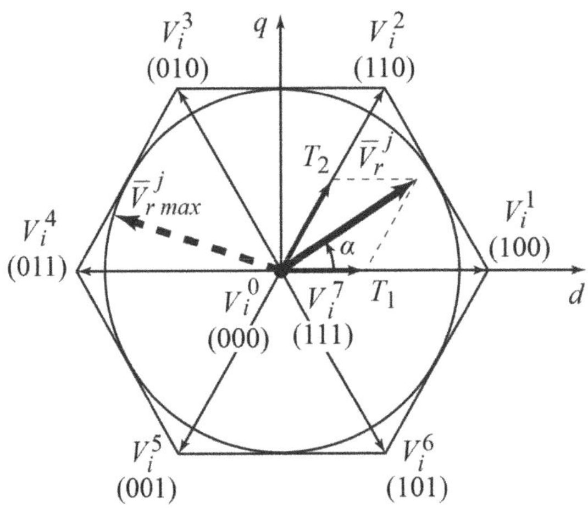

Figure B41. Various strategies of selecting the order of the vectors.

More complicated VECM strategies for the unbalanced operation of four-column ECM DC–AC commutator, acting as DC–AC inverters do exist. In these strategies the commutation vectors define a **three-dimensional** (3D) shape (a hexagonal prism in coordinates (Zhang *et al* 2002) or a dodecahedron in *a–b–c* coordinates (Perales *et al* 2003) rather than a **two-dimensional** (2D) hexagon.

B.7 Methods of PWM control: holor pulse width modulation (HOLPWM)

B.7.1 Introduction

The **electrical commutation matrixer** (ECM) DC–AC commutator, acting as the DC–AC inverter is used to convert DC power into AC power at the desired output voltage and frequency. The waveform of the output voltage depends on the commutation states of the electrical valves used in the ECM DC–AC commutator, acting as the DC–AC inverter. Major limitations and requirements of the ECM DC–AC commutator, acting as the DC–AC inverter are, harmonic contents, the commutation frequency, and the best utilisation of DC-link voltage.

Pulse width modulation (PWM) ECM DC–AC commutators, acting as DC–AC inverters, have been studied extensively during the past decades. In this method, a fixed DC input voltage is given to the ECM DC–AC commutator, acting as the DC–AC inverter and a controlled AC output voltage is obtained by adjusting the *ON* and *OFF* periods of the ECM components, i.e. electrical valves.

The most popular PWM techniques are the **sinusoidal PWM** (SINPWM), **vector PWM** (VECPWM) and a novel **holor PWM** (HOLPWM), first introduced by the authors of this textbook. With the development of **digital signal processors** (DSP), **holor modulation** (HOLM) has become one of the most important PWM methods for ECM DC–AC commutators, acting as DC–AC three-phase **voltage-source** (VS) DC–AC inverters. In this technique, the voltage holor concept is used to compute the duty cycle of the electrical valves. It is simply the digital implementation of PWM modulators. Most advanced features of HOLM are easy digital implementation and wide linear modulation range for output line-to-line voltages.

B.7.2 Holor PWM (HOLPWM)

The holor pulse width modulation (HOLPWM) method is a novel computation-intensive PWM method and it will possibly be the best one among all the PWM techniques for adjustable-velocity IEMD application. Because of its superior performance characteristics, it will find widespread application in the coming years. There are various variations of holor modulation (HOLM) that result in different quality and computational requirements. One major benefit is the reduction of total harmonic distortion (THD) created by the rapid commutation inherent to this PWM algorithm. The HOLPWM refers to a special commutation sequence of the upper three electrical valves of an ECM DC–AC commutator, acting as three-phase voltage-source DC–AC inverters used in applications such as a DC–AC commutator synchronous or asynchronous (induction) motor IEMD. It is a more sophisticated technique for generating sine waves that provides a higher voltage to the EM motor with lower total harmonic distortion.

In the SINPWM technique, instead of using a separate modulator for each of the three phases, the complex reference voltage holor is processed as a whole. Therefore, the interaction between the three EM motor phases is considered. HOLPWM generates less harmonic distortion in the output voltages and currents in the windings of the EM motor load and provides a more efficient use of the DC supply voltage in comparison with SINPWM techniques.

Since HOLPWM provides a constant commutation frequency; the commutation frequency can be adjusted easily. Although HOLPWM is more complicated than SINPWM, it may be implemented easily with modern DSP based control systems.

Voltage holors

Consider three-phase waveforms which are displaced by $2\pi/3$,

$$v_i^a(t) = V_m^a \sin \omega t;$$

$$v_i^b(t) = V_m^b \sin (\omega t - 2\pi/3);$$

$$v_i^c(t) = V_m^c \sin (\omega t + 2\pi/3);$$

these three voltage holors can be represented by one holor which is known as voltage space holor.

Voltage space holor is defined as the sum of three voltage holors, namely,

$$V_i^s = V_i^a + V_i^b + V_i^c = V^a \angle 0 + V^b \angle(-2\pi/3) + V^c \angle 2\pi/3 =$$

$$= \left\|\begin{matrix} V_c^a \\ V_s^a \end{matrix}\right\| + \left\|\begin{matrix} V_c^b \\ V_s^b \end{matrix}\right\| + \left\|\begin{matrix} V_c^c \\ V_s^c \end{matrix}\right\| = \left\|\begin{matrix} V^a\cos 0 \\ V^a\sin 0 \end{matrix}\right\| + \left\|\begin{matrix} V^b\cos(-2\pi/3) \\ V^b\sin(-2\pi/3) \end{matrix}\right\| + \left\|\begin{matrix} V^c\cos 2\pi/3 \\ V^c\sin 2\pi/3 \end{matrix}\right\| =$$

$$= V^a\left\|\begin{matrix} \cos 0 \\ \sin 0 \end{matrix}\right\| + V^b\left\|\begin{matrix} \cos(-2\pi/3) \\ \sin(-2\pi/3) \end{matrix}\right\| + V^c\left\|\begin{matrix} \cos 2\pi/3 \\ \sin 2\pi/3 \end{matrix}\right\| =$$

$$= \left\|\begin{matrix} 1 & -1/2 & -1/2 \\ 0 & \sqrt{3}/2 & -\sqrt{3}/2 \end{matrix}\right\| \left\|\begin{matrix} V^a \\ V^b \\ V^c \end{matrix}\right\| = \left\|\begin{matrix} V_c^s \\ V_s^s \end{matrix}\right\|.$$

where:

$$V_i^a = V^a \angle 0 = \left\|\begin{matrix} V_c^a \\ V_s^a \end{matrix}\right\| = \left\|\begin{matrix} V^a\cos 0 \\ V^a\sin 0 \end{matrix}\right\| = V^a\left\|\begin{matrix} \cos 0 \\ \sin 0 \end{matrix}\right\|;$$

$$V_i^b = V^b \angle(-2\pi/3) = \left\|\begin{matrix} V_c^b \\ V_s^b \end{matrix}\right\| = \left\|\begin{matrix} V^b\cos(-2\pi/3) \\ V^b\sin(-2\pi/3) \end{matrix}\right\| = V^b\left\|\begin{matrix} \cos(-2\pi/3) \\ \sin(-2\pi/3) \end{matrix}\right\|;$$

$$V_i^c = V^c \angle 2\pi/3 = \left\|\begin{matrix} V_c^c \\ V_s^c \end{matrix}\right\| = \left\|\begin{matrix} V^c\cos 2\pi/3 \\ V^c\sin 2\pi/3 \end{matrix}\right\| = V^c\left\|\begin{matrix} \cos 2\pi/3 \\ \sin 2\pi/3 \end{matrix}\right\|.$$

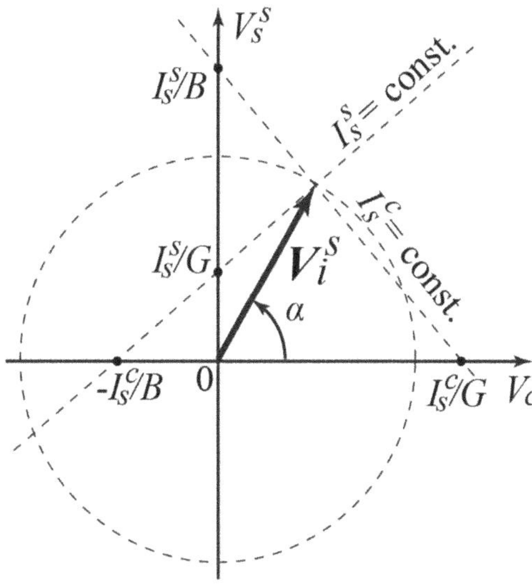

Figure B42. Rotating voltage space holor.

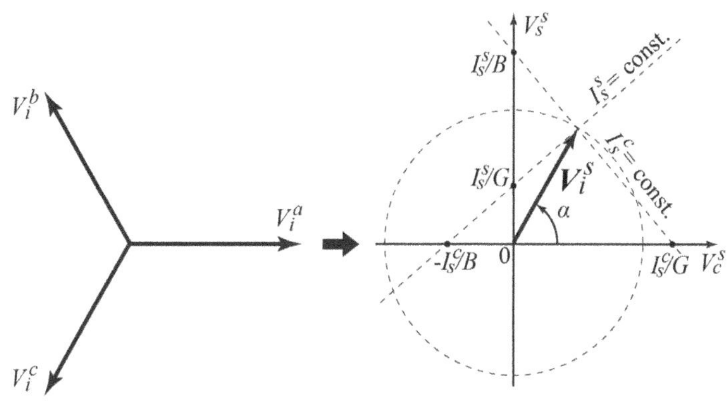

Figure B43. *a–b–c* to $V_c^s - V_s^s$ transformation.

Rotating voltage space holor $\boldsymbol{V}_i^s$ is one having a magnitude of $3/2V_{max}$ and that rotates in space at ω rad s^{-1} as shown in figure B42.

Rotating voltage space holor $\boldsymbol{V}_i^s$ can be represented in **two-dimensional** (2D) space by simply resolving *a–b–c* in to $V_c^s - V_s^s$ axis as shown in figure B43.

Therefore, the voltage space holor can also be written as

$$|\overline{V}_i^s| = \sqrt{(V_c^s)^2 + (V_s^s)^2};$$

$$\alpha = \tan^{-1}\left(\frac{V_s^s}{V_c^s}\right)$$

$$\| V_i^s \| = \left\| \begin{matrix} V_c^s \\ V_s^s \end{matrix} \right\| = \left\| \begin{matrix} 1 & -\frac{1}{2} & -\frac{1}{2} \\ 0 & \frac{\sqrt{3}}{2} & -\frac{\sqrt{3}}{2} \end{matrix} \right\| \left\| \begin{matrix} V_i^{an} \\ V_i^{bn} \\ V_i^{cn} \end{matrix} \right\|$$

Principle of holor PWM (HOLPWM)

Holor pulse with modulation (HOLPWM) aims to generate a voltage space holor that is close to the reference circle through the various commutation modes of the ECM DC–AC commutator, acting as three-phase DC–AC inverter.

Figure B44 is the structural and functional diagram of an ECM DC–AC commutator, acting as a VS DC–AC inverter physical model of the DC–AC commutator synchronous or asynchronous (induction) motor.

$\boldsymbol{C}_{ij}^{Pa}$, $\boldsymbol{C}_{ij}^{Pb}$, $\boldsymbol{C}_{ij}^{Pc}$, $\boldsymbol{C}_{ij}^{Na}$, $\boldsymbol{C}_{ij}^{Nb}$ to $\boldsymbol{C}_{ij}^{Nc}$ are the six commutation nodules, i.e. the six electrical valves that shape the output, which are controlled by the commutation variables $\boldsymbol{G}_{ij}^{Pa}$, $\boldsymbol{G}_{ij}^{Na}$, $\boldsymbol{G}_{ij}^{Pb}$, $\boldsymbol{G}_{ij}^{Nb}$, $\boldsymbol{G}_{ij}^{Pc}$ and $\boldsymbol{G}_{ij}^{Nc}$.

When an upper electrical valve is commuted (switched) *ON*, i.e. when *a, b* or *c* is 1, the corresponding lower electrical valve is commuted (switched) *OFF*, i.e. the corresponding *a′, b′ or c′* is 0. Therefore, the *ON* and *OFF* states of the upper electrical valves $\boldsymbol{C}_{ij}^{Pa}$, $\boldsymbol{C}_{ij}^{Pb}$and$\boldsymbol{C}_{ij}^{Pb}$ can be used to determine the output voltage.

Hence there are eight possible ECM DC–AC commutator states, i.e. (0, 0, 0), (0, 0, 1), (0, 1, 0), (0, 1, 1), (1, 0, 0), (1, 0, 1), (1, 1, 0), (1, 1, 1).

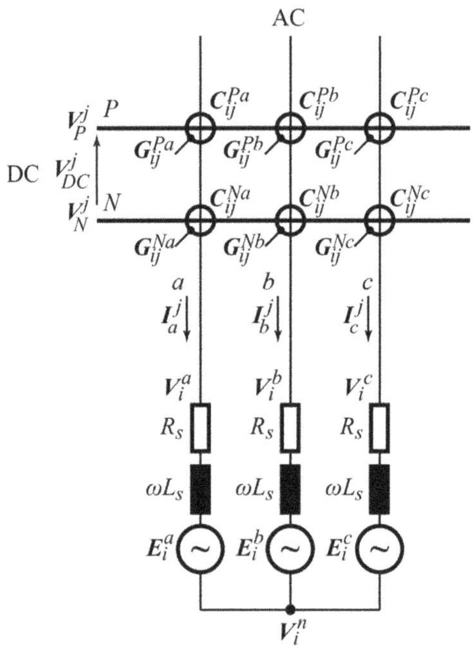

Figure B44. Voltage-source PWM DC–AC inverter physical model of the DC–AC commutator synchronous or asynchronous (induction) motor.

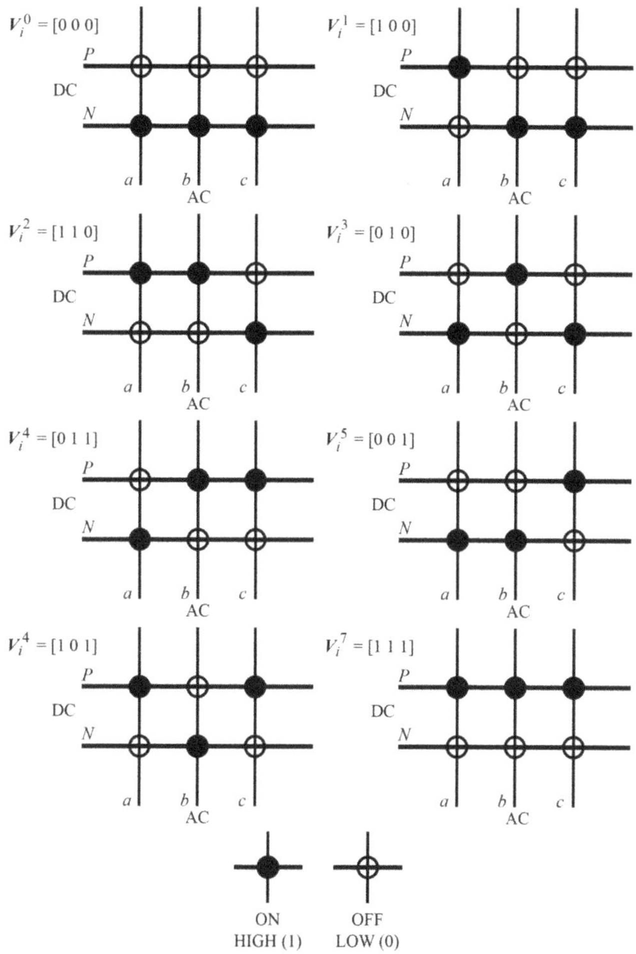

Figure B45. ECM DC–AC commutator, acting as a DC–AC inverter voltage holors (V_i^0 to V_i^7).

The ECM DC–AC commutator, acting as DC–AC inverter, has six commutation states when a voltage is applied to the EM motor and two states when the EM motor is shorted through the upper or lower electrical valves resulting in *NULL* voltages being applied to the EM motor (figure B45).

Consider an ECM DC–AC commutator, acting as a DC–AC inverter feeding a star connected load and centre point of the DC link is take as reference point as shown in figure B46.

The potential of point *a*, point *b* and point *c* with respect to the centre zero point of the DC link is known if the conducting states of the electrical valves are known. When the upper electrical valve is '*ON*', the potential of *a*, *b* and *c* is $\frac{1}{2}V_i^{DC}$ and when lower electrical valve is '*ON*', the potential of *a, b* and *c* is $-\frac{1}{2}V_i^{DC}$.

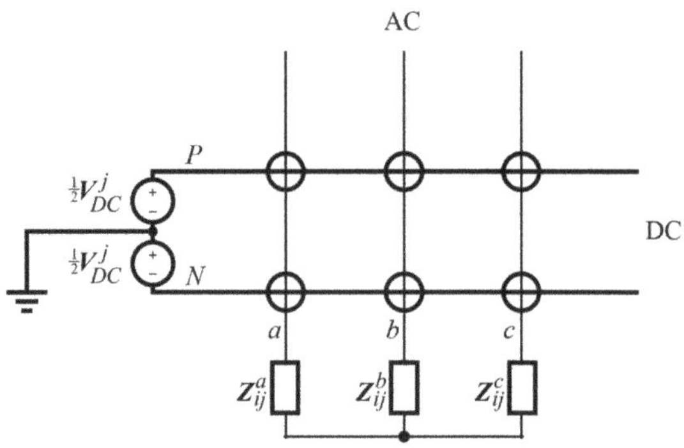

Figure B46. ECM DC–AC commutator, acting as a DC–AC inverter, feeding a star connected load.

$$V_i^{a0} = V_i^{an} + V_i^{n0};$$

$$V_i^{b0} = V_i^{bn} + V_i^{n0};$$

$$V_i^{c0} = V_i^{cn} + V_i^{n0};$$

and

$$V_i^{n0} = \frac{1}{3}[V_i^{a0} + V_i^{b0} + V_i^{c0}].$$

$$\begin{Vmatrix} V_i^{an} \\ V_i^{bn} \\ V_i^{cn} \end{Vmatrix} = \frac{1}{3}\begin{Vmatrix} 2 & -1 & -1 \\ -1 & 2 & -1 \\ -1 & -1 & 2 \end{Vmatrix}\begin{Vmatrix} V_i^{a0} \\ V_i^{b0} \\ V_i^{c0} \end{Vmatrix}.$$

Consider the commutation states, (0,0,0) and (1,1,1).

$$V_i^{an} = V_i^{bn} = V_i^{cn} = 0;$$

Hence,

$$V_c^s = V_s^s = 0.$$

Therefore,

$$V_i^s = 0\angle 0.$$

Now consider the commutation state (1, 0, 0),

$$V_i^{a0} = \frac{1}{2}V_i^{DC};$$

$$V_i^{b0} = V_i^{c0} = -\frac{1}{2} V_i^{DC};$$

$$V_i^{an} = \frac{2}{3} V_i^{DC};$$

$$V_i^{bn} = V_i^{cn} = -\frac{1}{2} V_i^{DC}.$$

Hence,

$$V_c^s = \frac{3}{2} V_i^{an} = V_i^{DC};$$

$$V_s^s = 0.$$

Therefore,

$$V_i^s = 0\angle 0.$$

Since (0, 1, 1) is the complementary of (1, 0, 0);

- For (0, 1, 1),

$$V_i^s = V_i^{DC} \angle \pi.$$

Similarly derive the magnitude and angle of space holor for all possible commutation states.

They are,

- For (0, 0, 0):

$$V_i^s = 0\angle 0 \rightarrow V_i^0$$

- For (1, 0, 0):

$$V_i^s = V_i^{DC} \angle 0 \rightarrow V_i^1$$

- For (1, 1, 0):

$$V_i^s = V_i^{DC} \angle \frac{1}{3}\pi \rightarrow V_i^2.$$

- For (0, 1, 0):

$$V_i^s = V_i^{DC}\angle\frac{2}{3}\pi \rightarrow V_i^3.$$

- For (0, 1, 1):

$$V_i^s = V_i^{DC}\angle\pi \rightarrow V_i^4.$$

- For (0, 0, 1):

$$V_i^s = V_i^{DC}\angle\frac{4}{3}\pi \rightarrow V_i^5.$$

- For (1, 0, 1):

$$V_i^s = V_i^{DC}\angle\frac{5}{3}\pi \rightarrow V_i^6.$$

- For (1, 1, 1):

$$V_i^s = V_i^{DC}\angle 0 \rightarrow V_i^7.$$

There are six *ACTIVE* (non-zero) voltage holors (V_i^1 to V_i^6) and two *NULL* (zero) holors (V_i^0 and V_i^7).

Table B4 summarises commutation holors along with the corresponding line to neutral voltage holor and line to line voltage holors, applied to the EM motor.

While plotting the eight voltage holors in the complex plane, the *ACTIVE* (non-zero) voltage holors form the axes of a hexagon as shown in figure B47. The angle

Table B4. Commutation holors, phase voltage holors and output line to line voltage holors.

	Commutating holors			Line to neutral voltage holors			Line to line voltage holors		
Voltage holors	a	b	c	V_i^{an}	V_i^{bn}	V_i^{cn}	V_i^{ab}	V_i^{bc}	V_i^{ca}
V_i^0	0	0	0	0	0	0	0	0	0
V_i^1	1	0	0	2/3	−1/3	−1/3	1	0	−1
V_i^2	1	1	0	1/3	1/3	−2/3	0	1	−1
V_i^3	0	1	0	−1/3	2/3	−1/3	−1	1	0
V_i^4	0	1	1	−2/3	1/3	1/3	−1	0	1
V_i^5	0	0	1	−1/3	−1/3	2/3	0	−1	1
V_i^6	1	0	1	1/3	−2/3	1/3	1	−1	0
V_i^7	1	1	1	0	0	0	0	0	0

(Note that the respective voltage holor should be multiplied by V_i^{DC}.)

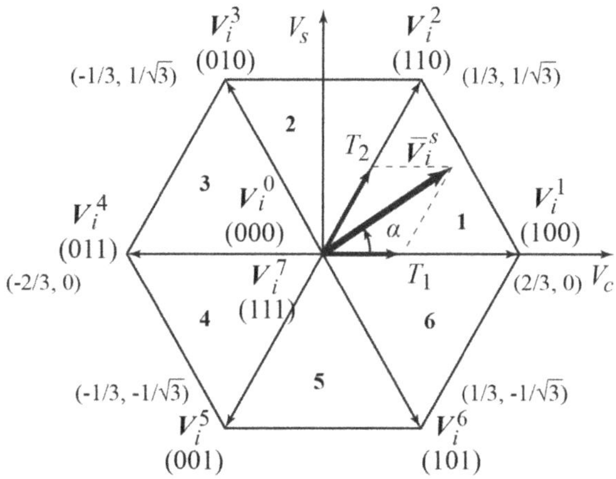

Figure B47. Fundamental commutation (switching) holors and sectors.

between any adjacent two *ACTIVE* voltage holors is $\frac{1}{3}\pi$ electrical radians. The *NULL* holors are at the origin and apply a *NULL* voltage holor to the EM motor. If the phase voltages are sinusoidal, the locus of the 'V_i^s' is a circle. The maximum value of V_i^s for which the locus is a circle is the radius of the inscribing circle, i.e. $\frac{\sqrt{3}}{2} V_i^{DC}$.

Realisation of holor PWM (HOLPWM)

The **holor PWM** (HOLPWM) is realised based on the following steps:

Step 1: Determine V_c^s, V_s^s, $\overline{V}_r^s$, and angle α

From equation (B.11), the, V_c^s, V_s^s, $\overline{V}_r^s$, and angle α can be determined as follows:

$$\| V_i^s \| = \left\| \begin{matrix} V_c^s \\ V_s^s \end{matrix} \right\| = \left\| \begin{matrix} 1 & -\frac{1}{2} & -\frac{1}{2} \\ 0 & \frac{\sqrt{3}}{2} & -\frac{\sqrt{3}}{2} \end{matrix} \right\| \left\| \begin{matrix} V_i^{an} \\ V_i^{bn} \\ V_i^{cn} \end{matrix} \right\|$$

$$|\overline{V}_i^s| = \sqrt{(V_c^s)^2 + (V_s^s)^2};$$

$$\alpha = \tan^{-1}\left(\frac{V_s^s}{V_c^s}\right)$$

Step 2: Determine time duration T_1, T_2, T_0

In the holor PWM technique, the required space holor is synthesised by two adjacent *ACTIVE* holors and a null holor.

- Commutation time duration at sector 1.

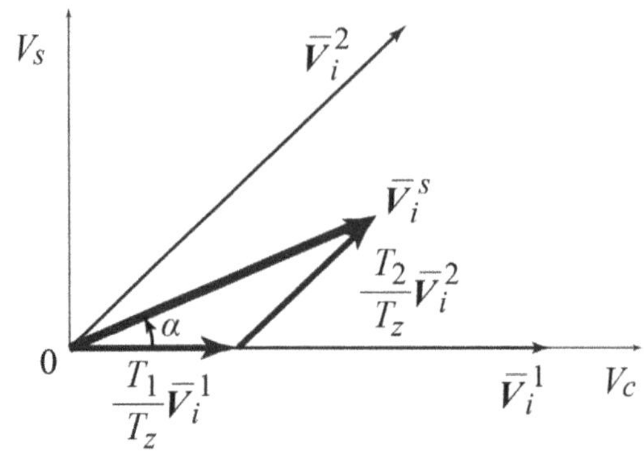

Figure B48. Reference holor as a combination of adjacent *ACTIVE* holors at sector 1.

From figure B48, the commutation time duration can be calculated as follows:

According to volt-second balance principle,

$$\int_0^{T_z} \overline{V}_r^s dt = \int_0^{T_1} \overline{V}_r^1 dt + \int_{T_1}^{T_1+T_2} \overline{V}_r^2 dt + \int_{T_1+T_2}^{T_z} \overline{V}_r^0 dt$$

$$T_z = |\overline{V}_r^s| \left\| \begin{matrix} \cos\alpha \\ \sin\alpha \end{matrix} \right\| = T_1 \overline{V}_i^1 \left\| \begin{matrix} 1 \\ 0 \end{matrix} \right\| + T_2 \overline{V}_i^2 \left\| \begin{matrix} \cos\frac{1}{3}\pi \\ \sin\frac{1}{3}\pi \end{matrix} \right\|$$

$$T_1 = T_z \frac{|\overline{V}_r^s|}{V_i^{DC}} \frac{\sin\left(\frac{1}{3}\pi - \alpha\right)}{\sin\frac{1}{3}\pi}$$

$$T_2 = T_z \frac{|\overline{V}_r^s|}{V_i^{DC}} \frac{\sin\alpha}{\sin\frac{1}{3}\pi}$$

where:

$$0 \leqslant \alpha \leqslant \frac{1}{3}\pi;$$

$$T_z = \frac{1}{2}T_s;$$

and

$$T_s = \frac{1}{f_s};$$

$$T_0 = T_z - (T_1 - T_2).$$

and

T_1 is the time for which V_i^1 is applied.
T_2 is the time for which V_i^2 is applied.
T_0 is the time for which null holor is applied.
T_s is the sampling time.
Similarly commutation time duration at any sector can be calculated.

Step 3: Determine the commutation time of each commutation nodulus, i.e. electrical valves (C_{ij}^{Pa}, C_{ij}^{Pb}, C_{ij}^{Pc}, C_{ij}^{Na}, C_{ij}^{Nb} to C_{ij}^{Nc}).

Figure B49 shows space holor PWM commutation patterns at each sector.

Based on figure B49, the commutation time at each sector is summarised in table B5, and it will be built in the physical model to implement HOLPWM.

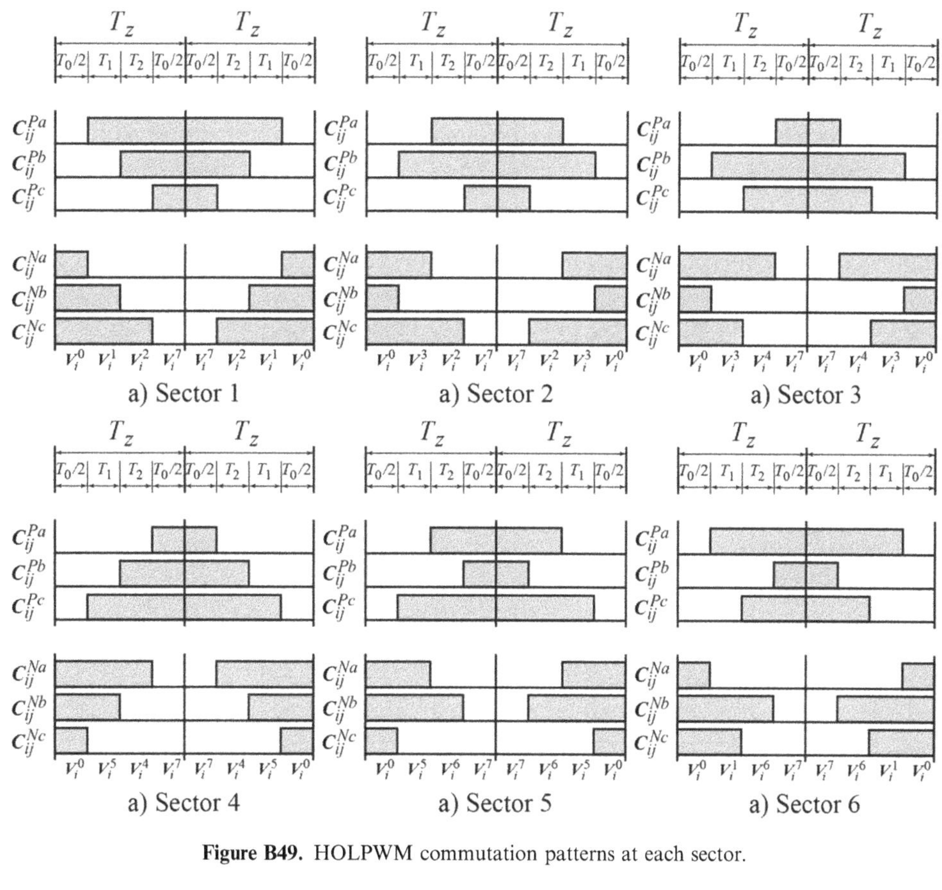

Figure B49. HOLPWM commutation patterns at each sector.

Table B5. Commutation time calculation at each sector.

Sector	Upper electrical valves	Lower electrical valves
1	$C_{ij}^{Pa}: T_1 + T_2 + T_0/2$	$C_{ij}^{na}: T_0/2$
	$C_{ij}^{Pb}: T_2 + T_0/2$	$C_{ij}^{nb}: T_1 + T_0/2$
	$C_{ij}^{Pc}: T_0/2$	$C_{ij}^{nc}: T_1 + T_2 + T_0/2$
2	$C_{ij}^{Pa}: T_1 + T_0/2$	$C_{ij}^{na}: T_2 + T_0/2$
	$C_{ij}^{Pb}: T_1 + T_2 + T_0/2$	$C_{ij}^{nb}: T_0/2$
	$C_{ij}^{Pc}: T_0/2$	$C_{ij}^{nc}: T_1 + T_2 + T_0/2$
3	$C_{ij}^{Pa}: T_0/2$	$C_{ij}^{na}: T_1 + T_2 + T_0/2$
	$C_{ij}^{Pb}: T_1 + T_2 + T_0/2$	$C_{ij}^{nb}: T_0/2$
	$C_{ij}^{Pc}: T_2 + T_0/2$	$C_{ij}^{nc}: T_1 + T_0/2$
4	$C_{ij}^{Pa}: T_0/2$	$C_{ij}^{na}: T_1 + T_2 + T_0/2$
	$C_{ij}^{Pb}: T_1 + T_0/2$	$C_{ij}^{nb}: T_2 + T_0/2$
	$C_{ij}^{Pc}: T_1 + T_2 + T_0/2$	$C_{ij}^{nc}: T_0/2$
5	$C_{ij}^{Pa}: T_2 + T_0/2$	$C_{ij}^{na}: T_1 + T_0/2$
	$C_{ij}^{Pb}: T_0/2$	$C_{ij}^{nb}: T_1 + T_2 + T_0/2$
	$C_{ij}^{Pc}: T_1 + T_2 + T_0/2$	$C_{ij}^{nc}: T_0/2$
6	$C_{ij}^{Pa}: T_1 + T_2 + T_0/2$	$C_{ij}^{na}: T_0/2$
	$C_{ij}^{Pb}: T_0/2$	$C_{ij}^{nb}: T_1 + T_2 + T_0/2$
	$C_{ij}^{Pc}: T_1 + T_0/2$	$C_{ij}^{nc}: T_2 + T_0/2$

B.7.3 Advantages of HOLPWM compared to SINPWM

I. Since the triplen order harmonics appeared in the phase-to-centre voltage of HOLPWM, it has higher modulation depth or index compared to the SINPWM. When the modulation depth or index increases the **total harmonic distortion** (THD) of the output voltage decreases. Hence HOLPWM has less current and torque harmonics than those of SINPWM.

II. For SINPWM

$$V_{\max} = V_{DC}/2.$$

III. For HOLPWM

$$V_{\max} = V_{DC}/\sqrt{3}.$$

where V_{DC} is DC-link voltage.

From this it is clear that HOLPWM can produce about 15% higher output voltage than SINPWM.

i. In SINPWM different phases may commute (switch) simultaneously. But in HOLPWM only one phase is switched at a time. Hence HOLPWM has reduced commutation losses compared to SINPWM.
ii. The SINPWM DC–AC inverter can be thought of as three separate driver stages which create each phase waveform independently. But **holor modulation** (HOLM) treats the ECM DC–AC commutator, acting as DC–AC inverter, as a single unit.

B.7.4 Conclusion

Holor modulation (HOLM) technique will become the most popular and important PWM technique for ECM AC–DC–AC or DC–AC commutators, acting as voltage-source DC–AC inverters for the angular-velocity and/or torque control of AC–DC–AC or DC–AC commutator synchronous or asynchronous (induction) motors. The modulation depth or index is higher for HOLPWM as compared to SINPWM. The current and torque harmonics produced are much less in the case of HOLPWM. In the case of HOLPWM the output voltage is about 15% more as compared to SINPWM. The HOLPWM technique utilises DC-link voltage more efficiently and generates less harmonic distortion in an ECM AC–DC–AC or DC–AC commutator, acting as a three-phase voltage-source DC–AC inverter. HOLPWM is very easy to implement. There are modern **digital signal processors** (DSP) with dedicated pins which give pulse width modulated waveforms using HOLM.

References

Brocker H W, Skudenly H C and Stanke G 1986 Analysis and realization of a pulse width modulator based on the voltage space vectors *Conf. Rec. IEEE-IAS Annu. Meeting* (Denver, CO) 244–51

Holmes D G and Lipo T A 2003 *Pulse Width Modulation for Power Converters Principles and Practice* (Piscataway, NJ: Institute of Electrical and Electronics Engineers)

Holtz J 1992 Pulse width modulation—A Survey *IEEE Trans. Ind. Electron.* **30** 410–20

Kazmierkowski M P, Krishnan R and Blaabjerg F 2002 *Control in Power Electronics: Selected Problems* (San Diego: Academic)

King F G 1974 A three-phase transistor class-B inverter with sinewave output and high efficiency, in *Inst. Elec. Eng. Conf. Publ. 123 Power Electronics, Power Semiconductors and Their Applications* 204–9

Perales M A, Prats M M, Portillo R, Mora J L, León J I and Franquelo L G 2003 Three-dimensional space vector modulation in abc coordinates for four-leg voltage source converters *IEEE Power Electron. Lett.* **1** 104–9

Zhang R, Prasad V, Boroyevich D and Lee F C 2002 Three-dimensional space vector modulation for four-leg voltage-source converters *IEEE Power Electron. Lett.* **17** 3

Zhou K and Wang D 2002 Relationship between space vector modulation and three phase carrier-based PWM: A comprehensive analysis *IEEE Trans. Ind. Electron.* **49** 186–96

IOP Publishing

Appendix C

Synthetic mathematical model of the abstract MMD electrical machine physical heterogeneous continuous dynamical hypersystem

C.1 Abstract MMD electrical machine physical heterogeneous continuous dynamical hypersystem

A synthetic mathematical model of the abstract MMD electrical machine physical heterogeneous continuous dynamical hypersystem, based on 'Hamilton–Ostrogradsky's variation principle of the stationary action', initiates an algorithm to formulate a real mathematical model of a physical heterogeneous dynamical hypersystem in the dynamical systems approach. This principle may be found in the work of an Irish mathematician William Rowan Hamilton (1885–1865), published in the years 1830–5. Hamilton assumes that the output dynamical system is 'scleronomic'; i.e. it is subjected only to the stationary constraints, in which the equation of time t does not appear explicitly. In the general case of non-stationary constraints, the principle was formulated and substantiated, by a Russian mathematician Mikhail Vasilevich Ostrogradsky (1891–1842) in 1845. In relation with the Hamilton's principle it is sometimes called the 'Hamilton–Ostrogradsky's principle'. In the 1760s a Swiss mathematician Leonhard Euler (1707–83) developed a way of laying out differential equations of dynamics based on the methods of variation calculus (Euler 1766—*Elementa Calculi Variatiorum*), and in the 1780s a French mathematician Joseph Louis de Lagrange (1736–813) developed a stacking of differential equations based on the energetic methods for mechanical homogenous continuous dynamical system (Lagrange 1788—*Mécanique Analytique*), of which both the differential equations of Euler and Lagrange are analogous so they may be

doi:10.1088/2053-2563/aae7d7ch11 © IOP Publishing Ltd 2019

applied in the case of general dynamical hypersystem of many **degrees of freedom** (DoF). In the 1870s the English physicist James Clark Maxwell (1831–79) developed a way of stacking the differential equations of dynamics based on Lagrange's energetic methods for an electromechanical continuous dynamical hyposystems (Maxwell 1873—*A Treatise on Electricity and Magnetism*), depending on a supplementation of the Lagrange's potential by an analogue of kinetic and potential energy, as well as on finding an analogue of the generalised force (Carvallo 1902, Liennard 1902, Maxwell 1873). By including in the textbook such fundamental concepts of dynamics for the methodology to create a mathematical model of an MMD electrical machine physical heterogeneous continuous dynamical hypersystem—like the MMD electrical machine physical continuous dynamical component (figure C1)—instead of using the concept of mechanical motion applied by Lagrange, as a material point, for mechanical homogeneous continuous dynamical systems, below we will leave the Euler–Lagrange second order differential equation of dynamics in the dynamical systems approach and hypermatrix notation.

The dynamical systems approach and holor-hypermatrix notation takes into account the work done by all the generalised forces acting on the abstract MMD electrical machine physical heterogeneous continuous dynamical hypersystem, regardless of whether they are conservative generalised forces, or not. In addition, the dynamical systems approach and hypermatrix notation make distributing the physical variables difficult by applying partial differentiation in many cases, the structure of the nonlinear MMD electrical machine physical heterogeneous continuous dynamical hypersystem denies this to each of its dynamical components and changes its dynamical states freely.

A number of conditions and restrictions can hamper their dynamical states. It is said then that on the dynamics of the MMD electrical machine physical dynamical

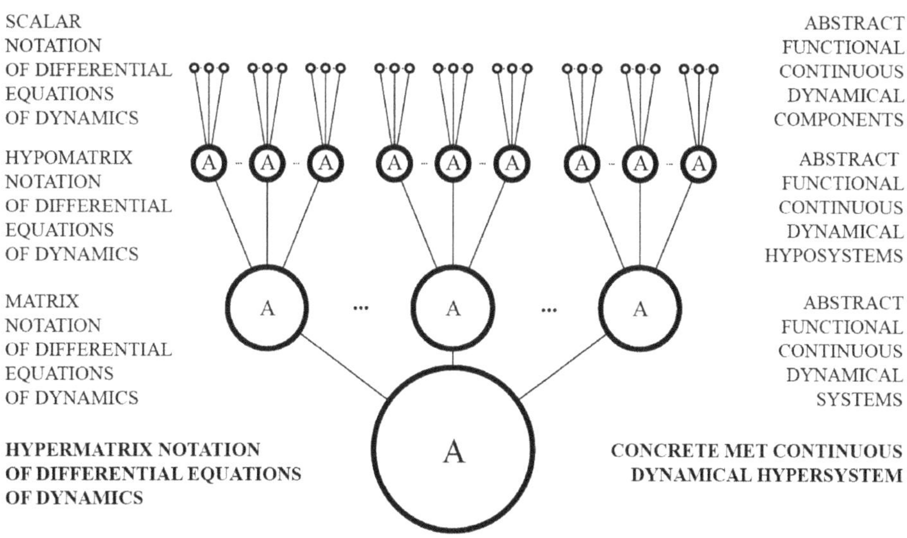

Figure C1. A hierarchical structure of a physical model of the abstract MMD electrical machine physical continuous dynamical hypersystem.

hypersystem are imposed constraints. A specific form of these constraints can be very varied, for example, the combination of two different MMD electrical machine physical continuous dynamical hyposystems of the **direct current** (DC) or **alternating current** (AC) commutator electrical machine by means of the arbitrary physical commutation matrixer (e.g. mechanical or electrical commutation matrixer)—see appendix A. If the constraints impose restrictions on the dynamical status of respective parts of the nonlinear abstract MMD electrical machine physical continuous dynamical hypersystem, they are called them the 'geometrical constraints'. Moreover, if they impose restrictions on the dynamics of the MMD electrical machine physical dynamical hypersystem, i.e. on the values of the generalised velocities of the respective parts of the MMD electrical machine physical dynamical hypersystem, then they are termed the 'kinematical constraints'. Not true, however, may be an inverse affirmation, i.e. the relationships between the generalised velocities of different respective parts of the MMD electrical machine physical dynamical hypersystem, and in particular the limit values of the generalised velocities may not have restrictions on the dynamical status of the MMD electrical machine physical continuous dynamical hypersystem. A convincing example of this is the arbitrary physical commutation matrixer. Relatively recently, it was noted the existence of constraints does not impose any restrictions on the dynamical state of the nonlinear abstract MMD electrical machine physical continuous dynamical hypersystem. Lagrange (1788) in *Mécanique Analytique* does not believe that such constraints may exist, noting that for any mechanical homogeneous continuous dynamical system using ordinary differential equations, resulting from the structure of the MMD electrical machine physical dynamical system, independent coordinates can be selected having independent variations. This over-sight became known after many years (Appell 1899a, 1899b, 1893, 1899d, 1903, Bobylev 1892, Beltrami, 1895, Tchaplygin 1897a, 1897b, Zhukovsky 1893) in considering rolling solid bodies without slipping after plane or complex surfaces. In 1894, the German physicist H Hertz (1857–94) introduced the division of the mechanical homogeneous continuous dynamical systems for holonomic and unholonomic (Hertz 1894). Nonlinear abstract MMD electrical machine physical heterogeneous continuous dynamical hypersystem with unholonomic constraints, no integral, which are not reduced to the geometrical constraints are called the 'unholonomic physical heterogeneous continuous dynamical hypersystem'.

C.2 Independent generalised coordinates

One of the basic concepts of the dynamics of nonlinear abstract MMD electrical machine physical heterogeneous continuous dynamical hypersystem is a concept of elementary 'abstract MMD electrical machine physical homogeneous continuous dynamical system'. Under this name, is meant the dynamical component elementary abstract MMD electrical machine physical homogeneous continuous dynamical system, whose physical, economic, social, etc, structure one can omit when creating a synthetic mathematical model of its dynamics.

It is understood that the possibility of such omission is dependent on the existing conditions of that or another problem. For instance, the planet can be seen as elementary astronomical homogeneous continuous dynamical system of the sunny homogeneous continuous dynamical hypersystem, in the investigation of its movement around the Sun, but one that cannot be seen in the investigation of their day and night movement.

The dynamical status of elementary abstract MMD electrical machine physical homogeneous continuous dynamical systems of nonlinear abstract MMD electrical machine physical heterogeneous continuous dynamical hypersystem, that is, their position in the generalised multi-dimensional spaces is determined by a hypermatrix of the independent generalised coordinates

$$\boldsymbol{q} = \left[||q_1||^T \, ||q_2||^T \, ||q_3||^T \ldots ||q_9||^T \right]^T$$

A derivative of the hypermatrix $\boldsymbol{q}$ over time

$$\dot{\boldsymbol{q}} = \frac{d\boldsymbol{q}}{dt} = \boldsymbol{v} \tag{C.1}$$

is a hypermatrix of the generalised velocities, and the second derivative

$$\ddot{\boldsymbol{q}} = \frac{d^2\boldsymbol{q}}{dt} = \frac{d\dot{\boldsymbol{q}}}{dt} = \boldsymbol{a} \tag{C.2}$$

is a hypermatrix of the generalised accelerations of the nonlinear MMD electrical machine physical heterogeneous continuous dynamical hypersystem.

To determine the dynamical status of the nonlinear MMD electrical machine physical heterogeneous continuous dynamical hypersystem consisting of N elementary MMD electrical machine physical homogeneous continuous dynamical systems in multi-dimensional spaces, should introduce N matrices of the independent generalised coordinates.

The total number of matrices of the independent generalised coordinates, which is necessary for introduction of the unique identification of the nonlinear MMD electrical machine physical continuous dynamical hypersystem, is simply the number of DoF. The hypomatrices of these matrices are independent generalised coordinates of elementary MMD electrical machine physical homogeneous continuous dynamical systems of the nonlinear MMD electrical machine physical heterogeneous continuous dynamical hypersystem.

The mathematical relationship relating the hypermatrix of generalised accelerations with the hypermatrix of the independent generalised coordinates and the hypermatrix of the generalised velocities was named the 'synthetic mathematical model' or the 'differential equation of dynamics' of the nonlinear MMD electrical machine physical heterogeneous continuous dynamical hypersystem. In relation to the hypermatrix, $\boldsymbol{q}$, it is the second-order differential equation in the hypermatrix notation, of which their successive integration allows, in principle, to determine these hypermatrices, i.e. trajectories of the dynamics of the nonlinear MMD electrical machine physical heterogeneous continuous dynamical hypersystem.

C.3 Unconstrained and constrained nonlinear MMD electrical machine physical heterogeneous continuous dynamical hypersystems—classification of the constraints

The subject of the work carried out by the authors in this textbook of the dynamical analytical studies is the dynamics of the nonlinear abstract MMD electrical machine physical heterogeneous continuous dynamical hypersystem, consisting of elementary MMD electrical machine physical homogeneous continuous dynamical systems of this MMD electrical machine physical dynamical hypersystem S_υ ($\upsilon = 1, 2, 3, \ldots, N$) relative to a set of matrices of the independent generalised coordinates. On matrices of the independent generalised coordinates (the generalised positions) and matrices of the generalised velocities of these elementary MMD electrical machine physical homogeneous continuous dynamical systems some restrictions called the 'constraints' are imposed.

The nonlinear MMD electrical machine physical heterogeneous continuous dynamical hypersystem with these types of constraints is called the 'unconstrained' one which is distinct from the 'constrained' MMD electrical machine physical dynamical hypersystems, which are not subject to any constraints. The constraints can be expressed in a hypermatrix notation by means of an equation[1]

$$\mathcal{F}(\boldsymbol{q}, \dot{\boldsymbol{q}}, t) = 0 \tag{C.3}$$

where after the left side—there is a time t, a hypermatrix of independent generalised coordinates $\boldsymbol{q}$ of all elementary nonlinear MMD electrical machine physical homogeneous continuous dynamical systems, S_υ of the nonlinear MMD electrical machine physical heterogeneous continuous dynamical hypersystem ($\upsilon = 1, 2, 3, \ldots, N$).

In the particular case, when the hypermatrix of the generalised velocities $\dot{\boldsymbol{q}}$ is not present in the equation of constraints (C.3), the constraints are called 'finite'. Analytically they may be represented with the hypermatrix notation as follows:

$$\mathcal{F}(\boldsymbol{q}, t) = 0 \tag{C.4}$$

In the general case, the constraints (C.3) are called 'differential'. In the case where there are finite constraints of a form (C.4) a nonlinear MMD electrical machine physical heterogeneous continuous dynamical hypersystem may not find itself in an arbitrary dynamical position, or take an arbitrary position in the multi-dimensional space in a given moment t. However, in the case when there are only the differential constraints, the MMD electrical machine physical dynamical hypersystem may impose in a given moment t to find itself in an arbitrary dynamical state (position) in the multi-dimensional space. However, in these dynamical states, the matrices of the generalised velocities of the elementary MMD electrical machine physical homogeneous continuous dynamical systems may not be arbitrary, because the differential constraints impose the restrictions on these matrices of the generalised velocities.

[1] Next $\mathcal{F}(\boldsymbol{q}, \dot{\boldsymbol{q}}, t) = 0$ is an abbreviated function designation $\mathcal{F}\left(\left[\|q_1\|^T \ldots \|q_\vartheta\|^T\right]^T, \left[\|\dot{q}_1\|^T \ldots \|\dot{q}_\vartheta\|^T\right]^T\right)$.

The finite constraints (C.4) are called 'stationary', in the case when in their equation time t does not appear explicitly.

The nonlinear MMD electrical machine physical heterogeneous continuous dynamical hypersystem, consisting of elementary MMD electrical machine physical homogeneous continuous dynamical systems of this MMD electrical machine physical dynamical hypersystem is called the 'holonomic MMD electrical machine physical dynamical hypersystem', if elementary MMD electrical machine physical heterogeneous continuous dynamical systems are not subject to any integral differential constraints.

The abbreviation of this kind is often used in this textbook. The function $\mathcal{F}$, as with all the functions found in the textbook assumes that they are continuous with those of their derivatives, which are found in the relevant chapters of the textbook. For example, the holonomic MMD electrical machine physical dynamical hypersystem is every one of the unconstrained nonlinear MMD electrical machine physical heterogeneous continuous dynamical hypersystem, consisting of elementary MMD electrical machine physical homogeneous continuous dynamical systems, as well as the constrained nonlinear MMD electrical machine physical heterogeneous continuous dynamical hypersystem with 'finite' and 'differential' constraints, but no integral. In the case of the holonomic nonlinear MMD electrical machine physical heterogeneous continuous dynamical hypersystem all constraints can be represented in a form of the finite one.

In the case when a nonlinear MMD electrical machine physical continuous dynamical hypersystem is subject to no integral differential constraints, this is called the 'unholonomic' MMD electrical machine physical dynamical hypersystem. No integral differential constraints are the so called 'unholonomic' one, while the integral differential constraints are the 'semi-holonomic' one.

The nonlinear MMD electrical machine physical heterogeneous continuous dynamical hypersystem is called the 'scleronomic' one, if it is subject only to stationary constraints. If not, the MMD electrical machine physical dynamical hypersystem is called the 'rheonomic' one.

Nonlinear MMD electrical machine physical heterogeneous continuous dynamical hypersystems, in which processing takes place in energy generation, conversion, dissipation or energy storage distribute on:

- conservative, in which the action of the external generalised forces change without loss into kinetic or potential energy;
- dissipative, in which the action of the generalised external forces dissipates into thermal energy.

C.4 Principle of stationary action

The most general law of dynamics of the conservative nonlinear MMD electrical machine physical heterogeneous continuous dynamical hypersystem, in this textbook is based on the variation principle of stationary action, which originates from the Hamilton–Ostrogradsky's principle of least action defined for conservative dynamical hypersystems after its generalisation on the dissipative dynamical hypersystems.

In accordance with this principle, each of the nonlinear abstract MMD electrical machine physical heterogeneous continuous dynamical hypersystems is characterised by some kinetic potential, that is, a function, called the 'Lagrange's energy function' (or the 'Lagrangian') which is equal to the surplus of the conservative kinetic co-energy $\mathcal{T}^*$ over the potential energy $\mathcal{U}$ of each of the nonlinear abstract MMD electrical machine physical heterogeneous continuous dynamical hypersystems, which was derived, for example, in the book written by White and Woodson (1959), but in the conventional (Newtonian Paradigm) systems approach and scalar notation, however, in the dynamical systems approach and hypermatrix notation is defined as:

$$\mathcal{L}(\boldsymbol{q}, \dot{\boldsymbol{q}}, t) \triangleq \mathcal{T}^*(\boldsymbol{q}, \dot{\boldsymbol{q}}, t) - \mathcal{U}(\boldsymbol{q}, \dot{\boldsymbol{q}}, t) \tag{C.5a}$$

or simply

$$\mathcal{L} \triangleq \mathcal{T}^* - \mathcal{U} \tag{C.5b}$$

moreover, a function called the 'virtual' or 'prepared' work, equal to the surplus of the dissipative kinetic co-energy (dissipation) $\mathcal{T}^*_R$ over the dissipative potential energy $\mathcal{U}_0$ of each of the nonlinear abstract MMD electrical machine physical heterogeneous continuous dynamical hypersystems, namely in the dynamical systems approach and hypermatrix notation is defined as:

$$\mathcal{A}(\boldsymbol{q}, \dot{\boldsymbol{q}}, t) \triangleq \mathcal{T}^*_R(\boldsymbol{q}, \dot{\boldsymbol{q}}, t) - \mathcal{U}_F(\boldsymbol{q}, \dot{\boldsymbol{q}}, t) \tag{C.6a}$$

or simply

$$\mathcal{A} \triangleq \mathcal{T}^*_R - \mathcal{U}_F \tag{C.6b}$$

Variables $\boldsymbol{q}, \dot{\boldsymbol{q}}, t$ by which are expressed Lagrange's energy function and virtual work, are called the 'Lagrange variables'. A set of the values of these variables has been characterised by a moment of time and correspond to the dynamical status of each of the nonlinear abstract MMD electrical machine physical heterogeneous continuous dynamical hypersystems, i.e. the generalised positions and generalised velocities of its elementary abstract MMD electrical machine physical heterogeneous continuous dynamical systems.

The conservative kinetic co-energy of any of the nonlinear abstract MMD electrical machine physical heterogeneous continuous dynamical hypersystems in the dynamical systems approach and the hypermatrix notation is defined as:

$$\mathcal{T}^* \triangleq \int_0^{\dot{q}} \boldsymbol{p}^T(\boldsymbol{q}, \dot{\boldsymbol{q}}) d\boldsymbol{q} \triangleq \int_0^{v} \boldsymbol{p}^T(\boldsymbol{q}, v) dv. \tag{C.7a}$$

where: $\boldsymbol{p}(\boldsymbol{q}, \dot{\boldsymbol{q}}) = \boldsymbol{p}(\boldsymbol{q}, v)$—the hypermatrix of conservative generalised momenta;

$\boldsymbol{K}(\boldsymbol{q}, \dot{\boldsymbol{q}}) = \boldsymbol{K}(\boldsymbol{q}, v) = \frac{\partial p}{\partial \dot{q}} = \frac{\partial p}{\partial v}$—the hypermatrix of conservative kinetic differential coefficients (generalised inertia).

The conservative kinetic co-energy of the arbitrary linear abstract MMD electrical machine physical heterogeneous continuous dynamical hypersystem in the dynamical systems approach and the hypermatrix notation is a homogeneous square form, namely:

$$\mathcal{T}^* \triangleq \frac{1}{2}\boldsymbol{q}^T\boldsymbol{K}(\boldsymbol{q}, \dot{\boldsymbol{q}})\boldsymbol{q} \triangleq \frac{1}{2}\boldsymbol{v}^T\boldsymbol{K}(\boldsymbol{q}, \boldsymbol{v})\boldsymbol{v} \tag{C.7b}$$

The conservative potential energy of the nonlinear abstract MMD electrical machine physical heterogeneous continuous dynamical hypersystem in the dynamical systems approach and the hypermatrix notation is defined as:

$$\mathcal{U} \triangleq \int_{\boldsymbol{0}}^{\boldsymbol{q}} \boldsymbol{f}d\boldsymbol{q} \tag{C.8a}$$

where: $\boldsymbol{f}$—the hypermatrix of conservative generalised forces;

$\boldsymbol{P}(\boldsymbol{q}, \dot{\boldsymbol{q}}) = \boldsymbol{P}(\boldsymbol{q}, \boldsymbol{v}) = \frac{\partial \boldsymbol{f}}{\partial \boldsymbol{q}}$—the hypermatrix of the conservative potential differential coefficients (generalised stiffnesses).

The dissipative potential energy of any linear MMD electrical machine physical heterogeneous continuous dynamical hypersystem in the dynamical systems approach and hypermatrix notation is a homogeneous square form, namely:

$$\mathcal{U} \triangleq \frac{1}{2}\boldsymbol{q}^T\boldsymbol{P}(\boldsymbol{q}, \dot{\boldsymbol{q}})\boldsymbol{q} \triangleq \frac{1}{2}\boldsymbol{q}^T\boldsymbol{P}(\boldsymbol{q}, \boldsymbol{v})\boldsymbol{q} \tag{C.8b}$$

The conservative kinetic co-energy of any of the nonlinear abstract MMD electrical machine physical heterogeneous continuous dynamical hypersystems in the dynamical systems approach and hypermatrix notation is defined as:

$$\mathcal{T}_R^* \triangleq \int_0^t \mathcal{R}dt \tag{C.9}$$

the integral of the 'Rayleigh function', i.e. the function of the thermal losses

$$\mathcal{R} \triangleq \mathcal{R}_V + \mathcal{R}_C \tag{C.10}$$

is the sum of the generalised losses on the viscous friction

$$\mathcal{R}_V \triangleq \int_{\boldsymbol{0}}^{\dot{\boldsymbol{q}}} \boldsymbol{p}_R(\dot{\boldsymbol{q}})d\dot{\boldsymbol{q}} \triangleq \int_0^{\boldsymbol{v}} \boldsymbol{p}_R(\boldsymbol{v})d\boldsymbol{v} \tag{C.11}$$

and generalised losses on the dry friction

$$\mathcal{R}_C \triangleq \int_{\boldsymbol{0}}^{\dot{\boldsymbol{q}}} \dot{\boldsymbol{q}}^T(\operatorname{sgn} \dot{\boldsymbol{q}}\, \Delta \boldsymbol{f})d\dot{\boldsymbol{q}} \triangleq \int_{\boldsymbol{0}}^{\boldsymbol{v}} \boldsymbol{v}^T(\operatorname{sgn} \boldsymbol{v}\, \Delta \boldsymbol{f})d\boldsymbol{v} \tag{C.12}$$

where: $\boldsymbol{p}_R(\dot{\boldsymbol{q}}) = \boldsymbol{p}_R(\boldsymbol{v})$—the hypermatrix of dissipative generalised momenta (dissipation);

$\boldsymbol{D}(\dot{\boldsymbol{q}}) = \boldsymbol{D}(\boldsymbol{v}) = \frac{\partial \boldsymbol{p}_R}{\partial \dot{\boldsymbol{q}}} = \frac{\partial \boldsymbol{p}_R}{\partial \boldsymbol{v}}$—the hypermatrix of dissipative kinetic differential coefficients of the thermal losses, caused by the viscous friction;

$\Delta\boldsymbol{f}$—the hypermatrix of dissipative generalised force-drops of thermal losses, caused by the dry friction.

The Rayleigh's dissipation function of any of the linear abstract MMD electrical machine physical heterogeneous continuous dynamical hypersystems is the sum of generalised losses on the viscous friction, establishing a homogeneous square form

$$\mathcal{R}_V \triangleq \frac{1}{2}\dot{\boldsymbol{q}}^T \boldsymbol{D}(\dot{\boldsymbol{q}})\dot{\boldsymbol{q}} \triangleq \frac{1}{2}\boldsymbol{v}^T \boldsymbol{D}(\boldsymbol{v})\boldsymbol{v} \tag{C.13}$$

and generalised losses on the dry friction

$$\mathcal{R}_C \triangleq \frac{1}{2}(\operatorname{sgn} \dot{\boldsymbol{q}}\ \Delta \boldsymbol{f})\dot{\boldsymbol{q}} \triangleq \frac{1}{2}(\operatorname{sgn} \boldsymbol{v}\ \Delta \boldsymbol{f})\boldsymbol{v} \tag{C.14}$$

where: $\operatorname{sgn} \dot{\boldsymbol{q}} = \left[\operatorname{sgn} \|\dot{q}_1\|^T \ldots \operatorname{sgn} \|\dot{q}_9\|^T\ \right]^T$;

$$\operatorname{sgn} \dot{\boldsymbol{q}} \triangleq \operatorname{sgn} \boldsymbol{v} \triangleq \begin{cases} +1 & v > 0; \\ -1 & v < 0. \end{cases}$$

The dissipative potential energy of any of the nonlinear abstract MMD electrical machine physical heterogeneous continuous dynamical hypersystem in the dynamical systems approach and hypermatrix notation assumed as:

$$\mathcal{U}_F \triangleq \int_{\boldsymbol{0}}^{\boldsymbol{q}} \boldsymbol{f}(t)^T\, d\boldsymbol{q} = \int_0^t \boldsymbol{f}(t)^T\, \dot{\boldsymbol{q}}\, dt = \int_0^t \boldsymbol{f}(t)^T\, \boldsymbol{v}\, dt \tag{C.15}$$

where: $\boldsymbol{f}(t)$—the hypermatrix of the dissipative generalised forces (dissipation) as a function of time.

If a nonlinear abstract MMD electrical machine physical heterogeneous continuous dynamical hypersystem transmits from the dynamical state described by a hypermatrix of the generalised coordinates

$$\boldsymbol{q} = \boldsymbol{q}^{(1)}$$

occurring in the time $t = t_1$, to the dynamical state described by a hypermatrix of the generalised coordinates

$$\boldsymbol{q} = \boldsymbol{q}^{(2)}$$

occurring in the time $t = t_2$, then this transition is done in such a way that the integral

$$\mathcal{I} \triangleq \int_{t_1}^{t_2} [\mathcal{L}(\boldsymbol{q}, \dot{\boldsymbol{q}}, t) + \mathcal{A}(\boldsymbol{q}, \dot{\boldsymbol{q}}, t)]dt = minimum \tag{C.16}$$

Let $\boldsymbol{q} = \boldsymbol{q}(t)$ be the hypermatrix of generalised coordinates for which the functional $\mathcal{I}$ has the smallest value. This means that $\mathcal{I}$ rose when changing the hypermatrix $\boldsymbol{q}(t)$ on any hypermatrix of the form

$$\boldsymbol{q}(t) + \delta\boldsymbol{q}(t) \tag{C.17}$$

where $\delta\boldsymbol{q}(t)$—the hypermatrix, where the matrices and hypomatrices are the elements of the hypermatrix $\boldsymbol{q}(t)$, which are small functions in the entire time period from t_1 to t_2 (called the 'variation' of the hypermatrix $\boldsymbol{q}(t)$); because at $t = t_1$ and $t = t_2$ the matrices, hypomatrices and elements of the hypermatrix (C.17) should assume one and the same values of the matrices, hypomatrices and components of the hypermatrix $\boldsymbol{q}^{(1)}$ and $\boldsymbol{q}^{(2)}$, it should be that both

$$\delta\boldsymbol{q}(t_1) = \mathbf{0} \tag{C.18a}$$

as well as

$$\delta\boldsymbol{q}(t_2) = \mathbf{0} \tag{C.18b}$$

A variation $\mathcal{I}$ at a permutation $\boldsymbol{q}$ into $\boldsymbol{q}(t) + \delta\boldsymbol{q}(t)$ may be obtained by differences

$$\int_{t_1}^{t_2} \mathcal{L}(\boldsymbol{q} + \delta\boldsymbol{q}, \dot{\boldsymbol{q}} + \boldsymbol{\delta}\dot{\boldsymbol{q}}, t)dt - \int_{t_1}^{t_2} \mathcal{L}(\boldsymbol{q}, \dot{\boldsymbol{q}}, t)dt$$

and

$$\int_{t_1}^{t_2} \mathcal{A}(\boldsymbol{q} + \delta\boldsymbol{q}, \dot{\boldsymbol{q}} + \boldsymbol{\delta}\dot{\boldsymbol{q}}, t)dt - \int_{t_1}^{t_2} \mathcal{A}(\boldsymbol{q}, \dot{\boldsymbol{q}}, t)dt$$

A distribution of these differences according to $\delta\boldsymbol{q}$ and $\boldsymbol{\delta}\dot{\boldsymbol{q}}$ (in the integral expressions) begins from the first-order terms.

A necessary condition of the integral minimisation $\mathcal{I}$ is a transformation of its all terms into zero; it is called a 'first variation' (or simply a variation of the integral). In this way, in unison with a symbolic of the variation calculus and matrix computation, the Hamilton–Ostrogradsky's principle of the stationary action for any nonlinear abstract MMD electrical machine physical heterogeneous continuous dynamical hypersystem can be written in a form:

$$\mathcal{I} \triangleq \delta \int_{t_1}^{t_2} [\mathcal{L}(\boldsymbol{q}, \dot{\boldsymbol{q}}, t) + \mathcal{A}(\boldsymbol{q}, \dot{\boldsymbol{q}}, t)]dt \tag{C.19}$$

or as a variation

$$\int_{t_1}^{t_2} \left(\frac{\partial \mathcal{L}}{\partial \boldsymbol{q}} \delta\boldsymbol{q} + \frac{\partial \mathcal{L}}{\partial \dot{\boldsymbol{q}}} \delta\dot{\boldsymbol{q}} + \frac{\partial \mathcal{A}}{\partial \boldsymbol{q}} \delta\boldsymbol{q} + \frac{\partial \mathcal{A}}{\partial \dot{\boldsymbol{q}}} \delta\dot{\boldsymbol{q}} \right) dt = 0.$$

Taking into consideration that $\dot{\boldsymbol{q}} = \frac{d}{dt}\delta\boldsymbol{q}$ and integrating by parts the second and fourth terms, one obtains

$$\begin{aligned} \delta\mathcal{I} = \frac{\partial \mathcal{L}}{\partial \boldsymbol{q}} \delta\boldsymbol{q} \Bigg|_{t_1}^{t_2} + \int_{t_1}^{t_1} \left(\frac{\partial \mathcal{L}}{\partial \boldsymbol{q}} \delta\boldsymbol{q} + \frac{d}{dt} \frac{\partial \mathcal{L}}{\partial \dot{\boldsymbol{q}}} \right) \delta\boldsymbol{q}\, dt + \\ + \frac{\partial \mathcal{A}}{\partial \boldsymbol{q}} \delta\boldsymbol{q} \Bigg|_{t_1}^{t_2} + \int_{t_1}^{t_1} \left(\frac{\partial \mathcal{A}}{\partial \boldsymbol{q}} \delta\boldsymbol{q} + \frac{d}{dt} \frac{\partial \mathcal{A}}{\partial \dot{\boldsymbol{q}}} \right) \delta\boldsymbol{q}\, dt = 0. \end{aligned} \tag{C.20}$$

After regarding the conditions (C.18), the first and third terms in the above expression disappear. Only the integrals remain, which should be equal to zero for any of the values of the matrices, hypomatrices and elements of the hypermatrix $\delta\boldsymbol{q}(t)$. This is possible only in this case, when the underintegral expression identically will be zero. In this way, one shall obtain the synthetic mathematical model of any nonlinear abstract MMD electrical machine physical heterogeneous continuous dynamical hypersystem, i.e. a differential equation of dynamics in the dynamical systems approach and hypermatrix notation

$$\left[\frac{d}{dt}\left(\frac{\partial \mathcal{L}}{\partial \dot{\boldsymbol{q}}}\right) - \frac{\partial \mathcal{L}}{\partial \boldsymbol{q}} + \frac{d}{dt}\left(\frac{\partial \mathcal{A}}{\partial \dot{\boldsymbol{q}}}\right) - \frac{\partial \mathcal{A}}{\partial \boldsymbol{q}}\right]\delta \boldsymbol{q} = 0. \tag{C.21a}$$

A differential equation of dynamics (C.21a), in the dynamical system approach in the methodology for a formation of the mathematical models of abstract MMD electrical machine physical heterogeneous continuous dynamical hypersystems, and the hypermatrix notation, is not known in the literature.

The authors decided to apply this hypermatrix notation, because it is comfortable for the dynamical systems approach in the methodology for a formation of the mathematical models. However, this differential equation in the non-matrix, i.e. scalar notation is well known and is called, in variation calculus, 'Euler's differential equation' (1766), and in analytical mechanics, 'Lagrange's differential equation' (1788) and it has a somewhat different form (White and Woodson 1959, Szklarski *et al* 1977, Fijalkowski 1987, 2016)

$$\frac{d}{dt}\left(\frac{\partial \mathcal{L}}{\partial \dot{q}_i}\right) - \frac{\partial \mathcal{L}}{\partial q_i} + f_D = f_E \tag{C.21b}$$

where: q_i, $\dot{q}_i$—the ith generalised coordinate and velocity;

i—the number of DoF;

$f_D = \frac{\partial \mathcal{R}}{\partial \dot{q}_i}$—the generalised dissipation forces;

$\mathcal{R}$—the Rayleigh function;

$f_E = f_E(t)$—the generalised external forces.

If the Lagrange function of energy and virtual work of an arbitrary nonlinear abstract MMD electrical machine physical heterogeneous continuous dynamical hypersystem are known then the Euler–Lagrange second-order differential equation of dynamics in the hypermatrix notation establishes a relationship between the relevant matrices of the hypermatrices of generalised accelerations, velocities, and coordinates, that is, it represents itself as a synthetic differential equation of the dynamics, and therefore a synthetic mathematical model of any nonlinear abstract MMD electrical machine physical heterogeneous continuous dynamical hypersystem.

Substituting the expressions (C.5) and (C.6) into the equation of dynamics (C.21*a*), one obtains a synthetic mathematical model of any nonlinear abstract MMD electrical machine physical heterogeneous continuous dynamical hypersystem in a more appropriate way for the computerised analytical studies of the dynamics, namely:

$$\left[\frac{d}{dt}\left(\frac{\partial \mathcal{T}^*}{\partial \dot{\boldsymbol{q}}}\right) - \frac{\partial \mathcal{T}^*}{\partial \boldsymbol{q}} + \frac{\partial \mathcal{U}}{\partial \boldsymbol{q}} + \frac{d}{dt}\left(\frac{\partial \mathcal{T}_R^*}{\partial \dot{\boldsymbol{q}}}\right) - \boldsymbol{f}^T\right]\delta \boldsymbol{q} = 0, \tag{C.22a}$$

or

$$\left[\frac{d}{dt}\left(\frac{\partial \mathcal{T}^*}{\partial v}\right) - \frac{\partial \mathcal{T}^*}{\partial \boldsymbol{q}} + \frac{\partial \mathcal{U}}{\partial \boldsymbol{q}} + \frac{d}{dt}\left(\frac{\partial \mathcal{T}_R^*}{\partial v}\right) - \boldsymbol{f}^T\right]\delta\boldsymbol{q}=0, \tag{C.22b}$$

where:

$$\boldsymbol{f}^T = (\mathrm{grad}\mathcal{A})^T = \frac{\partial \mathcal{A}}{\partial \boldsymbol{q}} = (\mathrm{grad}\mathcal{U}_F)^T = \frac{\partial \mathcal{U}_F}{\partial \boldsymbol{q}}$$

is the hypermatrix of generalised dissipative input (excitation) forces.

In the above Euler–Lagrange second-order differential equation of the dynamics, one can use the hypermatrix notation of generalised coordinates and velocities and scalar functions of the conservative kinetic co-energy $\mathcal{T}^*$ and conservative potential energy $\mathcal{U}$ as a homogeneous square form, where the partial derivatives in relation to the hypermatrices of the generalised coordinates and velocities form matrixes.

The reality of the subsistence of respective matrices of these hypomatrices must be verified in each case. Verification of that will ensure the completeness of the matrices and hypomatrices considerably acting in the abstract MMD electrical machine physical heterogeneous continuous dynamical hypersystem. This is the advantage of the accepted hypermatrix notation of differential equations of the dynamics of any nonlinear abstract MMD electrical machine physical heterogeneous continuous dynamical hypersystem and the consistently completed computation in this notation.

A synthetic mathematical model of any nonlinear abstract MMD electrical machine physical heterogeneous continuous dynamical hypersystem, establishes the Euler–Lagrange second-order differential equation of dynamics.

In the dynamical systems approach, it is written in the hypermatrix form, since the abstract MMD electrical machine physical heterogeneous continuous dynamical hypersystem is often a set of many MMD electrical machine physical homogeneous continuous dynamical systems, whose own mathematical models, establishing a set of the Euler–Lagrange differential equations are written in the matrix form.

Since each one is often a set of many MMD electrical machine physical heterogeneous continuous dynamical hyposystems whose mathematical models, forming a subset of the Euler–Lagrange second-order differential equations in the dynamical systems approach, are written in the hypomatrix form.

Since they are usually a few or more elements of the MMD electrical machine physical dynamical hyposystem, whose elementary mathematical models, constituting only the one single Euler–Lagrange second-order differential equation, are not written in the matrix form of a matrix, but in the scalar one (figure C1).

Taking into consideration equations (C.7)–(C.15), the Euler–Lagrange second-order differential equations of dynamics (C.22) in the dynamical systems approach can be written in the hypermatrix form as follows:

$$\left[\frac{d}{dt}[v^T\boldsymbol{K}(\boldsymbol{q}, v)] - \frac{1}{2}v^T\frac{\partial \boldsymbol{K}(\boldsymbol{q}, v)}{\partial \boldsymbol{q}}v + \boldsymbol{q}^T\boldsymbol{P}(\boldsymbol{q}, v) + v^T\boldsymbol{D}(\boldsymbol{q}, v) + \mathrm{sgn}\, v^T\Delta\boldsymbol{f} - \boldsymbol{f}^T\right] = 0 \tag{C.23a}$$

or simply

$$\left[\frac{d}{dt}(\boldsymbol{v}^T\boldsymbol{K}) - \frac{1}{2}\boldsymbol{v}^T\frac{\partial \boldsymbol{K}}{\partial \boldsymbol{q}}\boldsymbol{v} + \boldsymbol{q}^T\boldsymbol{P} + \boldsymbol{v}^T\boldsymbol{D} + \text{sgn}\,\boldsymbol{v}^T\Delta\boldsymbol{f} - \boldsymbol{f}^T\right]\delta\boldsymbol{q} = 0 \qquad (C.23b)$$

Assuming that $\boldsymbol{K}$, $\boldsymbol{P}$ and $\boldsymbol{D}$ are not the functions of $\boldsymbol{q}$ as well as $\dot{\boldsymbol{q}} = \boldsymbol{v}$, the Euler–Lagrange second-order differential equations of dynamics (C.23) in the dynamical systems approach can be written in the hypermatrix form as follows:

$$\left[\frac{d}{dt}\boldsymbol{v}^T\boldsymbol{K} + \boldsymbol{q}^T\boldsymbol{P} + \boldsymbol{v}^T\boldsymbol{D} + \text{sgn}\,\boldsymbol{v}^T\Delta\boldsymbol{f} - \boldsymbol{f}^T\right]\delta\boldsymbol{q} = 0. \qquad (C.24a)$$

or

$$\left[\frac{d}{dt}\boldsymbol{v}^T\boldsymbol{K} + \int_0^t \boldsymbol{v}^T dt\boldsymbol{P} + \boldsymbol{v}^T\boldsymbol{D} + \text{sgn}\,\boldsymbol{v}^T\Delta\boldsymbol{f} - \boldsymbol{f}^T\right]\delta\boldsymbol{q} = 0. \qquad (C.24b)$$

$$\left[\boldsymbol{v}^T(s)s\boldsymbol{K} + \boldsymbol{v}^T(s)\frac{\boldsymbol{P}}{s} + \boldsymbol{v}^T(s)\boldsymbol{D} + \text{sgn}\,\boldsymbol{v}^T(s)\Delta\boldsymbol{f}(s) - \boldsymbol{f}^T(s)\right]\delta\boldsymbol{q}(s) = 0. \qquad (C.25a)$$

or

$$\left[\boldsymbol{v}^T(s)\left(s\boldsymbol{K} + \frac{\boldsymbol{P}}{s} + \boldsymbol{D}\right) + \text{sgn}\,\boldsymbol{v}^T(s)\Delta\boldsymbol{f}(s) - \boldsymbol{f}^T(s)\right]\delta\boldsymbol{q}(s) = 0. \qquad (C.25b)$$

where: $s = \sigma + j\omega$—the Laplace operator.

After taking into account the notion of the 'impedance' $\boldsymbol{Z}(s)$ and/or 'admittance' $\boldsymbol{Y}$(s), which yields:

$$\boldsymbol{Z}(s) = \boldsymbol{D} + s\boldsymbol{K} + \frac{\boldsymbol{P}}{s}, \qquad (C.26a)$$

and

$$\boldsymbol{Y}(\text{s}) = \frac{1}{\boldsymbol{Z}(s)} = \frac{1}{\boldsymbol{D}} + \frac{1}{s\boldsymbol{K}} + \frac{s}{\boldsymbol{P}} = \boldsymbol{G} + \frac{\boldsymbol{B}}{s} + s\boldsymbol{S} \qquad (C.26b)$$

The Euler–Lagrange second-order differential equations of dynamics (C.25) in the dynamical systems approach can be written in the hypermatrix form as follows:

$$[\boldsymbol{v}^T(s)\boldsymbol{Z}(s) + \text{sgn}\,\boldsymbol{v}^T(s)\Delta\boldsymbol{f}(s) - \boldsymbol{f}^T(s)]\delta\boldsymbol{q}(s) = 0, \qquad (C.27a)$$

or

$$\left[\boldsymbol{v}^T(s)\frac{1}{\boldsymbol{Y}(s)} + \text{sgn}\,\boldsymbol{v}^T(s)\Delta\boldsymbol{f}(s) - \boldsymbol{f}^T(s)\right]\delta\boldsymbol{q}(s) = 0. \qquad (C.27b)$$

The mathematical representation of any measurable MMD electrical machine physical heterogeneous continuous dynamical hypersystems is greatly facilitated by representing it by a holor that fits the physical concept perfectly.

The term holor[2] is a generalisation of the more familiar mathematical concepts of tensors, vectors, matrices, complex numbers and scalars.

Rather than forcing all measurable MMD electrical machine physical heterogeneous continuous dynamical hypersystems to be imperfectly represented by a few families of mathematical quantities, the study of holors permits the development of a mathematical representation that fits the physical concept perfectly.

The set of numbers which specifies a holor are termed the merates. The physical quantity holor ought to satisfy two fundamental criteria. The holor must be defined in terms of physically realisable measurements that are significant at physical frequencies. The holor formulation is very simple and modern computer techniques make design of MMD electrical machine physical heterogeneous continuous dynamical hypersystems in terms of the generalised external forces holor (termed by the authors 'energy-potential-difference' holor, $\boldsymbol{F}_i$) and the generalised velocities holor (termed by the authors 'energy-flow' holor, $\boldsymbol{V}^j$) in **two-dimensional** (2D) spaces and generalised power holor, $\boldsymbol{P}^\kappa$ in a **three-dimensional** (3D) space a challenge of only moderate difficulty.

The mathematical model, represented by the Euler–Lagrange second-order differential equations of dynamics (C.27) of any MMD electrical machine physical heterogeneous continuous dynamical hypersystem with energy storage may originate from holor equations.

For that reason, the resolve of MMD electrical machine physical heterogeneous continuous dynamical hypersystems for a step-function or sinewave input-signal-holor $\boldsymbol{F}_i$ necessitates a solution of a set of holor equations

$$\left[\boldsymbol{V}^k\boldsymbol{Z}_{jk} + \operatorname{sgn} \boldsymbol{V}^k\Delta\boldsymbol{F}_{jk} - \boldsymbol{F}_k\right]\delta\boldsymbol{Q}^j = 0, \tag{C.28a}$$

or

$$\left[\boldsymbol{V}^k\frac{\delta_k^i}{\boldsymbol{Y}^{ij}} + \operatorname{sgn} \boldsymbol{V}^k\Delta\boldsymbol{F}_{jk} - \boldsymbol{F}_k\right]\delta\boldsymbol{Q}^j = 0. \tag{C.28b}$$

This appears for the physical and mathematical approaches presented in appendix C and chapter 7. One may consider if there is potential to model a physical heterogeneous continuous dynamical hypersystem by algebraic equations from which the holor equations can be created. To deal with this problem one must take into account the result of the generalised input-signal-holor $\boldsymbol{V}^j$ function.

[2] ***Definition from Wikipedia:*** A **holor** (/ˈhoʊlər'/; Greek ὅλος 'whole') is a mathematical entity that is made up of one or more independent quantities ('merates' as they are called in the theory of holors). Complex numbers, scalars, vectors, matrices, tensors, quaternions, and other hypercomplex numbers are kinds of holors. If proper index conventions are maintained then certain relations of holor algebra are consistent with that of real algebra; i.e. addition and uncontracted multiplication are both commutative and associative.

This type of driving holor function in its normal mathematical appearance is not conventionally realised by an MMD electrical machine physical heterogeneous continuous dynamical hypersystem. In spite of that, this generalised input-signal-holor function has a particular form (sine waveform) which symbolises the conventional generalised driving holor function applied in study.

After taking into account the notion of the 'impedance holor', $\boldsymbol{Z}_{ij}$ and/or 'admittance holor', $\boldsymbol{Y}^{ji}$, which yields:

$$\boldsymbol{Z}_{ij} = \begin{bmatrix} \boldsymbol{Z}_{cc} & -\boldsymbol{Z}_{cs} \\ \boldsymbol{Z}_{sc} & \boldsymbol{Z}_{ss} \end{bmatrix} = \begin{bmatrix} \boldsymbol{D} & -\boldsymbol{X} \\ \boldsymbol{X} & \boldsymbol{D} \end{bmatrix}, \tag{C.29a}$$

and

$$\boldsymbol{Y}^{ij} = \frac{\delta_k^i}{\boldsymbol{Z}_{jk}} = \begin{bmatrix} \boldsymbol{Y}^{ss} & \boldsymbol{Y}^{sc} \\ -\boldsymbol{Y}^{cs} & \boldsymbol{Y}^{cc} \end{bmatrix} = \begin{bmatrix} \boldsymbol{G} & \boldsymbol{B} \\ -\boldsymbol{B} & \boldsymbol{G} \end{bmatrix} = \frac{1}{(\boldsymbol{D}^\wedge 2)+(\boldsymbol{X}^\wedge 2)} \begin{bmatrix} \boldsymbol{D} & \boldsymbol{X} \\ -\boldsymbol{X} & \boldsymbol{D} \end{bmatrix}, \tag{C.29b}$$

where 'reactance', $\boldsymbol{X}$ and 'susceptance', $\boldsymbol{B}$ are as follows:

$$\boldsymbol{X} = \omega \boldsymbol{K} - \frac{1}{\omega}\boldsymbol{P}, \tag{C.30a}$$

$$\boldsymbol{B} = \omega \frac{1}{\boldsymbol{P}} - \frac{1}{\omega \boldsymbol{K}} = \omega \boldsymbol{S} - \frac{1}{\omega}\boldsymbol{H}, \tag{C.30b}$$

and

$$\boldsymbol{Y}^{ij}\boldsymbol{Z}_{jk} = \delta_k^i. \tag{C.30c}$$

The Euler–Lagrange second-order differential equations of dynamics (C.28) in the dynamical systems approach can be written in the hypermatrix holor form as follows:

$$\begin{aligned} &\left[\boldsymbol{V}^k \boldsymbol{Z}_{jk} + \operatorname{sgn} \boldsymbol{V}^k \Delta \boldsymbol{F}_{jk} - \boldsymbol{F}_k\right]\delta \boldsymbol{Q}^j = \\ &= \left\{[\boldsymbol{V}^c \ \ \boldsymbol{V}^s]\begin{bmatrix} \boldsymbol{D} & -\boldsymbol{X} \\ \boldsymbol{X} & \boldsymbol{D} \end{bmatrix} + \operatorname{sgn}\,[\boldsymbol{V}^c \ \ \boldsymbol{V}^s]\begin{bmatrix} \Delta \boldsymbol{F}_c & 0 \\ 0 & \Delta \boldsymbol{F}_s \end{bmatrix} - [\boldsymbol{F}_c \ \ \boldsymbol{F}_s]\right\}\delta\begin{bmatrix} \boldsymbol{Q}^c \\ \boldsymbol{Q}^s \end{bmatrix}, \end{aligned} \tag{C.31a}$$

or

$$\begin{aligned} &\left[\boldsymbol{V}^k \frac{\delta_k^i}{\boldsymbol{Y}^{ij}} + \operatorname{sgn} \boldsymbol{V}^k \Delta \boldsymbol{F}_{jk} - \boldsymbol{F}_k\right]\delta \boldsymbol{Q}^j = \\ &= \left\{[\boldsymbol{V}^c \ \ \boldsymbol{V}^s]\frac{1}{(\boldsymbol{G}\char`^2) + (\boldsymbol{B}\char`^2)}\begin{bmatrix} \boldsymbol{G} & -\boldsymbol{B} \\ \boldsymbol{B} & \boldsymbol{G} \end{bmatrix} + \operatorname{sgn}\,[\boldsymbol{V}^c \ \ \boldsymbol{V}^s]\begin{bmatrix} \Delta \boldsymbol{F}_c & 0 \\ 0 & \Delta \boldsymbol{F}_s \end{bmatrix} - [\boldsymbol{F}_c \ \ \boldsymbol{F}_s]\right\}\delta\begin{bmatrix} \boldsymbol{Q}^c \\ \boldsymbol{Q}^s \end{bmatrix} \end{aligned} \tag{C. 31b}$$

where:

$$\boldsymbol{F}_i = (\boldsymbol{F}_c, \boldsymbol{F}_s) = [\boldsymbol{F}_c \ \ \boldsymbol{F}_s]^T = \begin{bmatrix} \boldsymbol{F}_c \\ \boldsymbol{F}_s \end{bmatrix} = \begin{bmatrix} \boldsymbol{F}\cos\psi_F \\ \boldsymbol{F}\sin\psi_F \end{bmatrix} = \boldsymbol{F}\angle\psi_F, \tag{C.32a}$$

$$V^j = (V^c, V^s) = [V^c \quad V^s]^T = \begin{bmatrix} V^c \\ V^s \end{bmatrix} = \begin{bmatrix} V\cos\psi_V \\ V\sin\psi_V \end{bmatrix} = V\angle\psi_V, \qquad (C.32b)$$

and

$$\boldsymbol{F}^c = \boldsymbol{F}\cos\psi_F, \ \boldsymbol{F}^s = \boldsymbol{F}\sin\psi_F \qquad (C.32c)$$

$$V^c = V\cos\psi_V, \ V^s = V\sin\psi_V \qquad (C.32d)$$

$$\Delta\boldsymbol{F}_{ij} = \begin{bmatrix} \Delta\boldsymbol{F}_c & \boldsymbol{0} \\ \boldsymbol{0} & \Delta\boldsymbol{F}_s \end{bmatrix}. \qquad (C.32e)$$

References

Appell P 1893 *Traité de Mécanique Rationnelle 4 vols* (Paris: Gauthier-Villars) (In French)

Appell P 1899a *Les mouvements de roulement en dynamique Scientia* **4** Paris (In French)

Appell P 1899b Sur les mouvements de roulement, equations du movement analogu.htm es & celles de Lagrange. C.R., T. 129, Paris (In French)

Appell P 1899d Sur une forme gènèrale des èquations de la Dynamique *J. Reine Angew. Math.* **Bd. 121** Berlin (In French)

Appell P 1903 Remarques sur les systems non holonomes *J. Math. Pures Appl.* **9** 27–8 (In French)

Beltrami E 1895 Sulle equazioni diminiche di Lagrange Rendiconti dell'Istituto Lombardo, Ser. II, V. XXVIII, Fasc. XIV, pp. 745–752 (In Italian)

Bobylev D K 1892 O share s giroskopom vnutriy, katyashchemsya po gorizontalnoi ploskosti bez proskalzyvaniya (On a sphere with a gyroscope inside it rolling after horizontal plane without slipping) Mat, Sb. 16 3 (In Russian)

Carvallo E 1902 *L'Electricité deduite d'experience et remenée an pinxipe des travaux virtuels. Coll. Scientia*19 (Paris) (In French)

Euler L 1766 Elementa Calculi Variatiorum (The Elements of the Calculus Variations) Originally published in Novi Commentari Academiiae scientium Petropolitanae 10 51–93

Fijalkowski B 1987 *Modele matematyczne wybranych lotniczych i motoryzacyjnych mechano-elektro-termicznych dyskretnych nadsystemów dynamicznych (Mathematical Models of Selected Aviation and Automotive Continuous Dynamical Hyperystems), Monografia 53* (Krakow: Politechnika Krakowska imienia Tadeusza Kościuszki) 276 p (In Polish)

Fijalkowski B 2016 *Mechatronics: Dynamical Systems Approach and Theory of Holors* (Bristol, UK: IOP Publishing)

Hertz H 1894 Die Prinzipen der Mechanik *Gesammeite Werke 3* (In German)

Lagrange J L 1788 *Mécanique Analytique* (Paris: Chez la Veuve DESAINT, Libraire)

Liennard A 1902 Sur l'application des equations de Lagrange aux phenomènes èlectrodynamik et electromagnetiques *C.R., T.* 134 No. 163 (In French)

Maxwell J C 1873 *A Treatise of Electricity and Magnetism* (Oxford: Clarendon)

Szklarski L, Pelczewski W, Kolendowski J, Puchalka T and Komarzewska M 1963 *Dynamika ukladow nieholonomicznych (Dynamics of Electromechanical Systems)* (WARSZAWA: Komitet Elektrotechniki Polskiej Akademii Nauk, Panstwowe Wydawnictwo Naukowe) 259 p (In Polish)

Tchaplygin 1897a O dvizheniy tiazhalogo tela vrashcheniya na gorizontalnoy ploskosti (On a motion of the solid body on a horizontal plane) Trudy otdeleniya fizicheskih nauk obshchva liubiteley iestestvoznaniya 9 (In Russian)

Tchaplygin 1897b O nekotorom vozmozhnom obobshcheniy teoremy ploshchadey s primeneniem k zadache o kataniy sharov (On some possible restriction of a space theory with the application to the exercise on rolling spheres) Matem. Sb. 20 (In Russian)

White D C and Woodson H H 1959 *Electromechanics; Energy Conversion* (New York: Wiley)

Zhukovsky N E 1893 O giroskopicheskom share D K Bobyleva (On a gyroscopic sphere of D K Bobylev) Trudy ob-va lyubitieley iestiestvoznaniya, antropologii i etnografii 6 1 (In Russian)

Appendix D

Exemplary applications of the electrical commutation-matrixer commutators for DC and AC electrical machines

D.1 A 3 × 3, 3 × 5 or 5 × 5 ECM AC–AC commutator

This textbook, among others, especially concerns 3 × 3, 3 × 5 and/or 5 × 5 **electrical commutation matrixer** (ECM) AC–AC commutator IPM synchronous motors as alternative VS DC–AC inverter AC IMED (see figure D1).

The 3 × 3, 3 × 5 and 5 × 5 ECM commutators are direct ECM AC–AC commutators with no DC link. The lack of reactive components in the DC link is one of the salient advantages of the 3 × 3, 3 × 5 and 5 × 5 ECM commutators. Furthermore, the 3 × 3, 3 × 5 and/or 5 × 5 ECM features full four-quadrant operation and sinusoidal ECM-input current. The ECM-output-voltage is limited to 87% of the ECM-input voltage (Fijalkowski 1987a, 1987b, 1982, 1985a, 1985b, 1985c, 1988a, 1988b, 1988c, 1997).

The 3 × 3, 3 × 5 or 5 × 5 ECM AC–AC commutator needs 9, 15 or 25 bipolar electrical valves, respectively, to connect two three- or five-phase voltage systems in all possible combinations.

Most R&D work about 3 × 3, 3 × 5 and 5 × 5 ECMs has so far regarded the modulation and control of the ECM AC–AC commutator IPM synchronous motor.

The implemented 3 × 3 ECM AC–AC commutator shows a reduction of the reactive components at more than a factor 1 compared to a standard DC–AC/AC–DC commutator IPM synchronous motor (i.e. with a VSI DC–AC inverter) EM drive at the same power rating.

Power flow through the 3 × 3, 3 × 5 or 5 × 5 ECM AC–AC commutator is bipolar and it draws sinusoidal input current in generative as well as regenerative mode. The input

doi:10.1088/2053-2563/aae7d7ch12

© IOP Publishing Ltd 2019

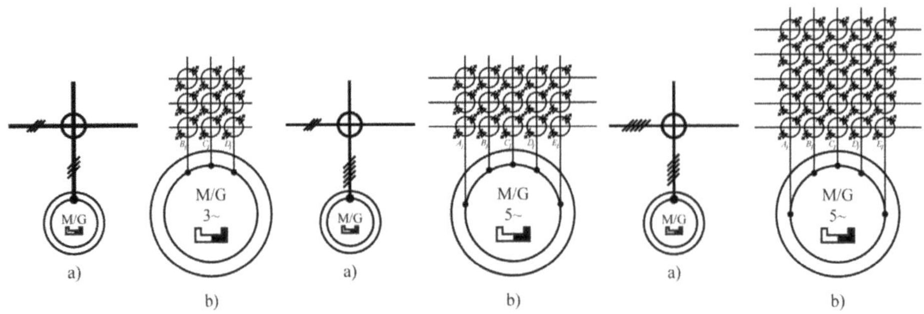

Figure D1. The 3 × 3, 3 × 5 and 5 × 5 ECM AC–AC commutator dynamotors for steered, motorised and/or generatorised wheels (SM&GW).

power–power factor is above 0.98. It may be concluded that the standard VS DC–AC inverter DC IEMD will be commonly used in the nearest future. For example, in civil and/or military **all-electric vehicles** (AEV) and **hybrid-electric vehicles** (HEV), mass and size reduction is critical. High-temperature operation is desirable to ease the problem of thermal management of the 3 × 3, 3 × 5 and/or 5 × 5 ECM AC–AC commutator IPM synchronous motors with single cooling heatsinks (Fijalkowski 1988b, 1988e, 1988g, 1997).

It is believed that the 3 × 3, 3 × 5 or 5 × 5 ECM AC–AC commutator IPM synchronous motor can have significant advantages over the conventional ECM AC–DC–AC commutator IPM synchronous motor in all of these areas since it is possible to eliminate the DC link capacitor especially if an AC primary power system is used.

An interesting example of the design and construction of a snubberless 150 kVA experimental 3 × 3 ECM AC–AC commutator induction dynamotor, built using 600 A, 1400 V continuous IGBTs has been presented by Wheeler *et al* (2004a, 2004b). The 150 kVA experimental 3 × 3 ECM AC–AC commutator is controlled with a microprocessor (*Infineon SAB80C 167*). Despite the relatively low performance control platform, it has been possible to achieve full closed loop vector control of the ECM AC–AC commutator asynchronous (induction) motor load. From some output-voltage and output-current waveforms of the 3 × 3 ECM AC–AC commutator asynchronous (induction) motor, it can be noted that the output power of the 3 × 3 ECM AC–AC commutator is 150 kVA. It is very important to minimise the parasitic inductance within the ECM AC–AC commutator. In this ECM AC–AC commutator, the parasitic inductance has been minimised using laminate input power planes with distributed input filter capacitance.

Another interesting example of the design and construction of a static Scherbius system with the 3 × 3 ECM AC–AC commutator wound-rotor asynchronous (induction) motor, which is used to control the rotor-side currents, has been presented by Altun (2005). Such a static Scherbius system, where the split power is taken from the rotor circuit is given to the grid through the 3 × 3 ECM AC–AC commutator.

The static Scherbius system combines a 3 × 3 ECM AC–AC commutator wound-rotor asynchronous (induction) motor in a split energy recovery EM propulsion mechatronic control system, as shown in figure D2.

This static Scherbius system is inherently capable of bilateral power flow and it also offers virtually sinusoidal input currents, without the harmonics usually

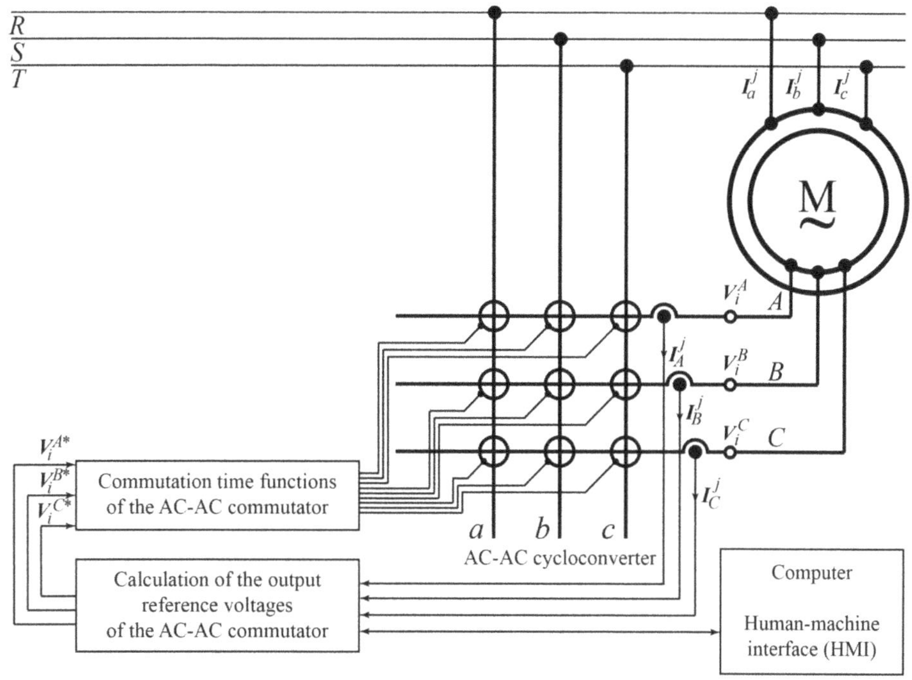

Figure D2. The static Scherbius system combines a 3 × 3 ECM AC–AC commutator wound-rotor asynchronous (induction) motor in a split energy recovery E-M propulsion mechatronic control system.

associated with present commercial ECM AC–AC and AC–DC–AC commutators, so-termed 'AC–AC or AC–DC–AC cycloconverters' and/or 'DC–AC inverters'.

In the static Scherbius system a 3 × 3 ECM AC–AC commutator has been used to extract the split energy from the rotor into the mains in an asynchronous region, and to supply the rotor in a supersynchronous region. It has been known that the use of the 3 × 3 ECM AC–AC commutator in the static Scherbius system has the advantages of single-stage power conversion, operation in both asynchronous and supersynchronous speed regions, sinusoidal voltage and current waveforms and simpler microelectronic gate-control drivers than other ECM AC–AC and/or AC–DC–AC commutators, so-termed 'AC–AC or AC–DC–AC cycloconverters' and/or 'DC–AC inverters' used instead.

At present, the static Scherbius system finds an important place in wind-power generation applications where a variable-speed, constant-frequency static Scherbius system produces electrical energy for a wide range of wind speeds.

D.2 A 3 × 2 ECM AC–DC commutator

A more complicated example of the 3 × 2 ECM AC–DC commutator synchronous generator (alternator), with a two heterogeneous anode and cathode commutation groups, i.e. the well-known, in older literature, three-phase bridge rectifier shown in figure D3 (Fijalkowski 1987a, 1987b).

There are three possible primary collectors (inputs), and the two secondary collectors (outputs) of the DC circuit, which give a 3 × 2 ECM.

Recently, due to the possibility of using ECMs in the ECM AC–AC and AC–DC/DC–AC commutator IPM synchronous motors two important actions may be used with great success, namely: contactless (sparkless) commutation and contactless (brushless) magnetoelectrical excitation.

D.3 A 3 × 3 ECM AC–AC commutator

The three-phase ECM AC–AC commutator supersynchronous motor (figure D4) with three-phase wound-rotor consists of the ECM commutator with a bipolar

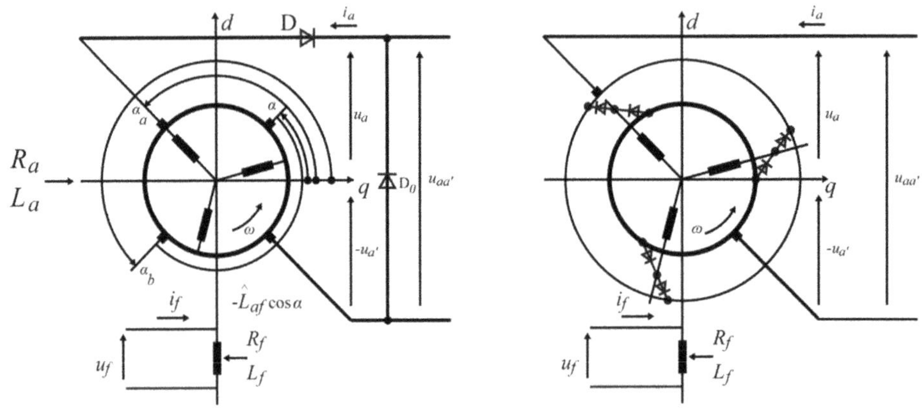

Figure D3. A 3 × 2 electrical commutation matrixer (ECM) AC–DC commutator generator (right) in comparison with a 3 × 2 mechanical commutation matrixer (MCM) AC–DC commutator generator (left).

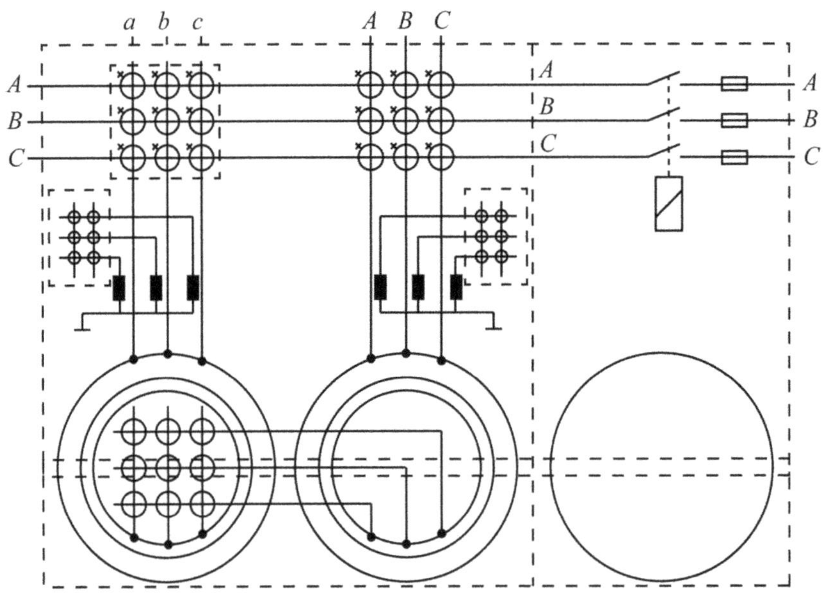

Figure D4. Three-phase ECM AC–AC commutator supersynchronous motor with a 3 × 3 ECM commutator and three-phase AC wound rotors.

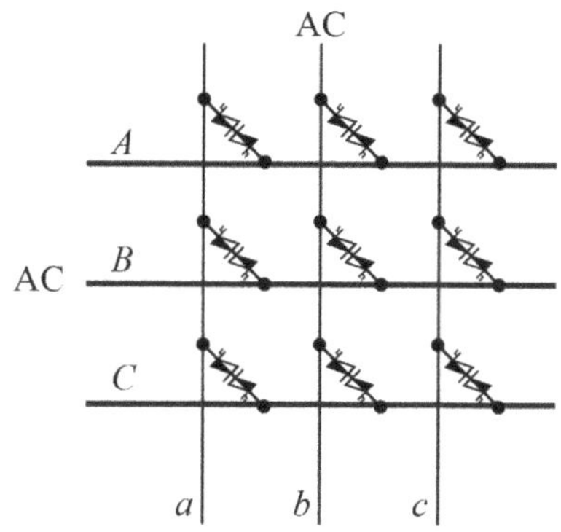

Figure D5. Three-phase 3 × 3 ECM AC–AC commutator with RB IGBT Six operating modes are possible in this MMD electrical machine.

ECM, realised on only nine *'continuous'* **reverse-blocking insulated-gate bipolar transistors** (RB IGBT), **metal-oxide semiconductor field-effect transistors** (MOSFET) or bipolar symmetrical trigistors, that is, **gate turn-off** (GTO) symmetrical thyristors (symistors) in 3 × 3 ECM connection (Fijalkowski 1988a, 1988b, 1988c, 1988d, 1988e, 1988f, 1988g). This 3 × 3 ECM AC–AC commutator (figure D5) makes the bilateral flow of secondary electrical energy possible as well during generating (absorbing) as during motoring (driving) with the supersynchronous angular velocity.

D.4 A single-phase ECM AC–AC and/or AC–DC/DC–AC commutator

The single-phase ECM AC–AC and/or AC–DC/DC–AC commutator synchronous (electromagnetically-excited) motor (figure D6) with two three-phase armature windings, one of which is wye-connected, and another delta-connected, consists of the 'non-rotary' main ECM commutator with two bipolar ECMs, realised on 12 'continuous' bipolar trigistors in 2s(2 × 3)/2s(3 × 2) ECM connection in series, and the 'rotary' auxiliary ECM commutator with the unipolar ECM, realised on six 'continuous' diodes in 2 × 3 ECM connection. It has a contactless (brushless) electromagnetical excitation.

The single-phase ECM AC–AC and/or AC–DC/DC–AC commutator synchronous (magnetoelectrically excited) motor (figure D7) consists of the ECM commutator with a single or two bipolar ECMs, realised on six or 12 'continuous' bipolar trigistors in 2 × 3 ECM connections or 2s(2 × 3) ECM connections in series, by means of **interior permanent-magnet** (IPM) poles, for example of alloys neodymium, iron and boron (Fijalkowski 1988a, 1988b, 1988c, 1988d, 1988e, 1988f, 1988g).

The single-phase ECM AC–AC and/or AC–DC/DC–AC commutator hyposynchronous (induction) dynamotor (figure D8) with a squirrel-cage rotor, consists of the ECM commutator with a single or two bipolar ECMs, realised on six or 12 'continuous' bipolar trigistors in 2 × 3 ECM connections or 2s(2 × 3) ECM connections in series (Fijalkowski 1988a, 1988b, 1988c, 1988d, 1988e, 1988f, 1988g).

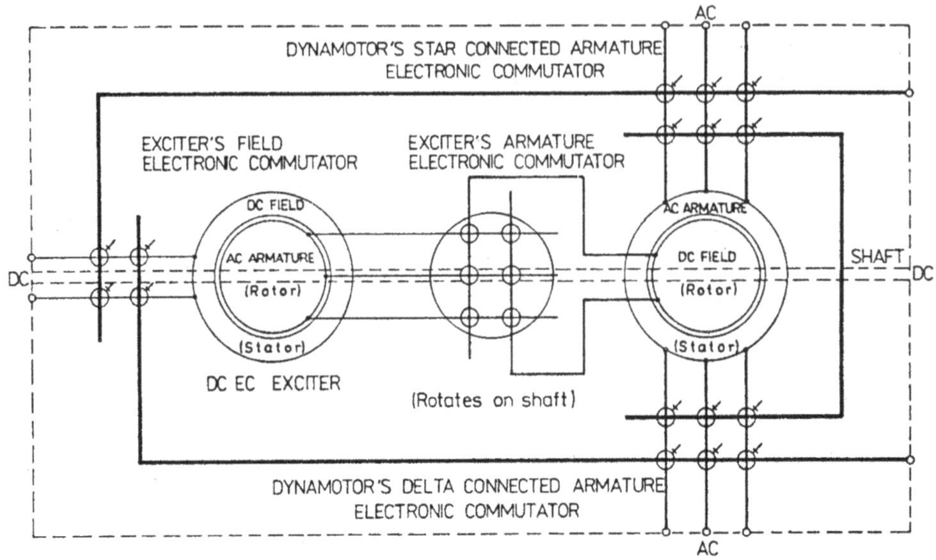

Figure D6. Single-phase ECM AC–AC and/or DC–AC/AC–DC commutator synchronous (electromagnetically-excited) motor with the 2s(2 × 3)/2s(3 × 2) ECM commutator.

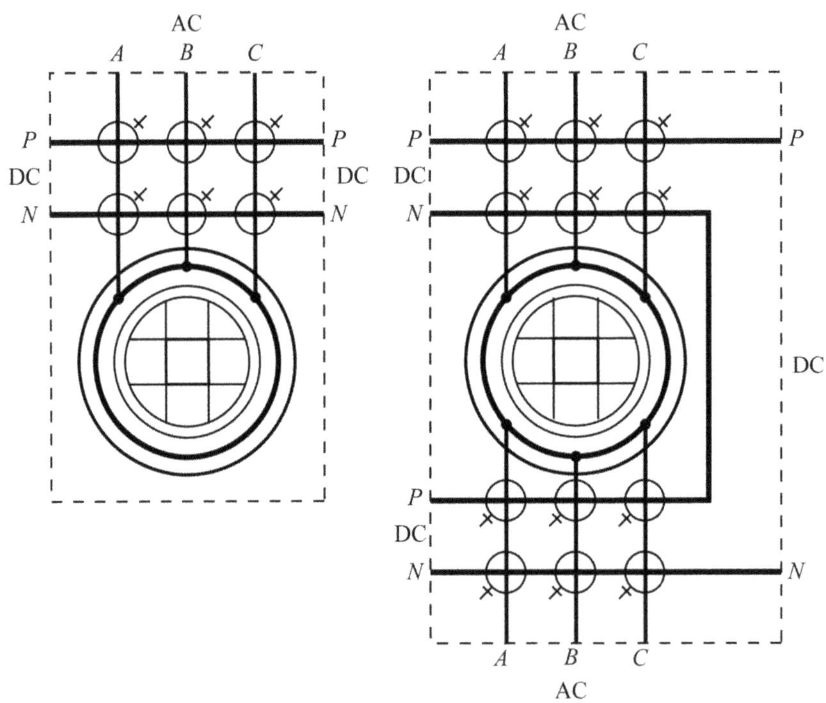

Figure D7. Single-phase ECM AC–AC and/or AC–DC/DC–AC commutator synchronous (magnetoelectrically-excited) dynamotor with the 2 × 3/3 × 2 or 2s(2 × 3)/2s(3 × 2) ECM commutator.

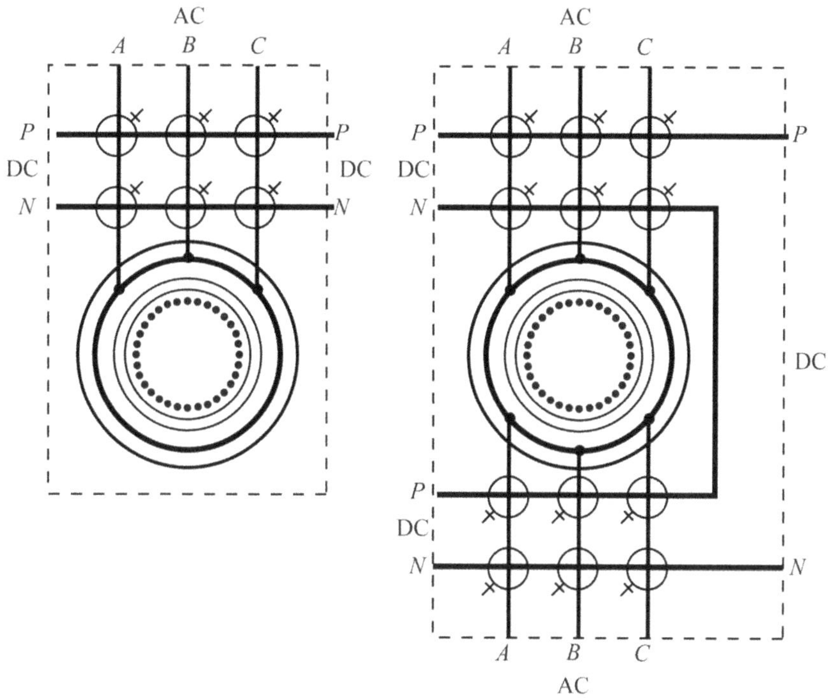

Figure D8. Single-phase ECM AC–AC and/or AC–DC/DC–AC commutator hyposynchronous (induction) dynamotor with the 2 × 3/3 × 2 or 2(2 × 3)/2(3 × 2) ECM commutator and squirrel-cage rotor.

D.5 A 2 × 3/3 × 2 ECM DC–AC/AC–DC commutator and 2 × 2 ECM DC–DC commutator

An advantage of the hybrid-electric **drive-by-wire** (DBW) **two-wheel-drive** (2WD) and/or **four-wheel-drive** (4WD) propulsion mechatronic control system is that a novel form of **external combustion engine** (ECE) or **internal combustion engine** (ICE) can be used, designed specifically for high efficiency constant engine-shaft angular velocity and power operation and minimum emission (Fijalkowski 1997).

The DBW 4WD propulsion mechatronical control system using E-M/M-E **steered, motorised and/or generatorised wheel** (SM&GW) with the ECM DC–AC/AC–DC commutator wheel-hub motors will eliminate the need for axles and drive shafts, which will allow a separate suspension for each independently-sprung SM&GW (figure D9). The rotating housing is made in the form of a wheel-hub, designed to fit standard rim sizes on the market. The ECM DC–AC/AC–DC commutator wheel-hub motor's transmission offers complete freedom in the design of the **hybrid-electric vehicle** (HEV).

A DBW 2WD and/or 4WD propulsion mechatronical control system may provide the HEV with estimable climbing ability (up to $\pi/4$ rad on firm surfaces). The two SM&GWs on each side of the HEV's hull-wheeled unit may have an independent **absorb-by-wire** (ABW) **four-wheel-absorbed** (4WA) suspension.

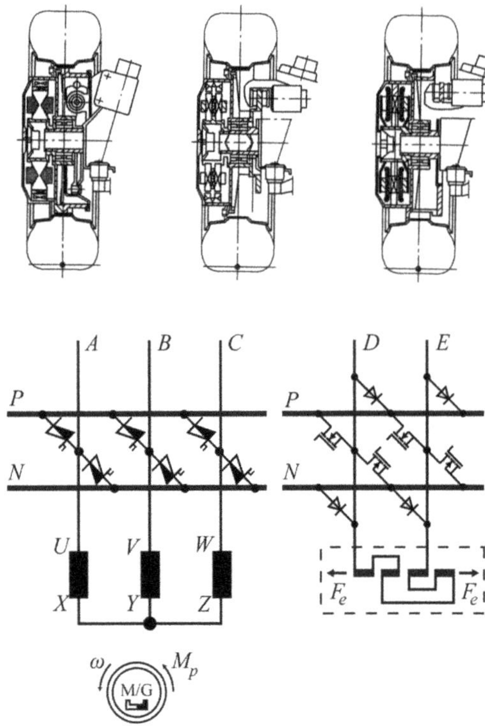

Figure D9. Principle layouts of three different independently-sprung, planetary-gearless steered, motorised and/or generatorised wheels (SM&GW) with 2 × 3/3 × 2 ECM DC–AC/AC–DC commutator in-wheel-hub motors and tubular, linear 2 × 2 ECM DC–DC commutator wheel-drum or wheel-disc brake-actuator motors.

D.6 A 2 × 5/5 × 2 ECM DC–AC/AC–DC commutator

The two front and the two rear SM&GWs (figure D10) are driven and/or absorbed individually by four 2 × 5/5 × 2 ECM DC–AC/AC–DC commutator reluctance and IPM magnetoelectrically-excited in-wheel-hub motors, respectively, and the angular velocity of each SM&GWs may be arbitrarily controlled by a driver-vehicle and terrain-vehicle real-time expert system, incorporating a mathematical-model following **fuzzy-logic** (FL) programmable and **neural-network** (NN) learning motion mechatronic control (Fijalkowski 1984a, 1984b, 1984c, 1984d, 1990, 2010).

D.7 A 2 × 3 ECM DC–AC/AC–DC commutator and a 2 × 2 DC–DC commutator

The authors considered a novel mobility and steerability concept of HEVs with E-M differentials. The use of these novel triad hybrid mobility and triad hybrid steerability enhancing concept DBW 2WD and/or 4WD propulsion mechatronical control systems for HEV (figure D11) opens up wide possibilities for improving fossil and non-fossil fuel economy, cutting initial and whole-life DBW 4WD propulsion costs, protecting the environment, and improving distribution of terrain

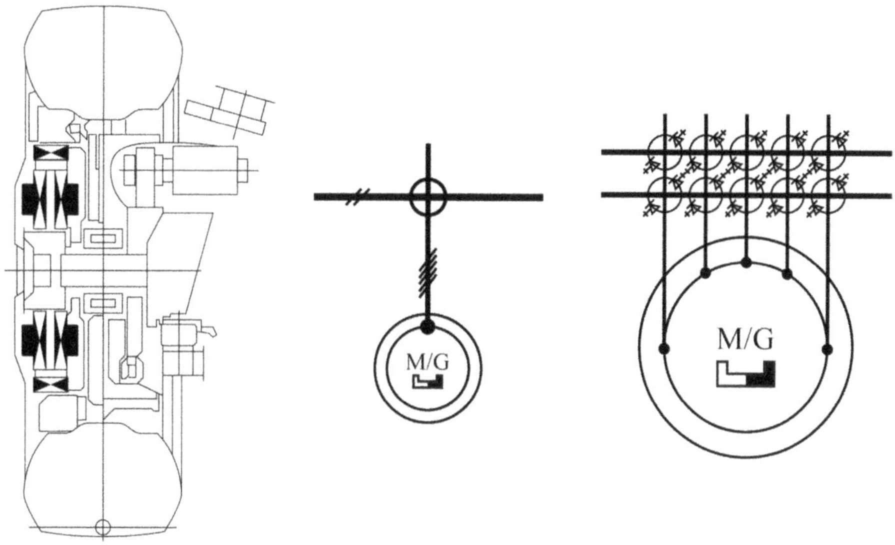

Figure D10. Layout of the steered, motorised and/or generatorised wheel (SM&GW)—(A) and the physical model of the ECM DC–AC/AC–DC commutator reluctance and IPM magnetoelectrically-excited in-wheel-hub motor/generator—(B).

thrust (gross tractive effort) as well as keeping both the net motion resistances of the DBW 4WD propulsion and SM&GW sinkages (rut depth) low not only by increasing the velocity of travel but also decreasing the rolling resistances of all the SM&GWs.

The ability of the HEVs to retain a sufficient level of mobility may exist, even if they have lost single or more SM&GWs. Using E-M differential torque and/or angular velocity controls of the outer SM&GWs as well as current and/or voltage controls of the inner SM&GWs, can increase the lateral motion control effect, especially when recovering from braking with the inner SM&GWs acting as the ECM AC–DC commutator in-wheel-hub generators, because the front gravitational forces on the HEVs become greater than the respective rear ones. At the same creep, this leads to greater horizontal (longitudinal and lateral) forces.

The DBW 4WD propulsion mechatronical control systems used on HEVs satisfy nearly all the same essential requirements as for the running gear systems used on conventional vehicles (Fijalkowski 1986, 1990, 2010), namely:

- To apply an E-M 'single-shaft' DBW 4WD propulsion mechatronical control system to a complete number of SM&GWs.
- To allow the outer side of the curve a positive propelling (driving) torque and to the inner side, a negative dispelling (braking) torque, achieving their maximum value for pivot-skid steering.
- To occupy the minimum volume within the space envelope of the HEV.
- To distribute the mass of the HEV over a relatively spacious ground surface or soil area.

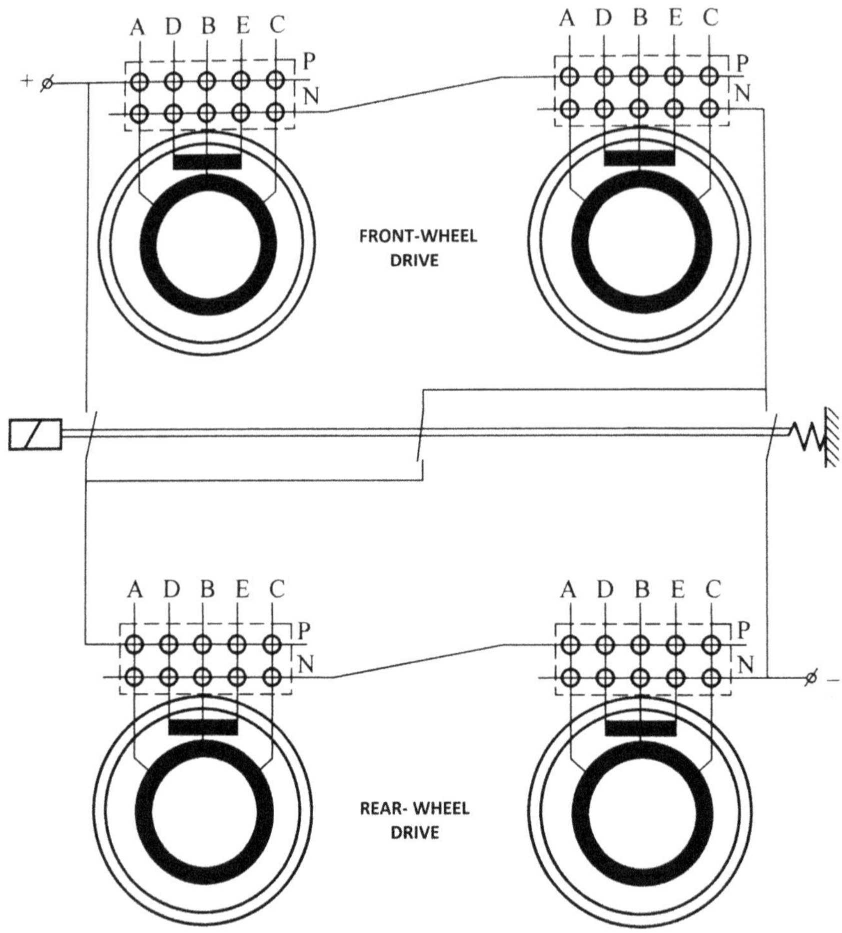

Figure D11. Elementary wiring connection for axleless, full-time DBW 4WD transmission arrangement with the torque proportioning front-wheel drive (FWD) and rear-wheel drive (RWD) as well as centre-wheel drive (CWD) electromechanical (E-M) differentials.

The requirement of the first feature may contribute to very good soil performance of HEVs. The third feature may tend to conflict with the requirement of the first feature.

D.8 A 2s(2 × 5)/2s(5 × 2) ECM DC–AC/AC–DC commutator

A principle layout of a particularly interesting planetary-gearless SM&GW with a wire-mesh tyre for the **Martian Roving Vehicle** (MRV) with the 2s(2 × 5)/2s(5 × 2) ECM DC–AC/AC–DC commutator reluctance and magnetoelectrically-excited (unwound MSI single ring-shape outer rotor and wound and IPM twin ring-shaped inner stator) in-wheel-hub motor/generator is shown in figure D12 (Fijalkowski 2001a, 2001b).

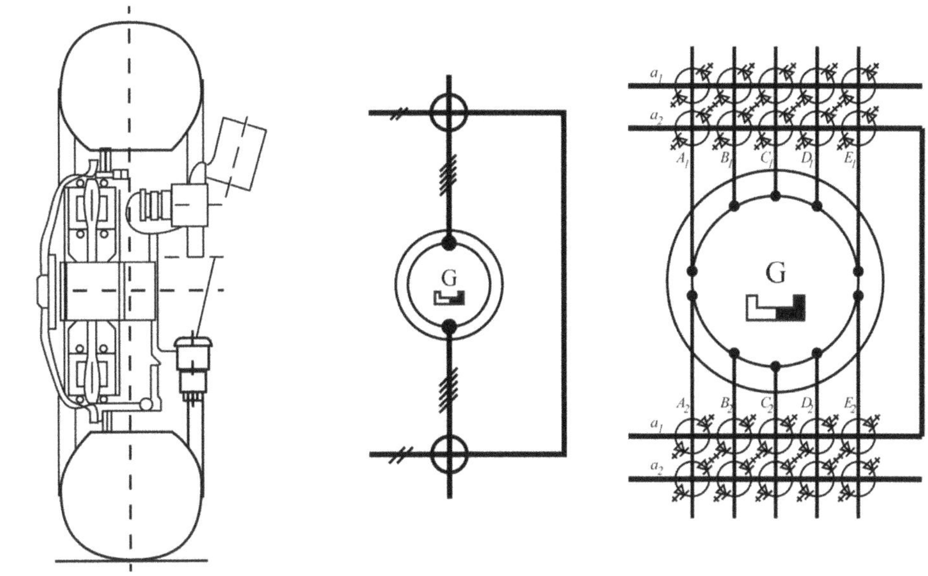

Figure D12. Planetary-gearless steered, motorised and/or generatorised wheel (SM&GW) with the 2s(2 × 5)/2s (5 × 2) ECM DC–AC/AC–DC commutator reluctance and magnetoelectrically-excited in-wheel-hub motor/generator.

It is evident that there are no windings on the mild soft iron (MSI) single ring-shaped outer rotor, no split rings, and no MCM DC–AC/AC–DC commutator with sliding copper-segments and their carbon brushes. Not evident is the nature of the twin ring-shaped inner-stator five-phase windings. If the twin ring-shaped inner stator is a six-pole, five-phase structure with a large number of 'teeth', then the single ring-shaped outer rotor will be also toothed and will be magnetised so that a south pole occupies one-half of the periphery while a north pole occupies the other half on both members affording a large number of opportunities for positional 'lock-up' of the single ring-shaped outer rotor.

Although the twin ring-shaped inner stator has only six poles and the single ring-shaped outer rotor has only three poles, the presence of 'teeth' on both members affords a large number of opportunities for positional 'lock-up' of the single ring-shaped outer rotor.

The 2s(2 × 5)/2s(5 × 2) ECM DC–AC/AC–DC commutator reluctance and magnetoelectrically-excited in-wheel-hub motor/generator with different pole numbers on the unwound MSI single ring-shaped outer rotor and the wound and IPM ring-shaped inner stator can provide traction if the stator five-phase winding phase-coils are sequentially energised. Here, the unwound MSI single ring-shaped outer rotor is made to 'chase' sequentially switched wound and IPM twin ring-shaped inner-stator's magnetic poles. As soon as magnetic alignment occurs, the attracting pole is de-energised. A variation of this basic scheme employs Hall-effect components to directly sense the position of ether the single ring-shaped outer rotor itself, or a suitably magnetised ring on the shaft.

D.9 Electrical commutation-matrixer commutators—A look into the future

The ECM commutator may represent the 'super-conductive ECM commutator' in which all-active components ('continuous' bipolar electrical valves) and reactive components (primary and secondary collectors, i.e., connections) of ECM are manufactured in the super- and/or semiconductor single wafer of amorphous material in the shape of the single chips (Fijalkowski 1988a, 1988b, 1988c, 1988d, 1988e, 1988f, 1988g).

In 1987, the Nobel Prize in the field of physics was awarded to Swiss physicist K A Müller and German physicist J G Bednorz, both working at IBM in Zürich. In early 1987, they discovered the properties of electrical-current conductivity (a rapid loss of resistance) in barium, lanthanum, copper and oxygen compounds at a temperature of 35 K. A little later American physicist P Chu, who followed the way indicated by Müller and Bednorz, went beyond a 'magic' limit of the nitrogen boiling point temperature of 77 K.

Currently, the electrical-current superconductivity phenomenon has already been registered even at the ambient temperature of the human natural environment. Ways of approaching the electrical-current superconductivity have changed completely. The amorphous materials may be—similarly to the mono- and polycrystalline—conductors (e.g. metals), non-conductors, as well as superconductors (Kazmerski 1980).

The electrical-current conductivity of amorphous germanium and silicon can be easily controlled by means of adequate doping, and the remaining properties of these materials are, in principle, analogous with those existing in mono- and polycrystalline equivalents. Thus, it is possible to perform the junctions of *p–n* type from amorphous germanium and silicon, which constitute the basic elements of the majority of semiconductor electrical valves.

Cheap and easy to attain, the amorphous material can also be used for the construction of active components (super- and/or semiconductor amorphous 'continuous' bipolar electrical valves) as reactive components (superconductor amorphous primary and secondary collectors, i.e. connections).

If the transistor revolution took place due to the discovery of properties of mono- and polycrystalline semiconductors, the successive overthrow in electronics will make use of semiconductor amorphous materials.

From the point of view of industrial EMD and traction technology, amorphous materials are characterised above all by the enormous heterogeneity of physical properties and by the fact that they require no complex technological processes.

Amorphous materials are usually manufactured in the shape of glasses by means of rapid quenching of the material from the liquid phase in the shape of thin films, which are vacuum evaporated on adequately quenched substrates. These technologies are distinctly competitive for the manufacturing technologies of mono- and polycrystalline materials.

The ECM AC–AC and AC–DC/DC–AC commutators with the bipolar ECMs, of which primary and secondary collectors are realised on ceramic superconductors, contribute to one more revolution in the history of both macroelectronics and ECM

AC–AC and AC–DC/DC–AC commutator dynamotors. Although the ECM AC–AC and AC–DC/DC–AC commutator with bipolar ECM, realised on 'continuous' bipolar ovonics is made of glassy material not having a crystalline structure, it is a super- and/or semiconductor static converter which provides a solution to the problem of its cooling.

The 'continuous' bipolar ovonics are characterised by the short time-lag of transition between the on state (conduction-state) of the electrical valve and its off state (non-conductive-state) of 0.1 μs. The ratio of the off state to the on state resistance yields 10^7 and more.

The triggering signal applied to the triggering gate may have an arbitrary polarisation. The 'continuous' bipolar ovonics can operate at a frequency below 500 kHz and at the ambient temperature range to 573 K (Ovshinsky 1958, 1970, Cohen *et al* 1969, Holmberg and Shaw 1974, Ovshinsky and Adler 1978, Kazmerski 1980).

The 'continuous' bipolar ovonics, under the influence of some control voltage, the value of which depends upon the chemical composition and dimensions of the glassy wafer, undergo metamorphosis from the resistor with resistance of $10^7\ \Omega$ into the conductor with resistance of $10^{-2}\ \Omega$.

Now for metamorphosis into a conductor, a relatively high electrical current starts to flow in the ovonic. As long as the electrical current is flowing, the glassy wafer remains a conductor.

It maintains this ability without any change even when the electrical current changes in a small range its instantaneous value. However, a single power pulse opposite to the primary one is sufficient for the amorphous conductor (without crystalline structure) to undergo metamorphosis into a resistor with a high resistance again.

Since the electrical current does not practically flow in this resistor, the heat is not released in it. However, if the ovonic metamorphoses into a conductor and conducts the electrical current, its resistance is so insignificant, that the quantity of heat released can be neglected.

In that way the dream of electronics scientists and engineers to make a great number of 'continuous' electrical valves on the single carrying wafer and to realise in practice uni- and/or bipolar ECMs for the ECM commutators becomes realisable.

The 2010 the Nobel Prize in Physics was awarded to the two researchers who performed the first experiments on graphene, a two-dimensional sheet of carbon atoms. The award, given to University of Manchester physicists Andre Geim and Konstantin Novoselov, recognises work that began less than a decade ago on a material that has since been used to make record-breaking transistors and stretchy electrodes.

Graphene is a one-atom-thick sheet of carbon atoms arranged in a honeycomb-like pattern. Graphene is considered to be the world's thinnest, strongest and most conductive material—of both electricity and heat. These properties are exciting researchers and businesses around the world as graphene has the potential to revolutionise entire industries in the fields of electricity, conductivity, energy generation, batteries, sensors and more.

For integrated circuits, graphene has a high carrier mobility, as well as low noise, allowing it to be used as the channel in a **field-effect transistor** (FET). Single sheets of graphene are hard to produce and even harder to make on an appropriate substrate.

Graphene exhibits a pronounced response to perpendicular external electric fields, potentially forming FET. The first top-gated FET (*ON–OFF* ratio of < 2) was demonstrated in 2007. Graphene nanoribbons may prove generally capable of replacing silicon as a semiconductor.

References

Altun H 2005 Application of a matrix converter in a slip energy recovery drive system *Eng. Model.* **18** 3–4 69–80

Cohen M H, Fritzsche H and Ovshinsky S R 1969 *Phys. Rev. Lett.* **22** 1065

Fijalkowski B T 1982 Trigistor frequency changer-controlled toothed gearless propulsion systems for electric and hybrid vehicles, *Proc. Europe's Int. Conf. on Electric Road Vehicle Systems Drive Electric Amsterdam, The Netherlands 25–28 October 1982* pp 554–67

Fijalkowski B 1984a Electronic commutator AC/DC motor-driven tracked all-terrain vehicles with extremely high mobility, *J. Terramechanics—Proc. 8th International ISTVS Conf. – The Performance of Off-road Vehicles and Machines, Churchill College, Cambridge University, Cambridge, England 5-11 August 1984* pp 1045–63

Fijalkowski B 1984b Development of electric propulsion for automotive vehicles in Poland, *Proc. 1984 EVS7: The Seventh Int. Electric Vehicle Symp., Versailles, France 26-29 June 1984* pp 378–83

Fijalkowski B 1984c City-bus hybrid-electric propulsion system with a brushless electronic-commutator DC dynamotor and continuously variable transmission, *Papers of 15th Meeting of Bus and Coach Experts, Budapest, Hungary 4-7 September 1984*

Fijalkowski B 1984d Power electronics propulsion systems for energy-saving automotive vehicles, *Proc. ISATA 84: Int. Symp. on Automotive Technology and Automation, Milan, Italy September 1984* **vol 1** pp 271–89

Fijalkowski B 1985a New concept MACRO- and MICRO-electronics cradled dynamometer systems for testing of combustion engines, *Proc. ISATA 85: Int. Symp. on Automotive Technology and Automation, Graz, Austria 23-27 September 1985* **vol 2** pp 587–616

Fijalkowski B T 1985b On the new concept hybrid and bimodal vehicles for the 1980s and 1990s, *Proc. Drive Electric Italy'85, Sorrento (Naples), Italy 1–4 October 1985* pp 4.04.2–8

Fijalkowski B T 1985c On the new concept MMD electrical machines with integral macro-electronic commutators—Development for the future, *Proc. First European Conf. on Power Electronics and Applications, Brussels, Belgium 16–18 October 1985* **vol 2** pp 3.377–84

Fijalkowski B 1986 Future hybrid electromechanical very advanced propulsion systems for civilian wheeled and tracked all-terrain vehicles with extremely high mobility, *Proc. 1986 EVS-8: The Eight International Electric Vehicle Symp., Washington, DC, 20-23 October 1986* pp 428–43

Fijalkowski B 1987a *Modele matematyczne wybranych lotniczych i motoryzacyjnych mechano-elektro-termicznych dyskretnych nadsystemów dynamicznych (Mathematical Models of Selected Aviation and Automotive Continuous Dynamical Hyperystems), Monografia* 53 (Krakow: Politechnika Krakowska imienia Tadeusza Kościuszki) p 276 (In Polish)

Fijalkowski B 1987b Choice of hybrid propulsion systems-wheeled city and urban vehicle; -tracked all-terrain vehicle. Part I *Electr. Veh. Dev.* **6** 113–7 142

Fijalkowski B 1988a Choice of hybrid propulsion systems-wheeled city and urban vehicle; -tracked all-terrain vehicle. Part II *Electr. Veh. Dev.* **7** 31–4

Fijalkowski B 1988b Odnochipnyi makrokommutator—Tehnologiia v stadii razvitiia – Uzhe nie teoriia, no eshche ne promyshlennost *Zbornik Prednasok z VIII celostatnej konferencje so zahranicnou ucastou Elektricke Pohony a Vykonova Elektronika* (I. DIEL, Kosice, Kosicka Bela, Czecho-Slovakia, 6.-9. 9.1988) pp 182–94 (In Russian)

Fijalkowski B 1988c Niekonwencjonalne uklady napedowe lokomotyw spalinowo-elektry-cznych z akumulacja energii – Spojrzenie w przyszlosc, *Materialy IV Konferencji Elektrotechnika, Elektronika i Automatyka w Transporcie Szyno-Wym Semtrak'88, Czesc II – Naped i sterowanie pojazdów trakcyjnych, Kraków – Zakopane* 28-30 wrzesnia 1988 pp 9–17 (In Polish)

Fijalkowski B 1988d Unconventional internal combustion engines for automotive vehicles, *Papers: Motor Vehicles and Motors'88* 3-4–5 October 1988 [also in *Motorna Vozila – Motori Saopstenja* (XIV-83) – *Kragujevac, November 1988* pp 265–74

Fijalkowski B 1988e Single-chip static-commutator—An indispensable module of new concept AC and /or DC dynamotors for automotive very advanced propulsion systems, *Proc. EVS9: The 9th Int. Electric Vehicle Symp., Toronto, Ontario, Canada 13-16 November 1988*, EVS88-031 pp 1–6

Fijalkowski B 1988f Automotive very advanced propulsion systems, *Proc. EVS9: the 9th Electric Vehicle Symp., Toronto, Ontario, Canada 13-16 November 1988*, EVS88-081 pp 1–9

Fijalkowski B 1988g Amorphous and polycrystalline semiconductor single-chip super-commutator. An indispensable module of new concept AC and/or DC electrical machines, *Proc. Int. Conf. and Intensive Tutorial Course on Semiconductor Materials, University of Delhi, New Delhi, India 8-16 December 1988*

Fijalkowski B T 1997 Intelligent automotive systems: development in full-time chassis motion spheres for intelligent vehicles *Advanced Vehicles and Infra-structure Systems—Computer Applications, Control and Automation* ed C O Nwagboso (New York: Wiley) pp 125–42

Fijalkowski B 1990 Very advanced propulsion spheres for high-speed tracked vehicles, *Proc. of the 10th Int. Conf. of the International Society for Terrain-Vehicle Systems (ISTVS), Kobe, Japan August 20-24, 1990* **vol III** pp 783–97

Fijalkowski B T 2001a Articulated triad Martian roving vehicle, *The Fourth International Mars Society Convention, Stanford University, Stanford, California August 23–26, 2001* (Lakewood, CO: The Mars Society Inc)

Fijalkowski B T 2001b A concept of the all-electric pressurized articulated triad martian roving vehicle *On To Mars 2* ed R M Zubrin and F Crossman (Burlington, ON: The Mars Society, Collector's Guide Publishing Inc)

Fijalkowski B T 2010 *Automotive Mechatronics: Operational and Practical Issues. — Volume I. International Series on Intelligent Control, and Automation Science and Engineering* vol 47 (Berlin: Springer)

Holmberg S N and Shaw M P 1974 *Proc. Int. Conf. Amorphous Liquid Semiconductors* 5th (London: Taylor and Francis)

Kazmerski L L (ed) 1980 *Polycrystalline and Amorphous Thin Films and Devices* (New York: Academic)

Ovshinsky S R 1958 *Phys. Rev. Lett.* **21** 1450

Ovshinsky S R 1970 *J. Non-Cryst. Solids.* **2** 99

Ovshinsky S R and Adler D 1978 *Contemp. Phys.* **10** 100

Wheeler P W, Clare J C, Empringham L, Bland M and Kerris K G 2004a Matrix converters *IEEE Ind. Appl. Mag.* **10** 59

Wheeler P W, Clare J C, Katsis D, Empringham L, Bland M and Podlesak T 2004b Design and construction of a 150 kVA matrix converter induction motor drive, *Proc of the 2nd IEE Int. Conf. on Power Electronics, Machines and Drives, Conference Publication No. 498, Edinburgh, UK* p 719

Glossary

AC–AC cycloconverter — An **electrical commutation matrixer** (ECM) AC–AC commutator, machine, device, or electrical dynamical system that changes directly AC power input voltage and frequency to AC power output voltage and frequency. Frequency changing in a PWM-type IEMD is performed by controlled bipolar electrical valves such as **reverse blocking** (RB) **insulated-gate bipolar transistors** (RB IGBTs) and **gate turn-off** (GTO) symistors.

AC–DC–AC frequency changer — An **electrical commutation matrixer** (ECM) AC–DC–AC commutator, machine, device, or electrical dynamical system that changes indirectly AC power input voltage and frequency to DC power output voltage and next DC power input voltage to AC power output voltage and frequency using AC–DC rectifier, DC link and DC–AC inverter. Frequency changing in a PWM-type IEMD is performed by uncontrolled unipolar electrical valves (diodes) for AC–DC rectification and controlled bipolar electrical valves such as **reverse blocking** (RB) **insulated-gate bipolar transistors** (IGBTs) and **gate turnoff** (GTO) thyristors for DC–AC inversion.

AC–DC rectifier — An **electrical commutation matrixer** (ECM) AC–DC commutator, machine, device, or electrical dynamical system that changes AC power to DC power. Rectification in a PWM-type IEMD is performed by uncontrolled unipolar electrical valves (diodes) or controlled uni- or bipolar electrical valves such as **reverse blocking** (RB) **insulated-gate bipolar transistors** (IGBTs) and **gate turnoff** (GTO) thyristors.

Actuator — The component of an open-loop or closed-loop mechatronical control system that connects the **electronic control unit** (ECU) with the process; the actuator consists of a **mechanical commutation matrixer** (MCM) or an **electrical commutation matrixer** (ECM) AC–AC, AC–DC–AC or DC–AC commutator motor and a final-control unit; positioning electrical signals are converted to mechanical output.

doi:10.1088/2053-2563/aae7d7ch13 © IOP Publishing Ltd 2019

All-electrical vehicle (AEV) Any ground automotive vehicle whose original source of energy is electrical power, such as an electrical automobile or electrical locomotive.

Angular velocity holor Defined as the rate of change of an angular displacement holor and is a vector quantity (more precisely, a pseudovector) which specifies the angular speed (rotational speed) of an object and the axis about which the object is rotating. The SI unit of angular velocity is *radians per second* (rad s^{-1}). Angular velocity holor is usually represented by the symbol omega ($\boldsymbol{\Omega}^{j}$). The sense of direction of the angular velocity holor is perpendicular to the plane of rotation, in a sense of direction that is usually specified by the 'right-hand rule'.

Artificial intelligence (AI) The synthetic intelligence exhibited by machines or software, and the domain of computer science that develops machines and software with human-like intelligence.

Application specific integrated circuit (ASIC) An **integrated circuit** (IC) designed for a custom requirement, frequently a gate array or single-chip microprocessor or programmable logic device.

Application specific integrated matrixer (ASIM) An **integrated matrixer** (IM) designed for a custom requirement, frequently a gate array or single-chip static-commutator or **electrical commutation matrixer** (ECM) AC–AC, AC–DC–AC, AC–DC/DC–AC or DC–DC commutator.

Cartesian dualism Defended by René Descartes, this states that there are two basic forms of substance: mental and material. According to his philosophy, the mental does not have extension in space, and the material cannot think. Also called as substance dualism it is credited with having given rise to much thought regarding the well-known mind–body dilemma.

Cartesian reductionism The idea of reductionism has existed since the ancient Greeks, René Descartes, a 17th century French philosopher, was the first to formally state the concept. He stated that the world was like a machine composed of many smaller parts, and that it could be understood by taking apart and studying the parts before learning how they all fit into the whole.

Chemo-electrical/electro-chemical storage battery (CH-E/E-Ch) Self-contained Ch-E/E-Ch cell/cells or accumulator system that converts chemical energy to electrical energy in a reversible process.

Complexity This is the property of a real-world dynamical hypersystem that is perceptible in the inability of any one formalism being able to denote all its properties. It necessitates that one finds particularly different means of interacting with physical heterogeneous continuous dynamical hypersystems. Particularly different in the sense that when one formulates adequate physical and mathematical models, the formal physical heterogeneous continuous dynamical hypersystems required to describe each particular aspect are not derivable from each other.

Computer-aided systems engineering (CASE) Tool designed to help a systems analyst complete development tasks.

Constraint A restriction on the natural **degrees of freedom** (DoF) of a physical heterogeneous continuous dynamical hypersystem, the number of constraints is the difference between the number of natural DoF and the number of actual DoF.

Cost function Can refer to: in mathematical optimisation, the loss function, a function to be minimised; in artificial neural networks, the function to return a number representing how well the neural network performed to map training examples to correct output.

Current holor A current holor, also known as electrical energy-transfer holor, is a flow of electrical charge. In **electrical commutation matrixers** (ECM) this charge is often carried by moving electrons in wires, i.e. ECM rows and columns. It can also be carried by ions in an electrolyte, or by both ions and electrons such as in a plasma. The SI unit for measuring a current is the ampere (A), which is the flow of electrical charges through a surface at the rate of one coulomb per second ($C\ s^{-1}$).

Cybernetics The study of human/machine interaction guided by the principle that numerous different types of systems can be studied according to principles of feedback, control, and communication. The field has a quantitative component, inherited from feedback control and information theory, but is primarily a qualitative, analytical tool—one might even say a philosophy of technology. Cybernetics is characterised by a tendency to universalise the notion of feedback, seeing it as the underlying principle of the technological world. The study of systems that are open to energy but closed to information and control systems that are tight is called cybernetics. Cybernetics takes as its domain the design or discovery and application of principles of regulation and communication. Cybernetics treats not things but ways of behaving.

Cyber-physical heterogeneous continuous complex and/or simple dynamical hypersystems Collections of various cyber-physical dynamical systems, hyposystems and components so connected or related as to perform a particular function not performable by the cyber-physical dynamical systems, hyposystems and components alone, and they must encompass computational (i.e. hardware and software) and physical homogeneous dynamical systems, hyposystems and components, seamlessly integrated and closely interacting to sense the changing state of the real world, These cyber-physical heterogeneous dynamical hypersystems combine distributed sensing, monitoring, actuation, and control networks.

d'Alambert principle The principle that the resultant of the external forces and the kinetic reaction acting on a body equals zero.

Darcy's law The law that the rate at which a fluid flows through a permeable substance per unit area is equal to the permeability, which is a property only of the substance through which the fluid is flowing, times the pressure drop per unit length of flow, divided by the viscosity of the fluid.

DC–AC inverter An **electrical commutation matrixer** (ECM) DC–AC commutator, machine, device, or electrical dynamical system that changes DC power to AC power. With regard to an IEMD, DC–AC inverter operation is carried out by controlled uni- or bipolar electrical valves (electronic switches) such as **reverse blocking** (RB) **insulated-gate bipolar transistors** (IGBTs) and **gate turnoff** (GTO) thyristors.

Deterministic mathematical model The physical heterogeneous continuous dynamical hypersystem is named such a model, description is possible in a form of the function relationships between inputs and outputs of the physical heterogeneous continuous dynamical hypersystem.

Developmental systems approach The developmental systems approach in the methodology of the excellence of physical models did not bring about the panacea on the routine for creating the physical model. This requires the considerably stronger impulse for the creation of the macro-conditions for computerised analytical studies of mathematical models.

Dynamical systems approach Offers a holistic—from whole to local approach for describing physical reality; produces a comprehensive image of physical reality; produces accurate predictions with straight-forward mathematics and clear logic.

Dynamical Universe (DU) A holistic description of the observable physical reality. It is a unifying theory converting space-time in variable coordinates into dynamic space in absolute coordinates. The **Dynamic Universe** (DU) theory relies on an overall zero-energy balance in space and the conservation of the total energy in inter-actions in space. Whole in the DU is not composed as the sum of elementary units —the multiplicity of elementary units is a result of diversification of whole. Many physical processes in the Universe are highly non-linear and coupled in a complex way. A deep understanding of the DU often requires detailed numerical simulations, coupled with sophisticated data analysis and visualisation.

Electrical commutation matrixer (ECM) AC–DC commutator This commutator is an ECM AC–DC rectifier for a rotary or transversal AC–DC commutator generator, the commutator electronically commutes (switches) the armature windings so that the resultant induced source AC armature phase-voltages always act with the same sense of voltage polarisation; this requires a reversal of the armature winding connection every π rad; the induced source AC armature phase-voltages are mechanically rectified to the induced source DC armature voltage via bipolar electrical valves, e.g. diodes, MOSFETs, IGBTs or MCTs.

Electrical commutation matrixer (ECM) DC–AC commutator This commutator is an ECM DC–AC inverter for a rotary or transversal DC–AC commutator motor or actuator, the commutator electronically commutes (switches) the armature windings so that the resultant torque or force always acts in the same sense of rotary or transverse direction, respectively; this requires a reversal of the armature winding connection every π rad; the DC armature supply is via bipolar electrical valves, e.g. MOSFETs, IGBTs or MCTs.

Electrical capacitance This is the ability of a body to store an electrical charge. Any object that can be electrically charged exhibits capacitance. A common form of energy storage device is a parallel-plate capacitor. In a parallel plate capacitor, capacitance is directly proportional to the surface area of the conductor plates and inversely proportional to the separation distance between the plates.

Electrical conductance The inverse quantity of electrical resistance is electrical conductance, the ease with which an electric current passes. The SI unit of electrical conductance is measured in siemens (S) or mhos (℧).

Electrical inductance In electromagnetism and electronics, inductance is the property of a conductor by which a change in current flowing through it 'induces' (generates) an electromotive force (induced voltage) in both the conductor itself (self-inductance) and in any nearby conductors (mutual inductance).

Electrical resistance This is the repulsion of a current within an electrical homogeneous continuous electrical system. It explains the relationship between voltage (amount of electrical pressure) and the current (flow of electricity), e.g. the electrical resistance of an electrical conductor is the opposition to the passage of an electric current through that conductor. Electrical resistance shares some conceptual parallels with the mechanical notion of friction. The SI unit of electrical resistance is the ohm (Ω).

Electrics The technology that generally deals with the application of electricity, electronics, and electromagnetism. This branch of physics first became distinct in the latter half of the 19th century after introduction of the electrical telegraph, the telephone, and electrical power distribution and use. Subsequently, broadcasting and recording media made electronics a part of everyday life.

Electro-mechanical (E-M) Pertaining to an **electro-mechanical** (E-M) motor or actuator or device, E-M heterogeneous continuous dynamical hypersystem, or E-M process which is electromagnetically or electrostatically actuated or controlled.

Engine A machine in which power is applied to do work by the conversion of various forms of energy into mechanical torque, force and motion.

Enthalpy The sum of the internal energy of a thermal homogeneous continuous dynamical system plus the product of the system's volume multiplied by the pressure exerted on the system by its surroundings. Also known as heat content, sensible heat, total heat.

Entropy Function of the state of a thermal homogeneous continuous dynamical system whose change in any differential reversible process is equal to the heat absorbed by the system from its surroundings divided by the absolute temperature of the system. Also known as thermal charge.

Eulerian coordinates Any system of coordinates in which properties of a fluid are assigned to points in space at a given time, without attempting to identify individual fluid parcels from one time to the next; a sequence of synoptic charts is a Eulerian representation of the data.

Eulerian correlation The correlation between the properties of a flow at various points in space at a single instant of time. Also known a synoptically correlation.

Eulerian equation A mathematical representation of the motions of a fluid in which the behaviour and the properties of the fluid are described at fixed points in a coordinate system.

Euler method A method of studying fluid motion and the mechanics of deformable bodies in which one considers volume elements at fixed locations in space, across which material flows, the Euler method is in contrast to the Lagrangian method.

Extended finite element method (XFEM) A numerical technique based on the **generalised finite element method** (GFEM) and the **partition of unity method** (PUM). It extends the classical **finite element method** (FEM) approach by enriching the solution space for solutions to differential equations with discontinuous functions.

External combustion engine (ICE) An engine that operates by the energy of external combustion of a fuel.

External energy-potential-difference holor An energy-potential-difference holor exerted on a physical heterogeneous continuous dynamical hypersystem or on some of physical homogeneous continuous dynamical systems, hyposystems and components by an agency outside the physical heterogeneous continuous dynamical hypersystem.

External work The work done by a physical heterogeneous continuous dynamical hyper-system in expanding against energy-potential-difference holors exerted from outside.

Final-control unit The second or last stage of an actuator to control mechanical output.

Finite elements method (FEM) A numerical technique for finding approximate solutions to boundary value problems for partial differential equations. It uses subdivision of a whole problem domain into simpler parts, called finite elements, and variational methods from the calculus of variations to solve the problem by minimising an associated error function.

Flow The forward continuous movement of a fluid, such as gases, vapours, or liquids, through closed or open channels or conduits.

Flow-rate Also known as rate of flow holor. (i) Time needed for a given quantity of flowable material to flow a measured distance. (ii) Weight or volume of flowable material flowing per unit time.

Flow resistance Any factor within a conduit or channel that impedes the flow of fluid, such as surface roughness or sudden bends, contractions, or expansions. Also known as viscosity.

Friction flow Fluid flow in which a significant amount of mechanical energy is dissipated into heat by the action of viscosity.

Friction torque The torque which is produced by frictional forces and opposes rotational motion, such as that associated with journal or sleeve bearings in machines.

Fluid A substance (a continuum) that deforms continuously the application of a shear, no matter how small the shear stress may be, and includes both liquids and gases. A fluid cannot sustain a shear when at rest. Liquids are nearly incompressible, but gases are highly compressible. Liquids are distinguished from gases by orders of magnitude differences in their density, absolute viscosity, and bulk modulus.

Fluid density The mass of a fluid per unit volume.

Fluid dynamics. The science of fluids in motion.

Fluid mechanics The science concerned with fluids, either at rest or in motion, dealing with pressures, velocities, and accelerations in the fluid, including fluid deformation and compression or expansion.

Fluid statics The determination of pressure intensities and forces exerted by fluids at rest.

Fluidical capacitance The volume of fluid in a channel can change just because of a change in pressure: this is either due to fluid compressibility or channel elasticity. Fluidical capacitors are one of two types of fluidical energy storing dynamical components in fluidical homogeneous continuous dynamical systems. Capacitance in a fluidical homogeneous continuous dynamical system comes from continuous **fluido-fluidical** (F-F) accumulators but also from the fluid itself if it is compliant. Fluid compliance is essential to consider in pneumatical homogeneous continuous dynamical systems, but generally does not play a significant role in physical models of hydraulical homogeneous continuous dynamical systems unless there is significant trapped air causing spongy behaviour. The capacitance of the fluid is captured by its bulk modulus property. The idea of capacitance of fluid trapped in a cylinder can be expanded to estimate the capacitance of a plug of fluid in a hose or pipe, which in turn can be used in a dynamic model. One application of a physical model involving fluid compression is to understand the **hydro-mechanical** (H-M) hammer, which is impact loading caused by sudden changes in flow, such as when a fluidical valve is commuted (switched) from on to off. One has already seen that a fluid itself, whether a liquid or gas, exhibits fluid compliance due to its compressibility. Certain fluidical machines or devices may also introduce flow compliance into a fluidical homogeneous continuous dynamical system, even if the fluids were absolutely incompressible.

Fluidical equation of motion One of the set of fluidodynamical equations representing the application of **Newton's Second Law** (NSL) of motion to a fluidical homogeneous continuous dynamical system, the total acceleration of an individual fluid particle is equated to the sum of the forces acting on the particle within the fluid.

Fluidical inertance or inductance The second type of energy-storing dynamical component is fluidical inertance or inductance. In mechanical homogeneous continuous dynamical systems, mass and rotary inertia often dominate dynamical-system behaviour and must be physically and mathematically modelled. In fluidical homogeneous continuous dynamical systems, the inertia of the fluid is generally insignificant and usually ignored in dynamical system physical models. The reason is that in hydraulically homogeneous continuous dynamical systems, pressures are so high that inertial forces can be neglected and in pneumatically homogeneous continuous dynamical systems the mass of air is so low that inertial forces can also be neglected. When analysing high frequency behaviour of a fluidical homogeneous continuous dynamical system, for example with sudden *ON/OFF* commutation (switching) of fluidical valves that causes transients in fluid flow, fluidical inertance should be included in the physical model.

Fluidical resistance The force exerted by a gas or liquid opposing the motion of a body through it. Like mechanical friction and electrical resistance this fluidical dynamical component performs an energy-dissipation function. The dissipation of fluidical energy into thermal energy (heat) occurs in all fluidical machines (F-M motors, MF pumps and compressors) or devices to some extent. Thus, a fluidical resistor is any dynamical component that resists flow. Another way of looking at fluidical resistors is that they are any dynamical component that causes a pressure drop when fluid flows through the component. Fluidical resistors include fluidical valves, filters, hoses, pipes and fittings.

Fluidical friction Conversion of mechanical energy in fluid flow into thermal energy.

Fluidics The technology of using the flow characteristics of liquid or gas to operate a fluidical homogeneous continuous dynamical system. One of the newest of the control technologies, fluidics has in recent years come to compete with mechanical and electrical homogeneous continuous dynamical systems.

Fluido-mechanical (F-M) Pertaining to a **fluido-mechanical** (F-M) motor or actuator or device, F-M heterogeneous continuous dynamical hypersystem, or F-M process which is fluido-dynamically or fluido-statically actuated or controlled.

Force holor That influence on a mechanical homogeneous continuous dynamical system which causes it to accelerate, quantitatively it is a mechanical energy-potential-difference holor, equal to the mechanical homogeneous continuous dynamical system's time rate of change of momentum holor.

Fuzzy-logic (FL) Software design based upon a reasoning nodal rather than fixed mathematical algorithm, a fuzzy logic design allows the mechatronic engineer to participate in the software design because the fuzzy language is linguistic and built upon easy-to-comprehend fundamentals.

Generalised admittance holor A measure of how easily a physical heterogeneous continuous dynamical hypersystem will allow an energy-transfer holor to flow. It is defined as the holor inverse of a generalized impedance holor, that is the holor reciprocal of a generalised impedance holor. Thus, a generalised admittance holor is the energy-transfer holor-to-energy-potential-difference holor ratio, and it conventionally carries SI units of siemens (S), formerly called mhos (℧).

Generalised coordinates A set of variables used to specify the position orientation of a physical heterogeneous continuous dynamical hypersystem, in principle defined in terms of Cartesian coordinates of the physical heterogeneous continuous dynamical hyposystem's particles and of the time to some convenient manner, the number of such coordinates equals the number of **degrees of freedom** (DoF) of the physical heterogeneous continuous dynamical hyposystem. Also known as Lagrangian coordinates.

Generalised equation of motion Equation which specifies the coordinates of particles as functions of time. Also known as a differential equation of dynamics, or one of several such equations, from which the coordinates of particles as functions of time can be obtained if the initial positions and velocities of the particles are known.

The generalised energy-potential-difference holor corresponding to a generalised coordinate is the ratio of the virtual work done in

	an infinitesimal virtual displacement, which alters those coordinates and no other, to the coordinate.
Generalised energy-potential-difference holor	
Generalised energy-transfer holor	The derivative with respect to time of one of the generalised coordinates of a particle. Also known as Lagrangian generalised energy-transfer holor.
Generalised finite element method (GFEM)	A direct extension of the **standard finite element method** (SFEM, or FEM), which makes possible the accurate solution of physical and engineering problems in complex domains which may be practically impossible to solve using the FEM.
Generalised impedance holor	The measure of the opposition that a physical heterogeneous continuous dynamical hypersystem presents to an energy-transfer holor when an energy-potential-difference holor is applied. In quantitative terms, it is the holor ratio of the energy-potential-difference holor to the energy-transfer holor in a physical heterogeneous continuous dynamical hypersystem. The impedance holor extends the concept of the resistance holor to physical heterogeneous continuous dynamical hypersystems, and possesses both magnitude and phase, unlike the resistance holor, which has only magnitude. When a physical heterogeneous discrete dynamical hypersystem is driven with a step function, there is no distinction between the impedance holor and the resistance holor; the latter can be thought of as the impedance holor with a zero phase angle. The magnitude of the impedance holor is the holor ratio of the energy-potential-difference holor amplitude to the energy-transfer holor amplitude. The phase of the impedance is the phase shift by which the energy-transfer holor lags the energy-potential difference holor. In general, generalised impedance will be a holor, with the same SI units as the physical resistance holor, for which the SI unit is the ohm (Ω).
Generalised momentum	If q_j (j = 1, 2, 3,...) are generalised coordinates of a classical physical heterogeneous continuous dynamical hypersystem, and $\mathscr{L}$ is its Lagrangian, the momentum conjugate to q_j is $p_j = \partial\mathscr{L}/\partial q_j$. Also known as canonical momentum, conjugate momentum.
Graphene	A remarkable substance with a multitude of astonishing properties which repeatedly earn it the title 'wonder material'. Graphene is the thinnest material known to man at one atom thick, and also incredibly strong—about 200 times stronger than steel. On top of that, graphene is an excellent conductor of heat and electricity and has interesting light absorption abilities. It is truly a material that could change the world, with unlimited potential for integration in almost any industry.
Hamilton–Ostrogradsky's principle	A variation principle which states that the path of a conservative physical heterogeneous continuous dynamical hypersystem in configuration space between two configurations is such that the integral of the Lagrangian function over time is a minimum or maximum relative to near paths between the same end points and taking the same time.
Hamilton's equations of motion	A set of first-order, highly symmetrical equations describing the motion of a classical dynamical system, namely $\dot{q}_j = \partial H/\partial p_j$, $p_j = -\partial H/\partial q_j$ here q_j (j = 1, 2, ...) are generalised

coordinates of the system, p_j is the momentum conjugate to q_j and H is the Hamiltonian. Also known as canonical equations of motion.

Hamiltonian function A function of the generalised coordinates and momenta of a physical heterogeneous continuous dynamical hypersystem, equal in value to the sum over the coordinates of the product of the generalised momentum corresponding to the coordinate, and the coordinate's time derivative, minus the Lagrangian of the physical heterogeneous continuous dynamical hypersystem; it is numerically equal to the total energy of the Lagrangian and does not depend on time explicitly, the equations of motion of the physical heterogeneous continuous dynamical hypersystem are determined by the functional dependence of the Hamiltonian on the generalised coordinates and momenta.

Heat Thermal energy transmitted due to a temperature difference between the source from which the energy is coming and a sink towards which the energy is going, other types of energy in transit are called work.

Heat conduction The flow of thermal energy through a substance from a higher to a lower temperature region. Also known as heat diffusion.

Heat convection The transfer of thermal energy by actual physical movement from one location to another of a substance to which thermal energy is stored. Also known as thermal convection.

Heat flow Heat thought of as thermal energy flowing from one substance to another, quantitatively, the amount of heat transferred in a unit time. Also known as heat transmission.

Heat radiation Electromagnetic radiation generated by the thermal motion of charged particles in matter, also known as thermal radiation. All matter with a temperature greater than absolute zero emits thermal radiation. When the temperature of the body is greater than absolute zero, interatomic collisions cause the kinetic energy of the atoms or molecules to change. Thermal radiation is one of the fundamental mechanisms of heat transfer. Heat transfer through radiation takes place in the form of electromagnetic waves mainly in the infrared region. Radiation emitted by a body is a consequence of thermal agitation of its composing molecules.

Heat transfer The exchange of thermal energy between physical heterogeneous continuous dynamical hypersystems depending on the temperature and pressure, by dissipating heat. Thermal homogeneous discrete dynamical systems which are not isolated may decrease in entropy. Most objects emit infrared thermal radiation near room temperature. The fundamental modes of heat transfer are conduction or diffusion, convection, advection and radiation.

Holism Asserts that everything exists in relationship, in a context of connection and meaning—and that any change or event causes realignment, however slight, throughout the entire pattern. The whole is greater than the sum of its parts means that the whole is comprised of a pattern of relationships that are not contained by the parts but ultimately define them. Holism stands in stark opposition to the method of reductionism, which holds that analysis, dissection, and strict definition are the tools for understanding reality. Holism asserts that phenomena can never be fully understood in isolation; it asserts that reductionism can only give us a partial view of anything it dissects. Holism is difficult to pin down precisely, because by its very nature it embraces paradox, mystery, and contradiction. The holistic world view has its roots in the new physics, systems thinking and ecology, holism and perennial philosophy.

Holonomic constraints An integrable set of differential equations of dynamics which describe the restrictions on the motion of a physical heterogeneous continuous dynamical hypersystem, a function relating several variables in the form $f(x_1, \ldots x_n) = 0$, in optimisation of physical problems.

Holonomic physical heterogeneous continuous dynamical hypersystem A physical heterogeneous continuous dynamical hypersystem in which the constraints are such that the original coordinates can be expressed in terms of independent coordinates and possibly also the time.

Holor A holor (hoʊlər; Greek ὅλος 'whole') is a mathematical entity that is made up of one or more independent quantities ('merates') as they are called in the theory of holors. Complex numbers, scalars, vectors, matrices, tensors, quaternions, and other hypercomplex numbers are kinds of holors.

Hybrid-electrical vehicle (HEV) An automotive vehicle that uses two or more distinct power sources to move the vehicle. However, one of them ought to be a secondary power source with a possibility of energy storage. The term most commonly refers to **hybrid electrical vehicles** (HEV), which combine an **internal combustion engine** (ICE) or **external combustion engine** (ECE) and one or more **electro-mechanical/mechano-electrical** (E-M/ME) dynamotors. However, other mechanisms to capture and utilise energy are included. A hybridisation of conventional automotive vehicle and electrical vehicle combines an ICE or ECE propulsion system with an electrical propulsion system. The presence of the electrical powertrain is intended to achieve either better **specific fuel consumption** (SFC) than a conventional automotive vehicle or better performance.

Internal combustion engine (ICE) A prime mover in which the fuel is burned within the engine and the products of combustion serve as thermodynamic fluid, as with gasoline and diesel engines. An engine that operates by the energy of internal combustion of a fuel.

Kinetic energy The energy which a physical heterogeneous continuous dynamical hypersystem possesses because of the motion in classical physics equal to one-half of the physical heterogeneous continuous dynamical hypersystem's mass and/or capacitance times the square of its energy-transfer variable.

Kirchhoff's first law (KFL) This law is also known as Kirchhoff's point rule, or Kirchhoff's junction rule (or nodal rule). The principle of conservation of electrical charge implies that: at any commutation node (junction) in an **electrical commutation matrixer** (ECM), the holor sum of current holors flowing into that commutation node is equal to the holor sum of current holors flowing out of that commutation node, or: the holor sum of current holors in an ECM of conductors (rows and columns) meeting at a point is zero.

Kirchhoff's second law (KSL) This law is also known as Kirchhoff's loop rule or as Kirchhoff's mesh rule. The principle of conservation of energy implies that the holor sum of the electrical energy-potential-difference holors (voltage holors) around any closed **electrical commutation matrixer** (ECM) is zero, or: more simply, the holor sum of the **electromotive force** (emf) holors in any closed loop is equivalent to the holor sum of the potential-drop holors in that loop, or: the holor sum of the products of the resistances of the ECM conductors (rows and columns) and the current holors in them in a closed loop is equal to the total emf holor available in that loop.

Lagrange–Hamilton theory The formalised study of continuous systems in terms of field variables where a Lagrangian density function and Hamiltonian density function are introduced to produce equations of motion.

Lagrange's equations Equation of dynamics of a physical heterogeneous continuous dynamical hypersystem for which a classical (non-quantum mechanical) physical description is possible, and which relate the kinetic energy of the physical heterogeneous continuous dynamical hypersystem to the generalised coordinates, the generalised energy-potential-difference variables, and the time. Also known as Lagrangian equations of dynamics.

Lagrangian The difference between the kinetic energy and the potential energy of a physical heterogeneous continuous dynamical hypersystem, expressed as a function of generalised coordinates and energy-transfer variables from which Lagrange's equations can be derived. Also known as kinetic potential Lagrange function.

Lagrangian density For a dynamical system of fields or continuous media, a function of the fields, of their time and space derivatives, and the coordination and time, whose integral over space is the Lagrangian.

Lagrangian function The function which measures the difference between kinetic and potential energy of a physical heterogeneous continuous dynamical hypersystem.

Lagrangian method A method of studying of fluid motion and the mechanics of deformable bodies in which one considers volume elements which are carried along with the fluid or body, and across whose boundaries material does not flow, in contrast to Euler method.

Machine A combination of rigid or resistant bodies having definite motions and capable of performing useful work.

Magnetic reluctance A concept used in the analysis of magnetic homogenous continuous dynamical systems. It is analogous to electrical resistance in an electrical homogeneous continuous dynamical system, but rather than dissipating electrical energy it stores magnetic energy. In likeness to the way an electrical field causes an electrical current to follow the path of least resistance, a magnetic field causes magnetic flux to follow the path of least magnetic reluctance. It is a scalar, extensive quantity, akin to electrical resistance. The units for magnetic reluctance are inverse henries, (H^{-1}). Also known as magnetic resistance.

Magnetic resistance See magnetic reluctance.

Magnetomotive force In physics, the **magnetomotive force**, (mmf) is a quantity appearing in the equation for the magnetic flux in a magnetic homogeneous continuous dynamical system, sometimes known as Hopkinson's law: $M = R\Phi$; where Φ is the magnetic flux and R is the magnetic reluctance of the homogeneous continuous dynamical system. It can be seen that the mmf plays a role in this equation analogous to the voltage V in Ohm's law: $V = RI$. Magnetomotive force is analogous to **electromotive force**, (emf), i.e. the difference in electrical potential, or voltage, between the terminals of a source of electricity, e.g. a Ch-E/E-Ch storage battery from which no current is being drawn, since it is the cause of magnetic flux in a magnetic homogeneous continuous dynamical system; i.e. (i) $M = NI$; where N is the number of turns in the coil and I is the electrical current through the magnetic homogeneous continuous dynamical system; (ii) $M = R\Phi$; where Φ is the magnetic flux and R is the magnetic reluctance; (iii) $M = Hl$; where H is the magnetising force (the strength of the magnetising field) and l is the mean length of a solenoid or the circumference of a toroid

Mass-flow holor Also known as mass-transfer holor and bulk-flow holor, this is the measure of the movement of material matter. In physics, mass flow occurs in open fluidical homogeneous continuous dynamical systems and is often measured as occurring when moving across a certain boundary characterised by its cross-sectional area and a flow-rate holor. In fluidics it may also be a flow of fluids in a tube or vessel of a certain diameter. A bulk transfer of particles of matter in a characterised type of flow is also known as bulk flow.

Mass-velocity holor The mass flow-rate holor of a fluid divided by the cross-sectional area of the enclosing chamber or conduit. The SI unit is kilogram per second and square metres (kg s^{-1} m^{-2}).

Mathematical model The physical heterogeneous continuous dynamical hypersystem is adequate to a physical process, functional and structural homogeneous continuous dynamical systems, hypo-systems and components and superimposed constraints.

Mechanical commutation matrixer (MCM) AC–DC commutator This commutator is a **mechanical commutation matrixer** (MCM) AC–DC rectifier for a rotary AC–DC commutator generator, the commutator **mechano-electrically** (ME) commutes (switches) the armature windings so that the resultant induced source AC armature phase-voltages always act with the same sense of voltage polarisation; this requires a reversal of the armature winding connection every π rad; the induced source AC armature phase-voltages are mechano-electronically rectified to the induced source DC armature voltage via bipolar mechano-electrical valves, i.e. commutator segments that contact the carbon brushes.

Mechanical commutation matrixer (MCM) DC–AC commutator This commutator is a mechano-electrical commutation matrixer (MCM) DC–AC inverter for a rotary DC–AC commutator motor or actuator, the commutator mechano-electrically commutes (switches) the armature windings so that the resultant torque always acts in the same sense of rotary direction; this requires a reversal of the armature winding connection every π rad; the DC armature supply is via bipolar mechano-electrical contact valves, i.e. carbon brushes that contact the commutator bars or segments.

Mechanics (Greek μηχανική) is concerned with the behavior of physical bodies when subjected to forces or displacements, and the subsequent effects of the bodies on their environment. It is a branch of classical physics that deals with particles that are either at rest or are moving with velocities significantly less than the speed of light. It can also be defined as a branch of science which deals with the motion of and forces on objects.

Mechanism The part of a machine which contains two or more pieces so arranged that the motion of one compels the motion of the others.

Mechano-electrical (ME) Pertaining to a mechano-electrical (ME) generator or device, ME heterogeneous continuous dynamical hypersystem. or ME process which is mechanically driven or controlled.

Mechano-fluidical (MF) Pertaining to a mechano-fluidical (MF) pump or compressor or device, MF heterogeneous continuous dynamical hypersystem. or MF process which is fluidically driven or controlled.

Microcomputer unit (MCU) A semiconductor device that has a **central processing unit** (CPU), memory, and input/output (I/O) capability on the same chip.

Motion A continuous change of position of a body.

Mechano-fluidical (MF) pump A machine that draws a fluid into itself through an entrance port and forces the fluid out through an exhaust port.

Mechano-pneumatical (M-P) compressor A **mechano-pneumatical** (M-P) machine that increases the pressure of a gas or vapour by increasing the gas density and delivering the gas or vapour against the connected pneumatical homogeneous continuous dynamical system resistance.

Newtonian mechanics The mechanical homogeneous continuous dynamical system based upon Newton's laws of motion in which mass and energy are considered as separate, conservative, mechanical properties, in contrast to their treatment in relativistic mechanics.

Newton's first law The law that a particle not subjected to external forces remains at rest or moves with constant velocity in a straight line. Also known as the first law of motion, Galileo's law of inertia.

Newtonian paradigm Newton gave us three 'laws of motion', which were intended to describe the motion of the planets. It turned out that these laws could be applied in a seemingly perfectly general way. This broader application has been the foundation of the modern scientific method and will be referred to here as the Newtonian paradigm. In the centre of this paradigm is 'dynamics'. Dynamics is the way the laws of motion get applied.

Newton's second law The law that the acceleration of a particle is directly proportional to the resultant external force acting on the particle and is inversely proportional to the mass of the particle. Also known as the second law of motion.

Newton's third law The law that, if two particles interact, the force exerted by the first particle on the second particle (called the action force) is equal in magnitude and opposite in sense of direction to the force exerted by the second particle on the first particle (called the reaction force). Also known as the law of action and reaction, the third law of motion.

Partition of unity method (PUM) This can be regarded as a generalisation of classic **finite element methods** (FEM). Instead of a mesh, the construction of a global approximation space and the evaluation of the discretisation use a much more versatile set of patches, the so-called cover. The method enables the inclusion of arbitrary local function spaces and is particularly well suited for adaptive refinements with respect to spatial resolution and reproducing quality of the function space.

Physical heterogeneous continuous complex andlor simple dynamical hypersystems These have collections of various physical dynamical systems, hyposystems and components so connected or related as to perform a particular function not performable by the dynamical systems, hyposystems and components alone.

Physical model The physical model of the physical heterogeneous continuous dynamical hypersystem is a model in which passing physical processes are equivalent each other when described by mathematical relationships.

Pneumatics Fluid statics and behaviour in closed pneumatical homogeneous continuous dynamical systems when the fluid is a gas.

Potential energy The capacity to do the work that a physical heterogeneous continuous dynamical hypersystem has by virtue of its position or configuration.

Pressure-drop holor The difference in pressure holor between two points in a fluidical homogeneous continuous dynamical system, usually caused by frictional resistance to a fluid flowing through a conduit, filter media, or other flow-conducting, fluidical homogeneous continuous dynamical system.

Pressure holor A type of stress which is exerted uniformly in all senses of direction; its measure is the force holor exerted per unit area.

Prime mover The component of a power plant that transforms energy from the thermal or the fluidical form to the mechanical form.

Principle of virtual work The principle that the total work done by all energy-potential-difference variables acting on a physical heterogeneous continuous dynamical hypersystem in static equilibrium is zero for any infinitesimal generalised coordinate from equilibrium which is consistent with the constraints of the physical heterogeneous continuous dynamical hypersystem. Also known as virtual work principle.

Protocol The rules governing the exchange of information (data) between networked components.

Pulse-width modulation (PWM) The precise and timely creation of negative and positive waveform edges to achieve a waveform with a specific frequency and duty cycle.

Semicustom MCU A **microcontroller unit** (MCU) that incorporates normal MCU elements plus application specified peripheral devices such as higher-power port outputs, special times units, etc, mixed semiconductor technologies, such as high-density CMOS (HCMOS) and bipolar analog, are available in a semicustom MCU, generally HCMOS is limited to 10 V_{DC}, whereas bipolar-analog is suitable to 60 V_{DC}.

Specific fuel consumption (SFC) The mass flow rate of fuel required to produce a unit of power or thrust, e.g. kilograms per kilowatt-hour (kg kW^{-1}).

Split This term reflects the difference between an AC asynchronous (induction) motor's synchronous velocity and its rotor velocity. Split ratio in percent is equal to (synchronous velocity – rotor velocity)/(synchronous velocity 100).

Statistical mathematical model The physical heterogeneous continuous dynamical hypersystem is such a model, it is formulated taking into consideration the statistical parameters and the probability expansion function.

Statistical mechanics A branch of physics that studies, using probability theory, the average behaviour of a mechanical homogeneous continuous dynamical system where the state of the system is uncertain. Statistical mechanics is a collection of mathematical tools that are used to fill this disconnection between the laws of mechanics and the practical experience of incomplete knowledge.

Synchronous velocity In regard to AC synchronous or asynchronous (induction) motors, the rotational velocity of the stator's rotating magnetic field is called the synchronous velocity, which is equal to (in rpm): [120] f (the line frequency in cps)] [divided by] p (the number of magnetic poles).

Systems thinking approach This involves shifting one's attention from the dynamical parts to the whole, from objects to relationships, from structures to processes, from hierarchies to networks. It also includes shifts of emphasis from the rational to the intuitive, from analysis to synthesis, from linear to non-linear thinking. Contrary to reductionism, systems thinking is a style of thought and reasoning that seeks to understand a whole dynamical hypersystem by examining the physical heterogeneous continuous dynamical hypersystem as a whole instead of disassembling and studying the dynamical parts, i.e. physical homogeneous continuous dynamical systems, hyposystems and components. While some scientists prefer to use one of the two styles of thinking to the exclusion of the other, it is more common to use whichever style fits a given situation. Quite simply, some situations call for systems thinking, while others require a closer look at the dynamical parts of the physical heterogeneous continuous dynamical hypersystem are better suited for reductionism. Summing up, a system thinking is an approach to integration that is based on the belief that the dynamical parts, i.e. physical homogeneous continuous dynamical systems, hyposystems and components of a physical heterogeneous continuous dynamical hypersystem will act differently when isolated from the hypersystem's environment or other parts of the hypersystem. Standing in contrast to positivist and reductionist thinking, systems thinking sets out to view physical heterogeneous continuous dynamical hypersystems in a holistic manner. Consistent with systems philosophy, systems thinking concerns an understanding of a physical heterogeneous continuous dynamical hypersystem by examining the linkages and interactions between the physical homogeneous continuous dynamical systems, hyposystems and components that comprise the whole of the physical heterogeneous continuous dynamical hypersystem. When one encounters situations which are complex and messy, then systems thinking can help one to understand the situation systemically. This helps one to see the full picture—from which one may identify multiple leverage points that can be addressed to support constructive change. It also helps one to see the connectivity between physical homogeneous continuous dynamical systems, hyposystems and components in the situation, so as to support joined-up actions.

Theoretical reductionism This states that all theories in a field are part of a larger theory with a broader scope. In theory, this supports the idea of the existence of a 'grand unified theory' of physics that combines quantum physics with other observed phenomena.

Thermal capacitance The ratio of the entropy added to a body to the resulting rise in temperature.

Thermal conductance The amount of heat transmitted by a material divided by the difference in temperature of the surfaces of the material.

Thermal conductivity The heat flow through a surface per unit area per unit time, divided by the negative of the rate of change of temperature with distance to a direction perpendicular to the surface. Also known as heat conductivity.

Thermal inductance The product of temperature difference and time divided by entropy flow.

Thermal energy potential difference The difference between the thermodynamic temperatures of two points.

Thermal resistance A measure of a body's ability to prevent heat from flowing through it, equal to the difference between the temperatures of opposite faces of the body divided by the rate of heat flow. Also known as heat resistance.

Thermal resistivity The reciprocal of the thermal conductivity.

Thermics Also known as thermal physics is the combined study of thermodynamics, statistical mechanics, and kinetic theory. This umbrella-subject is typically designed for physics students and functions to provide a general introduction to each of three core heat-related subjects. Others, however, define thermics loosely as a summation of only thermodynamics and statistical mechanics.

Thermodynamics A branch of physics concerned with heat and temperature and their relation to energy and work. It defines macroscopic variables, such as internal energy, entropy, and pressure, which partly describe a body of matter or radiation. It states that the behaviour of those variables is subject to general constraints that are common to all materials, not the peculiar properties of particular materials. These general constraints are expressed in the four laws of thermodynamics.

Torque holor (i) For a single force holor, the cross product of a holor from some reference point to the point of application of the force holor with the force holor itself. Also known as moment of force holor, rotation moment holor. (ii) For several force holors, the holor sum of the torque holors (first definition) associated with each of the force holors.

Transfer-function holor The mathematical relationship between the output holor of a mechatronical control system and its output holor for a linear system; it is the output holor divided by the input holor under conditions of zero initial energy stored. Also known as transmittance holor.

Transfer-matrix holor The generalisation of the concept of a transfer-function holor to multi-variable mechatronical control system; it is the matrix whose product with the vector representing the input holors yields the vector representing the output holors.

Unholonomic physical heterogeneous continuous dynamical hypersystem A physical heterogeneous continuous dynamical hypersystem which is subjected to restraints of such a nature that the physical heterogeneous continuous dynamical hypersystem cannot be described by independent coordinates, examples are **generalised commutation matrixer** (GCM) commutators, e.g. in a mechanical homogeneous continuous dynamical systems are a rolling hoop, or an ice skate which must point along its path.

Vapour A gas at a temperature below the critical temperature, so that it can be liquefied by compression, without lowering the temperature.

Velocity holor The time rate of change of position of a body; it is a vector quantity having sense of direction and magnitude. Also known as linear velocity holor.

Voltage holor Voltage holor, also known as electrical energy-potential-difference holor, electrical tension holor or electrical pressure holor (denoted $\boldsymbol{V}_i$) and measured in SI units of electrical potential: volts (V), or joules per coulomb (J C^{-1}) is the electrical energy-potential-difference holor between two **electrical commutation matrixer's** (ECM) separated terminals, i.e. between ECM rows and columns, or the difference in electrical potential energy of a unit charge transported between two ECM separated terminals.

Volumetric flow-rate holor In fluidics, in particular fluid dynamics and hydrometry, the volumetric flow-rate holor, (also known as volume flow-rate holor, rate of fluid-flow holor or volume velocity holor) is the volume of fluid which passes through a given surface per unit time. The SI unit is cubic meters per second (m^3 s^{-1}). It is usually represented by the symbol $\boldsymbol{Q}^j$. Volumetric flow-rate holor should not be confused with volumetric flux holor, as defined by Darcy's law and represented by the symbol $\boldsymbol{q}^j$, with SI units of $m^3/(m^2$ s), that is, (m s^{-1}). The integration of a flux over an area gives the volumetric flow rate.

Lightning Source UK Ltd.
Milton Keynes UK
UKHW050638100419
340794UK00003BA/47/P

9 780750 320467